WORLD HEALTH ORGANIZATION

INTERNATIONAL AGENCY FOR RESEARCH ON CANCER

IARC MONOGRAPHS

ON THE

EVALUATION OF CARCINOGENIC RISKS TO HUMANS

Man-made Vitreous Fibres

VOLUME 81

This publication represents the views and expert opinions
of an IARC Working Group on the
Evaluation of Carcinogenic Risks to Humans,
which met in Lyon,

9–16 October 2001

2002

IARC MONOGRAPHS

In 1969, the International Agency for Research on Cancer (IARC) initiated a programme on the evaluation of the carcinogenic risk of chemicals to humans involving the production of critically evaluated monographs on individual chemicals. The programme was subsequently expanded to include evaluations of carcinogenic risks associated with exposures to complex mixtures, life-style factors and biological and physical agents, as well as those in specific occupations.

The objective of the programme is to elaborate and publish in the form of monographs critical reviews of data on carcinogenicity for agents to which humans are known to be exposed and on specific exposure situations; to evaluate these data in terms of human risk with the help of international working groups of experts in chemical carcinogenesis and related fields; and to indicate where additional research efforts are needed.

The lists of IARC evaluations are regularly updated and are available on Internet: http://monographs.iarc.fr/

This project was supported by Cooperative Agreement 5 UO1 CA33193 awarded by the United States National Cancer Institute, Department of Health and Human Services. Additional support has been provided since 1993 by the United States National Institute of Environmental Health Sciences.

This publication was made possible in part by a Cooperative Agreement between the United States Environmental Protection Agency, Office of Research and Development (US EPA, ORD) and IARC and does not necessarily express the views of the US EPA, ORD. This project was funded in part by the European Commission, Directorate-General EMPL (Employment and Social Affairs), Health, Safety and Hygiene at Work Unit.

©International Agency for Research on Cancer, 2002

Distributed by IARC*Press* (Fax: +33 4 72 73 83 02; E-mail: press@iarc.fr)
and by the World Health Organization Marketing and Dissemination, 1211 Geneva 27
(Fax: +41 22 791 4857; E-mail: publications@who.int)

Publications of the World Health Organization enjoy copyright protection in accordance with the provisions of Protocol 2 of the Universal Copyright Convention.

All rights reserved. Application for rights of reproduction or translation, in part or *in toto*, should be made to the International Agency for Research on Cancer.

IARC Library Cataloguing in Publication Data

Man-made vitreous fibres /

    IARC Working Group on the Evaluation of Carcinogenic Risks to Humans (2002 : Lyon, France)

    (IARC monographs on the evaluation of carcinogenic risks to humans ; 81)

    1. Carcinogens – congresses   2. Fibres – congresses   I. IARC Working Group on the Evaluation of Carcinogenic Risks to Humans   II. Series

    ISBN 978-9-2832128-12   (NLM Classification: W1)
    ISSN 1017-1606

PRINTED IN FRANCE

CONTENTS

NOTE TO THE READER 1

LIST OF PARTICIPANTS 3

PREAMBLE 9
   1. Background 9
   2. Objective and Scope 9
   3. Selection of Topics for Monographs 10
   4. Data for Monographs 11
   5. The Working Group 11
   6. Working Procedures 11
   7. Exposure Data 12
   8. Studies of Cancer in Humans 14
   9. Studies of Cancer in Experimental Animals 17
  10. Other Data Relevant to an Evaluation of Carcinogenicity
       and its Mechanisms 20
  11. Summary of Data Reported 22
  12. Evaluation 23
  13. References 27

GENERAL REMARKS ON MAN-MADE VITREOUS FIBRES 33
   1. Composition, production and use 34
   2. Toxicity 35
   3. Chronic inflammation, fibrosis and cancer 35
   4. Studies of cancer in humans 35
   5. Studies of cancer in experimental animals 36
   6. Administration to experimental animals by inhalation 37
   7. Administration to experimental animals by intraperitoneal injection 37
   8. References 38

MAN-MADE VITREOUS FIBRES 41
   1. Exposure data 43
      1.1 Chemical and physical data 43
          1.1.1 Nomenclature and general description 43
              (a) Categorization 43

1.1.2 Chemical and physical properties 45
    (a) Chemical properties 45
        (i) Continuous glass filament 48
        (ii) Glass wool 52
        (iii) Rock (stone) and slag wool 52
        (iv) Refractory ceramic fibres 53
        (v) Newly developed fibres 54
    (b) Physical properties 54
        (i) Fibre diameter 54
        (ii) Fibre length 57
        (iii) Fibre density 58
        (iv) Fibre coatings and binders 58
        (v) Structural changes 59
1.1.3 Analysis 59
    (a) Principle 59
    (b) Sampling 60
    (c) Gravimetric analysis 60
    (d) Fibre counting 60
        (i) Counting criteria 61
        (ii) Fibre identification 61
        (iii) Detection limit 62
        (iv) Quality assurance 63
    (e) Surface deposition 63
    (f) Bulk material 63
1.2 Production 64
  1.2.1 History and production levels 64
    (a) Continuous glass filament 64
    (b) Glass wool, rock (stone) wool and slag wool 64
    (c) Refractory ceramic fibres 66
    (d) Newly developed fibres 67
  1.2.2 Production methods 67
    (a) Continuous glass filament 70
    (b) Glass wool 71
    (c) Rock (stone) wool 72
    (d) Slag wool 73
    (e) Refractory ceramic fibres 74
    (f) Newly developed fibres 75
1.3 Uses 75
  1.3.1 Continuous glass filament 75
  1.3.2 Glass wool, rock (stone) wool and slag wool 76
    (a) Thermal and acoustic insulation, fire protection 76
    (b) Other uses 78

CONTENTS

    (c) Special-purpose glass fibres 78
   1.3.3 Refractory ceramic fibres 78
   1.3.4 Newly developed fibres 80
 1.4 Occurrence 80
   1.4.1 Occupational exposure 80
    (a) Exposure in production plants 82
     (i) United States of America 82
     (ii) Europe 98
     (iii) Other studies 106
    (b) Exposure to compounds other than MMVFs in production
     plants 108
    (c) Exposure of workers during the installation and removal of
     MMVF insulation products 109
    (d) Exposure in residential, commercial and public buildings .. 122
   1.4.2 Environmental occurrence 125
 1.5 Regulations and guidelines 127
2. Studies of cancer in humans 133
 2.1 Glass wool 133
   2.1.1 United States University of Pittsburgh cohort 140
    (a) Cohort studies 140
    (b) Nested case–control studies 143
   2.1.2 European glass fibre cohort 145
    (a) Cohort studies 145
    (b) Nested case–control study 147
   2.1.3 Other cohort studies 147
 2.2 Continuous glass filament 148
    (a) United States University of Pittsburgh cohort 148
    (b) European cohort 152
    (c) United States Georgetown University cohort 152
    (d) Canadian cohort 153
 2.3 Rock (stone) wool and slag wool 154
   2.3.1 Cohort studies 154
    (a) United States University of Pittsburgh cohort 154
    (b) European study 165
 2.4 Refractory ceramic fibres 169
   2.4.1 Cohort study 169
   2.4.2 Case–control study 169
 2.5 MMVF (not otherwise specified) 169
   2.5.1 Cohort studies 175
   2.5.2 Population-based case–control studies 176
    (a) Lung cancer 176

  (b) Mesothelioma...177
  (c) Cancer at sites other than the lung and pleura.....................178
 2.5.3 Registry-based studies 178
3. Studies of cancer in experimental animals 181
 3.1 Continuous glass filament...181
  3.1.1 Intraperitoneal injection 181
 3.2 Glass wool..181
  3.2.1 Inhalation exposure..181
   (a) Rat 181
   (b) Hamster 190
  3.2.2 Intraperitoneal injection 191
   (a) Rat 191
   (b) Hamster 193
 3.3 Special-purpose glass fibres 193
  3.3.1 Inhalation exposure..193
   (a) Rat 193
   (b) Hamster 195
   (c) Guinea-pig 196
  3.3.2 Intraperitoneal injection 196
  3.3.3 Intratracheal instillation 201
   (a) Rat 201
   (b) Hamster 201
  3.3.4 Intrapleural injection...202
   (a) Mouse...202
   (b) Rat 203
 3.4 Rock (stone) wool 204
  3.4.1 Inhalation exposure..204
  3.4.2 Intratracheal instillation 208
   (a) Rat 208
   (b) Hamster 210
  3.4.3 Intraperitoneal injection 210
  3.4.4 Intrapleural injection...222
 3.5 Slag wool 223
  3.5.1 Inhalation exposure..223
  3.5.2 Intraperitoneal injection 224
   (a) Rat 224
   (b) Hamster 225
  3.5.3 Intrapleural injection...226
 3.6 Refractory ceramic fibres 226
  3.6.1 Inhalation exposure..226
   (a) Rat 226
   (b) Hamster 233

       3.6.2 Intratracheal instillation 234
             (a) Rat 234
             (b) Hamster 235
       3.6.3 Intrapleural injection..235
       3.6.4 Intraperitoneal injection 236
             (a) Rat 236
             (b) Hamster 238
   3.7 Newly developed fibres 238
       3.7.1 Inhalation exposure...238
       3.7.2 Intraperitoneal injection 239
4. Other data relevant to an evaluation of carcinogenicity
   and its mechanisms 241
   4.1 Deposition, retention and clearance 241
       4.1.1 Deposition..242
       4.1.2 Clearance and retention 247
             (a) Studies in animals 251
             (b) Studies in humans 253
       4.1.3 Fibre biopersistence, concepts and definition 255
             (a) Fibre biopersistence studies..257
             (b) Fibre biopersistence and pathogenicity 263
       4.1.4 Fibre dissolution 265
             (a) In-vitro dissolution...265
                 (i)   Cell-free systems 265
                 (ii) Mechanisms of in-vitro fibre degradation 268
                 (iii) Relationship between in-vivo retention of fibres,
                       in-vitro dissolution rates and fibre composition 268
                 (iv) Cellular in-vitro systems 269
             (b) In-vivo solubility 270
   4.2 Toxic effects in humans 272
       4.2.1 Adverse health effects other than respiratory cancer 272
             (a) Mortality data..272
             (b) Morbidity data 275
                 (i)   Pneumoconiosis 275
                 (ii) Pleural fibrosis and related findings 278
                 (iii) Symptoms and changes in lung function.....................282
             (c) Other respiratory findings 287
       4.2.2 Other toxic effects...288
             (a) Mortality data..288
             (b) Morbidity data 288
   4.3 Toxic effects in experimental systems 288
       4.3.1 Continuous glass filament and glass wool 289
             (a) Inflammation and fibrosis 290

                (b) Cell toxicity 293
                (c) Cell activation 294
                (d) Other effects..294
            4.3.2 Rock (stone) and slag wool 295
                (a) Inflammation and fibrosis 295
                (b) Cell activation 297
                (c) Other effects..297
            4.3.3 Refractory ceramic fibres 297
                (a) Inflammation 297
                (b) Fibrosis..298
                (c) Cell proliferation 299
                (d) Cellular toxicity 299
                (e) Cell activation 300
                (f) Oxidant generation 301
                (g) Other effects..301
        4.4 Effects on gene expression 301
            4.4.1 Continuous glass filament and glass wool 301
            4.4.2 Rock (stone) wool, slag wool and refractory ceramic fibres 306
                (a) Pulmonary epithelial cells 306
                (b) Mesothelial cells 307
        4.5 Genetic and related effects 307
            4.5.1 Continuous glass filament and glass wool 307
            4.5.2 Rock (stone) wool and slag wool 314
            4.5.3 Refractory ceramic fibres 318
    5. Summary of data reported and evaluation 327
        5.1 Exposure data 327
        5.2 Human carcinogenicity data 328
        5.3 Animal carcinogenicity data 331
        5.4 Other relevant data 334
        5.5 Evaluation 338
    6. References..341

LIST OF ABBREVIATIONS USED IN THIS VOLUME 375

GLOSSARY...379

SUPPLEMENTARY CORRIGENDUM TO VOLUMES 1–80 383

CUMULATIVE INDEX TO THE *MONOGRAPHS* SERIES..................................385

NOTE TO THE READER

The term 'carcinogenic risk' in the *IARC Monographs* series is taken to mean the probability that exposure to an agent will lead to cancer in humans.

Inclusion of an agent in the *Monographs* does not imply that it is a carcinogen, only that the published data have been examined. Equally, the fact that an agent has not yet been evaluated in a monograph does not mean that it is not carcinogenic.

The evaluations of carcinogenic risk are made by international working groups of independent scientists and are qualitative in nature. No recommendation is given for regulation or legislation.

Anyone who is aware of published data that may alter the evaluation of the carcinogenic risk of an agent to humans is encouraged to make this information available to the Unit of Carcinogen Identification and Evaluation, International Agency for Research on Cancer, 150 cours Albert Thomas, 69372 Lyon Cedex 08, France, in order that the agent may be considered for re-evaluation by a future Working Group.

Although every effort is made to prepare the monographs as accurately as possible, mistakes may occur. Readers are requested to communicate any errors to the Unit of Carcinogen Identification and Evaluation, so that corrections can be reported in future volumes.

IARC WORKING GROUP ON THE EVALUATION OF CARCINOGENIC RISKS TO HUMANS: MAN-MADE VITREOUS FIBRES

Lyon, 9–16 October 2001

LIST OF PARTICIPANTS

Members

A. Andersen, The Cancer Registry of Norway, Institute for Epidemiological Cancer Research, Montebello, 0310 Oslo, Norway

C. Axten, Health Risks Solutions, LLC, 1606 Maddux Lane, McLean, VA 22101-3200, USA

D.M. Bernstein, 40 chemin de la Petite-Boissière, 1208 Geneva, Switzerland

P. Brochard, Outpatient Department of Occupational Pathology, Pellegrin Hospital, Place Amélie Raba-Léon, 33076 Bordeaux Cédex, France

V. Castranova, Pathology and Physiology Research Branch, Health Effects Laboratory Division, National Institute for Occupational Safety and Health, 1095 Willowdale Road, Morgantown, WV 26505, USA

K. Donaldson, School of Life Sciences, Napier University, 10 Colinton Road, Edinburgh EH10 5DT, United Kingdom

P. Dumortier, Pneumology Unit – Mineralogy Laboratory, Erasmus Hospital, 808 route de Lennik, 1070 Brussels, Belgium

J.I. Everitt, CIIT Centers for Health Research, 6 Davis Drive, PO Box 12137, Research Triangle Park, NC 27709, USA

P. Gustavsson, Department of Occupational Health, Karolinska Hospital, 171 76 Stockholm, Sweden

T.W. Hesterberg, Johns Manville Corporation, 10100 W. Ute Avenue (80127), PO Box 635005, Littleton, CO 80162-5005, USA

M.C. Jaurand, INSERM EMI-9909, University of Medicine – Paris XII, 8 rue du Général Sarrail, 94010 Créteil Cédex, France

A.B. Kane, Department of Biology and Laboratory Medicine, Division of Biology and Medicine, School of Medicine, Brown University, Box G-B511, 171 Meeting Street, Providence, RI 02912, USA (*Chairperson*)

G.M. Marsh, Department of Biostatistics, Graduate School of Public Health, University of Pittsburgh, A410 Crabtree Hall, 130 DeSoto Street, Pittsburgh, PA 15261, USA

Y. Morimoto, Department of Occupational Pneumology, Institute of Industrial Ecological Sciences, University of Occupational and Environmental Health, 1-1 Iseigaoka, Yahata-nishi, Kitakyushu City 807-8555, Japan

H. Muhle, Fraunhofer Institute of Toxicology and Aerosol Research, Nikolai-Fuchs Strasse 1, 30625 Hannover, Germany

G. Oberdörster, Department of Environmental Medicine, University of Rochester Medical Center, 573 Elmwood Avenue, Rochester, NY 14642, USA

S. Olin, ILSI Risk Science Institute, One Thomas Circle, 9th Floor, Washington, DC 20005-5802, USA

K.M. Savolainen, Department of Industrial Hygiene and Toxicology, Finnish Institute of Occupational Health, Topeliuksenkatu 41 aA, 00250 Helsinki, Finland

T. Schneider, National Institute of Occupational Health, Lerso Parkalle 105, 2100 Copenhagen, Denmark

Representatives/Observers[1]

Representative of the European Ceramic Fibre Industries Association (ECFIA)
B.C. Brown, 4 Bramble Close, Uppingham, Rutland LE15 9PH, United Kingdom

Representative of the Italian National Toxicology Commission
B. Fubini, Department of Chemistry IFM and Interdepartmental Centre 'G. Scansetti' for Studies on Asbestos and Other Toxic Particulates, Università degli Studi di Torino, Via Pietro Giuria 7, 10125 Torino, Italy

Representative of the North American Insulation Manufacturers' Association (NAIMA)
J.G. Hadley, Owens Corning, Science and Technology Center, 2790 Columbus Road, Route 16, Granville, OH 43023-1200, USA

Observer, Technical Resources International
T. Junghans, Technical Resources International Inc., 6500 Rock Spring Drive, Suite 650, Bethesda, MD 20817-1197, USA

Representative of the European Insulation Manufacturers' Association (EURIMA)
O. Kamstrup, Rockwool International A/S, Hovedgaden 584, 2640/Hedehusene, Denmark

Representative of the Refractory Ceramic Fibers Coalition (RCFC)
L.D. Maxim, Everest Consulting Associates, Inc., 15 N. Main Street, Cranbury, NJ 08512-3203, USA

Representative of the Joint European Medical Research Board
C.E. Rossiter, Joint European Medical Research Board, 902 Helderberg Village, Private Bag X19, Somerset West 7129, South Africa

Representative of the European Commission
K. Ziegler-Skylakakis, European Commission, DG Employment D/5, Bâtiment Jean Monnet, Plateau du Kirchberg, 2920 Luxembourg

IARC Secretariat
R.A. Baan, Unit of Carcinogen Identification and Evaluation
P. Boffetta, Unit of Environmental Cancer Epidemiology
P. Buffler[2], IARC Visiting Scientist Award
D. Evered, Office of the Director
M. Friesen, Unit of Nutrition and Cancer
Y. Grosse, Unit of Carcinogen Identification and Evaluation (*Responsible Officer*)
J.-C. Hung, Unit of Environmental Cancer Epidemiology
S. Kaplan, Bern, Switzerland (*Editor*)
N. Napalkov[3], Office of the Director (Consultant)
C. Partensky, Unit of Carcinogen Identification and Evaluation
J.M. Rice, Unit of Carcinogen Identification and Evaluation (*Head of Programme*)
L. Stayner, Unit of Carcinogen Identification and Evaluation
K. Straif, Unit of Carcinogen Identification and Evaluation
E. Suonio, Unit of Carcinogen Identification and Evaluation

Technical assistance
S. Egraz
B. Kajo
M. Lézère
J. Mitchell
E. Perez

---

[1] The following individuals had been scheduled to participate as observers representing their governmental organizations but ultimately were unable to attend: P. Infante, US Occupational Safety and Health Administration; D. Longfellow, US National Cancer Institute; R. Zumwalde, US National Institute of Occupational Safety and Health.

[2] Present address: School of Public Health, University of California, Warren Hall, Room 140, Berkeley, CA 94720, USA

[3] Present address: Director Emeritus, Petrov Institute of Oncology, Pesochny-2, 197758 St Petersburg, Russia

PREAMBLE

IARC MONOGRAPHS PROGRAMME ON THE EVALUATION OF CARCINOGENIC RISKS TO HUMANS

PREAMBLE

1. BACKGROUND

In 1969, the International Agency for Research on Cancer (IARC) initiated a programme to evaluate the carcinogenic risk of chemicals to humans and to produce monographs on individual chemicals. The *Monographs* programme has since been expanded to include consideration of exposures to complex mixtures of chemicals (which occur, for example, in some occupations and as a result of human habits) and of exposures to other agents, such as radiation and viruses. With Supplement 6 (IARC, 1987a), the title of the series was modified from *IARC Monographs on the Evaluation of the Carcinogenic Risk of Chemicals to Humans* to *IARC Monographs on the Evaluation of Carcinogenic Risks to Humans*, in order to reflect the widened scope of the programme.

The criteria established in 1971 to evaluate carcinogenic risk to humans were adopted by the working groups whose deliberations resulted in the first 16 volumes of the *IARC Monographs series*. Those criteria were subsequently updated by further ad-hoc working groups (IARC, 1977, 1978, 1979, 1982, 1983, 1987b, 1988, 1991a; Vainio *et al.*, 1992).

2. OBJECTIVE AND SCOPE

The objective of the programme is to prepare, with the help of international working groups of experts, and to publish in the form of monographs, critical reviews and evaluations of evidence on the carcinogenicity of a wide range of human exposures. The *Monographs* may also indicate where additional research efforts are needed.

The *Monographs* represent the first step in carcinogenic risk assessment, which involves examination of all relevant information in order to assess the strength of the available evidence that certain exposures could alter the incidence of cancer in humans. The second step is quantitative risk estimation. Detailed, quantitative evaluations of epidemiological data may be made in the *Monographs*, but without extrapolation beyond the range of the data available. Quantitative extrapolation from experimental data to the human situation is not undertaken.

The term 'carcinogen' is used in these monographs to denote an exposure that is capable of increasing the incidence of malignant neoplasms; the induction of benign neoplasms may in some circumstances (see p. 19) contribute to the judgement that the exposure is carcinogenic. The terms 'neoplasm' and 'tumour' are used interchangeably.

Some epidemiological and experimental studies indicate that different agents may act at different stages in the carcinogenic process, and several mechanisms may be involved. The aim of the *Monographs* has been, from their inception, to evaluate evidence of carcinogenicity at any stage in the carcinogenesis process, independently of the underlying mechanisms. Information on mechanisms may, however, be used in making the overall evaluation (IARC, 1991a; Vainio *et al.*, 1992; see also pp. 25–27).

The *Monographs* may assist national and international authorities in making risk assessments and in formulating decisions concerning any necessary preventive measures. The evaluations of IARC working groups are scientific, qualitative judgements about the evidence for or against carcinogenicity provided by the available data. These evaluations represent only one part of the body of information on which regulatory measures may be based. Other components of regulatory decisions vary from one situation to another and from country to country, responding to different socioeconomic and national priorities. **Therefore, no recommendation is given with regard to regulation or legislation, which are the responsibility of individual governments and/or other international organizations.**

The *IARC Monographs* are recognized as an authoritative source of information on the carcinogenicity of a wide range of human exposures. A survey of users in 1988 indicated that the *Monographs* are consulted by various agencies in 57 countries. About 2500 copies of each volume are printed, for distribution to governments, regulatory bodies and interested scientists. The Monographs are also available from IARC*Press* in Lyon and via the Distribution and Sales Service of the World Health Organization in Geneva.

3. SELECTION OF TOPICS FOR MONOGRAPHS

Topics are selected on the basis of two main criteria: (a) there is evidence of human exposure, and (b) there is some evidence or suspicion of carcinogenicity. The term 'agent' is used to include individual chemical compounds, groups of related chemical compounds, physical agents (such as radiation) and biological factors (such as viruses). Exposures to mixtures of agents may occur in occupational exposures and as a result of personal and cultural habits (like smoking and dietary practices). Chemical analogues and compounds with biological or physical characteristics similar to those of suspected carcinogens may also be considered, even in the absence of data on a possible carcinogenic effect in humans or experimental animals.

The scientific literature is surveyed for published data relevant to an assessment of carcinogenicity. The IARC information bulletins on agents being tested for carcinogenicity (IARC, 1973–1996) and directories of on-going research in cancer epidemiology (IARC, 1976–1996) often indicate exposures that may be scheduled for future meetings. Ad-hoc working groups convened by IARC in 1984, 1989, 1991, 1993 and

1998 gave recommendations as to which agents should be evaluated in the IARC Monographs series (IARC, 1984, 1989, 1991b, 1993, 1998a,b).

As significant new data on subjects on which monographs have already been prepared become available, re-evaluations are made at subsequent meetings, and revised monographs are published.

4. DATA FOR MONOGRAPHS

The *Monographs* do not necessarily cite all the literature concerning the subject of an evaluation. Only those data considered by the Working Group to be relevant to making the evaluation are included.

With regard to biological and epidemiological data, only reports that have been published or accepted for publication in the openly available scientific literature are reviewed by the working groups. In certain instances, government agency reports that have undergone peer review and are widely available are considered. Exceptions may be made on an ad-hoc basis to include unpublished reports that are in their final form and publicly available, if their inclusion is considered pertinent to making a final evaluation (see pp. 25–27). In the sections on chemical and physical properties, on analysis, on production and use and on occurrence, unpublished sources of information may be used.

5. THE WORKING GROUP

Reviews and evaluations are formulated by a working group of experts. The tasks of the group are: (i) to ascertain that all appropriate data have been collected; (ii) to select the data relevant for the evaluation on the basis of scientific merit; (iii) to prepare accurate summaries of the data to enable the reader to follow the reasoning of the Working Group; (iv) to evaluate the results of epidemiological and experimental studies on cancer; (v) to evaluate data relevant to the understanding of mechanism of action; and (vi) to make an overall evaluation of the carcinogenicity of the exposure to humans.

Working Group participants who contributed to the considerations and evaluations within a particular volume are listed, with their addresses, at the beginning of each publication. Each participant who is a member of a working group serves as an individual scientist and not as a representative of any organization, government or industry. In addition, nominees of national and international agencies and industrial associations may be invited as observers.

6. WORKING PROCEDURES

Approximately one year in advance of a meeting of a working group, the topics of the monographs are announced and participants are selected by IARC staff in consultation with other experts. Subsequently, relevant biological and epidemiological data are collected by the Carcinogen Identification and Evaluation Unit of IARC from recognized sources of information on carcinogenesis, including data storage and retrieval systems such as MEDLINE and TOXLINE.

For chemicals and some complex mixtures, the major collection of data and the preparation of first drafts of the sections on chemical and physical properties, on analysis, on production and use and on occurrence are carried out under a separate contract funded by the United States National Cancer Institute. Representatives from industrial associations may assist in the preparation of sections on production and use. Information on production and trade is obtained from governmental and trade publications and, in some cases, by direct contact with industries. Separate production data on some agents may not be available because their publication could disclose confidential information. Information on uses may be obtained from published sources but is often complemented by direct contact with manufacturers. Efforts are made to supplement this information with data from other national and international sources.

Six months before the meeting, the material obtained is sent to meeting participants, or is used by IARC staff, to prepare sections for the first drafts of monographs. The first drafts are compiled by IARC staff and sent before the meeting to all participants of the Working Group for review.

The Working Group meets in Lyon for seven to eight days to discuss and finalize the texts of the monographs and to formulate the evaluations. After the meeting, the master copy of each monograph is verified by consulting the original literature, edited and prepared for publication. The aim is to publish monographs within six months of the Working Group meeting.

The available studies are summarized by the Working Group, with particular regard to the qualitative aspects discussed below. In general, numerical findings are indicated as they appear in the original report; units are converted when necessary for easier comparison. The Working Group may conduct additional analyses of the published data and use them in their assessment of the evidence; the results of such supplementary analyses are given in square brackets. When an important aspect of a study, directly impinging on its interpretation, should be brought to the attention of the reader, a comment is given in square brackets.

7. EXPOSURE DATA

Sections that indicate the extent of past and present human exposure, the sources of exposure, the people most likely to be exposed and the factors that contribute to the exposure are included at the beginning of each monograph.

Most monographs on individual chemicals, groups of chemicals or complex mixtures include sections on chemical and physical data, on analysis, on production and use and on occurrence. In monographs on, for example, physical agents, occupational exposures and cultural habits, other sections may be included, such as: historical perspectives, description of an industry or habit, chemistry of the complex mixture or taxonomy. Monographs on biological agents have sections on structure and biology, methods of detection, epidemiology of infection and clinical disease other than cancer.

For chemical exposures, the Chemical Abstracts Services Registry Number, the latest Chemical Abstracts Primary Name and the IUPAC Systematic Name are recorded; other synonyms are given, but the list is not necessarily comprehensive. For biological agents, taxonomy and structure are described, and the degree of variability is given, when applicable.

Information on chemical and physical properties and, in particular, data relevant to identification, occurrence and biological activity are included. For biological agents, mode of replication, life cycle, target cells, persistence and latency and host response are given. A description of technical products of chemicals includes trade names, relevant specifications and available information on composition and impurities. Some of the trade names given may be those of mixtures in which the agent being evaluated is only one of the ingredients.

The purpose of the section on analysis or detection is to give the reader an overview of current methods, with emphasis on those widely used for regulatory purposes. Methods for monitoring human exposure are also given, when available. No critical evaluation or recommendation of any of the methods is meant or implied. The IARC published a series of volumes, *Environmental Carcinogens: Methods of Analysis and Exposure Measurement* (IARC, 1978–93), that describe validated methods for analysing a wide variety of chemicals and mixtures. For biological agents, methods of detection and exposure assessment are described, including their sensitivity, specificity and reproducibility.

The dates of first synthesis and of first commercial production of a chemical or mixture are provided; for agents which do not occur naturally, this information may allow a reasonable estimate to be made of the date before which no human exposure to the agent could have occurred. The dates of first reported occurrence of an exposure are also provided. In addition, methods of synthesis used in past and present commercial production and different methods of production which may give rise to different impurities are described.

Data on production, international trade and uses are obtained for representative regions, which usually include Europe, Japan and the United States of America. It should not, however, be inferred that those areas or nations are necessarily the sole or major sources or users of the agent. Some identified uses may not be current or major applications, and the coverage is not necessarily comprehensive. In the case of drugs, mention of their therapeutic uses does not necessarily represent current practice, nor does it imply judgement as to their therapeutic efficacy.

Information on the occurrence of an agent or mixture in the environment is obtained from data derived from the monitoring and surveillance of levels in occupational environments, air, water, soil, foods and animal and human tissues. When available, data on the generation, persistence and bioaccumulation of the agent are also included. In the case of mixtures, industries, occupations or processes, information is given about all agents present. For processes, industries and occupations, a historical description is also given, noting variations in chemical composition, physical properties and levels of occupational exposure with time and place. For biological agents, the epidemiology of infection is described.

Statements concerning regulations and guidelines (e.g., pesticide registrations, maximal levels permitted in foods, occupational exposure limits) are included for some countries as indications of potential exposures, but they may not reflect the most recent situation, since such limits are continuously reviewed and modified. The absence of information on regulatory status for a country should not be taken to imply that that country does not have regulations with regard to the exposure. For biological agents, legislation and control, including vaccines and therapy, are described.

8. STUDIES OF CANCER IN HUMANS

(a) Types of studies considered

Three types of epidemiological studies of cancer contribute to the assessment of carcinogenicity in humans—cohort studies, case–control studies and correlation (or ecological) studies. Rarely, results from randomized trials may be available. Case series and case reports of cancer in humans may also be reviewed.

Cohort and case–control studies relate the exposures under study to the occurrence of cancer in individuals and provide an estimate of relative risk (ratio of incidence or mortality in those exposed to incidence or mortality in those not exposed) as the main measure of association.

In correlation studies, the units of investigation are usually whole populations (e.g. in particular geographical areas or at particular times), and cancer frequency is related to a summary measure of the exposure of the population to the agent, mixture or exposure circumstance under study. Because individual exposure is not documented, however, a causal relationship is less easy to infer from correlation studies than from cohort and case–control studies. Case reports generally arise from a suspicion, based on clinical experience, that the concurrence of two events—that is, a particular exposure and occurrence of a cancer—has happened rather more frequently than would be expected by chance. Case reports usually lack complete ascertainment of cases in any population, definition or enumeration of the population at risk and estimation of the expected number of cases in the absence of exposure. The uncertainties surrounding interpretation of case reports and correlation studies make them inadequate, except in rare instances, to form the sole basis for inferring a causal relationship. When taken together with case–control and cohort studies, however, relevant case reports or correlation studies may add materially to the judgement that a causal relationship is present.

Epidemiological studies of benign neoplasms, presumed preneoplastic lesions and other end-points thought to be relevant to cancer are also reviewed by working groups. They may, in some instances, strengthen inferences drawn from studies of cancer itself.

(b) *Quality of studies considered*

The Monographs are not intended to summarize all published studies. Those that are judged to be inadequate or irrelevant to the evaluation are generally omitted. They may be mentioned briefly, particularly when the information is considered to be a useful supplement to that in other reports or when they provide the only data available. Their inclusion does not imply acceptance of the adequacy of the study design or of the analysis and interpretation of the results, and limitations are clearly outlined in square brackets at the end of the study description.

It is necessary to take into account the possible roles of bias, confounding and chance in the interpretation of epidemiological studies. By 'bias' is meant the operation of factors in study design or execution that lead erroneously to a stronger or weaker association than in fact exists between disease and an agent, mixture or exposure circumstance. By 'confounding' is meant a situation in which the relationship with disease is made to appear stronger or weaker than it truly is as a result of an association between the apparent causal factor and another factor that is associated with either an increase or decrease in the incidence of the disease. In evaluating the extent to which these factors have been minimized in an individual study, working groups consider a number of aspects of design and analysis as described in the report of the study. Most of these considerations apply equally to case–control, cohort and correlation studies. Lack of clarity of any of these aspects in the reporting of a study can decrease its credibility and the weight given to it in the final evaluation of the exposure.

Firstly, the study population, disease (or diseases) and exposure should have been well defined by the authors. Cases of disease in the study population should have been identified in a way that was independent of the exposure of interest, and exposure should have been assessed in a way that was not related to disease status.

Secondly, the authors should have taken account in the study design and analysis of other variables that can influence the risk of disease and may have been related to the exposure of interest. Potential confounding by such variables should have been dealt with either in the design of the study, such as by matching, or in the analysis, by statistical adjustment. In cohort studies, comparisons with local rates of disease may be more appropriate than those with national rates. Internal comparisons of disease frequency among individuals at different levels of exposure should also have been made in the study.

Thirdly, the authors should have reported the basic data on which the conclusions are founded, even if sophisticated statistical analyses were employed. At the very least, they should have given the numbers of exposed and unexposed cases and controls in a case–control study and the numbers of cases observed and expected in a cohort study. Further tabulations by time since exposure began and other temporal factors are also important. In a cohort study, data on all cancer sites and all causes of death should have been given, to reveal the possibility of reporting bias. In a case–control study, the effects of investigated factors other than the exposure of interest should have been reported.

Finally, the statistical methods used to obtain estimates of relative risk, absolute rates of cancer, confidence intervals and significance tests, and to adjust for confounding should have been clearly stated by the authors. The methods used should preferably have been the generally accepted techniques that have been refined since the mid-1970s. These methods have been reviewed for case–control studies (Breslow & Day, 1980) and for cohort studies (Breslow & Day, 1987).

(c) Inferences about mechanism of action

Detailed analyses of both relative and absolute risks in relation to temporal variables, such as age at first exposure, time since first exposure, duration of exposure, cumulative exposure and time since exposure ceased, are reviewed and summarized when available. The analysis of temporal relationships can be useful in formulating models of carcinogenesis. In particular, such analyses may suggest whether a carcinogen acts early or late in the process of carcinogenesis, although at best they allow only indirect inferences about the mechanism of action. Special attention is given to measurements of biological markers of carcinogen exposure or action, such as DNA or protein adducts, as well as markers of early steps in the carcinogenic process, such as proto-oncogene mutation, when these are incorporated into epidemiological studies focused on cancer incidence or mortality. Such measurements may allow inferences to be made about putative mechanisms of action (IARC, 1991a; Vainio et al., 1992).

(d) Criteria for causality

After the individual epidemiological studies of cancer have been summarized and the quality assessed, a judgement is made concerning the strength of evidence that the agent, mixture or exposure circumstance in question is carcinogenic for humans. In making its judgement, the Working Group considers several criteria for causality. A strong association (a large relative risk) is more likely to indicate causality than a weak association, although it is recognized that relative risks of small magnitude do not imply lack of causality and may be important if the disease is common. Associations that are replicated in several studies of the same design or using different epidemiological approaches or under different circumstances of exposure are more likely to represent a causal relationship than isolated observations from single studies. If there are inconsistent results among investigations, possible reasons are sought (such as differences in amount of exposure), and results of studies judged to be of high quality are given more weight than those of studies judged to be methodologically less sound. When suspicion of carcinogenicity arises largely from a single study, these data are not combined with those from later studies in any subsequent reassessment of the strength of the evidence.

If the risk of the disease in question increases with the amount of exposure, this is considered to be a strong indication of causality, although absence of a graded response is not necessarily evidence against a causal relationship. Demonstration of a decline in risk after cessation of or reduction in exposure in individuals or in whole populations also supports a causal interpretation of the findings.

Although a carcinogen may act upon more than one target, the specificity of an association (an increased occurrence of cancer at one anatomical site or of one morphological type) adds plausibility to a causal relationship, particularly when excess cancer occurrence is limited to one morphological type within the same organ.

Although rarely available, results from randomized trials showing different rates among exposed and unexposed individuals provide particularly strong evidence for causality.

When several epidemiological studies show little or no indication of an association between an exposure and cancer, the judgement may be made that, in the aggregate, they show evidence of lack of carcinogenicity. Such a judgement requires first of all that the studies giving rise to it meet, to a sufficient degree, the standards of design and analysis described above. Specifically, the possibility that bias, confounding or misclassification of exposure or outcome could explain the observed results should be considered and excluded with reasonable certainty. In addition, all studies that are judged to be methodologically sound should be consistent with a relative risk of unity for any observed level of exposure and, when considered together, should provide a pooled estimate of relative risk which is at or near unity and has a narrow confidence interval, due to sufficient population size. Moreover, no individual study nor the pooled results of all the studies should show any consistent tendency for the relative risk of cancer to increase with increasing level of exposure. It is important to note that evidence of lack of carcinogenicity obtained in this way from several epidemiological studies can apply only to the type(s) of cancer studied and to dose levels and intervals between first exposure and observation of disease that are the same as or less than those observed in all the studies. Experience with human cancer indicates that, in some cases, the period from first exposure to the development of clinical cancer is seldom less than 20 years; latent periods substantially shorter than 30 years cannot provide evidence for lack of carcinogenicity.

9. STUDIES OF CANCER IN EXPERIMENTAL ANIMALS

All known human carcinogens that have been studied adequately in experimental animals have produced positive results in one or more animal species (Wilbourn *et al.*, 1986; Tomatis *et al.*, 1989). For several agents (aflatoxins, 4-aminobiphenyl, azathioprine, betel quid with tobacco, bischloromethyl ether and chloromethyl methyl ether (technical grade), chlorambucil, chlornaphazine, ciclosporin, coal-tar pitches, coal-tars, combined oral contraceptives, cyclophosphamide, diethylstilboestrol, melphalan, 8-methoxypsoralen plus ultraviolet A radiation, mustard gas, myleran, 2-naphthylamine, nonsteroidal estrogens, estrogen replacement therapy/steroidal estrogens, solar radiation, thiotepa and vinyl chloride), carcinogenicity in experimental animals was established or highly suspected before epidemiological studies confirmed their carcinogenicity in humans (Vainio *et al.*, 1995). Although this association cannot establish that all agents and mixtures that cause cancer in experimental animals also cause cancer in humans, nevertheless, **in the absence of adequate data on humans, it is biologically plausible and prudent to regard agents and mixtures for which there is *sufficient evidence* (see p. 24) of carcinogenicity in experimental animals as if they presented a carcinogenic risk to humans**. The possibility that a given agent may cause cancer through a species-specific mechanism which does not operate in humans (see p. 27) should also be taken into consideration.

The nature and extent of impurities or contaminants present in the chemical or mixture being evaluated are given when available. Animal strain, sex, numbers per group, age at start of treatment and survival are reported.

Other types of studies summarized include: experiments in which the agent or mixture was administered in conjunction with known carcinogens or factors that modify carcinogenic effects; studies in which the end-point was not cancer but a defined precancerous lesion; and experiments on the carcinogenicity of known metabolites and derivatives.

For experimental studies of mixtures, consideration is given to the possibility of changes in the physicochemical properties of the test substance during collection, storage, extraction, concentration and delivery. Chemical and toxicological interactions of the components of mixtures may result in nonlinear dose–response relationships.

An assessment is made as to the relevance to human exposure of samples tested in experimental animals, which may involve consideration of: (i) physical and chemical characteristics, (ii) constituent substances that indicate the presence of a class of substances, (iii) the results of tests for genetic and related effects, including studies on DNA adduct formation, proto-oncogene mutation and expression and suppressor gene inactivation. The relevance of results obtained, for example, with animal viruses analogous to the virus being evaluated in the monograph must also be considered. They may provide biological and mechanistic information relevant to the understanding of the process of carcinogenesis in humans and may strengthen the plausibility of a conclusion that the biological agent under evaluation is carcinogenic in humans.

(*a*)   *Qualitative aspects*

An assessment of carcinogenicity involves several considerations of qualitative importance, including (i) the experimental conditions under which the test was performed, including route and schedule of exposure, species, strain, sex, age, duration of follow-up; (ii) the consistency of the results, for example, across species and target organ(s); (iii) the spectrum of neoplastic response, from preneoplastic lesions and benign tumours to malignant neoplasms; and (iv) the possible role of modifying factors.

As mentioned earlier (p. 11), the *Monographs* are not intended to summarize all published studies. Those studies in experimental animals that are inadequate (e.g., too short a duration, too few animals, poor survival; see below) or are judged irrelevant to the evaluation are generally omitted. Guidelines for conducting adequate long-term carcinogenicity experiments have been outlined (e.g. Montesano *et al.*, 1986).

Considerations of importance to the Working Group in the interpretation and evaluation of a particular study include: (i) how clearly the agent was defined and, in the case of mixtures, how adequately the sample characterization was reported; (ii) whether the dose was adequately monitored, particularly in inhalation experiments; (iii) whether the doses and duration of treatment were appropriate and whether the survival of treated animals was similar to that of controls; (iv) whether there were adequate numbers of animals per group; (v) whether animals of each sex were used; (vi) whether animals were allocated randomly to groups; (vii) whether the duration of observation was adequate; and (viii) whether the data were adequately reported. If available, recent data on the incidence of specific tumours in historical controls, as well as in concurrent controls, should be taken into account in the evaluation of tumour response.

When benign tumours occur together with and originate from the same cell type in an organ or tissue as malignant tumours in a particular study and appear to represent a stage in the progression to malignancy, it may be valid to combine them in assessing tumour incidence (Huff *et al.*, 1989). The occurrence of lesions presumed to be preneoplastic may in certain instances aid in assessing the biological plausibility of any neoplastic response observed. If an agent or mixture induces only benign neoplasms that appear to be end-points that do not readily progress to malignancy, it should nevertheless be suspected of being a carcinogen and requires further investigation.

(b)   Quantitative aspects

The probability that tumours will occur may depend on the species, sex, strain and age of the animal, the dose of the carcinogen and the route and length of exposure. Evidence of an increased incidence of neoplasms with increased level of exposure strengthens the inference of a causal association between the exposure and the development of neoplasms.

The form of the dose–response relationship can vary widely, depending on the particular agent under study and the target organ. Both DNA damage and increased cell division are important aspects of carcinogenesis, and cell proliferation is a strong determinant of dose–response relationships for some carcinogens (Cohen & Ellwein, 1990). Since many chemicals require metabolic activation before being converted into their reactive intermediates, both metabolic and pharmacokinetic aspects are important in determining the dose–response pattern. Saturation of steps such as absorption, activation, inactivation and elimination may produce nonlinearity in the dose–response relationship, as could saturation of processes such as DNA repair (Hoel *et al.*, 1983; Gart *et al.*, 1986).

(c) *Statistical analysis of long-term experiments in animals*

Factors considered by the Working Group include the adequacy of the information given for each treatment group: (i) the number of animals studied and the number examined histologically, (ii) the number of animals with a given tumour type and (iii) length of survival. The statistical methods used should be clearly stated and should be the generally accepted techniques refined for this purpose (Peto *et al.*, 1980; Gart *et al.*, 1986). When there is no difference in survival between control and treatment groups, the Working Group usually compares the proportions of animals developing each tumour type in each of the groups. Otherwise, consideration is given as to whether or not appropriate adjustments have been made for differences in survival. These adjustments can include: comparisons of the proportions of tumour-bearing animals among the effective number of animals (alive at the time the first tumour is discovered), in the case where most differences in survival occur before tumours appear; life-table methods, when tumours are visible or when they may be considered 'fatal' because mortality rapidly follows tumour development; and the Mantel-Haenszel test or logistic regression, when occult tumours do not affect the animals' risk of dying but are 'incidental' findings at autopsy.

In practice, classifying tumours as fatal or incidental may be difficult. Several survival-adjusted methods have been developed that do not require this distinction (Gart *et al.*, 1986), although they have not been fully evaluated.

10. OTHER DATA RELEVANT TO AN EVALUATION OF CARCINOGENICITY AND ITS MECHANISMS

In coming to an overall evaluation of carcinogenicity in humans (see pp. 25–27), the Working Group also considers related data. The nature of the information selected for the summary depends on the agent being considered.

For chemicals and complex mixtures of chemicals such as those in some occupational situations or involving cultural habits (e.g. tobacco smoking), the other data considered to be relevant are divided into those on absorption, distribution, metabolism and excretion; toxic effects; reproductive and developmental effects; and genetic and related effects.

Concise information is given on absorption, distribution (including placental transfer) and excretion in both humans and experimental animals. Kinetic factors that may affect the dose–response relationship, such as saturation of uptake, protein binding, metabolic activation, detoxification and DNA repair processes, are mentioned. Studies that indicate the metabolic fate of the agent in humans and in experimental animals are summarized briefly, and comparisons of data on humans and on animals are made when possible. Comparative information on the relationship between exposure and the dose that reaches the target site may be of particular importance for extrapolation between species. Data are given on acute and chronic toxic effects (other than cancer), such as organ toxicity, increased cell proliferation, immunotoxicity and endocrine effects. The presence and toxicological significance of cellular receptors is described. Effects on reproduction, teratogenicity, fetotoxicity and embryotoxicity are also summarized briefly.

Tests of genetic and related effects are described in view of the relevance of gene mutation and chromosomal damage to carcinogenesis (Vainio et al., 1992; McGregor et al., 1999). The adequacy of the reporting of sample characterization is considered and, where necessary, commented upon; with regard to complex mixtures, such comments are similar to those described for animal carcinogenicity tests on p. 18. The available data are interpreted critically by phylogenetic group according to the end-points detected, which may include DNA damage, gene mutation, sister chromatid exchange, micronucleus formation, chromosomal aberrations, aneuploidy and cell transformation. The concentrations employed are given, and mention is made of whether use of an exogenous metabolic system *in vitro* affected the test result. These data are given as listings of test systems, data and references. The Genetic and Related Effects data presented in the *Monographs* are also available in the form of Graphic Activity Profiles (GAP) prepared in collaboration with the United States Environmental Protection Agency (EPA) (see also Waters et al., 1987) using software for personal computers that are Microsoft Windows® compatible. The EPA/IARC GAP software and database may be downloaded free of charge from *www.epa.gov/gapdb*.

Positive results in tests using prokaryotes, lower eukaryotes, plants, insects and cultured mammalian cells suggest that genetic and related effects could occur in mammals. Results from such tests may also give information about the types of genetic effect produced and about the involvement of metabolic activation. Some end-points described are clearly genetic in nature (e.g., gene mutations and chromosomal aberrations), while others are to a greater or lesser degree associated with genetic effects (e.g. unscheduled DNA synthesis). In-vitro tests for tumour-promoting activity and for cell transformation may be sensitive to changes that are not necessarily the result of genetic alterations but that may have specific relevance to the process of carcinogenesis. A critical appraisal of these tests has been published (Montesano et al., 1986).

Genetic or other activity manifest in experimental mammals and humans is regarded as being of greater relevance than that in other organisms. The demonstration that an agent or mixture can induce gene and chromosomal mutations in whole mammals indicates that it may have carcinogenic activity, although this activity may not be detectably expressed in any or all species. Relative potency in tests for mutagenicity and related effects is not a reliable indicator of carcinogenic potency. Negative results in tests for mutagenicity in selected tissues from animals treated *in vivo* provide less weight, partly because they do not exclude the possibility of an effect in tissues other than those examined. Moreover, negative results in short-term tests with genetic end-points cannot be considered to provide evidence to rule out carcinogenicity of agents or mixtures that act through other mechanisms (e.g. receptor-mediated effects, cellular toxicity with regenerative proliferation, peroxisome proliferation) (Vainio et al., 1992). Factors that may lead to misleading results in short-term tests have been discussed in detail elsewhere (Montesano *et al.*, 1986).

When available, data relevant to mechanisms of carcinogenesis that do not involve structural changes at the level of the gene are also described.

The adequacy of epidemiological studies of reproductive outcome and genetic and related effects in humans is evaluated by the same criteria as are applied to epidemiological studies of cancer.

Structure–activity relationships that may be relevant to an evaluation of the carcinogenicity of an agent are also described.

For biological agents—viruses, bacteria and parasites—other data relevant to carcinogenicity include descriptions of the pathology of infection, molecular biology (integration and expression of viruses, and any genetic alterations seen in human tumours) and other observations, which might include cellular and tissue responses to infection, immune response and the presence of tumour markers.

11. SUMMARY OF DATA REPORTED

In this section, the relevant epidemiological and experimental data are summarized. Only reports, other than in abstract form, that meet the criteria outlined on p. 11 are considered for evaluating carcinogenicity. Inadequate studies are generally not summarized: such studies are usually identified by a square-bracketed comment in the preceding text.

(a) Exposure

Human exposure to chemicals and complex mixtures is summarized on the basis of elements such as production, use, occurrence in the environment and determinations in human tissues and body fluids. Quantitative data are given when available. Exposure to biological agents is described in terms of transmission and prevalence of infection.

(b) Carcinogenicity in humans

Results of epidemiological studies that are considered to be pertinent to an assessment of human carcinogenicity are summarized. When relevant, case reports and correlation studies are also summarized.

(c) Carcinogenicity in experimental animals

Data relevant to an evaluation of carcinogenicity in animals are summarized. For each animal species and route of administration, it is stated whether an increased incidence of neoplasms or preneoplastic lesions was observed, and the tumour sites are indicated. If the agent or mixture produced tumours after prenatal exposure or in single-dose experiments, this is also indicated. Negative findings are also summarized. Dose–response and other quantitative data may be given when available.

(d)   *Other data relevant to an evaluation of carcinogenicity and its mechanisms*

Data on biological effects in humans that are of particular relevance are summarized. These may include toxicological, kinetic and metabolic considerations and evidence of DNA binding, persistence of DNA lesions or genetic damage in exposed humans. Toxicological information, such as that on cytotoxicity and regeneration, receptor binding and hormonal and immunological effects, and data on kinetics and metabolism in experimental animals are given when considered relevant to the possible mechanism of the carcinogenic action of the agent. The results of tests for genetic and related effects are summarized for whole mammals, cultured mammalian cells and nonmammalian systems.

When available, comparisons of such data for humans and for animals, and particularly animals that have developed cancer, are described.

Structure–activity relationships are mentioned when relevant.

For the agent, mixture or exposure circumstance being evaluated, the available data on end-points or other phenomena relevant to mechanisms of carcinogenesis from studies in humans, experimental animals and tissue and cell test systems are summarized within one or more of the following descriptive dimensions:

(i) Evidence of genotoxicity (structural changes at the level of the gene): for example, structure–activity considerations, adduct formation, mutagenicity (effect on specific genes), chromosomal mutation/aneuploidy

(ii) Evidence of effects on the expression of relevant genes (functional changes at the intracellular level): for example, alterations to the structure or quantity of the product of a proto-oncogene or tumour-suppressor gene, alterations to metabolic activation/inactivation/DNA repair

(iii) Evidence of relevant effects on cell behaviour (morphological or behavioural changes at the cellular or tissue level): for example, induction of mitogenesis, compensatory cell proliferation, preneoplasia and hyperplasia, survival of premalignant or malignant cells (immortalization, immunosuppression), effects on metastatic potential

(iv) Evidence from dose and time relationships of carcinogenic effects and interactions between agents: for example, early/late stage, as inferred from epidemiological studies; initiation/promotion/progression/malignant conversion, as defined in animal carcinogenicity experiments; toxicokinetics

These dimensions are not mutually exclusive, and an agent may fall within more than one of them. Thus, for example, the action of an agent on the expression of relevant genes could be summarized under both the first and second dimensions, even if it were known with reasonable certainty that those effects resulted from genotoxicity.

12. EVALUATION

Evaluations of the strength of the evidence for carcinogenicity arising from human and experimental animal data are made, using standard terms.

It is recognized that the criteria for these evaluations, described below, cannot encompass all of the factors that may be relevant to an evaluation of carcinogenicity. In considering all of the relevant scientific data, the Working Group may assign the agent, mixture or exposure circumstance to a higher or lower category than a strict interpretation of these criteria would indicate.

(a) Degrees of evidence for carcinogenicity in humans and in experimental animals and supporting evidence

These categories refer only to the strength of the evidence that an exposure is carcinogenic and not to the extent of its carcinogenic activity (potency) nor to the mechanisms involved. A classification may change as new information becomes available.

An evaluation of degree of evidence, whether for a single agent or a mixture, is limited to the materials tested, as defined physically, chemically or biologically. When the agents evaluated are considered by the Working Group to be sufficiently closely related, they may be grouped together for the purpose of a single evaluation of degree of evidence.

(i) Carcinogenicity in humans

The applicability of an evaluation of the carcinogenicity of a mixture, process, occupation or industry on the basis of evidence from epidemiological studies depends on the variability over time and place of the mixtures, processes, occupations and industries. The Working Group seeks to identify the specific exposure, process or activity which is considered most likely to be responsible for any excess risk. The evaluation is focused as narrowly as the available data on exposure and other aspects permit.

The evidence relevant to carcinogenicity from studies in humans is classified into one of the following categories:

Sufficient evidence of carcinogenicity: The Working Group considers that a causal relationship has been established between exposure to the agent, mixture or exposure circumstance and human cancer. That is, a positive relationship has been observed between the exposure and cancer in studies in which chance, bias and confounding could be ruled out with reasonable confidence.

Limited evidence of carcinogenicity: A positive association has been observed between exposure to the agent, mixture or exposure circumstance and cancer for which a causal interpretation is considered by the Working Group to be credible, but chance, bias or confounding could not be ruled out with reasonable confidence.

Inadequate evidence of carcinogenicity: The available studies are of insufficient quality, consistency or statistical power to permit a conclusion regarding the presence or absence of a causal association between exposure and cancer, or no data on cancer in humans are available.

Evidence suggesting lack of carcinogenicity: There are several adequate studies covering the full range of levels of exposure that human beings are known to encounter, which are mutually consistent in not showing a positive association between exposure to the agent, mixture or exposure circumstance and any studied cancer at any observed level of exposure. A conclusion of 'evidence suggesting lack of carcinogenicity' is inevitably limited to the cancer sites, conditions and levels of exposure and length of observation covered by the available studies. In addition, the possibility of a very small risk at the levels of exposure studied can never be excluded.

In some instances, the above categories may be used to classify the degree of evidence related to carcinogenicity in specific organs or tissues.

(ii) *Carcinogenicity in experimental animals*

The evidence relevant to carcinogenicity in experimental animals is classified into one of the following categories:

Sufficient evidence of carcinogenicity: The Working Group considers that a causal relationship has been established between the agent or mixture and an increased incidence of malignant neoplasms or of an appropriate combination of benign and malignant neoplasms in (a) two or more species of animals or (b) in two or more independent studies in one species carried out at different times or in different laboratories or under different protocols.

Exceptionally, a single study in one species might be considered to provide sufficient evidence of carcinogenicity when malignant neoplasms occur to an unusual degree with regard to incidence, site, type of tumour or age at onset.

Limited evidence of carcinogenicity: The data suggest a carcinogenic effect but are limited for making a definitive evaluation because, e.g. (a) the evidence of carcinogenicity is restricted to a single experiment; or (b) there are unresolved questions regarding the adequacy of the design, conduct or interpretation of the study; or (c) the agent or mixture increases the incidence only of benign neoplasms or lesions of uncertain neoplastic potential, or of certain neoplasms which may occur spontaneously in high incidences in certain strains.

Inadequate evidence of carcinogenicity: The studies cannot be interpreted as showing either the presence or absence of a carcinogenic effect because of major qualitative or quantitative limitations, or no data on cancer in experimental animals are available.

Evidence suggesting lack of carcinogenicity: Adequate studies involving at least two species are available which show that, within the limits of the tests used, the agent or mixture is not carcinogenic. A conclusion of evidence suggesting lack of carcinogenicity is inevitably limited to the species, tumour sites and levels of exposure studied.

(b) *Other data relevant to the evaluation of carcinogenicity and its mechanisms*

Other evidence judged to be relevant to an evaluation of carcinogenicity and of sufficient importance to affect the overall evaluation is then described. This may include data on preneoplastic lesions, tumour pathology, genetic and related effects, structure–activity relationships, metabolism and pharmacokinetics, physicochemical parameters and analogous biological agents.

Data relevant to mechanisms of the carcinogenic action are also evaluated. The strength of the evidence that any carcinogenic effect observed is due to a particular mechanism is assessed, using terms such as weak, moderate or strong. Then, the Working Group assesses if that particular mechanism is likely to be operative in humans. The strongest indications that a particular mechanism operates in humans come from data on humans or biological specimens obtained from exposed humans. The data may be considered to be especially relevant if they show that the agent in question has caused changes in exposed humans that are on the causal pathway to carcinogenesis. Such data may, however, never become available, because it is at least conceivable that certain compounds may be kept from human use solely on the basis of evidence of their toxicity and/or carcinogenicity in experimental systems.

For complex exposures, including occupational and industrial exposures, the chemical composition and the potential contribution of carcinogens known to be present are considered by the Working Group in its overall evaluation of human carcinogenicity. The Working Group also determines the extent to which the materials tested in experimental systems are related to those to which humans are exposed.

(c)   *Overall evaluation*

Finally, the body of evidence is considered as a whole, in order to reach an overall evaluation of the carcinogenicity to humans of an agent, mixture or circumstance of exposure.

An evaluation may be made for a group of chemical compounds that have been evaluated by the Working Group. In addition, when supporting data indicate that other, related compounds for which there is no direct evidence of capacity to induce cancer in humans or in animals may also be carcinogenic, a statement describing the rationale for this conclusion is added to the evaluation narrative; an additional evaluation may be made for this broader group of compounds if the strength of the evidence warrants it.

The agent, mixture or exposure circumstance is described according to the wording of one of the following categories, and the designated group is given. The categorization of an agent, mixture or exposure circumstance is a matter of scientific judgement, reflecting the strength of the evidence derived from studies in humans and in experimental animals and from other relevant data.

Group 1 —The agent (mixture) is carcinogenic to humans.
The exposure circumstance entails exposures that are carcinogenic to humans.

This category is used when there is *sufficient evidence* of carcinogenicity in humans. Exceptionally, an agent (mixture) may be placed in this category when evidence of carcinogenicity in humans is less than sufficient but there is *sufficient evidence* of carcinogenicity in experimental animals and strong evidence in exposed humans that the agent (mixture) acts through a relevant mechanism of carcinogenicity.

Group 2

This category includes agents, mixtures and exposure circumstances for which, at one extreme, the degree of evidence of carcinogenicity in humans is almost sufficient, as well as those for which, at the other extreme, there are no human data but for which there is evidence of carcinogenicity in experimental animals. Agents, mixtures and exposure circumstances are assigned to either group 2A (probably carcinogenic to humans) or group 2B (possibly carcinogenic to humans) on the basis of epidemiological and experimental evidence of carcinogenicity and other relevant data.

Group 2A—The agent (mixture) is probably carcinogenic to humans.
The exposure circumstance entails exposures that are probably carcinogenic to humans.

This category is used when there is *limited evidence* of carcinogenicity in humans and *sufficient evidence* of carcinogenicity in experimental animals. In some cases, an agent (mixture) may be classified in this category when there is *inadequate evidence* of carcinogenicity in humans, *sufficient evidence* of carcinogenicity in experimental animals and strong evidence that the carcinogenesis is mediated by a mechanism that also operates in humans. Exceptionally, an agent, mixture or exposure circumstance may be classified in this category solely on the basis of *limited evidence* of carcinogenicity in humans.

Group 2B—The agent (mixture) is possibly carcinogenic to humans.
The exposure circumstance entails exposures that are possibly carcinogenic to humans.

This category is used for agents, mixtures and exposure circumstances for which there is *limited evidence* of carcinogenicity in humans and less than *sufficient evidence* of carcinogenicity in experimental animals. It may also be used when there is *inadequate evidence* of carcinogenicity in humans but there is *sufficient evidence* of carcinogenicity in experimental animals. In some instances, an agent, mixture or exposure circumstance for which there is *inadequate evidence* of carcinogenicity in humans but *limited evidence* of carcinogenicity in experimental animals together with supporting evidence from other relevant data may be placed in this group.

Group 3—The agent (mixture or exposure circumstance) is not classifiable as to its carcinogenicity to humans.

This category is used most commonly for agents, mixtures and exposure circumstances for which the *evidence of carcinogenicity* is *inadequate* in humans and *inadequate* or *limited* in experimental animals.

Exceptionally, agents (mixtures) for which the *evidence of carcinogenicity* is *inadequate* in humans but *sufficient* in experimental animals may be placed in this category when there is strong evidence that the mechanism of carcinogenicity in experimental animals does not operate in humans.

Agents, mixtures and exposure circumstances that do not fall into any other group are also placed in this category.

Group 4—The agent (mixture) is probably not carcinogenic to humans.

This category is used for agents or mixtures for which there is *evidence suggesting lack of carcinogenicity* in humans and in experimental animals. In some instances, agents or mixtures for which there is *inadequate evidence* of carcinogenicity in humans but *evidence suggesting lack of carcinogenicity* in experimental animals, consistently and strongly supported by a broad range of other relevant data, may be classified in this group.

13. REFERENCES

Breslow, N.E. & Day, N.E. (1980) *Statistical Methods in Cancer Research*, Vol. 1, *The Analysis of Case–Control Studies* (IARC Scientific Publications No. 32), Lyon, IARC*Press*

Breslow, N.E. & Day, N.E. (1987) *Statistical Methods in Cancer Research*, Vol. 2, *The Design and Analysis of Cohort Studies* (IARC Scientific Publications No. 82), Lyon, IARC*Press*

Cohen, S.M. & Ellwein, L.B. (1990) Cell proliferation in carcinogenesis. *Science*, **249**, 1007–1011

Gart, J.J., Krewski, D., Lee, P.N., Tarone, R.E. & Wahrendorf, J. (1986) *Statistical Methods in Cancer Research*, Vol. 3, *The Design and Analysis of Long-term Animal Experiments* (IARC Scientific Publications No. 79), Lyon, IARC*Press*

Hoel, D.G., Kaplan, N.L. & Anderson, M.W. (1983) Implication of nonlinear kinetics on risk estimation in carcinogenesis. *Science*, **219**, 1032–1037

Huff, J.E., Eustis, S.L. & Haseman, J.K. (1989) Occurrence and relevance of chemically induced benign neoplasms in long-term carcinogenicity studies. *Cancer Metastasis Rev.*, **8**, 1–21

IARC (1973–1996) *Information Bulletin on the Survey of Chemicals Being Tested for Carcinogenicity/Directory of Agents Being Tested for Carcinogenicity*, Numbers 1–17, Lyon, IARC*Press*

IARC (1976–1996), Lyon, IARC*Press*

    Directory of On-going Research in Cancer Epidemiology 1976. Edited by C.S. Muir & G. Wagner

    Directory of On-going Research in Cancer Epidemiology 1977 (IARC Scientific Publications No. 17). Edited by C.S. Muir & G. Wagner

    Directory of On-going Research in Cancer Epidemiology 1978 (IARC Scientific Publications No. 26). Edited by C.S. Muir & G. Wagner

    Directory of On-going Research in Cancer Epidemiology 1979 (IARC Scientific Publications No. 28). Edited by C.S. Muir & G. Wagner

    Directory of On-going Research in Cancer Epidemiology 1980 (IARC Scientific Publications No. 35). Edited by C.S. Muir & G. Wagner

Directory of On-going Research in Cancer Epidemiology 1981 (IARC Scientific Publications No. 38). Edited by C.S. Muir & G. Wagner

Directory of On-going Research in Cancer Epidemiology 1982 (IARC Scientific Publications No. 46). Edited by C.S. Muir & G. Wagner

Directory of On-going Research in Cancer Epidemiology 1983 (IARC Scientific Publications No. 50). Edited by C.S. Muir & G. Wagner

Directory of On-going Research in Cancer Epidemiology 1984 (IARC Scientific Publications No. 62). Edited by C.S. Muir & G. Wagner

Directory of On-going Research in Cancer Epidemiology 1985 (IARC Scientific Publications No. 69). Edited by C.S. Muir & G. Wagner

Directory of On-going Research in Cancer Epidemiology 1986 (IARC Scientific Publications No. 80). Edited by C.S. Muir & G. Wagner

Directory of On-going Research in Cancer Epidemiology 1987 (IARC Scientific Publications No. 86). Edited by D.M. Parkin & J. Wahrendorf

Directory of On-going Research in Cancer Epidemiology 1988 (IARC Scientific Publications No. 93). Edited by M. Coleman & J. Wahrendorf

Directory of On-going Research in Cancer Epidemiology 1989/90 (IARC Scientific Publications No. 101). Edited by M. Coleman & J. Wahrendorf

Directory of On-going Research in Cancer Epidemiology 1991 (IARC Scientific Publications No.110). Edited by M. Coleman & J. Wahrendorf

Directory of On-going Research in Cancer Epidemiology 1992 (IARC Scientific Publications No. 117). Edited by M. Coleman, J. Wahrendorf & E. Démaret

Directory of On-going Research in Cancer Epidemiology 1994 (IARC Scientific Publications No. 130). Edited by R. Sankaranarayanan, J. Wahrendorf & E. Démaret

Directory of On-going Research in Cancer Epidemiology 1996 (IARC Scientific Publications No. 137). Edited by R. Sankaranarayanan, J. Wahrendorf & E. Démaret

IARC (1977) *IARC Monographs Programme on the Evaluation of the Carcinogenic Risk of Chemicals to Humans*. Preamble (IARC intern. tech. Rep. No. 77/002)

IARC (1978) *Chemicals with Sufficient Evidence of Carcinogenicity in Experimental Animals—IARC Monographs Volumes 1–17* (IARC intern. tech. Rep. No. 78/003)

IARC (1978–1993) *Environmental Carcinogens. Methods of Analysis and Exposure Measurement*, Lyon, IARC*Press*

Vol. 1. Analysis of Volatile Nitrosamines in Food (IARC Scientific Publications No. 18). Edited by R. Preussmann, M. Castegnaro, E.A. Walker & A.E. Wasserman (1978)

Vol. 2. Methods for the Measurement of Vinyl Chloride in Poly(vinyl chloride), Air, Water and Foodstuffs (IARC Scientific Publications No. 22). Edited by D.C.M. Squirrell & W. Thain (1978)

Vol. 3. Analysis of Polycyclic Aromatic Hydrocarbons in Environmental Samples (IARC Scientific Publications No. 29). Edited by M. Castegnaro, P. Bogovski, H. Kunte & E.A. Walker (1979)

Vol. 4. Some Aromatic Amines and Azo Dyes in the General and Industrial Environment (IARC Scientific Publications No. 40). Edited by L. Fishbein, M. Castegnaro, I.K. O'Neill & H. Bartsch (1981)

Vol. 5. Some Mycotoxins (IARC Scientific Publications No. 44). Edited by L. Stoloff, M. Castegnaro, P. Scott, I.K. O'Neill & H. Bartsch (1983)

Vol. 6. N-Nitroso Compounds (IARC Scientific Publications No. 45). Edited by R. Preussmann, I.K. O'Neill, G. Eisenbrand, B. Spiegelhalder & H. Bartsch (1983)

Vol. 7. Some Volatile Halogenated Hydrocarbons (IARC Scientific Publications No. 68). Edited by L. Fishbein & I.K. O'Neill (1985)

Vol. 8. Some Metals: As, Be, Cd, Cr, Ni, Pb, Se, Zn (IARC Scientific Publications No. 71). Edited by I.K. O'Neill, P. Schuller & L. Fishbein (1986)

Vol. 9. Passive Smoking (IARC Scientific Publications No. 81). Edited by I.K. O'Neill, K.D. Brunnemann, B. Dodet & D. Hoffmann (1987)

*Vol. 10. Benzene and Alkylated Benzenes (*IARC Scientific Publications No. 85). Edited by L. Fishbein & I.K. O'Neill (1988)

Vol. 11. Polychlorinated Dioxins and Dibenzofurans (IARC Scientific Publications No. 108). Edited by C. Rappe, H.R. Buser, B. Dodet & I.K. O'Neill (1991)

Vol. 12. Indoor Air (IARC Scientific Publications No. 109). Edited by B. Seifert, H. van de Wiel, B. Dodet & I.K. O'Neill (1993)

IARC (1979) *Criteria to Select Chemicals for* IARC Monographs (IARC intern. tech. Rep. No. 79/003)

IARC (1982) *IARC Monographs on the Evaluation of the Carcinogenic Risk of Chemicals to Humans,* Supplement 4, *Chemicals, Industrial Processes and Industries Associated with Cancer in Humans* (IARC Monographs, Volumes 1 to 29), Lyon, IARC*Press*

IARC (1983) *Approaches to Classifying Chemical Carcinogens According to Mechanism of Action* (IARC intern. tech. Rep. No. 83/001)

IARC (1984) *Chemicals and Exposures to Complex Mixtures Recommended for Evaluation in IARC Monographs and Chemicals and Complex Mixtures Recommended for Long-term Carcinogenicity Testing* (IARC intern. tech. Rep. No. 84/002)

IARC (1987a) *IARC Monographs on the Evaluation of Carcinogenic Risks to Humans,* Supplement 6, *Genetic and Related Effects: An Updating of Selected* IARC Monographs *from Volumes 1 to 42,* Lyon, IARC*Press*

IARC (1987b) *IARC Monographs on the Evaluation of Carcinogenic Risks to Humans,* Supplement 7, *Overall Evaluations of Carcinogenicity: An Updating of* IARC Monographs *Volumes 1 to 42,* Lyon, IARC*Press*

IARC (1988) *Report of an IARC Working Group to Review the Approaches and Processes Used to Evaluate the Carcinogenicity of Mixtures and Groups of Chemicals* (IARC intern. tech. Rep. No. 88/002)

IARC (1989) *Chemicals, Groups of Chemicals, Mixtures and Exposure Circumstances to be Evaluated in Future IARC Monographs, Report of an ad hoc Working Group* (IARC intern. tech. Rep. No. 89/004)

IARC (1991a) *A Consensus Report of an IARC Monographs Working Group on the Use of Mechanisms of Carcinogenesis in Risk Identification* (IARC intern. tech. Rep. No. 91/002)

IARC (1991b) *Report of an ad-hoc* IARC Monographs *Advisory Group on Viruses and Other Biological Agents Such as Parasites* (IARC intern. tech. Rep. No. 91/001)

IARC (1993) *Chemicals, Groups of Chemicals, Complex Mixtures, Physical and Biological Agents and Exposure Circumstances to be Evaluated in Future* IARC Monographs, *Report of an ad-hoc Working Group* (IARC intern. Rep. No. 93/005)

IARC (1998a) *Report of an ad-hoc* IARC Monographs *Advisory Group on Physical Agents* (IARC Internal Report No. 98/002)

IARC (1998b) *Report of an ad-hoc IARC Monographs Advisory Group on Priorities for Future Evaluations* (IARC Internal Report No. 98/004)

McGregor, D.B., Rice, J.M. & Venitt, S., eds (1999) *The Use of Short and Medium-term Tests for Carcinogens and Data on Genetic Effects in Carcinogenic Hazard Evaluation* (IARC Scientific Publications No. 146), Lyon, IARC*Press*

Montesano, R., Bartsch, H., Vainio, H., Wilbourn, J. & Yamasaki, H., eds (1986) *Long-term and Short-term Assays for Carcinogenesis—A Critical Appraisal* (IARC Scientific Publications No. 83), Lyon, IARC*Press*

Peto, R., Pike, M.C., Day, N.E., Gray, R.G., Lee, P.N., Parish, S., Peto, J., Richards, S. & Wahrendorf, J. (1980) Guidelines for simple, sensitive significance tests for carcinogenic effects in long-term animal experiments. In: *IARC Monographs on the Evaluation of the Carcinogenic Risk of Chemicals to Humans*, Supplement 2, *Long-term and Short-term Screening Assays for Carcinogens: A Critical Appraisal*, Lyon, IARC*Press*, pp. 311–426

Tomatis, L., Aitio, A., Wilbourn, J. & Shuker, L. (1989) Human carcinogens so far identified. *Jpn. J. Cancer Res.*, **80**, 795–807

Vainio, H., Magee, P.N., McGregor, D.B. & McMichael, A.J., eds (1992) *Mechanisms of Carcinogenesis in Risk Identification* (IARC Scientific Publications No. 116), Lyon, IARC*Press*

Vainio, H., Wilbourn, J.D., Sasco, A.J., Partensky, C., Gaudin, N., Heseltine, E. & Eragne, I. (1995) Identification of human carcinogenic risk in IARC Monographs. *Bull. Cancer*, **82**, 339–348 (in French)

Waters, M.D., Stack, H.F., Brady, A.L., Lohman, P.H.M., Haroun, L. & Vainio, H. (1987) Appendix 1. Activity profiles for genetic and related tests. In: *IARC Monographs on the Evaluation of Carcinogenic Risks to Humans*, Suppl. 6, *Genetic and Related Effects: An Updating of Selected IARC Monographs from Volumes 1 to 42*, Lyon, IARC*Press*, pp. 687–696

Wilbourn, J., Haroun, L., Heseltine, E., Kaldor, J., Partensky, C. & Vainio, H. (1986) Response of experimental animals to human carcinogens: an analysis based upon the IARC Monographs Programme. *Carcinogenesis*, **7**, 1853–1863

GENERAL REMARKS ON MAN-MADE VITREOUS FIBRES

This eighty-first volume of *IARC Monographs* considers certain man-made vitreous (glass-like) fibres of highly variable composition that are widely used for thermal and acoustical insulation and to a lesser extent for other purposes. The generic term, man-made vitreous fibres (MMVFs), denotes non-crystalline, fibrous inorganic substances (silicates) made primarily from rock, slag, glass or other processed minerals. These fibres, also called man-made mineral fibres, include glass fibres (used in glass wool and continuous glass filament), rock (stone)/slag wool and refractory ceramic fibres. Rock (stone) wool, slag wool and glass wool are used extensively in thermal and acoustical insulation, typically in buildings, vehicles and appliances. The refractory ceramic fibres are designed for high-temperature applications, mainly in industrial settings. Continuous glass filament is used primarily in reinforced composite materials for the insulation, electronics and construction industries. These substances were evaluated by a previous IARC Working Group (IARC, 1988) (Table 1). Since these evaluations, new data have become available, which have been incorporated into the monograph and were taken into consideration in the present evaluations.

Man-made vitreous fibres have some physical similarities to asbestos, in particular, their fibrous character which gives them the same aerodynamic properties and leads to their deposition throughout the respiratory tract. Unlike amphibole asbestos, however, they are synthetic and amorphous, and generally have a lower biopersistence in lung tissues. Also, unlike serpentine asbestos, they tend to break transversely rather than cleaving along the fibre axis.

Inhaled asbestos fibres can cause two quite different malignancies in humans: malignant mesothelioma, which arises from the lining of the body cavities, and carcinoma of the lung, which arises from pulmonary epithelial cells (IARC, 1987). Epidemiological studies of human populations exposed to MMVFs have therefore focused on these two types of cancer.

The mechanisms of carcinogenesis by inhaled fibres and the use of data on these mechanisms in the identification of carcinogenic hazard have been reviewed by Kane *et al.* (1996).

Table 1. Previous evaluations[a] of agents (names as used in Volume 43) considered in this volume

Agent[b]	Evidence for carcinogenicity		Overall evaluation of carcinogenicity to humans
	Human	Animal	
Glasswool	I	S	2B
Glass filaments	I	I	3
Rockwool	L	L	2B
Slagwool		I	2B
Ceramic fibres	I (no data)	S	2B

S, sufficient evidence; L, limited evidence; I, inadequate evidence; Group 2B, possibly carcinogenic to humans; Group 3, cannot be evaluated as to its carcinogenicity to humans
[a] *IARC Monographs* Volume 43 (IARC, 1988)
[b] See section 1.1.1(a).

1. Composition, production and use

The compositions of individual MMVF products were historically driven by production technology, the availability of raw materials and, more importantly, the intended use and the temperature ranges over which the products were designed to operate. During the period 1940–1980, changes in product formulation were introduced as production methods were improved or alternative raw materials became available. Most of these changes represented minor modifications to basic product formulations, but more significant changes took place in the early 1990s. In recognition of concerns over the possible adverse health effects of the fibres released from MMVF products and in response to governmental regulations, some manufacturers altered the chemical compositions of their products to enhance their solubility in biological systems. Other manufacturers developed completely new products (e.g. alkaline earth silicate wools and high-alumina, low-silica wools) to achieve the same effect. These products have become commercially available so recently that no relevant epidemiological data have yet been published.

A large experimental database is available on many fibre compositions, although inevitably the number of epidemiological studies on fibres is limited. For new fibres with compositions that differ considerably from those of the older fibre types, studies of toxicity and determinations of biopersistence are required for evaluations of possible inhalation hazards.

2. Toxicity

The end-points used in short-term toxicity studies range from inflammation in experimental animals *in vivo* to cytotoxicity and cell activation *in vitro*. In-vitro assays vary in duration from hours to a few days at most. During long-term residence in the lung, some non-biopersistent fibres undergo changes that act to dissolve, shorten or otherwise decrease the biological activity of the long fibres. This decrease in biological activity would not be detected in short-term assays and it would be difficult to extrapolate these assays to predict long-term effects.

3. Chronic inflammation, fibrosis and cancer

Chronic inflammation and increased turnover of epithelial cells are features of human cancers that are associated with chronic infections in the liver, gastric mucosa and colon (IARC, 1999). Chronic or persistent inflammation, especially in the lung, is frequently accompanied by progressive fibrosis in humans with idiopathic pulmonary fibrosis (reviewed by Samet, 2000); a sevenfold increase in incidence of lung cancer was reported in a recent cohort study of a population with idiopathic pulmonary fibrosis, although confounding by cigarette smoking could not be ruled out (Hubbard *et al.*, 1999).

In chronic inhalation assays of particulate materials in rodents, chronic inflammation and fibrosis almost always precede the development of lung cancer (Davis & Cowie, 1990). Chronic inflammation may contribute to the initiation, promotion and progression of tumours by several mechanisms. Firstly, inflammatory cells release reactive oxygen and nitrogen species that may lead to DNA damage in adjacent parenchymal cells. Secondly, inflammatory cells may release mediators such as cytokines, growth factors and proteases that may alter proliferation, differentiation and migration of preneoplastic cells (reviewed by Coussens & Werb, 2001). Activated fibroblasts may play a role in tumour progression by increasing turnover of the extracellular matrix which may also alter the adhesion, differentiation, proliferation and motility of epithelial cells. Active fibrosis is often accompanied by angiogenesis that may provide a favourable local environment for growth and invasion of developing tumours (reviewed by Tlsty, 2001). Although the experimental evidence for these processes in the pathogenesis of human lung cancer is currently limited, these proposed mechanistic links between chronic inflammation, fibrosis and cancer provide a plausible biological mechanism for lung carcinogenesis by fibres.

4. Studies of cancer in humans

Since the publication of the previous *IARC Monographs* on MMVFs (IARC, 1988), there have been substantial improvements in the quality of the epidemiological information available for the evaluation of the carcinogenicity of glass fibres, continuous glass filament and rock (stone)/slag wool. The new investigations have addressed the limitations of the earlier cohort studies of workers exposed to MMVFs from the United States of America and Europe, particularly concerning the lack of adjustments in these studies of lung cancer risk to take into account concomitant risk factors such as smoking and other sources of occupational exposure.

These studies, like all epidemiological investigations, have limitations that must be borne in mind when interpreting their results. Although the methods of exposure assessment used in these studies are far better than in most, there is still the potential for exposure misclassification. Most notably these studies were not able to examine fully the risks to workers who were exposed to the more durable fibres, which appear to be more hazardous based on toxicological studies. Information on smoking and on the other potential confounders that were adjusted for in these studies was also subject to measurement error, which may have influenced the validity of the adjustments made. Underascertainment and misclassification of mesothelioma were possible in these studies, since they relied primarily upon information from death certificates. Finally, although these studies were very large by epidemiological standards, their sensitivity may have been limited by the low concentrations of fibres to which a large proportion of the study population was exposed.

There is some concern that workers in industries that use or remove products containing MMVFs (e.g. construction workers), may have experienced higher, but perhaps more intermittent exposure. The data available to evaluate risks for cancer from exposure to MMVFs in these workers are very limited.

The results of studies on mortality among workers in the refractory ceramic fibre industry have also been published since the last IARC Monograph. However, the epidemiological data for refractory ceramic fibres are still very limited. Radiographic evidence indicating pleural plaques has been reported for refractory ceramic fibres workers. Although the prognostic significance of pleural plaques is unclear, such plaques are common in workers exposed to asbestos.

5. Studies of cancer in experimental animals

The carcinogenicity of fibres in experimental animals has been studied using very different routes of administration, i.e. inhalation, intratracheal instillation or intracavitary injection. There is no general agreement on which of these routes of administration best predicts human cancer risk, but it is known that intraperitoneal injection allows high doses of fibres to reach the target organ.

Muhle and Pott (2000) analysed studies of asbestos inhalation and concluded that the rat inhalation model is not sufficiently sensitive to predict the cancer risk presented by fibre types other than asbestos for humans and proposed that the intraperitoneal injection test be used instead. In contrast, Maxim and McConnell (2001) reported that well-conducted inhalation studies of carcinogenicity are very sensitive and that rats may be more sensitive than humans in detecting the carcinogenic potential of MMVFs.

In a recent statistical analysis of the available data from studies that used intraperitoneal injection, chronic inhalation and measures of biopersistence, Bernstein *et al.* (2001a, b) showed that the studies that used intraperitoneal injection provide a ranking comparable to that obtained in studies of carcinogenicity following chronic inhalation of fibres of similar biopersistence and length.

6. Administration to experimental animals by inhalation

Before 1989, a number of inhalation studies on rodents had been conducted to evaluate the biological effects of the different types of MMVF. The results of many of these studies were negative for fibrosis and tumorigenesis even when relatively durable fibres were tested. For example, different results were obtained in two studies of refractory ceramic fibres — one study reported both fibrosis and tumorigenesis while the other reported neither fibrosis nor tumours in rats and mesothelioma in only 2% of hamsters. Many earlier studies did not appreciate the importance of fibre diameter in the respirability of fibres in the rat. In addition, fibres were often ground before administration; this procedure significantly shortened their length and often resulted in exposure of the test animals to primarily short fibres. Thus, it is not surprising that some inhalation studies of amphibole asbestos reported no tumours. The contradictory results of these studies led to a better understanding of the importance of respirability and length of fibres and to the development of new study designs (Hesterberg *et al.*, 1993; Bernstein *et al.*, 1995).

More recent inhalation studies in rodents have addressed the technological limitations of the earlier studies using test fibres prepared by new size-separation methods. Such fibres are respirable by rats and long enough to be biologically active, with nominal dimensions of 1×20 μm. An aerosolization system has been designed to create uniform, high concentrations of airborne fibres without destroying the biologically important long–thin fibre geometry.

In the chronic inhalation studies of MMVFs reviewed in section 3, the Working Group has clearly noted those studies that they considered to be 'well-conducted long-term inhalation studies' which meet the criteria summarized above.

7. Administration to experimental animals by intraperitoneal injection

The potential of asbestos fibres to produce mesothelioma was first demonstrated in animals by the implantation or injection of fibres into the pleural cavity of rats (Wagner, 1963; Wagner & Berry, 1969). Subsequently, Stanton and Wrench (1972) showed by implantation in the pleural cavity, and Pott and Friedrichs (1972) and Pott (1978) by injection into the peritoneal cavity, that fibre shape was important and that fibres can produce tumours if they are sufficiently long, thin and durable. Since then the intraperitoneal injection route has been used more often than pleural implantation due to its relative simplicity.

The intraperitoneal test, in which fibres are injected directly into the intraperitoneal cavity, bypasses the natural route of exposure. Because the lung is bypassed, the natural mechanisms by which the lung removes, dissolves or breaks fibres, thereby reducing or eliminating potential exposure of the pleural cavity, do not operate. Therefore the intraperitoneal test has no physiologically imposed maximum dose to which the animals can be exposed. The intraperitoneal test can indicate whether a fibre should be classified as a carcinogen if a proper positive control is used.

8. References

Bernstein, D.M., Thevenaz, P., Fleissner, H., Anderson, R., Hesterberg, T.W. & Mast, R. (1995) Evaluation of the oncogenic potential of man-made vitreous fibres: The inhalation model. *Ann. occup. Hyg.*, **39**, 661–672

Bernstein, D.M., Riego Sintes, J.M., Ersboell, B.K. & Kunert, J. (2001a) Biopersistence of synthetic mineral fibers as a predictor of chronic inhalation toxicity in rats. *Inhal. Toxicol.*, **13**, 823–849

Bernstein, D.M., Riego Sintes, J.M., Ersboell, B.K. & Kunert, J. (2001b) Biopersistence of synthetic mineral fibers as a predictor of chronic intraperitoneal injection tumor response in rats. *Inhal. Toxicol.*, **13**, 851–875

Coussens, L.M. & Werb, Z. (2001) Inflammatory cells and cancer: think different! *J. exp. Med.*, **6**, F23–F26

Davis, J.M.G. & Cowie, H.A. (1990) The relationship between fibrosis and cancer in experimental animals exposed to asbestos and other fibers. *Environ. Health Perspect.*, **88**, 305–309

Hesterberg, T.W., Miiller, W.C., McConnell, E.E., Chevalier, J., Hadley, J., Bernstein, D.M., Thevenaz, P. & Anderson, R. (1993) Chronic inhalation toxicity of size-separated glass fibers in Fischer 344 rats. *Fundam. appl. Toxicol.*, **20**, 464–476

Hubbard, R., Venn, A., Lewis, S. & Britton, J. (1999) Lung cancer and cryptogenic fibrosing: A population based cohort study. *Am. J. respir. crit. Care Med.*, **161**, 5–8

IARC (1987) *IARC Monographs on the Evaluation of Carcinogenic Risks to Humans*, Suppl. 7, *Overall Evaluations of Carcinogenicity: An Updating of* IARC *Monographs Volumes 1 to 42*, Lyon, IARCPress, pp. 106–116

IARC (1988) *IARC Monographs on the Evaluation of Carcinogenic Risks to Humans*, Vol. 43, *Man-made Mineral Fibres and Radon*, Lyon, IARCPress, pp. 33–171

IARC (1999) *IARC Monographs on the Evaluation of Carcinogenic Risks to Humans*, Vol. 74, *Surgical Implants and Other Foreign Bodies*, Lyon, IARCPress, pp. 313–322

Kane, A.B., Boffetta, P., Saracci, R. & Wilbourn, J.D., eds (1996) *Mechanisms of Fibre Carcinogenesis* (IARC Scientific Publications No. 140), Lyon, IARCPress

Maxim, L.D. & McConnell, E.E. (2001) Interspecies comparisons of the toxicity of asbestos and synthetic vitreous fibers: A weight-of-the-evidence approach. *Regul. Toxicol. Pharmacol.*, **33**, 1–24

Muhle, H. & Pott, F. (2000) Asbestos as reference material for fibre-induced cancer. *Arch. occup. environ. Health*, **73**, 53–59

Pott, F. (1978) Some aspects on the dosimetry of the carcinogenic potency of asbestos and other fibrous dusts. *Staub-Reinhalt. Luft*, **38**, 486–490

Pott, F. & Friedrichs, K.H. (1972) [Tumours in rats after i.p. injection of fibrous dust.] *Naturwissenschaften*, **59**, 318 (in German)

Samet, J.M. (2000) Does idiopathic pulmonary fibrosis increase lung cancer risk? *Am. J. respir. crit. Care Med.*, **161**, 1–2

Stanton, M.F. & Wrench, C. (1972) Mechanisms of mesothelioma induction with asbestos and fibrous glass. *J. natl Cancer Inst.*, **48**, 797–821

Tlsty, T.D. (2001) Stromal cells can contribute oncogenic signals. *Seminars Cancer Biol.*, **11**, 97–104

Wagner, J.C. (1963) Asbestosis in experimental animals. *Br. J. ind. Med.*, **20**, 1–12

Wagner, J.C. & Berry, G. (1969) Mesotheliomas in rats following inoculation with asbestos. *Br. J. Cancer*, **23**, 567–581

MAN-MADE VITREOUS FIBRES

1. Exposure Data

1.1 Chemical and physical data

1.1.1 *Nomenclature and general description*

Man-made vitreous fibre (MMVF) is a generic name used to describe an inorganic fibrous material manufactured primarily from glass, rock, minerals, slag and processed inorganic oxides. The MMVFs produced are non-crystalline (glassy, vitreous, amorphous). Other names for MMVFs include manufactured vitreous fibres, man-made mineral fibres (MMMF), machine-made mineral fibres and synthetic vitreous fibres. Continuous glass filament is sometimes also referred to as 'glass textile fibre'.

The term 'mineral wool' has been used in the USA to describe only rock (stone) wool and slag wool. In Europe, 'mineral wool' also includes glass wool. In this monograph, the terms rock (stone) wool, slag wool and glass wool are used rather than mineral wool, whenever possible.

Man-made vitreous fibres are manufactured by a variety of processes based on the attenuation of a thin stream of molten inorganic oxides at high temperatures. All commercially important MMVFs are silica-based and contain various amounts of other inorganic oxides. The non-silica components typically include, but are not limited to, oxides of alkaline earths, alkalis, aluminium, boron, iron and zirconium. These additional oxides may be constituents of the raw materials used to make the fibres, or they may be added to enhance the manufacturing process or the product performance.

Depending on the process of fibre formation, MMVFs are produced either as wool, which is a mass of tangled, discontinuous fibres of variable lengths and diameters, or as filaments, which are continuous fibres (of indeterminate length) with diameters having ranges that are more uniform and typically thicker than those of wool.

(a) Categorization

The previous *IARC Monographs* on man-made mineral fibres (IARC, 1988) grouped fibres into five categories based loosely on raw materials, production process and/or product application. These categories were: glass filament, glass wool, rock wool, slag wool and ceramic fibres.

To reflect developments in the industry, the categories have been expanded and modified somewhat in this monograph, as depicted in Figure 1. The present monograph

Figure 1. Categories of MMVFs

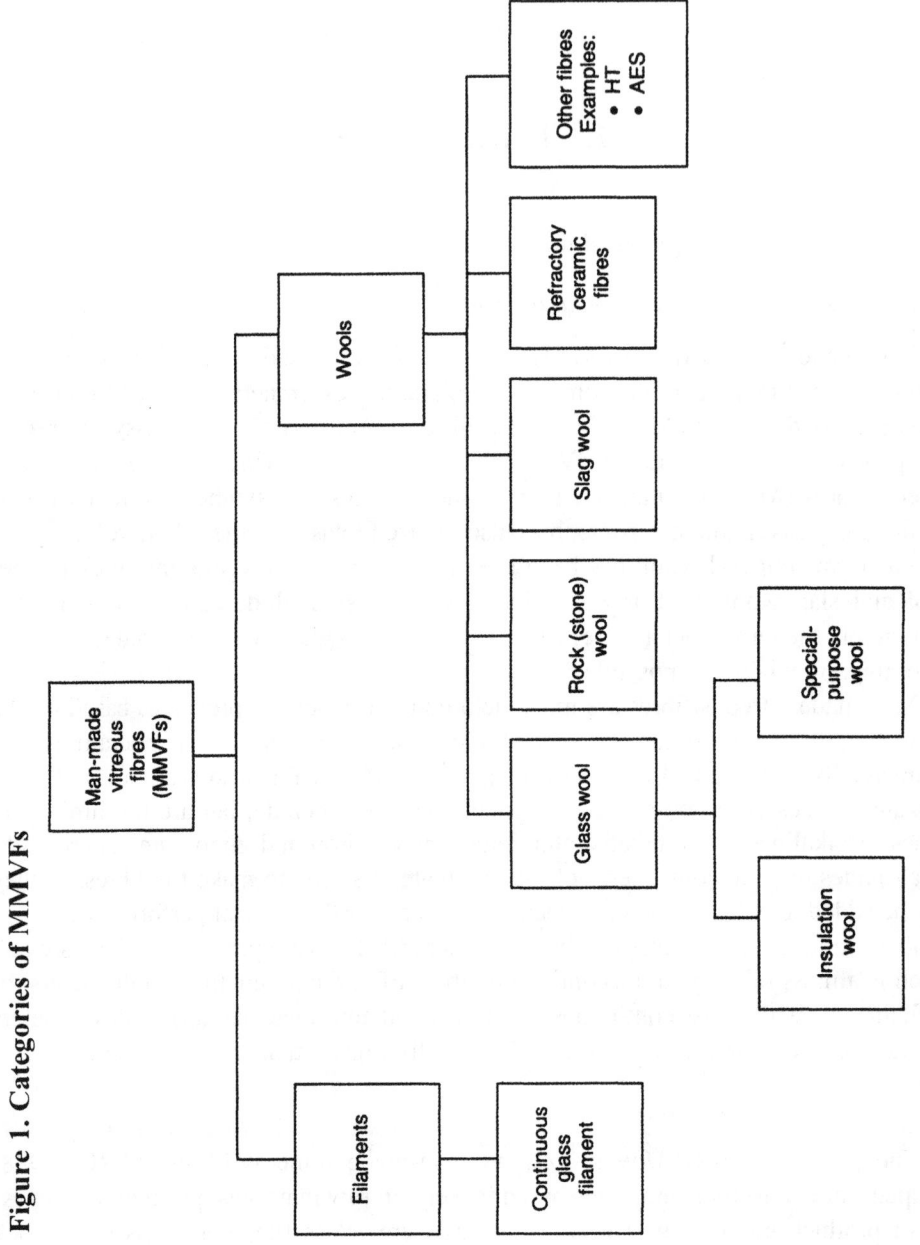

Within each of these categories, there are commercial products representing a range of compositions and durabilities.
AES, alkaline earth silicate wools; HT, high-alumina, low-silica wools

includes only vitreous fibres, whereas the 1988 monograph also included some crystalline ceramic fibres (e.g. silicon carbide) used in high-temperature applications. In this monograph, the MMVFs evaluated in the ceramic fibre category are wool-type fibres known as refractory ceramic fibres.

Certain characteristics of MMVFs such as their respirability and biopersistence in the lung have been the focus of increased attention and research in recent years, and a number of new fibres with reduced biopersistence have been developed. These fibre characteristics are discussed in detail in section 4.1, and the regulatory categorization systems adopted by the European Union and by several other countries, based on biopersistence and respirability, are described in section 1.5. The development of the newer fibres is recognized in Figure 1 under the category 'other fibres'. Examples of more biosoluble fibres include the alkaline earth silicate wools and the high-alumina, low-silica wools. Other newly developed fibres may be less biosoluble and more biopersistent.

Finally, the generic names of several categories of MMVFs have been updated in the current monograph (i.e. glass filament → continuous glass filament; glasswool → glass wool; rockwool → rock (stone) wool; slagwool → slag wool; ceramic fibres → refractory ceramic fibres (see also General Remarks on Man-made Vitreous Fibres).

1.1.2 *Chemical and physical properties*

(*a*) *Chemical properties*

The MMVFs have a broad variety of chemical compositions as shown in Table 1. Within each traditional category of MMVF, the composition of the fibres may vary substantially. Several factors account for the compositional variability of MMVFs:

End-use: The end-use of each product requires fibres to have specific chemical and physical characteristics. For example, 'continuous glass filament' includes eight distinctly different fibre types. Each type has its own formulation with a narrow range of variability. The formulations differ considerably from one another because each type is designed for a specific set of performance criteria, such as high strength, high electrical resistivity or resistance to attack by various chemical agents. Similarly, the grades of refractory ceramic fibres designed for high-temperature (1000–1460 °C) end-uses may have a high alumina and zirconia content (Maxim *et al.*, 1999a; EIPPCB, 2000; Mast *et al.*, 2000a).

Manufacturing requirements: Variations in manufacturing processes and in the availability of raw materials are responsible for much of the compositional variation in glass, rock (stone) and slag wools.

Biopersistence considerations: MMVFs have historically been made with a range of compositions and associated durabilities. Since it was recognized that fibre biopersistence affects the potential effects on respiratory health and that fibre chemistry is an important determinant of biopersistence, the industry has introduced some new,

Table 1. Typical chemical compositional ranges for classes of MMVFs expressed as oxide mass percentage (for the major constituents found in most commercially important MMVFs)

	Continuous glass filament[a]	Glass wool[b,c] Insulation wool	Glass wool Special-purpose fibre	Rock (stone) wool[d]	Slag wool[b,c]	Refractory ceramic fibres[c]	Alkaline earth silicate wool[e]	High-alumina, low-silica wool[f]
SiO_2	52–75	55–70	54–69	43–50	38–52	47–54	50–82	33–43
Al_2O_3	0–30	0–7	3–15	6–15	5–16	35–51	<2	18–24
CaO	0–25	5–13	0–21	10–25	20–43	<1		
MgO	0–10	0–5	0–4.5	6–16	4–14	<1		
MgO + CaO	0–35	5–18	0–25.5	16–41	24–57		18–43	23–33
BaO	0–1	0–3	0–5.5					
ZnO	0–5		0–4.5					
Na_2O		13–18	0–16	1–3.5	0–1	<1	<1	
K_2O		0–2.5	0–15	0.5–2	0.3–2	<1	<1	
$Na_2O + K_2O$	0–21	12–20.5			0.3–3			1–10
B_2O_3	0–24	0–12	4–11	<1	<1		<1	
Fe_2O_3[g]	0–5	0–5	0–0.4		0–5	0–1		
FeO				3–8				3–9
TiO_2	0–12	0–0.5	0–8	0.5–3.5	0.3–1	0–2		0.5–3
ZrO_2	0–18		0–4			0–17	0–6	
$Al_2O_3 + TiO_2 + ZrO_2$								
P_2O_5				<1	0–0.5		<6	
F_2	0–5	0–1.5	0–2					

Table 1 (contd)

	Continuous glass filament[a]	Glass wool[b,c]		Rock (stone) wool[d]	Slag wool[b,c]	Refractory ceramic fibres[c]	Alkaline earth silicate wool[e]	High-alumina, low-silica wool[f]
		Insulation wool	Special-purpose fibre					
S					0–2			
SO_3		0–0.5						
Li_2O	0–1.5	0–0.5						

[a] Hartman et al. (1996)
[b] EIPPCB (2000) (includes the new less biopersistent fibre-types)
[c] TIMA (1993)
[d] Guldberg et al. (2000)
[e] ECFIA & RCFC (2001)
[f] Guldberg et al. (2002) (HT fibre; CAS No. 287922-11-6)
[g] Total iron expressed as Fe_2O_3

less biopersistent fibre compositions. To accomplish this, the industry has extended the traditional compositional ranges of MMVFs in several ways:
— by increasing the content of alkali oxides and borate in glass wools;
— by substituting alumina for silica or alkaline earth oxides for alumina in rock wools; and
— by developing high-temperature-resistant compositions based on the alkaline earth silicate (AES) wools as an alternative to the aluminosilicate compositions of refractory ceramic fibres, in some applications.

The raw materials that have been commonly used to make MMVFs are listed in Table 2. The sources for the raw materials include the following (TIMA, 1993):

Mined: materials mined or quarried from the earth that have received only minimal physical processing to ensure the required particle size and reasonable chemical homogeneity.

Processed: mined materials that have received minimal chemical processing such as thermal treatment to remove water or carbonate.

Recycled: materials that are by-products of the manufacture of MMVFs or other manufacturing processes.

Manufactured: relatively pure manufactured chemical compounds.

(i) Continuous glass filament

Most of the continuous glass filament produced worldwide has an E-glass composition (Table 3), developed solely for electrical applications and more than 98% of all continuous glass filament currently produced is of this type (EIPCCB, 2000; APFE, 2001). E-glass is a calcium-aluminosilicate glass, in which the alkali oxides of sodium and potassium are maintained at low concentrations (< 2 weight %) to achieve acceptable electrical properties. Boron oxide is often a major additive, but in recent years, alternative formulations of E-glass without boron oxide have been developed to reduce emissions of boron compounds into the air during production and to lower the cost of raw material. These boron-free formulations are used in applications other than printed circuit boards or aerospace materials (Hartman *et al.*, 1996).

Other types of glass are also produced as continuous filament. These are used in applications that require specific properties such as high mechanical strength, increased temperature resistance, improved resistance to corrosion, resistance to alkali in cement or low dielectric properties (see Table 3). C-glass is resistant to acids and is used in composites that come into contact with mineral acids and as a reinforcement material in bituminous roofing sheet. AR-glass is used for cement reinforcement and differs from other glasses in that it contains zirconium oxide, which provides resistance to corrosion by alkalis. S-glass is a high-strength glass developed in the 1960s for applications such as rocket motor cases. S-glass is difficult and costly to make and is therefore limited to highly technological uses (Loewenstein, 1993; Hartman *et al.*, 1996; APFE, 2001).

Table 2. Raw materials commonly used in the manufacture of MMVFs

Raw material	Desired element	Source	Continuous glass filament	Glass wool - Insulation wool	Glass wool - Special-purpose fibre	Rock (stone) wool	Slag wool	Refractory ceramic fibre	AES wool	High-alumina, low-silica wool
Anorthosite	Al, Si	Mined								x
Basalt	Si, Ca, Mg, Fe	Mined				x	x			x
Bauxite	Al	Mined					x			x
Colemanite	B	Mined	x		x					
Dolomite	Ca, Mg	Mined	x	x	x	x	x		x	x
Fluorspar	F	Mined	x		x					
Kaolin clay	Al	Mined	x	x	x					
Limestone	Ca	Mined	x	x	x	x	x	x		
Nepheline syenite	Al	Mined		x		x	x			x
Silica sand	Si	Mined	x	x	x	x	x	x	x	x
Ulexite	B	Mined		x						
Wollastonite	Ca, Si	Mined			x				x	
Zircon sand	Zr, Si	Mined			x			x	x	
Briquettes (artificial stones)	Si, Al, Ca, Mg	Processed								x
Burned dolomite	Ca, Mg	Processed	x	x	x				x	
Granite	Si, Al	Processed				x	x			x
Alu-dross	Al	Recycled								x
Blast furnace slag	Si, Al, Ca, Mg	Recycled				x	x			x
Converter slag	Fe, Ca	Recycled								x
Cullet[a]	Si, Ca, Mg, Na, B	Recycled		x						
Ladle slag	Al, Fe, Ca	Recycled								
Alumina	Al	Manufactured		x	x			x		x
Borax (5 H_2O)	B	Manufactured		x						x
Magnesia	Mg	Manufactured							x	
Magnesite	Mg	Manufactured	x		x					

Table 2 (contd)

Raw material	Desired element	Source	Continuous glass filament	Glass wool		Rock (stone) wool	Slag wool	Refractory ceramic fibre	AES wool	High-alumina, low-silica wool
				Insulation wool	Special-purpose fibre					
Manganese dioxide	Oxidizing power	Manufactured		x						
Sodium nitrate	Oxidizing power	Manufactured		x						
Sodium carbonate	Na	Manufactured		x						
Sodium sulfate	Oxidizing power	Manufactured	x	x						
Zirconia	Zr	Manufactured	x		x			x		

From TIMA (1993)
AES, alkaline earth silicate
[a]Cullet (broken or waste glass) includes purchased recycled cullet and cullet recycled from the same manufacturing plant.

Table 3. Typical chemical compositional ranges of representative continuous glass filament expressed as oxide (wt%)

	E-glass (high electrical resistivity)	ECR-glass (high electrical resistivity, corrosion resistant)	C-glass (acid resistant)	D-glass (low dielectric constant)	R-glass (high strength)	AR-glass (alkali resistant)	S-glass (high strength, high-temperature resistant)
SiO_2	52–56	54–62	64–68	72–75	55–65	55–75	64–66
Al_2O_3	12–16	9–15	3–5	0–1	15–30	0–5	24–25
CaO	16–25	17–25	11–15	0–1	9–25	1–10	0–0.1
MgO	0–5	0–4	2–4		3–8		9–10
BaO			0–1				
ZnO		2–5					
$Na_2O + K_2O$	0–2	0–2	7–10	0–4	0–1	11–21	0–0.2
Li_2O						0–1.5	
B_2O_3	5–10		4–6	21–24		0–8	
Fe_2O_3 [a]	0–0.8	0–0.8	0–0.8	0–0.3		0–5	0–0.1
TiO_2	0–1.5	0–4				0–12	
ZrO_2						1–18	
F_2	0–1				0–0.3	0–5	

From Hartman et al. (1996)
[a] Total iron expressed as Fe_2O_3

(ii) *Glass wool*

In principle, many different chemical elements could be present in glass. However, in commercial glass manufacturing, the number of oxides used is limited by their cost. Almost all of the glass products manufactured have silicon dioxide, silica (SiO_2), as the single largest oxide ingredient, measured by weight or volume, in the final composition. To form a glass, a glass-forming compound, or glass former, is required. A glass former is a compound that, in its pure form, can be melted and quenched into the glassy state. In principle, the oxide glass formers can be boric oxide (B_2O_3), phosphorus pentoxide (P_2O_5) or even germanium dioxide (GeO_2), but SiO_2 is the major commercial glass former because it is readily available in a variety of inexpensive forms that can be mixed and processed into a glass (TIMA, 1993).

Although SiO_2 is the principal ingredient, it is necessary to modify the composition using other oxides, commonly referred to as either intermediate oxides or modifiers. There is no sharp distinction between the intermediates and modifiers. However, oxides such as aluminium oxide, alumina (Al_2O_3), titanium dioxide, titania (TiO_2) and zinc oxide (ZnO) are often classified as intermediates, while oxides such as magnesium oxide, magnesia (MgO), lithium oxide, lithia (Li_2O), barium oxide, baria (BaO), calcium oxide, calcia (CaO), sodium oxide, soda (Na_2O) and potassium oxide (K_2O) are usually classified as modifiers. Sometimes, the modifiers are called fluxes, while the intermediate oxides are referred to as stabilizers (TIMA, 1993).

Glasses containing a large fraction of fluxes permit reaction of the raw materials to occur at relatively low temperatures, but such glasses tend to have lower chemical resistance. As an example, a sodium silicate glass with a large fraction of sodium oxide is soluble in water and, in fact, such compositions are manufactured as soluble silicates or water-soluble glasses. The intermediate oxides help impart to a silicate glass a higher degree of chemical resistance, and they control, together with the fluxes, the viscous character of the melt, which is especially important in fiberization (TIMA, 1993).

Most glass wool has been used for a variety of insulation applications. An additional category has been used to group those glass fibres produced by flame attenuation for special applications. This category, termed 'special-purpose fibres' in Figure 1 and Table 1, includes, for example, fibres such as E-glass and 475-glass used for high-efficiency air filtration media, acid battery separators and certain fine-diameter glass fibres.

(iii) *Rock (stone) and slag wool*

Typical modern rock (stone) and slag wools are composed of calcium magnesium aluminium silicate glass. They are produced by melting a mixture of various slags and/or rock raw materials in a coke-fired cupola. Alternatively, they can be melted in an electric or gas-heated furnace. For rock (stone) wool, the procedure is carried out using a mixture of various natural and synthetic rock sources to yield the desired composition. In the manufacturing of both rock (stone) and slag wool, one raw material is normally the main component, and other materials are added to make up for a particular deficiency in that raw material. If, for example, the main component is too rich in acid oxides such as silica, then limestone or a slag rich in calcium oxide is added. In slag wool production, iron-ore blast-furnace slag is the primary component, while in rock (stone) wool production, basalt is usually the primary raw material. In rock (stone) and slag wool produced from materials melted in a cupola with coke as fuel, all the iron oxide is reduced to ferrous oxide (FeO). During the spinning process, a surface layer may form in which the iron is oxidized to ferric oxide (Fe_2O_3). Typically 8–15% of the iron is oxidized to ferric oxide. In an electric furnace melting basalt, up to 50% of the iron is in the form of ferric oxide which is more evenly distributed throughout the entire fibre volume than after heating in a coke-fired cupola (TIMA, 1993).

The production of slag wool in Europe began in the 1880s using slags of various types and continued until the mid-1940s. After the Second World War, most plants began using rock rather than slag as the raw material and currently most European plants continue to melt rock. In the USA, the production of rock (stone) wool dominated from about 1900 until the late 1930s when several of the rock (stone) wool plants converted to iron-ore blast-furnace slag, a waste-product in the production of pig iron because the use of slag was more economical. While the use of slags other than iron-ore blast-furnace slag was once quite widespread, this is no longer the case. Slag formed during the reduction of iron ore to pig iron is now the primary raw material used in the USA to make slag wool. It accounts for 70–90% of the weight of the raw materials that make up the slag wool. Since the mid-1970s, the slag wool industry in the USA has relied entirely on blast-furnace slag with small amounts of additives such as phosphate-smelter slag and natural materials like silica gravel, limestone, nepheline syenite and, for certain dark coloured wools, small amounts of an essentially arsenic-free copper slag. The rock (stone) wool plants use basaltic rock, limestone, clay and feldspar, together with the additives mentioned above (TIMA, 1993).

(iv) *Refractory ceramic fibres*

Refractory ceramic fibres are produced by melting a combination of alumina (Al_2O_3) and silica (SiO_2) in approximately equal proportions or by melting kaolin clay. Other oxides, such as zirconium dioxide (ZrO_2), boric oxide (B_2O_3), titanium oxide (TiO_2) and chrome oxide (Cr_2O_3) are sometimes added to alter the properties of the resulting fibres (TIMA, 1993). For example, the chemical composition is one of the factors that determine the maximum feasible end-use temperature. As for all MMVFs, the fibre length, diameter and bulk density — controllable to some degree by the manufacturing method and chemical composition — also affect key physical properties of the refractory ceramic fibres, e.g. the thermal conductivity (Everest Consulting Associates, 1996).

The basic composition of refractory ceramic fibres has not changed appreciably since their initial formulation in the 1940s (Environmental Resources Management, 1995), but modifications to the composition such as raising the content of alumina and the addition of zirconium dioxide and other materials create fibres that tolerate higher maximum end-use temperatures.

(v) *Newly developed fibres*

In recent years, the industry has developed newer fibres that have similar properties to older products, but are more biosoluble. Some examples of these newly developed fibres are the alkaline earth silicate (AES) wools and high-alumina, low-silica wools. Producers of refractory ceramic fibres and other MMVFs have developed new fibre compositions designed to withstand high end-use temperatures, but with significantly lower biopersistence than the older types. Although these new fibres can be produced in the same furnaces as are used to manufacture refractory ceramic fibres, their chemistry differs substantially from that of refractory ceramic fibres; they are new fibres rather than a modification or hybrid of refractory ceramic fibres. These new products, termed AES fibres, were first commercialized in 1991. They are wool-like products composed of alkaline earth oxides (calcium oxide + magnesium oxide) in the range of 18%–43% by weight, silica (SiO_2) in the range of 50%–80% by weight, and alumina + titania + zirconia < 6% by weight. Traces of other elements are also present (ECFIA–RCFC, 2001).

Another product introduced in the early 1990s is the high-alumina, low-silica stone wool (known as HT wool). The traditional raw materials for the production of rock (stone) wool are the rock types basalt or diabase (dolerite) in a mixture with the fluxing agents limestone or dolomite. Briquettes or form stones (artificial rocks) often bound together by cement can now be used instead of natural rocks. The briquettes make it possible to use raw materials that have a higher melting point than the melt temperature of 1500–1550 °C normally used and allow the inclusion of fine-grained high-melting temperature raw materials such as quartz sand, olivine sand and bauxite in the melt composition (Guldberg *et al.*, 2002).

(*b*) *Physical properties*

(i) *Fibre diameter*

The distribution of fibre diameters in MMVFs varies with the fibre type and the manufacturing process employed. Because they are amorphous (i.e. non-crystalline), MMVFs do not have cleavage planes that cause them to split lengthwise into fibres with smaller diameters. Rather, MMVFs break across the fibre, resulting in fibres which are of the same diameter as the original fibre but shorter, together with a small amount of dust (Assuncao & Corn, 1975).

Continuous glass filament is produced by a continuous process of drawing through the calibrated holes of the bushings at constant speed, thus leading to a very narrow variation in the filament diameter. In any given product, the diameter of the fibres differs little from the mean or nominal diameter. The standard deviation of the diameter in continuous filament products is typically less than 10% of the nominal diameter. Filaments are divided into 19 classes by a letter designation from B to U corresponding to a range of mean diameters as shown in Table 4. The nominal diameter of continuous glass filaments ranges from 5–25 µm. The majority of the filaments produced have a diameter of 9 µm or more, corresponding to G filaments and above (APFE, 2001). Small quantities of C filaments are produced in North America, and small quantities of B filaments are produced in Japan. Table 5 lists the fibres that have been tested in animal carcinogenicity studies.

The post-production processing of continuous glass filament does not cause any change in diameter. However, in a recent study, examination of dust from highly chopped and pulverized continuous glass filament by microscopy demonstrated the presence of small amounts of respirable dust particles, a small number of which had aspect ratios equal to or greater than 3:1. These elongated particles have been called 'shards' (APFE, 2001).

The fibre formation processes used to manufacture wools produce fibres with diameters that vary much more within a given wool product than within a continuous glass filament product. The diameters within a vitreous wool product have an approximately log-normal distribution. Nearly all wool products have average diameters of 3–10 µm. For example, in a wool product with an average fibre diameter

Table 4. Letter designations for continuous glass filaments

Size designation	Range of mean diameters (µm)
B	3.30–4.05
C	4.06–4.82
D	4.83–5.83
DE	5.84–6.85[a]
E	6.35–7.61[a]
F	7.62–8.88
G	8.89–10.15
H	10.16–11.42
J	11.43–12.69
K	12.70–13.96
L	13.97–15.23
M	15.24–16.50
N	16.51–17.77
P	17.78–19.04
Q	19.05–20.31
R	20.32–21.58
S	21.59–22.85
T	22.86–24.12
U	24.13–25.40

From ASTM (2000)
[a] Mixture of ranges

Table 5. List of fibres tested in the animal carcinogenicity studies performed after 1987 and reported in this monograph

Fibre	Reference
Fibres A, C, F, G, H	Lambré et al. (1998)
B-01-0.9 glass wool, B-09-0.6 glass wool, B-09-2.0 glass wool	Roller et al. (1996)
B-20-0.6 slag wool	Roller et al. (1996)
B-20-2.0 (experimental rock (stone) wool)	Davis et al. (1996a); Roller et al. (1996)
Bayer B1, B2, B3 (glass wools)	Pott et al. (1991)
104E glass fibre	Cullen et al. (2000)
JM 475	Pott (1989); Pott et al. (1989)
JM 753	Roller et al. (1996)
M-stone wool	Davis et al. (1996a); Roller et al. (1996)
R-stone wool-E3	Roller et al. (1996)
MMVF10, MMVF11	Hesterberg et al. (1993)
MMVF10a	Hesterberg et al. (1997, 1999); McConnell et al. (1999)
MMVF21 (rock (stone) wool)	Mc Connell et al. (1994)
MMVF22	Roller et al. (1996)
MMVF33	McConnell et al. (1999)
MMVF34 (or HT fibre or HT stone wool)	Kamstrup et al. (2001, 2002)
X-607 (or AES)	Hesterberg et al. (1998a)
RCF1, RCF2, RCF3, RCF4	Mast et al. (1995a)

of approximately 5 µm, the diameters of individual fibres may range from less than 1 µm to > 20 µm. In addition to fibres, some wool fibre formation processes can produce a considerable number of large, rounded particles approximately 60 µm or larger in diameter, which are termed 'shot' (TIMA, 1993).

The preparation of a bulk fibre sample for measurement by microscopy typically breaks the fibres into shorter lengths. Under these circumstances, it is not meaningful to report the number of fibres in various ranges of diameters. Instead, either the total length of fibres falling within each diameter range is measured, or fibres are sampled for measurement of their diameters in proportion to their lengths (the intercept method) (Schneider et al., 1983; Koenig et al., 1993; TIMA, 1993). The mean, geometric mean or median fibre diameter determined in this way is referred to as a 'length-weighted' (or 'accumulated length') mean, geometric mean or median diameter. Table 6 lists typical length-weighted mean fibre diameter and standard deviation and the shot content for various wool products. It is apparent from Table 6 that the average fibre diameter is a function of the manufacturing process and not an inherent property of a particular type of fibre (TIMA, 1993).

If, for a given sample, the number of fibres per diameter interval is determined, the corresponding median diameter will be smaller than the length-weighted median diameter, since thicker fibres tend to be longer. For this reason, the median diameter of

Table 6. Fibre length-weighted diameter and shot content for various wool products[a]

Product	Fibre diameter (μm)		Shot content (percentage by weight)
	Average	Standard deviation	
Glass wools	7.7	4.2	2
	5.8	4.7	0
	5.6	3.3	1
	5.3	3.2	5
	4.7	2.0	0
	4.0	2.4	1
	3.4	2.0	1
	3.3	2.7	0
	1.2	1.0	0
	0.6	0.5	0
Rock (stone) and slag wools	5.3	3.8	16
	4.5	4.1	49
	4.4	2.7	43
	4.0	2.4	24
	4.0	3.1	45
	4.0	3.1	39
	3.9	3.2	55
	3.5	3.5	51
	2.4	2.3	50
Refractory ceramic fibres	3.8	2.8	20
	2.4	2.2	43
Alkaline earth silicate wools[b]	2.2	2.5	35
	3.0	2.4	40
High-alumina, low-silica wool[c]	4.6	–	30

[a] From TIMA (1993)
[b] From ECFIA & RCFC (2001)
[c] From Knudsen et al. (1996)

the fibres in an airborne fibre cloud is smaller than the length-weighted median diameter of the fibres in the product from which it originates: settling and ventilation further shift the distribution towards lower diameters. For example, for nominal diameters of, say, 8 μm, the typical count median diameter for the airborne fibres could not exceed 3–5.5 μm, even at the moment of dispersion (Schneider et al., 1983).

(ii) *Fibre length*

As with diameter, the fibre length varies according to the manufacturing process. Continuous glass filaments are produced by a continuous drawing process that results in extremely long fibres (typically several metres). During post-fibrization processing, however, fibres may be broken either intentionally or inadvertently. Thus, fibre lengths in continuous glass filament products are highly dependent on the nature of such processing. For example, the typical length of chopped strands is 3.0–4.5 mm, while that of chopped rovings is 2.5–3.0 cm (TIMA, 1993).

Wool fibres are manufactured as discontinuous fibres. In an analysis of the fibres found in glass wool insulation, most of them were found to be several centimetres long. The mean length of fibres in other wool products is variable, ranging from several centimetres to < 1 cm (TIMA, 1993). Fibres with lengths less than the 250 μm upper limit for respirability (Timbrell, 1965) are certainly present in most, if not all, wool products, and probably also in continuous filament products as a result of the various post-fibrization processes. There are no good methods to quantify the number of fibres of length < 250 μm within MMVF products (TIMA, 1993). Furthermore, the ability of fibres to become airborne depends strongly on the degree to which they are immobilized in the product by binder, other additives and facing, and on the way they are handled. Thus, the best way to evaluate fibre length in relation to health effects is to analyse the airborne dust generated during the manufacturing and handling of the MMVF product (see section 1.4).

(iii) *Fibre density*

Unlike fibre length and diameter, fibre density does not vary widely between MMVFs. The four traditional classes of MMVF range in density from 2.1–2.9 (Table 7). Fibre density, length and diameter are the critical properties that determine the aerodynamic behaviour of MMVFs and their respirability (see section 4.1.1).

Table 7. Density ranges for MMVFs

Fibre	Density (g/cm^3)
Continuous glass filament	2.1–2.7[a]
Glass wool	2.4–2.6[a]
Rock (stone) and slag wool	2.7–2.9[a]
Refractory ceramic fibres	2.6–2.7[a]
Alkaline earth silicate wools	2.6[b]
High-alumina, low-silica (HT) wools	2.8[c]

[a] TIMA (1993) and Hartman *et al.* (1996)
[b] ECFIA & RCFC (2001)
[c] Hesterberg *et al.* (1998b)

(iv) *Fibre coatings and binders*

During the drawing of continuous filament glass fibres, an aqueous polymer emulsion or solution is usually applied to each filament. This coating material is referred to as binder or size and serves to: protect the filaments from their own abrasion during further processing and handling, and ensure good adhesion of the glass fibre to the resin during polymer reinforcements. The quantity of binder on the filaments is typically in the range of 0.5%–1.5% by mass. The coating material applied varies depending on the end-use of the product. Typical coating components include: film formers such as polyvinyl acetate, starch, polyurethane and epoxy resins; coupling agents such as organofunctional silanes; pH modifiers such as acetic acid, hydrochloric acid and ammonium salts; and lubricants such as mineral oils and surfactants (EIPPCB, 2000).

MMVF wools may contain other types of additive. Oils and other lubricants may be added to wools during processing to reduce dust generation from the product. An organic binder may be applied to wools immediately after fibrization in order to hold the fibres together in a spongy mass. This binder is usually a phenol–formaldehyde resin in aqueous solution, which, after drying and curing, tends to concentrate at fibre junctions, but also partially coats the individual fibres. In rock (stone) and slag wools the binders account for up to 10% of the mass of the final product. Other additives applied to wools may include antistatic agents, extenders and stabilizers, and inhibitors of microorganisms (TIMA, 1993). In recent years, alternatives to phenol–formaldehyde resins such as melamine and acrylic resins have been used.

The binder content of insulation wool products is typically quite low, but for high-density products may range up to 25% by mass. In some products no binder is applied. Such binder-free products are designed either for an application in which integrity of the wool fibres is not necessary or in cases where that integrity is achieved by other means, such as encapsulation in a plastic sheath. Typically, some lubricant is sprayed on these fibres immediately after fibrization to protect them from mechanical damage during processing and subsequent use (TIMA, 1993).

(v) *Structural changes*

Man-made vitreous fibres are noncrystalline and remain vitreous when used at temperatures below 500 °C. At higher temperatures, they flow, melt or crystallize depending on their composition. High-silica and low-alkali metal oxide compositions such as refractory ceramic fibres, AES wools, and some rock (stone) wools will start to crystallize at temperatures above 900 °C. The crystalline phases produced will depend on composition and temperature. Longer exposure times are required for fibre devitrification at lower temperatures (Brown *et al.*, 1992; Laskowski *et al.*, 1994).

1.1.3 *Analysis*

(a) *Principle*

Dust samples are collected by drawing a measured quantity of air through a filter. For determination of the concentration of airborne dust, the collected dust is weighed. For the examination of fibres by optical microscopy, the filter is rendered transparent and mounted on a microscope slide. The numbers of fibres on a measured area of the filter are then counted visually using phase-contrast optical microscopy (PCOM) following a set of counting rules. Samples for scanning electron microscopy (SEM) are taken on filters that have a smooth surface suitable for direct examination. After sampling, part of the filter is cut out, mounted on a specimen stub and coated with a thin layer of gold (WHO, 1985) or platinum (Yamato *et al.*, 1998). For transmission electron microscopy (TEM), special preparation methods are used. Widely used methods of measurement, as described in WHO (1996), represent a consensus reached by experts from 14 countries, NIOSH (1994) and earlier versions of these methods. Since the sampler design specifications for optical microscopy and counting criteria have all been improved and standardized, the fibre concentrations assessed by the old and new methods may not be directly comparable.

(*b*)   *Sampling*

The sampler is mounted on a worker in the breathing zone or placed in a fixed location where exposure is to be characterized. A range of filter diameters and filter-holder designs has been used. The most recent sampling methods (NIOSH, 1994; WHO, 1996) use a 25-mm diameter filter placed in an open-faced holder with a 50-mm electrically conductive extension cowl.

Fibres depositing inside the cowl are not considered in the NIOSH 7400 procedure, and WHO (1996) states that fibres rinsed from the cowl should be disregarded when calculating fibre concentration. Jacob *et al.* (1992) found average cowl losses of 25% (standard deviation (SD), 12%) for sampling during the installation of residential glass-fibre insulation. During manufacturing operations involving glass wool, the average loss was 27% (SD, 13%) (Jacob *et al.*, 1993). For refractory ceramic fibres, Cornett *et al.* (1989) found an average cowl deposition of 17% (SD, 12%).

Fibre mass and fibre concentration have been determined from separate samples (e.g. Schneider, 1979a) or from single samples used in turn for weighing and fibre counting (e.g. Ottery *et al.*, 1984). Fibre counting requires that the dust be uniformly distributed across the filter, and therefore open-faced filters are used. However, for proper sampling of inhalable and respirable dust, the samplers used must conform to the specifications given in international standards (CEN, 1993; ISO, 1995).

(*c*)   *Gravimetric analysis*

Gravimetric analysis measures the total mass of dust in a volume of air. The collection filter is desiccated, conditioned to the relative humidity of the weighing room and weighed before sampling and together with the collected dust after sampling. The concentration of airborne particles, expressed as mg/m^3, is calculated from the sampling rate and the weight gain of the filter (Ottery *et al.*, 1984).

(*d*)   *Fibre counting*

Fibre counting methods are used to determine the concentration of airborne fibres.

(i) *Counting criteria*

The filter is rendered optically transparent and the fibres present within a specified number of randomly selected areas are counted using a PCOM at a magnification of × 500. The total number of fibres on the filter is calculated to give the concentration of airborne fibres. Since the visibility of the thinnest fibres is dependent on the optical parameters of the microscope and on the refractive index of the filter medium, these parameters are also specified. The performance of the microscope in terms of visibility of thin fibres is assessed using a standard test slide. The microscopy techniques are based on those commonly used for the monitoring of asbestos.

A countable fibre, as defined by WHO (1996), is any particle that has a length > 5 µm, a length:diameter ratio larger than 3:1 and a fibre diameter < 3 µm (often referred to as WHO fibres).

A fibre as defined by the NIOSH 'B' rules (1994) is any particle that has a length > 5 µm, a length:width aspect ratio equal to or greater than 5:1 and a diameter < 3 µm.

Investigators in the USA typically used the NIOSH 'A' rule and its predecessor P&CAM (Physical & Chemical Analytical Method) 239 (Taylor, 1977) until the late 1980s. The 'A' rule has no upper diameter bound and the aspect ratio lower limit is 3:1. Thus, the 'B' rule is bound to give lower results than the 'A' rule because some fibres are excluded. The difference will depend on the bivariate fibre size distribution. Breysse *et al.* (1999) found that this difference can be significant. For samples of airborne glass and rock (stone) wool fibres, the 'A' rule gave approximately 70% higher results than the 'B' rule. For loose insulation wool without binder, applied by blowing, the difference was only 8%. For refractory ceramic fibres, the difference was 33%. However, Buchta *et al.* (1998) found no difference in the density of refractory ceramic fibres measured by the two counting rules.

The criteria for counting fibres that are branching or crossing or that are attached to other particles are marginally different between WHO (1996) and NIOSH (1994). Laboratories in the USA using the NIOSH 'B' rule counted fibre densities on average lower by 27% than European laboratories using the WHO method to count the same set of pre-mounted slides (Breysse *et al.*, 1994). Maxim *et al.* (1997) compared the NIOSH 'B' rule with the WHO rule and found that use of the NIOSH 'B' rules resulted in counts that were approximately 95% of those obtained using the WHO rules.

Criteria for counting fibres that are not completely within the counting field are also specified (NIOSH, 1994; WHO, 1996). Some of the previously used criteria have overestimated the number of long fibres (Schneider, 1979b).

(ii) *Fibre identification*

The workplace atmosphere may be contaminated by fibres other than MMVFs since, for example, fibres of cellulose, organic textiles and gypsum are ubiquitous. The presence of particles that fulfil the counting criteria, but are not MMVFs, has been acknowledged by several investigators. Jacob *et al.* (1993) studied the removal of pipe and ceiling insulation and, using the NIOSH 'B' rule, determined the arithmetic mean concentration of fibres to be 0.13 fibre/cm^3. Of the fibres present, 0.042 fibre/cm^3 were MMVFs (identified using morphology and polarized light microscopy (PLM)). In a study of workers in the prefabricated wooden house industry in Sweden, MMVFs constituted only about 25% of the total number of fibres in air samples (Plato *et al.*, 1995a).

Switala *et al.* (1994) measured MMVF concentrations near emission sources and used dispersion modelling, supplemented by PLM if it was not readily apparent that a fibre was an MMVF. When assessing exposure in non-industrial environments and exposure of the general public, care should be taken to discriminate between different types of fibre and better criteria for identifying MMVFs are required (Schneider *et al.*, 1996). The work of Draeger *et al.* (1998), Rödelsperger *et al.* (1998) and data quoted by Höfert and Rödelsperger (1998), indicate that the parallel edge criterion shows promise as a replacement for the chemical composition criterion used in the past as an inclusion criterion for MMVFs in addition to analysis of elemental composition.

NIOSH (1994) specifically states that it is incumbent on all laboratories to report all fibres meeting the counting criteria. For assessing exposure to asbestos, however, the method states explicitly that if serious contamination from other fibres occurs, PLM can be used to eliminate non-crystalline fibres of diameter > 1 μm. WHO (1996) provides guidance on the application of PLM for identifying many types of fibre with diameters > 1 μm, such as cellulose fibres, many synthetic organic fibres and asbestos fibres. The use of SEM with energy dispersive X-ray analysis (EDXA) can distinguish between various types of MMVF. Transmission electron microscopy with EDXA and electron diffraction is generally considered to be the most definitive method available for providing both chemical and structural information on fibres down to 0.01 μm in diameter (WHO, 1996). The VDI (1994) method for non-industrial environments is based on SEM and EDXA. The VDI method provides guidelines on how to attribute MMVFs to specific bulk MMVFs present in a building, according to a classification scheme based on the presence and relative intensity of characteristic energy peaks in the EDXA spectrum. Such fibres are termed 'product fibres'.

While some attention has been paid to the effect on fibre counts of changing from the NIOSH 'A' to the 'B' rule, much less attention has been paid to the differences caused by using various approaches to exclude non-MMVFs.

(iii) *Detection limit*

In optical microscopy, the practical limit of detection is about 0.2 μm. For conventional MMVFs, this is no great disadvantage, since they are mostly more than 1 μm in diameter. The median diameter of some airborne microfibres and other special-purpose fibres, however, can range from 0.1–0.3 μm, and therefore a substantial proportion of such fibres would not be detected using optical microscopy (Rood & Streeter, 1985). Furthermore, some of these fibre types may have a refractive index close to that of the filter medium, further increasing the difficulty in detecting them. In this case, special sample preparation methods should be used (Health and Safety Executive, 1988).

For the NIOSH 7400 method (B counting rules), the quantitative working range is 0.04–0.5 fibre/cm^3 for a 1000-L air sample. The detection limit depends on the sample volume and quantity of interfering dust, and is < 0.01 fibre/cm^3 for atmospheres free of interferences (NIOSH, 1994). Using PLM, fibre concentrations below 0.001 fibre/cm^3 have been measured (Schneider, 1986).

For routine analysis, SEM allows good visualization of fibres as thin as 0.05 μm (WHO, 1996). The percentage of visible fibres starts to decrease as fibre width decreases below 0.3 μm. These figures vary according to the substrate, type of instrument and display mode (Kauffer *et al.*, 1993).

(iv) *Quality assurance*

During their exposure survey for the European epidemiological study (see section 2), Cherrie *et al.* (1988) found that use of the WHO PCOM method, combined with participation in interlaboratory workshops and slide exchanges, reduced the difference between participating laboratories from a factor of 3 to 1.4. Their fibre counting level increased by a factor of 3 and the exposure data have been corrected accordingly. For another participating laboratory, the increase in fibre counting was by a factor of 4.5.

(e) *Surface deposition*

Particles deposited on the skin or on horizontal surfaces in buildings can be removed using sticky foils (e.g. adhesive tape) and analysed by optical microscopy (Cuypers *et al.*, 1975; Schneider, 1986). Mucous threads and clumps from the inner corner of the eye can be used to estimate particle deposition in the eyes (Schneider & Stokholm, 1981).

(f) *Bulk material*

The diameter distribution of bulk material can be uniquely characterized by using the length-weighted (also termed 'accumulated length') diameter distribution. By this method, a bulk sample is heated to remove organic binder and oil and is comminuted. A small fraction is prepared for analysis by optical or electron microscopy. The length of the fibres within each given diameter interval (but not the total fibre number) is determined. The median is called the nominal diameter of the material. This distribution can also be determined using the intercept method (Schneider & Holst, 1983) by which diameters are measured for all fibres intercepting a line. Since the probability of intercepting the line is proportional to fibre length, this gives the length-weighted distribution.

To quantify the fibre diameter distribution of a wool product, typically a large number of fibres from the product are measured individually, using either optical microscopy or SEM. If the average fibre diameter is > 1 μm, the two techniques give comparable results. If it is appreciably less, the resolution of the optical microscope is insufficient, and the SEM results are more accurate (TIMA, 1993).

Analytical methods for MMVFs in bulk material have been developed by the industry and are based on optical microscopy (Koenig *et al.*, 1993) and specific techniques for refractory ceramic fibres are based on SEM (Alexander *et al.*, 1997).

1.2 Production

1.2.1 *History and production levels*

Most of the MMVF produced worldwide is used as insulation. For at least the last decade, glass wool, rock (stone) wool and slag wool have together met just over half of the world demand for insulation, with the remainder consisting of foamed plastics such as polyurethanes and polystyrenes and other minor products (cellulose, perlite, vermiculite, etc.). About 75% of the world's insulation material is produced and used in North America and Europe (the Freedonia Group, 2001).

World demand for glass, rock (stone) and slag wool insulation for selected years and by region is presented in Table 8. Approximately 88% of glass wool and 80% of rock (stone) and slag wool are used in the construction of residential and commercial buildings, and 12% of glass wool and 20% of rock (stone) and slag wool are used in industrial applications, including heating, ventilation and air conditioning, household appliances and transportation (the Freedonia Group, 2001).

(a) *Continuous glass filament*

The production of continuous glass filament which began in the 1930s (TIMA, 1993; Vetrotex, 2001) is one of the smallest segments of the glass industry in terms of tonnage. The USA is the biggest producer accounting for over 40% of worldwide output; Europe and Asia each produce 20–25% of the total. Twenty-six furnaces were in operation in the European Union in 1997, producing 475 000 tonnes of continuous glass filament (Table 9) (EIPPCB, 2000).

(b) *Glass wool, rock (stone) wool and slag wool*

Insulation products made from rock (stone) and slag wool were first produced around 1840 in Wales (Mohr & Rowe, 1978). By 1885, commercial manufacturing plants for rock (stone) wool were also operating in England and later spread to Germany and the USA (IARC, 1988). Glass wool was not produced commercially until the late 1930s and early 1940s (Mohr & Rowe, 1978).

Although some rock (stone) and slag wool plants were already operating in the USA and throughout Europe by the 1900s, the industry did not begin to grow until after the First World War. By 1928, at least eight plants operated in the USA. By 1939, the number of glass, rock (stone) and slag wool plants in the USA had increased to more than 25 (IARC, 1988).

Table 8. World demand for insulation

	1989	1994	1999
World demand for insulation[a] (thousand tonnes)	8564	9060	10 683
Rock (stone) and slag wool (thousand tonnes)	2641	2738	3090
Glass wool (thousand tonnes)	2646	2800	3493

Demand for rock (stone)/slag and glass wool insulation by region

	Rock (stone)/ slag	Glass	Rock (stone)/ slag	Glass	Rock (stone)/ slag	Glass
North America	223 (8.4%)	1508 (56.9%)	208 (7.6%)	1542 (55.1%)	236 (7.6%)	1915 (54.8%)
Western Europe	1195 (45.2%)	690 (26%)	1259 (46%)	757 (27%)	1374 (44.5%)	812 (23.2%)
Rest of the world	1223 (46.3%)	450 (17%)	1271 (46.4%)	500 (17.9%)	1480 (47.9%)	770 (22%)

From the Freedonia Group (2001)

[a] All types of insulation, including rock (stone), slag and glass wools, foamed plastics and other minor products

Table 9. Glass filament production in the European Union in 1997

	No. of plants	No. of furnaces	Percentage of total European Union production
Northern Europe	5	10	43
Finland	1	3	
Germany	1	1	
Netherlands	1	2	
United Kingdom	2	4	
Central and southern Europe	7	16	57
Belgium	2	7	
France	2	4	
Italy	2	3	
Spain	1	2	
Total	12	26	475 000 tonnes

From EIPPCB (2000)

Glass fibre manufacturers opened new markets such as textile manufacturing, while rock (stone) wool and slag wool manufacturers continued to supply the thermal insulation market. The number of rock (stone) wool and slag wool plants in the USA peaked at between 80 and 90 in the 1950s and then declined as glass wool began to be used more in thermal insulation. In Europe, rock (stone) wool plants predominated until the mid-1970s when glass wool use increased (IARC, 1988). Table 10 gives the number of plants in various areas in 2000, and the volumes of MMVFs produced by these plants are presented in Table 11.

(c) Refractory ceramic fibres

Refractory ceramic fibres were first produced in the USA in the 1940s for the aerospace industry. The commercial importance of refractory ceramic fibres increased during the 1970s when rising energy costs created a strong demand for efficient refractory insulating products (Schupp, 1990; Maxim *et al.*, 1994).

Although refractory ceramic fibre is a refractory insulating product with a defined market niche, the annual volume produced worldwide is relatively small (1–2% of the total production of MMVFs) (National Research Council, 2000). Table 12 shows estimates of capacity for refractory ceramic fibre production by region and by country for the year 1990 (Monopolies and Mergers Commission, 1991). In 2000, the world market for refractory ceramic fibres was estimated to be 150 000–200 000 tonnes per year; the market was divided approximately equally between the Americas, Europe and the rest of the world (NAIMA/EURIMA, 2001).

Table 10. Number of plants manufacturing glass wool, rock (stone) wool and slag wool in selected regions

Location	Glass wool	Rock (stone) wool and slag wool
Australia and New Zealand	4	1
Canada	8	3
Europe	30	30[a]
Japan and China	6	8
USA	32	12

From NAIMA/EURIMA (2001)
[a] Includes manufacture of high-alumina, low-silica wools

Table 11. Volume of glass wool, rock (stone) wool and slag wool manufactured in selected regions (thousand tonnes)

Location	Year	Glass wool	Rock (stone) wool and slag wool
Australia and New Zealand	2000	42	6
Europe	2000	1300	1200[a]
Japan and China	1999	223	342
USA	2000	1950	746

From NAIMA/EURIMA (2001)
[a] Includes manufacture of high-alumina, low-silica wools

(d)   *Newly developed fibres*

In recent years, newer fibres, often with reduced biopersistence, have been developed for specific purposes. Examples include the high-alumina, low-silica (HT) wools, more than 1 million tonnes of which were produced for the European market in 2000 (Rockwool International, 2001) and the alkaline earth silicate (AES) wools, 10–20 thousand tonnes of which were produced in Europe in 2000 (ECFIA & RCFC, 2001).

1.2.2   *Production methods*

Figure 2 is a generic representation of the processes involved in producing MMVFs.

The technological changes that have occurred in the MMVF industry had the potential to affect the distribution of diameters in the bulk material, the propensity of the product to release fibres, the production rate and the extent of manual handling of

Table 12. Estimated production capacity for refractory ceramic fibres by region and country in 1990

Region	Country	Production capacity (thousand tonnes)	Percentage of subtotal	Percentage of total
East Asia and Oceania	Australia	3.10	11.0	2.0
	India	3.50	12.4	2.2
	Japan	17.50	62.1	11.0
	Malaysia	0.90	3.2	0.6
	Republic of Korea	2.30	8.2	1.4
	Taiwan, China	0.90	3.2	0.6
Subtotal		28.20		17.77
Europe	Former Czechoslovakia	1.60	3.0	1.0
	France	21.50	39.8	13.6
	Germany	7.90	14.6	5.0
	Hungary	1.20	2.2	0.8
	Italy	1.50	2.8	0.9
	Poland	2.20	4.1	1.4
	Spain	1.60	3.0	1.0
	United Kingdom	15.00	27.8	9.5
	Former Yugoslavia	1.50	2.8	0.9
Subtotal		54.00		34.0
USA and Canada	Canada	2.00	4.3	1.3
	Puerto Rico	2.75	6.0	1.7
	USA	41.40	89.6	26.1
Subtotal		46.15		29.1
Latin America	Brazil	4.75	51.1	3.0
	Mexico	3.80	40.9	2.4
	Venezuela	0.75	8.1	0.5
Subtotal		9.30		5.9
Other	China[a]	7.50	35.7	4.7
	South Africa	3.50	16.7	2.2
	Former Soviet Union[a]	10.00	47.6	6.3
Subtotal		21.0		13.2
Total		158.65		100.0

Secondary source: Monopolies and Mergers Commission (1991) citing Morgan Crucible estimates

[a] Countries for which these estimates are less reliable

Figure 2. Schematic of a typical MMVF manufacturing plant

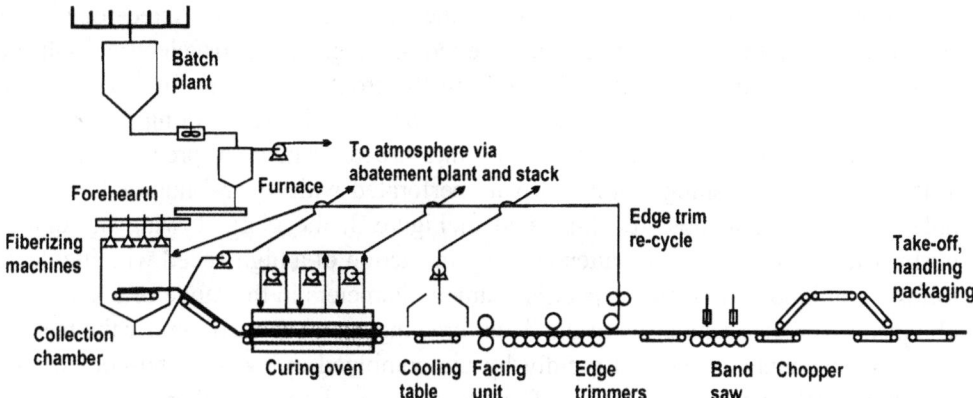

From EIPPCB (2000)

the products. Such changes may have resulted in qualitative differences in the concentrations of airborne fibres and may thus have affected exposure levels. For epidemiological purposes, three technological phases in the history of the European insulation wool manufacturing industry have been defined between the 1930s and the 1970s (Dodgson *et al.*, 1987a):

(i) the early phase, corresponding to periods when MMVFs were manufactured by discontinuous batch production and/or oil was not added during production;

(ii) the late phase, corresponding to modern production techniques including the addition of oil; and

(iii) the intermediate phase which included all the mixed production processes excluded by definition from (i) and/or (ii).

The timing of the three phases differs between the ten manufacturing plants surveyed by Dodgson *et al.* (1987a), so that for the industry as a whole, the early phase spans the period from 1933–68, the intermediate phase from 1940–69, and the late phase from 1951–78 (the last year included in the study).

The history of the production of glass, rock (stone) and slag wool and refractory ceramic fibres in the USA has essentially paralleled the development of the fiberization process. The steam-blown process that was initially used to produce MMVFs was quickly replaced by the flame attenuation process in the mid-1940s. The spinning process was introduced in the mid-1950s and was further enhanced in the rotary process that was introduced in the late 1950s and remains the primary means of fibre production today (Mohr & Rowe, 1978; TIMA, 1993). Lubricating oils and mineral oils were used in the manufacturing process from its inception. The early binder formulations contained inorganic minerals (e.g. clay) and petroleum-based mixtures, but these were rapidly replaced in the mid 1940s and early 1950s by the phenol–formaldehyde system which has been used predominantly ever since (TIMA, 1993).

(a) *Continuous glass filament*

To produce continuous glass filament, the raw materials are mixed together, or supplied as pre-formed marbles, and melted in a gas-fired or electrically-heated furnace. The resulting glass then flows from the front end of the furnace through a series of refractory-lined, gas-heated channels to the forehearths. Along the bottom of each forehearth are bushings, complex box-like structures made of precious metal, and at the base of the bushing is a metal plate perforated with several hundred calibrated holes called bushing tips. As illustrated in Figure 3, the glass flowing through the bushing tips is drawn out and attenuated by the action of a high-speed winding device to form continuous filaments. Specific filament diameters in the range of 5–24 μm are obtained by precise regulation of the linear drawing speed. Directly under the bushing, the glass filaments are cooled rapidly by the combined effect of water-cooled metal fins, high airflow and water sprays. The filaments are drawn together and pass over a roller or belt, which applies a protecting and lubricating coating, called size.

Figure 3. Continuous drawing process for manufacturing textile fibres

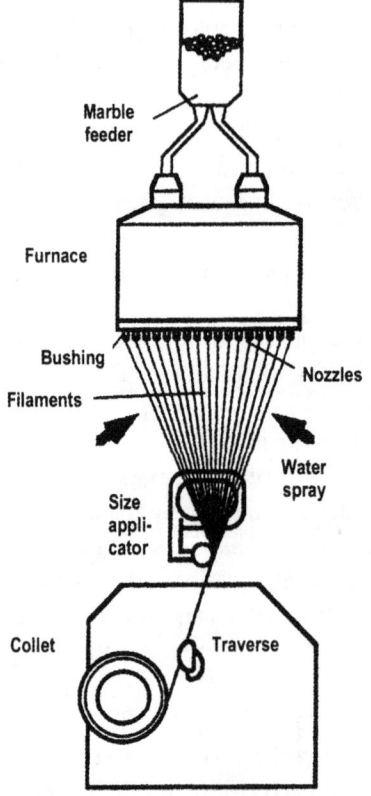

From TIMA (1993)

The coated filaments are gathered together into bundles called strands that are further processed (e.g. by chopping, twisting, milling and/or the application of binder), depending on the type of reinforcement being made as end-use. The typical products of these processes include continuous and chopped strand mat, yarns, roofing or surface tissue, rovings and roving cloths (EIPPCB, 2000).

(b)   *Glass wool*

During the production of glass wool by the rotary process, raw materials are blended and melted in an electrically heated furnace, an oxy-gas furnace or a traditional refractory-lined, gas-fired recuperative furnace. As depicted in Figure 4, a stream of molten glass flows from the furnace along a heated refractory-lined forehearth and pours through a number of single-orifice bushings into specially designed rotary centrifugal spinners. Primary fiberizing is the result of the centrifugal action of the rotating spinner; fibres are further attenuated by hot gases from a circular burner. This forms a veil of randomly interlaced fibres with a range of lengths and diameters. The veil passes through a ring of binder sprays that deposit a phenolic resin binder and mineral oil onto the fibres to provide integrity, resilience, durability and handling quality to the finished product (EIPPCB, 2000).

The resin-coated fibre is drawn under suction onto a moving conveyor to form a mat of fibres. This mat passes through a gas-fired oven, which dries the product and

Figure 4. Rotary process for manufacturing glass wool fibres

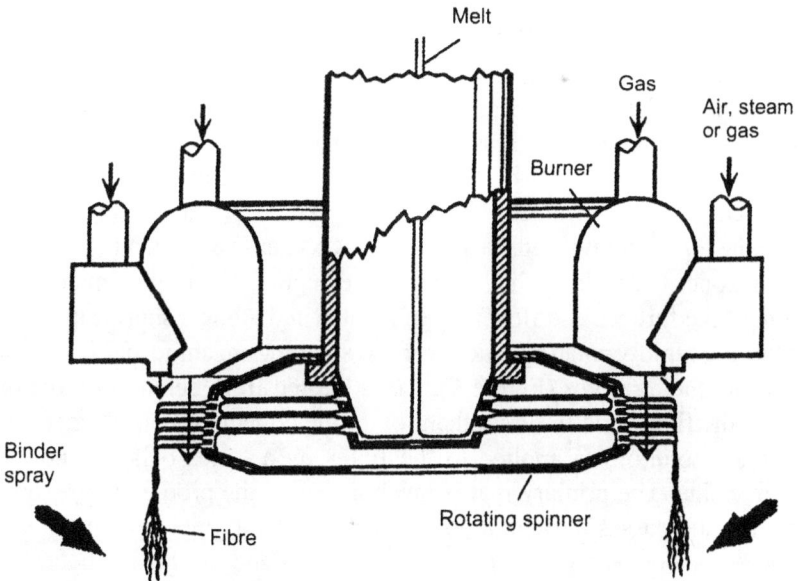

From TIMA (1993)

cures the binder. The product is then air-cooled and cut to size before packaging. Fibres trimmed from the edges can be granulated and blown back into the fibre veil, or combined with surplus product to form a loose wool. Some glass wools are produced without oven curing, (e.g. microwave-cured, hot-pressed, uncured or binder-free products). Also, certain laminated products are made by the application of a coating (e.g. aluminium foil or glass tissue) which is applied with an adhesive (EIPPCB, 2000).

A range of secondary products can be made from manufactured glass fibre. These include granulated insulation wool for installation by blowing, packaged uncured wool for supply to customers for further processing, and laminated or faced products. Pipe insulation is a significant secondary product that is usually manufactured by diverting uncured wool from the main process for press moulding and curing (EIPPCB, 2000).

Special-purpose fibres

Glass wool can also be made by the flame attenuation (pot and marble) process, which is more commonly used for the production of special-purpose fibres. This is a two-step process as shown in Figure 5. First, a coarse primary filament is drawn from a viscous melt. The coarse fibre is then remelted and attenuated into many finer fibres using a high-temperature gas flame, normally mounted at right angles to the primary fibre. Fibres are usually propelled by the high-velocity gases through a forming tube, where they are sprayed with a binder, and then to a moving collection chain where they deposit and tangle, producing a mat which is further processed into a variety of special application and filtration products (TIMA, 1993).

(c) *Rock (stone) wool*

The first step in the manufacture of rock (stone) wool is to melt a combination of alumino-silicate rock (usually basalt), blast furnace slag and limestone or dolomite. The batch may also contain recycled process or product waste. The most common melting apparatus is the coke-fired hot-blast cupola. The cupola consists of a cylindrical steel mantle which may be refractory-lined and closed in at the bottom. The cupola is charged to the top with raw materials and coke. Oxygen-enriched air is injected into the combustion zone, 1–2 m from the bottom of the cupola. The molten material gathers in the bottom of the furnace and flows out of a notch and along a short trough from where it falls onto the rapidly rotating wheels of the spinning machine and is thrown off in a fine spray, producing fibres (Figure 6). Air is blasted from behind the rotating wheels to attenuate the fibres and to direct them on to the collection belt. Binder (an aqueous phenolic resin solution) is applied to the fibres by a series of spray nozzles on the spinning machine. The primary mat is layered to give the product the required weight per unit area, and passes through an oven, which sets the thickness of the mat, dries it and cures the binder. The product is then air-cooled and cut to size before packaging (EIPPCB, 2000).

Figure 5. Flame attenuation process for manufacturing special-purpose glass fibres

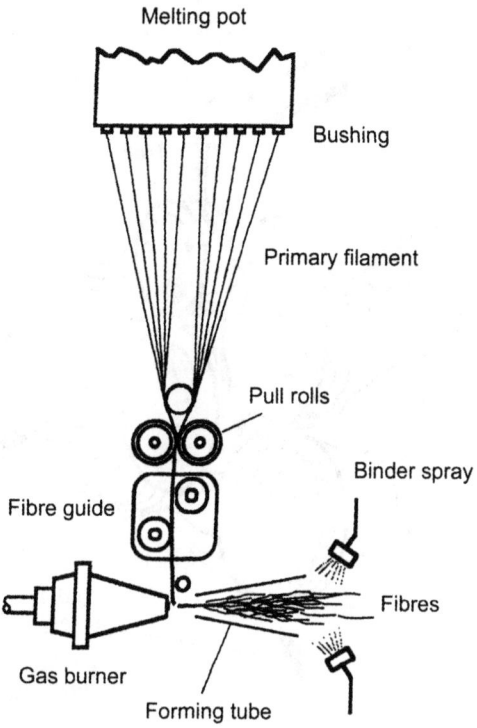

From TIMA (1993)

Pipe insulation and some secondary products may be manufactured by the process described for glass wool (EIPPCB, 2000).

Rock (stone) wool can also be produced using immersed electric arc furnaces (graphite electrodes) and flame furnaces. The subsequent operations including fibrization are the same as described above (EIPPCB, 2000).

(d) Slag wool

The production of slag wool is similar to that of rock (stone) wool. Slag wool is made by melting the primary component, blast furnace slag, with a combination of inorganic additives. The batch may also contain recycled process or product waste. As with rock (stone) wool, the most common melting apparatus is the coke-fired cupola, but immersed electric arc furnaces and flame furnaces may also be used. The subsequent process operations including fibrization are the same as for the production of rock (stone) wool (EIPPCB, 2000).

Figure 6. Wheel centrifuge or spinning process for manufacturing rock (stone) and slag wool and some refractory ceramic fibres

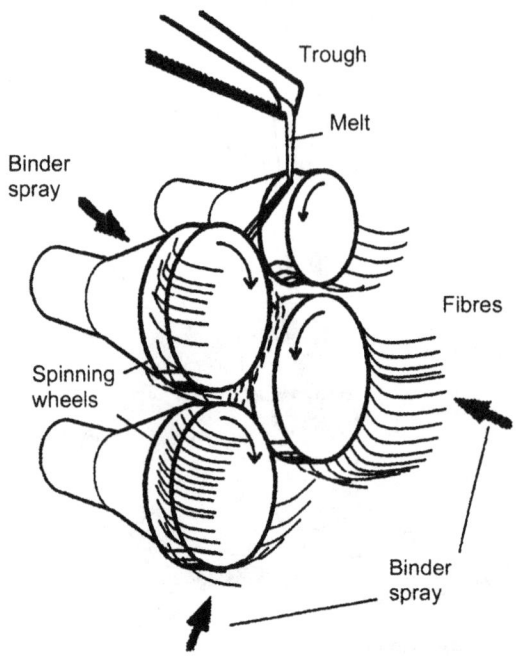

From TIMA (1993)

(e) *Refractory ceramic fibres*

Refractory ceramic fibres are produced by melting a combination of alumina and silica in approximately equal proportions at temperatures up to 2000 °C or in the USA by melting kaolin clay together with several trace ingredients. The molten mixture is made into fibre either by blowing an air stream on to the molten material flowing from an orifice at the bottom of the melting furnace (the blowing process; Figure 7), or by directing the molten material into a series of spinning wheels (the spinning process; Figure 6). The fibres are either collected directly as bulk fibre, or further processed into a blanket by a needling process. Although refractory ceramic fibres are sold in a variety of forms, all start with the production of either bulk or blanket material, termed 'primary refractory ceramic fibre production' (TIMA, 1993; EIPPCB, 2000).

The physical properties of the refractory ceramic fibre produced vary according to the method of fibrization. Spun fibre, for example, usually has a higher average diameter than blown fibre (TIMA, 1993).

The processing of refractory ceramic fibres begins with the fibre in either bulk or blanket form. Bulk material may be used directly, but is usually used as a feedstock for other processes. Bulk material can be converted into paper, board, felt, textiles, vacuum-formed products or dispersed in a fluid or solid matrix for other applications.

Figure 7. Fibre blowing process for manufacturing refractory ceramic fibres

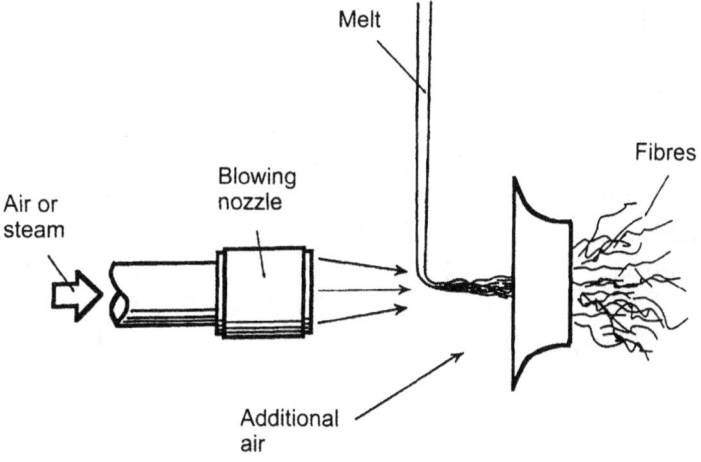

From TIMA (1993)

The blanket form is often used directly (e.g. as a furnace insulation material), but may be fabricated into modules, gaskets and other products. These 'downstream' processing operations may be carried out in the plants of the primary producers or in plants operated by other firms (EIPPCB, 2000).

(f)   Newly developed fibres

Alkaline earth silicate (AES) wools are made using the same equipment that is used to manufacture refractory ceramic fibres, although the different compositions of refractory ceramic fibres and AES wools result from very different operating conditions for furnacing and fibrization.

The high-alumina, low-silica fibres are produced by processes similar to those described for rock (stone) wool, but briquettes (artificial stones) containing fine-grained raw material with a high melting point such as quartz sand, olivine sand and bauxite are used as starting materials (Guldberg *et al*., 2002).

1.3   Uses

1.3.1   *Continuous glass filament*

Continuous glass filament products are used in a broad variety of applications, but their main end-use (approximately 75%) is as reinforcements in composites with thermosetting or thermoplastic resins. The main markets for composite materials are the automotive and other transport industries, the electrical and electronics industry and the building industry. Around 50% out of the sector output goes into the building and automotive/transport industries (EIPPCB, 2000).

Chopped strand mats are used to reinforce thermoplastics in the construction of boat hulls and decks, vehicle bodies, sheeting and storage tanks. Continuous strand mats are used in laminate production, when press moulding is employed, and to improve the appearance and strength of the laminates. Rovings have a variety of uses. They may be chopped to make chopped strands, woven into roving cloth or wound onto a mould for making convex-shaped composites such as nose cones for aeroplanes. Rovings are also used to reinforce plastic parts and in applications that require electrical insulation. In addition to being used as a reinforcement in thermoplastics, chopped fibres are used as a reinforcement in roof mat, which is commonly used to cover concrete or wooden roofs. Glass fibre mat is also used as a reinforcement in vinyl floor tiles and sheet linoleum floor covering (Loewenstein, 1993). Glass fibre yarns are used in the manufacture of glass cloth and heavy-duty cord for tyre reinforcement. The main market for glass textiles is the electronics industry where they are used in the production of high-quality printed circuit boards (EIPPCB, 2000). Other important uses are for aeroplane structures and for fireproof textiles, such as draperies and emergency protective clothing (IARC, 1988).

1.3.2 Glass wool, rock (stone) wool and slag wool

Figure 8 shows the main uses of glass, rock (stone) and slag wools.

(a)   Thermal and acoustic insulation, fire protection

Whatever the method employed to manufacture glass, rock (stone) or slag wools, the small-diameter, stiff, tangled fibres of these vitreous wools form a spongy mass, in which millions of small air pockets are trapped. These air pockets create an effective barrier against the transmission of both heat and sound energy. Therefore, glass, rock (stone) and slag wools provide effective thermal insulation for buildings (i.e. they help to keep buildings warm in the winter and cool in the summer). Many different areas of the home may be protected thermally: ceilings, side walls, perimeters of slabs, floors and other areas. Sound reduction is also an important use of glass, rock (stone) or slag wools, not only in buildings, but also in appliances, machinery and air-handling systems.

The insulation of homes, other buildings and industrial processes against heat loss and heat gain represents the largest single use for glass, rock (stone) and slag wools; up to 70% of industry output is for these applications (EIPPCB, 2000). Vitreous wools can be blown into structural spaces, such as in walls and attics. Bulk rock (stone) wool and glass fibre rovings are incorporated into ceiling tiles to provide fire resistance and thermal and sound insulation. Batts, blankets and semi-rigid boards made of glass, rock (stone) or slag wool fibres are used in both residential and commercial buildings. Pipe and board insulation is used extensively in industrial processes.

Sound absorption: Glass, rock (stone) or slag wool insulation in partitions, floors and ceilings significantly reduces sound transmission.

Figure 8. Main uses of glass, rock (stone) and slag wools

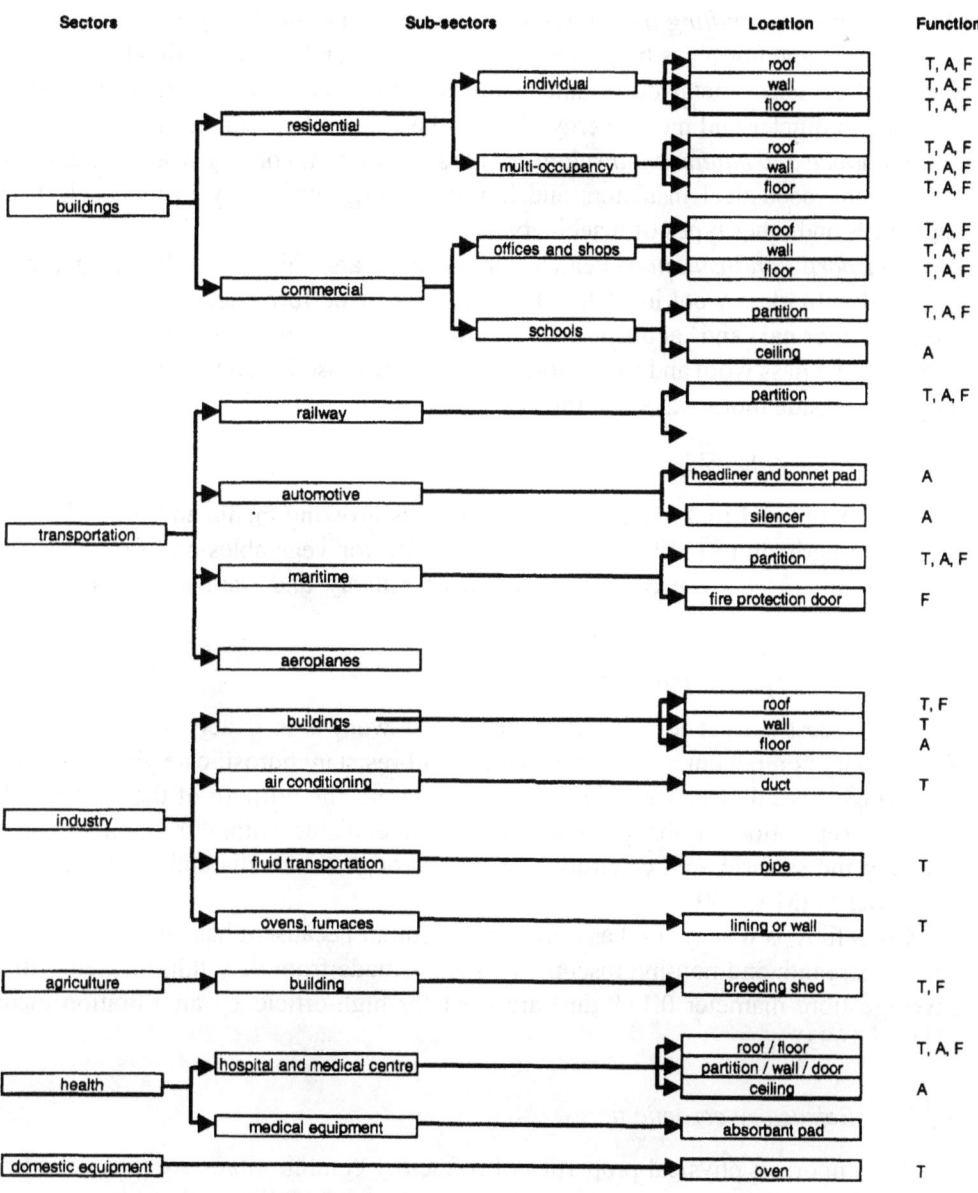

Modified from various industry sources.
T, thermal; A, acoustic; F, fire protection

Ceiling tiles and wall panels: Ceiling tiles and wall panels that serve as acoustical and thermal insulators can be made from glass, rock (stone) or slag wools.

Pipe and air-handling insulation: Glass, rock (stone) or slag wool can be used to insulate cold and hot pipes both indoors and outdoors and in many climates. They are also used on sheet-metal ducts and plenums for thermal and acoustical insulation, resulting in quieter and more energy efficient heating and air conditioning systems.

Appliance and equipment insulation: Glass, rock (stone) or slag wools are effective thermal and acoustical insulators and improve energy efficiency in many electrical appliances and other types of machinery.

Transportation insulation: Vehicles or carriers (cars, ships, aircraft and spacecraft are fitted with glass wool insulation to enhance their performance and to provide the required thermal and acoustic conditions for the goods or passengers being transported. Glass wool and rock (stone) wool are also used in sound-absorbent barrier panels alongside motorways and railways.

(b) *Other uses*

Glass wool and rock (stone) wool are used as growing media and for soil conditioning in agriculture (EIPPCB, 2000), especially for vegetables and flowers. Rock (stone) wool mats are used for insulation of railway and tramway tracks against vibration.

(c) *Special-purpose glass fibres*

The largest market for special-purpose glass wool is in battery separator media. The primary component of such media is an acid-resistant borosilicate glass fibre with an average fibre diameter ranging from 0.75–3 μm. The purpose of the media is the physical separation of the positive and negative plates within the battery, while enabling the sulfuric acid electrolyte to pass through the media and the filtration of impurities (TIMA, 1993).

Glass fibre is widely used as a filtration medium because it has unique properties, such as strength and non-hygroscopicity. Papers made from very thin-diameter fibres (average fibre diameter 0.1–1 μm) are used for high-efficiency air filtration media (TIMA, 1993).

1.3.3 *Refractory ceramic fibres*

The important physical properties of refractory ceramic fibres are their relatively high (1000–1460 °C) maximum use temperature (EIPPCB, 2000), low thermal conductivity, very low thermal mass (low bulk heat capacity), resistance to thermal shock and low density (light weight). Refractory ceramic fibres are produced and sold in a variety of physical forms, including bulk, blanket, felt, modules, board, vacuum-formed shapes, mixes, cements, putties and textile. The main applications and product characteristics vary with the product form (Table 13).

Table 13. Major applications of refractory ceramic fibres

Product form	Major applications	Product characteristics
Blanket, felt, modules	Furnace, kiln lining; firewall protection; high-temperature gaskets and seals for expansion joints; turbines; insulation wraps	Low thermal conductivity; flexibility; strength
Bulk	Fill-in packaging material for expansion joints; furnace base seals; conversion to other forms	Resilience
Board	Expansion joints; furnace, kiln back-up insulation; furnace lining; heat shields	Resilience, rigidity, resistance to high air velocity
Paper	Thermal, electrical insulators; refractory backup; ingot mould linings	Low thermal conductivity, dielectrical properties
Vacuum-formed shapes	Furnace linings (boards), insulation for special foundry components	Variable
Mixes, textiles (cloth, rope, sleeving) and miscellaneous	Patching refractory cracks and fissures; composite insulation for space firings and launchings; cloth (furnace curtains, welding curtains and blankets); sleeving (tube protection, cable insulation)	Resilience, insulating properties

From ECA (1996)

Refractory ceramic fibres are used chiefly in industrial applications; they have few consumer uses. In the USA, the largest percentage of the demand for refractory ceramic fibres (63%) in 1990 was for furnace lining and related applications (ECA, 1996). According to the estimates of the MMVF industry in the USA, six key sectors of the economy: ceramics, glass, forging, hydrocarbon processing, aluminium and steel consume 70–80% of the refractory ceramic fibres used for furnace lining and related applications.

Table 14 provides estimates of the percentages of the European refractory ceramic fibre market accounted for by various applications (ERM, 1995); furnace applications are the largest single market (50%). Data on markets and applications for refractory ceramic fibres in other areas of the world are not readily available, but they are believed to be similar to those in Europe and the USA (Horie, 1987).

Examples of applications of refractory ceramic fibres include: appliances (water combustion chambers, hearth products, fireplace logs, stove tops), automotive uses (catalytic converters, brake pads, air bags, heat shields), chemical applications (ethylene furnace insulation, reformer insulation, crude heaters), fire protection (temperature-resistant door lining, chimney liners, expansion joints), iron and steel incinerators (ladle preheat stands and covers, continuous casters, reheat furnaces, coke ovens), non-ferrous

Table 14. The European market for refractory ceramic fibres by application

Application	Proportion of the market (per cent)
Furnace, heater and kiln linings	50
Appliances	20
Metal processing (excluding furnace linings)	10
Automotive uses	5
Fire protection	5
General industrial insulation	10

From ERM (1995), citing estimates prepared by ECFIA

metal incinerators (soaking pit covers, melting furnaces), power generation (co-generation systems, turbine exhaust duct work, heat recovery steam generators, various types of boiler), aerospace and defence (heat shields) (Horie, 1987; ICF, Inc., 1991; ERM, 1995; ECA, 1996).

1.3.4 Newly developed fibres

Alkaline earth silicate wools generally have the same market applications as refractory ceramic fibres. However, although AES wools tolerate continuous service temperatures above those of rock (stone) wool and slag wool, they are below those tolerated by refractory ceramic fibres and therefore AES wools cannot be substituted for refractory ceramic fibres in all applications.

High-alumina, low-silica wools are now widely used in the same applications as rock (stone) wool.

1.4 Occurrence

1.4.1 Occupational exposure

The extent to which exposure to MMVFs occurs during their manufacture, fabrication, installation and end-use has been the subject of reviews and reports for some time. The European CAREX database (Kauppinen *et al.*, 2000) estimated that approximately 930 000 workers were exposed to glass wool and/or continuous glass filament, and 62 000 to refractory ceramic fibres in the European Union in 1990–93.

In the late 1960s, exposure to MMVFs was evaluated as total dust (gravimetric) concentrations (i.e. mg/m^3) as a result of the then existing regulatory requirements and the limitations of the sampling and analytical equipment available at the time (see section 1.1.3). However, the critical evaluation of exposure data collected using this methodology (Corn & Sansone, 1974; Corn *et al.*, 1976; Esmen *et al.*, 1978, 1979a,b;

Head & Wagg, 1980) revealed that for a given exposure reported in mg/m^3, considerable variation in the concentration of airborne fibres was possible depending on the mean aerodynamic diameter of the fibres being evaluated, and the presence of other particulate matter in the environments being studied.

Thus, sampling and evaluation of exposure to MMVFs came to rely on fibre counts made using optical and electron microscopy (Taylor, 1977; Eller, 1984; WHO, 1985). These became the conventional means of reporting the extent of the exposure of individuals to fibres in occupational environments. These methods were based on an explicit definition of what constituted a fibre (see section 1.1.3) and reported the results in fibres/cc (i.e. fibres/cm^3 or fibres/mL); they had the advantage not only of standardizing the approach to fibre monitoring, but also producing results in terms of 'respirable' fibres only. This is significant because, as discussed more fully in section 4.1, airborne fibres with diameters of 3 µm or less are currently thought to be the most likely to have adverse effects on human health.

Considerable quantities of data were obtained using these standardized methods for the evaluation and reporting of exposure to MMVFs in the MMVF manufacturing and installation industries. These data gave relatively consistent results from one MMVF product type to another irrespective of the geographical location of the manufacturing or installation site. For this reason, the MMVF exposure data presented here report the results in fibres/cm^3 only.

Certain factors must, however, be considered when interpreting the results of individual studies or comparing the results of two or more of the studies described below. The length of the fibre sampling period (i.e. full-shift or short-term) as well as the sample type (i.e. personal or area) is significant. Similarly important are the fibre counting methods and the type of microscopy (i.e. PCOM, TEM or SEM) used for sample analysis. As noted above, fibre counting methods and microscopy techniques have evolved over time, and the results obtained using the NIOSH P & CAM 239 method (Taylor, 1977) in the 1970s and early 1980s are not directly comparable to those obtained using the WHO method and the NIOSH 7400 (B counting rules) method (Eller, 1984) from the late 1980s onwards (WHO, 1996).

The manner in which the data on MMVFs is presented or summarized in each study is also significant when interpreting the results. Arithmetic means or medians with standard deviations and sample ranges have been reported in some studies while geometric means with geometric standard deviations were given in others. In common with most other types of occupational exposure, the concentrations of airborne fibres are better described by a log-normal than a normal distribution (Leidel et al., 1977). In contrast to normally distributed data sets which tend to follow traditional Gaussian (i.e. bell shaped) distributions, log-normal data sets are generally positively skewed (long 'tail' to the right) indicating a larger probability of very high individual exposure concentrations than would be the case for normally distributed data sets. Geometric means are the preferred method of expressing central tendency for log-normally distributed data sets, and are usually smaller and more similar to arithmetic medians than to arithmetic means for a given set of data. Each study presented in section 2 should thus be carefully interpreted to determine the influence of each of these factors, individually or in combination, on the results.

Exposure is normally determined principally by counting airborne fibres, but for the complete characterization of fibre sizes the joint values of diameter and length of each fibre should also be determined. The size distribution of a population of fibres characterized in this way can be summarized by count geometric mean, GM, and geometric standard deviation, GSD, of diameter, D, and length, L, and the correlation between $\log_e(D)$ and $\log_e(L)$, CORR. If the distribution is bivariate log-normal, these five parameters will completely specify the size distribution (Schneider, 2000). Table 15 presents the data on the sizes of airborne fibres analysed by electron microscopy in several studies.

(a) Exposure in production plants

(i) United States of America

Among the earliest evaluations of the exposure of workers in MMVF production facilities in the USA was a survey conducted by Johnson *et al.* (1969) of five glass fibre factories, four of which manufactured insulation material and one that produced textile (i.e. continuous glass filament) fibres. The results were reported in terms of both gravimetric (i.e. mg/m^3) and airborne fibre (i.e. $fibre/cm^3$) concentrations. The concentrations of airborne fibres longer than 5 μm ranged from 0.0–0.97 fibre/cm^3 in the insulation plants and from 0.0–1.97 fibre/cm^3 in the textile fibre plant. From the results presented in Table 16, the authors of the survey concluded that exposure was low when compared with that measured for similar operations performed in the asbestos processing industry at the time.

In a follow-up of this work, Dement (1975) surveyed concentrations of airborne fibres in four glass fibre facilities that manufactured 'large-diameter' (i.e. most fibre diameters > 1 μm) insulation products and six facilities that produced and/or used 'small-diameter' (i.e. most fibre diameters < 1 μm) fibre products. One facility that manufactured glass-fibre reinforced products was also surveyed. The results of the analysis of air samples taken at the plants where 'large-diameter' glass fibre insulation was produced were classified according to the fibre-forming method and fabrication operations within each plant. The mean and range for concentrations of airborne fibres is shown in Table 17. The highest mean concentration of fibres measured in these facilities was 0.20 fibre/cm^3 in the 'all other operations' category of plant C. The results of the analysis of air samples taken at the facilities that produced and used 'small-diameter' glass fibre were grouped according to product type and fabrication operations within each plant. The mean and range for concentrations of airborne fibres measured at these plants is shown in Table 18. In contrast to the results seen in the plants where 'large-diameter' glass-fibre insulation was produced, the mean concentrations of airborne fibres from 'small-diameter' glass fibre bulk-handling operations ranged from

Table 15. Bivariate size distributions of airborne fibres in various occupational settings summarized by various parameters

Reference	Type	Method	GM (D)	GSD (D)	GM (L)	GSD (L)	CORR
Cherrie et al. (1987)	Simulation of early rock (stone) wool production	SEM	0.3–0.5	1.9–2.7	7.0–9.0	2.2–3.0	0.4–0.6
Schneider et al. (1985)	Use of rock (stone) wool	SEM	1.2	2.7	22	4.0	0.7
	Use of glass wool	SEM	0.75	2.8	16	3.5	0.7
Breysse et al. (2001)	Use of glass wool	SEM	0.8–1.9	1.4–1.9	9.5–30	1.4–2.5	0.2–0.7
	Use of rock (stone) wool	SEM	1.6–1.9	1.6–1.9	19	1.7–2.7	0.4–0.6
Plato et al. (1995a)	House prefabrication: glass wool	SEM	0.91–1.2	1.7–1.8	9.2–9.3	2.3–2.5	–
	House prefabrication: rock (stone) wool	SEM	1.3–1.7	1.9	12–17	2.5–2.8	–
Lees et al. (1993)	Installation of MMVF batts	SEM	0.9–1.3	2.2	22–37	2.8–2.9	0.5–0.6
	Installation of loose MMVF with binder	SEM	1.0–2.0	1.8–2.2	30–50	2.3–2.6	0.4–0.6
	Installation of loose MMVF wool without binder	SEM	0.60	1.9	14–15	2.4–2.6	0.5–0.6
Maxim et al. (2000a)	RCF production and use	TEM[a]	0.84	2.1	14	2.5	0.4
Hori et al. (1993)	RCF, factory A	SEM-micrographs	0.96–1.2	1.7–1.9	12–19	2.4–2.6	–
	RCF, factory B	SEM-micrographs	0.86	1.9–2.0	11–13	2.4–2.6	–

GM, geometric mean; D, diameter; GSD, geometric standard deviation; L, length; CORR, correlation between $\log_e(D)$ and $\log_e(L)$; –, not determined; RCF, refractory ceramic fibre; SEM, scanning electron microscopy; TEM, transmission electron microscopy
[a] For analytical method, see Mast et al. (2000a)

Table 16. Fibre concentrations (fibre/cm³) by plant and operation

Operations	Plant no.[a]	Total fibres		Fibre length > 5 µm		Fibre length > 10 µm	
		Mean	Range	Mean	Range	Mean	Range
Mixing and melting	1	–	–	–	–	–	–
	2	3.64	3.64	0.97	0.97	0.54	0.54
	3	0.66	0.41–1.03	0.16	0.10–0.26	0.08	0.02–0.16
	4	0.30	0.08–0.67	0.10	0.02–0.25	0.04	0–0.07
	1–4[b]	1.53	0.08–3.64	0.41	0.02–0.97	0.22	0–0.54
	5	0.09	0.09	0.04	0.04	0	0
Forming	1	–	–	–	–	–	–
	2	0.41	0.04–2.95	0.12	0–0.56	0.08	0–0.35
	3	0.15	0.02–0.45	0.04	0–0.14	0.02	0–0.09
	4	0.19	0.07–0.31	0.07	0.01–0.19	0.03	0–0.06
	1–4[b]	0.25	0.02–2.95	0.08	0–0.56	0.04	0–0.35
	5	0.10	0–0.19	0.02	0–0.04	0.01	0–0.04
Spinning, twisting and waste recovery	5	0.72	0.03–12.67[c]	0.11	0–1.97	0.01	0–0.06

From Johnson et al. (1969) (personal samples; 2–7 h sampling times; analysis by phase-contrast optical microscopy including fibres < 5 µm in diameter)

[a] Plants 1–4 are insulation manufacturing plants; Plant 5 is a textile manufacturing plant.
[b] Combined results for plants 1–4
[c] [The Working Group noted that this range is not reliable because of the difficulty of discriminating between vitreous fibres and other particles with the methodology used.]

Table 17. Concentrations of airborne fibres (fibre/cm^3) in wool insulation facilities manufacturing 'large-diameter' (> 1 μm) glass fibres

Forming method or operation	Insulation plant			
	A	B	C	D
Centrifugal-formed building insulation				
Mean	0.07	0.08	0.09	0.09
Range	0.04–0.13	0.00–0.18	0.08–0.12	0.01–0.83
Number of samples	9	19	4	22
Centrifugal-formed appliance insulation				
Mean	–	0.04	–	0.06
Range	–	0.01–0.11	–	0.02–0.09
Number of samples	–	13	–	22
Flame-attenuated insulation				
Mean	0.04	–	0.14	–
Range	0.02–0.06	–	0.04–0.24	–
Number of samples	8	–	2	–
Pipe insulation				
Mean	0.07	–	0.14	–
Range	0.03–0.12	–	0.06–0.27	–
Number of samples	10	–	6	–
Scrap reclamation				
Mean	0.06	0.07	0.10	0.07
Range	0.03–0.15	0.02–0.47	0.08–0.13	0.01–0.14
Number of samples	10	4	5	7
All other operations				
Mean	0.07	0.04	0.20	–
Range	0.01–0.13	0.01–0.08	0.04–0.26	–
Number of samples	10	12	4	–

From Dement (1975) (personal and stationary samples; analysed by PCOM)

1.0–21.9 fibres/cm^3; the highest fibre concentration observed was 44.1 fibres/cm^3 in plant G.

In the plant that produced glass-fibre reinforced plastics, the concentrations of airborne fibres ranged from 0.02–0.10 fibre/cm^3 with a mean concentration of 0.07 fibre/cm^3 for lamination operations and 0.03 fibre/cm^3 for cutting and grinding operations. Dement (1975) commented that the nominal diameter of the fibres produced was a principal factor in determining the concentrations of airborne fibres present.

The largest body of data on exposure of production workers in the USA was collected in support of an epidemiological study of the MMVF industry (Corn &

Table 18. Concentrations of airborne fibres (fibre/cm³) in operations producing or using 'small-diameter' (< 1 μm) glass fibres

Operations	Plant					
	Bulk fibre production		Paper manufacture		Aircraft insulation fabrication	
	C[a]	E	F[b]	G[b]	H	P[b]
Bulk fibre handling						
Mean	1.0	9.7	5.8	21.9	1.2	14.1
Range	0.1–1.7	0.9–33.6	4.7–6.9	8.9–44.1	0.4–3.1	3.2–24.4
Number of samples	5	54	2	3	13	3
Fabrication and finishing						
Mean	–	5.3	1.9	10.6	0.8	2.1
Range	–	0.3–14.3	1.6–2.1	–	0.2–4.4	–
Number of samples	–	24	2	1	15	1

From Dement (1975) (personal and stationary samples; analysed by PCOM)
[a] In addition to producing 'large-diameter' wool insulation, plant C had several lines that produced 'small-diameter' fibres.
[b] Only limited surveys were conducted in these facilities.

Sansone, 1974; Corn *et al.*, 1976; Esmen *et al.*, 1978, 1979a) (see IARC, 1988). This study encompassed 16 glass wool, continuous glass filament, rock (stone) wool and slag wool plants. A summary of the concentrations of airborne fibres, measured using PCOM with counting criteria equivalent to those of the NIOSH A rule combined with TEM, is presented in Table 19. The mean fibre concentrations varied widely from plant to plant ranging from 0.001 fibre/cm³ in the forming area of plant 10 to 1.56 fibres/cm³ in the manufacturing area of plant 15. Esmen *et al.* (1978, 1979a) attributed these differences to the type of MMVF produced as well as to differences in manufacturing operations. However, most specifically, the concentrations of airborne fibres observed correlated closely to the nominal diameter of the fibres manufactured, as reported previously by Dement (1975).

Estimates of exposure to respirable fibres by plant during the manufacture of glass fibre (glass wool and continuous glass filament) were recently published by Smith *et al.* (2001). These estimates, shown in Table 20, were used in the epidemiological analysis of cause-specific mortality conducted by Marsh *et al.* (2001a) (see section 2).

When data from rock (stone) wool and slag wool manufacturing plants were evaluated separately, concentrations of airborne fibres were found to be slightly higher than those measured in glass-fibre manufacturing plants (Corn *et al.*, 1976; Esmen *et al.*, 1978). The average concentrations of airborne fibres varied from 0.01–0.42 fibre/cm³ in one plant that produced slag wool and from 0.20–1.4 fibre/cm³ in one that produced rock (stone) wool (see Table 21). Corn *et al.* (1976) suggested that the higher concentrations

Table 19. Concentrations of airborne fibres (fibre/cm^3), as determined by optical microscopy, in 16 MMVF plants in the USA

Plant no.	Forming[a]		Production[b]		Manufacturing[c]		Maintenance[d]		Quality control[e]		Shipping[f]		Overall	
	Mean	SD	Mean	SD	Mean	SD	Mean	SD	Mean	SD	Mean	SD	Mean	SD
1	0.002	0.001	0.38	0.32	0.03	0.02	0.02	0.02	0.07	0.10	0.01	0.001	0.01	0.25
2	0.07	0.03	0.17	0.14	0.12	0.11	0.08	0.05	0.19	0.16	0.07	0.06	0.11	0.12
3	–	–	0.02	0.02	–	–	0.07	0.18	–	–	0.005	0.01	0.04	0.10
4	0.01	0.004	0.07	0.12	0.04	0.05	0.03	0.02	0.01	0.01	0.02	0.01	0.04	0.08
5	0.02	0.01	0.03	0.02	0.03	0.02	0.02	0.01	0.03	–	0.03	0.01	0.02	0.02
6	0.05	0.10	0.01	0.01	0.009	0.01	0.01	0.03	0.01	0.02	0.005	0.004	0.01	0.03
7	0.15	0.03	0.24	0.12	0.43	0.32	0.44	0.37	–	–	0.15	0.17	0.34	0.30
8	–	–	0.03	0.02	0.04	0.03	0.01	0.01	–	–	0.01	0.01	0.02	0.02
9	0.02	0.02	0.01	0.01	0.02	0.07	0.01	0.006	–	–	0.004	0.002	0.02	0.01
10	0.001	0.001	0.003	0.004	0.004	0.004	0.002	0.003	0.003	0.003	0.002	0.002	0.002	0.003
11	0.09	0.11	0.05	0.03	0.04	0.03	0.04	0.04	0.08	0.08	0.03	0.02	0.05	0.05
12	0.01	0.01	0.02	0.03	0.01	0.004	0.01	0.02	0.01	0.003	0.007	0.005	0.01	0.02
13	0.58	–	0.08	0.06	0.11	0.17	0.09	0.08	–	–	0.03	0.02	0.10	0.10
14	0.01	0.01	0.04	0.09	0.05	0.05	0.05	0.13	–	–	0.03	0.03	0.04	0.03
15	0.19	0.22	0.92	1.02	1.56	3.79	0.11	0.10	0.89	0.33	0.10	0.09	0.78	2.1
16	0.02	0.01	0.02	0.02	0.05	0.03	0.07	0.23	0.04	–	0.02	0.01	0.04	0.12

From Esmen et al. (1979a) (personal samples; 7–8-h sampling times; analysed by PCOM, using the equivalent of NIOSH A rules, combined with TEM) (Esmen et al., 1978)

[a] Forming: all hot-end workers, cupola operators, batch mixers, transfer operators and charging operators
[b] Production: cold-end workers in direct contact with fibres, but not involved in cutting, sawing, sanding or finishing operations
[c] Manufacturing: workers involved in general manufacturing operations such as trimming, sawing, cutting, finishing, painting finished boards and handling boxed and/or packaged goods
[d] Maintenance: maintenance workers who repair production machinery and do general work in the production area as needed, including sweeping floors and cleaning dust collectors and machinery
[e] Quality control: workers who sample products and ascertain product quality
[f] Shipping (transport of packaged material): fork-lift truck operators and shipping yard operators

Table 20. Summary statistics[a] for measures of exposure to respirable fibres, by plant, total fibre-glass-worker cohort, 1946–87[b,c]

Statistics	Plant										All plants
	1	2	4	5	6	9	10	11	14	15	
RFib-AIE (fibres/cm^3)											
Median	0.039	0.002	0.050	0.001	0.167	0.040	0.036	0.096	0.046	0.018	0.035
Mean	0.053	0.003	0.055	0.001	0.211	0.048	0.350	0.096	0.049	0.034	0.073
Coefficient of variation[d]	0.854	1.447	0.579	2.955	0.990	0.747	2.109	0.588	0.732	1.404	2.796
RFib-Cum (fibre/cm^3–months)											
Median	2.638	0.101	2.466	0.086	6.382	1.833	1.168	5.500	1.839	0.892	1.441
Mean	7.116	0.211	5.041	0.129	13.408	4.998	23.483	10.273	5.032	2.788	6.080
Coefficient of variation[d]	1.324	1.629	1.277	1.285	1.346	1.534	2.882	1.085	1.624	1.856	3.078

[a] Computed through December 31, 1987, over the entire work history, for exposed workers only
[b] RFib, respirable fibre; Cum, cumulative exposure; AIE, average intensity of exposure
[c] Marsh et al. (2001a)
[d] Coefficient of variation = standard deviation divided by mean

Table 21. Concentrations (fibre/cm³) of total airborne fibres in dust zones of one rock (stone) wool and one slag wool production plant in the USA

Dust zone	No. of samples	Total fibres (fibre/cm³)	
		Average	Range
Rock (stone) wool			
Warehouse	3	1.4	1.1–1.7
Mixing/Fourdrinier[a] ovens	3	0.14	0.13–0.18
Panel finishing	12	0.40	0.13–1.3
Fibre forming	10	0.20	0.07–0.65
Erection and repair	13	0.24	0.04–1.1
Tile finishing	22	0.31	0.10–0.74
All samples	63	0.34	0.04–1.7
Slag wool			
Maintenance	15	0.08	0.01–0.24
Block production	8	0.05	0.02–0.11
Blanket line	5	0.05	0.02–0.09
Boiler room	2	0.05	0.04–0.07
Yard	2	0.09	0.05–0.13
Ceramic block	7	0.42	0.11–0.95
Shipping	3	0.04	0.02–0.06
Main plant	11	0.01	0.006–0.58
Mould formation	19	0.03	0.005–0.08
All samples	72	0.10	0.005–0.95

From Corn *et al.* (1976) (personal samples; 8-h sampling times; analysed by PCOM, using the equivalent of NIOSH A rules, combined with TEM)
[a] Fourdrinier, a machine which forms wet board of approximately 50% fibre

of airborne fibres seen in these plants were due to the larger percentage of fibres with diameters < 3 µm compared to that in the glass fibre production operations surveyed.

Jacob *et al.* (1993) evaluated the concentrations of airborne fibres associated with several production operations in glass wool manufacturing plants in the USA. As indicated in Table 22, some variation in fibre concentrations was observed between the different operations, but the overall concentrations measured were low (i.e. < 0.20 fibre/cm³) and the arithmetic mean concentration ranged from 0.022–0.29 fibre/cm³ for the 14 manufacturing operations evaluated. Typically half or fewer of the airborne fibres counted were identified as glass fibres.

The North American Insulation Manufacturers' Association (NAIMA), an association of producers of glass fibre and rock (stone) and slag wool in the USA, recently published a large database of airborne fibre concentrations measured during manufacturing and installation operations during 1984–2000 (Marchant *et al.*, 2002). The

Table 22. Arithmetic mean concentration of airborne fibres (fibre/cm^3) associated with various manufacturing operations involving glass wool insulation

	Total fibres[a]	Total glass fibres[b]	Respirable fibres[c]	Respirable glass fibres[d]
Equipment insulation fabrication[e]				
Arithmetic mean	0.026	0.009	0.007	0.001
95% confidence limits[f]	0.022–0.031	0.007–0.012	0.004–0.009	0.0004–0.002
95th percentile[f]	0.034	0.016	0.010	0.003
Number of samples	8	8	4	4
Moulding media				
Arithmetic mean	0.028	0.012	0.012	0.008
95% confidence limits	0.021–0.039	0.007–0.019	0.007–0.018	0.004–0.014
95th percentile	0.083	0.048	0.042	0.033
Number of samples	31	31	16	16
Fabrication, press operator, packer				
Arithmetic mean	0.20	0.14	0.087	0.071
95% confidence limits	0.17–0.23	0.11–0.17	0.072–0.102	0.059–0.083
95th percentile	0.32	0.24	0.12	0.11
Number of samples	24	24	12	12
Fabrication, other				
Arithmetic mean	0.038	0.015	0.007	0.002
95% confidence limits	0.028–0.051	0.011–0.020	0.005–0.009	0.0008–0.004
95th percentile	0.064	0.031	0.010	0.005
Number of samples	12	12	6	6
Metal building insulation[h]				
Arithmetic mean	0.043	0.017	0.017	0.009
95% confidence limits	0.038–0.049	0.014–0.020	0.014–0.022	0.006–0.012
95th percentile	0.076	0.032	0.028	0.018
Number of samples	31	31	16	16
Manufactured housing[i]				
Arithmetic mean	0.19	0.080	0.032	0.019
95% confidence limits	0.12–0.31	0.05–0.124	0.014–0.052	0.010–0.028
95th percentile	0.34	0.18	0.074	0.040
Number of samples	20	20	8	8
Pipe insulation				
Arithmetic mean	0.20	0.052	0.055	0.020
95% confidence limits	0.13–0.27	0.030–0.081	0.035–0.072	0.013–0.027
95th percentile	0.46	0.15	0.091	0.037
Number of samples	15	15	8	8
Range assembly				
Arithmetic mean	0.058	0.039	0.029	0.023
95% confidence limits	0.043–0.080	0.027–0.055	0.015–0.044	0.011–0.037
95th percentile	0.113	0.095	0.071	0.062
Number of samples	25	25	11	11

Table 22 (contd)

	Total fibres[a]	Total glass fibres[b]	Respirable fibres[c]	Respirable glass fibres[d]
Range assembly, wool unloader only				
Arithmetic mean	0.16	0.078	0.12	0.10
Values[j]	0.18, 0.13	0.083, 0.072		
Number of samples	2	2	1	1
Duct assembly				
Arithmetic mean	0.038	0.009	0.008	0.005
95% confidence limits	0.027–0.055	0.007–0.010	0.007–0.009	0.005–0.006
95th percentile	0.078	0.012	0.009	0.007
Number of samples	12	12	6	6
Duct board installation				
Arithmetic mean	0.022	0.002	0.006	0.002
95% confidence limits	0.019–0.026	0.001–0.002	0.004–0.007	0.001–0.003
95th percentile	0.042	0.006	0.009	0.005
Number of samples	29	29	15	15
Water heater assembly				
Arithmetic mean	0.047	0.030	0.022	0.018
95% confidence limits	0.034–0.058	0.022–0.040	0.013–0.034	0.010–0.028
95th percentile	0.074	0.054	0.044	0.036
Number of samples	11	11	6	6
Flex duct assembly				
Arithmetic mean	0.073	0.037	0.029	0.022
95% confidence limits	0.062–0.087	0.029–0.047	0.022–0.035	0.016–0.028
95th percentile	0.16	0.11	0.063	0.055
Number of samples	62	62	31	31
Removal of pipe and ceiling insulation				
Arithmetic mean	0.29	0.10	0.13	0.042
95% confidence limits	0.20–0.41	0.06–0.14	0.10–0.16	0.023–0.059
95th percentile	0.77	0.25	0.25	0.085
Number of samples	14	14	14	14

From Jacob *et al.* (1993) (personal and area samples; task weighted; analysed by NIOSH Method 7400 — A and B counting rules)

[a] Total fibres, NIOSH 7400 A rules, both filters and cowls
[b] Total glass fibres, NIOSH 7400 A rules, both filters and cowls, glass fibres only
[c] Respirable fibres, NIOSH 7400 B rules, filters and cowls
[d] Respirable glass fibres, NIOSH 7400 B rules, filters and cowls, glass fibres only
[e] A roll of light-density glass fibre insulation is fed through a coater that applies a facing and it is rolled up again.
[f] 95% confidence limits on the arithmetic mean
[g] 95th percentile of the data (95% of all measurements were less than this value)
[h] A roll of light-density glass fibre insulation is unrolled, sometimes cut into a thinner blanket, a coated facer is applied, and the insulation is rolled up again.
[i] Prefabricated housing
[j] When only two samples were measured, the measured values are presented.

database was established as a component of a voluntary workplace safety programme (designated the Health and Safety Partnership Program [HSPP]) developed by NAIMA after discussion with representatives from the US Occupational Safety and Health Administration (OSHA). These data, presented in Tables 23–26, demonstrate that the concentrations of airborne fibres measured during the production of glass and slag and/or rock (stone) wool are generally below 1 fibre/cm^3, which is a voluntary permissible 8-h time-weighted average (TWA) exposure limit, except in those areas where small-diameter fibres (< 1 μm in diameter) are produced (i.e. manufacture of aircraft insulation and separator and filtration media).

The concentrations of fibres measured during the production of refractory ceramic fibres in the USA were higher than those measured in glass wool and continuous glass filament plants, but were comparable with those measured in rock (stone) wool and slag wool production plants (Esmen et al., 1979b). These data are presented in Table 27. Considerable variation in the concentrations of airborne fibres was noted between plants and within operations in individual plants. Approximately 90% of the airborne fibres in the three facilities evaluated were determined to be respirable (i.e. < 3 μm in diameter), as determined by PCOM and electron microscopy.

Additional monitoring of exposure to fibres was conducted in plants that produced refractory ceramic fibres in the USA from the late 1970s to the early 1990s (Corn et al., 1992). However, much of this monitoring was sporadic, conducted by different consultants or institutions, focused on unit operations or processes of particular interest and emphasized monitoring of internal facilities rather than plants operated by secondary users. The estimates of exposure during the production of refractory ceramic fibres and related operations were discussed in detail by Rice et al. (1996, 1997).

The development of the unified Product Stewardship Program (PSP) by the industry trade association, the Refractory Ceramic Fiber Coalition (RCFC), in 1990 and the signing of a voluntary consent agreement with the US Environmental Protection Agency in 1993, brought greater conformity to methods for fibre sampling, counting and analysis of refractory ceramic fibres; more data were collected from consumers and end-users and increased emphasis was placed on random sampling. The data collected for this programme are presented in Table 28 (Maxim et al., 1994, 1997, 1999a, 2000a).

These data demonstrate that exposure to refractory ceramic fibres varies widely from one functional job category to another. Most mean time-weighted average (TWA) concentrations of refractory ceramic fibres recorded during manufacture and processing operations are currently below the recommended exposure limit set by the US MMVF industry of 0.5 fibre/cm^3; higher mean TWA concentrations were measured during product finishing and removal operations (Maxim et al., 1997). In addition, the analysis of industry data on a calendar time-period basis indicates that the average concentrations of airborne refractory ceramic fibres during manufacture and processing operations declined from 1.2 fibre/cm^3 in 1990 to less than 0.4 fibre/cm^3 in 2000 (Maxim et al., 2000a).

Table 23. Glass wool manufacture: exposure to glass wool fibres (fibre/cm³) by product type

Product type	Sample size	Exposure (fibre/cm³)			
		Mean	Standard deviation	Median	Range
Air handling products	12	0.03	0.03	0.02	0.01–0.13
Aircraft insulation	67	0.19	0.36	0.06	0.01–2.29
Appliance insulation	28	0.12	0.29	0.03	0.01–1.30
Automotive insulation	102	0.02	0.03	0.01	0.01–0.18
Separator and filtration media	376	0.80	0.84	0.51	0.01–4.63
Blowing wool with binder	71	0.04	0.03	0.03	0.01–0.02
Blowing wool without binder	53	0.11	0.12	0.08	0.01–0.49
High density board	14	0.02	0.02	0.01	0.01–0.09
Pipe insulation	114	0.05	0.10	0.02	0.01–0.70
Insulation batts and blankets	472	0.05	0.09	0.02	0.01–0.97
Other[a]	339	0.07	0.18	0.02	0.01–2.30

From Marchant et al. (2002) (personal samples; > 90% of the samples in the database were collected and analysed by NIOSH method 7400 B counting rules; sampling times greater than 4 h)

[a] Includes acoustical panels and non–specified industrial and commercial products

Table 24. Glass wool manufacture: exposure to glass wool fibres (fibre/cm³) by job description

Job description	Sample size	Exposure (fibre/cm³)			
		Mean	Standard deviation	Median	Range
Scrap baler/compactor	29	0.05	0.05	0.04	0.01–0.25
Batch/binder mixer	40	0.18	0.33	0.04	0.01–1.30
Cutting/hot press mould	109	0.04	0.12	0.01	0.01–0.88
Forming	289	0.11	0.23	0.02	0.01–2.30
General labourer/maintenance	62	0.11	0.33	0.02	0.01–2.29
Packaging	890	0.34	0.67	0.04	0.01–4.63
Quality control/research	75	0.18	0.23	0.09	0.01–1.20
Sewing/laminating/assembly	91	0.08	0.11	0.03	0.01–0.62
Shipping/receiving	53	0.01	0.01	0.01	0.01–0.06
Other[a]	10	0.11	0.20	0.05	0.01–0.66

From Marchant et al. (2002) (personal samples; sampling times greater than 4 h; analysed by NIOSH method 7400 B counting rules)

[a] Includes administration and blowing wool chopper operator or nodulator

Table 25. Slag and/or rock (stone) wool manufacture and installation: exposure to wool fibres (fibre/cm^3) by product type

Product type	Sample size	Exposure (fibre/cm^3)			
		Mean	Standard deviation	Median	Range
Manufacturing					
Ceiling panel/tile	412	0.20	0.19	0.15	0.01–1.41
Other manufacturing[a]	17	0.06	0.04	0.05	0.01–0.15
Installation					
Ceiling panel/tile	33	0.23	0.21	0.17	0.02–0.82
Spray-on fire-proofing	15	0.08	0.10	0.05	0.02–0.42
Insulation batt and blanket	12	0.09	0.04	0.08	0.04–0.16
Other installation[a]	14	0.11	0.11	0.06	0.02–0.4

From Marchant et al. (2002) (personal samples; sampling times greater than 4 h; analysed by NIOSH method 7400, B counting rules)
[a] Includes air-handling products, appliance insulation, blowing wool with binder, cavity loose-fill insulation, pipe insulation and safety blanket and board.

Table 26. Slag and/or rock (stone) wool manufacture and installation: exposure to wool fibres (fibre/cm^3) by job description

Job description	Sample size	Exposure (fibre/cm^3)			
		Mean	Standard deviation	Median	Range
Manufacturing					
Supervisory	17	0.13	0.11	0.10	0.01–0.40
Forming	162	0.24	0.22	0.18	0.01–1.41
Maintenance	79	0.18	0.16	0.14	0.01–0.79
Packaging	62	0.25	0.20	0.23	0.01–1.00
Quality control	20	0.21	0.21	0.16	0.01–0.80
Shipping/receiving	55	0.14	0.14	0.08	0.01–0.57
Other manufacturing[a]	34	0.09	0.10	0.05	0.01–0.42
Installation					
Installers	65	0.16	0.17	0.10	0.02–0.82
Other installation[a]	9	0.09	0.12	0.05	0.02–0.40

From Marchant et al. (2002) (personal samples; sampling times greater than 4 h; analysed by NIOSH method 7400 B counting rules)
[a] Includes assembly, cutting and sawing with power tools, vehicle driver production, warehousing, feeder and general labourer.

Table 27. Total concentrations of airborne fibres and concentrations of respirable fibres in dust zones of three refractory ceramic fibre production plants in the USA

Dust zone	Total respirable fibres[a] (fibre/cm^3)	Total fibres[a] (fibre/cm^3)	Total fibres[b] (fibre/cm^3)	Respirable fraction[c]
Plant A (total suspended particulate matter)	2.6	3.3	2.6	0.79
Finishing	2.1	2.6	1.9	0.82
CVF	4.2	5.2	4.3	0.80
Lines 1 and 2	0.94	1.1	0.73	0.83
Lines 3 and 4	0.08	0.09	0.04	0.89
OEM	6.9	8.7	7.6	0.79
Maintenance	0.50	0.64	0.52	0.79
GFA	0.53	0.80	0.74	0.66
Shipping	0.27	0.34	0.22	0.78
Quality control	0.11	0.15	0.11	0.71
Plant B (total suspended particulate matter)	1.4	1.5	0.63	0.92
Textile	0.88	1.1	0.62	0.79
Maintenance	0.95	1.0	0.27	0.96
Furnace	1.5	1.6	0.60	0.96
Process	2.4	2.6	1.1	0.95
Quality control	0.62	0.68	0.33	0.92
Plant C (total suspended particulate matter)	0.21	0.23	0.05	0.91
Maintenance	0.12	0.12	0.01	0.98
Fiberizing	0.22	0.23	0.04	0.96
Felting	0.20	0.24	0.10	0.82
Pressing	0.23	0.26	0.08	0.89
Finishing	0.26	0.28	0.06	0.93
Fibre cleaning	0.06	0.07	0.01	0.94
Mixing	0.02	0.03	0.01	0.93
Shipping	0.04	0.05	0.03	0.84
Job centre	0.22	0.23	0.04	0.94

From Esmen *et al.* (1979b)

[a] As determined by both electron and optical microscopy, including fibres < 5 μm in diameter
[b] As determined by optical microscopy only
[c] Total concentration of respirable fibres/total concentration of fibres
CVF, mixing of bulk fibre with colloidal silica and vacuum formed; OEM, trimming of some products from CVF process with hand saws, drilling and packaging; GFA, cutting by hand of blankets from line 2 into specific shapes

Table 28. Concentrations (fibre/cm^3) of refractory ceramic fibres during manufacturing, processing and use in the USA (1993–96)

Functional job category	Data set	Fibre concentration (TWA) (fibre/cm^3)					
		Sample size	Minimum	Maximum	Arithmetic mean of sample	Median of sample	Standard deviation of sample
Assembly	Internal	207	0.005	2.2	0.30	0.22	0.31
	External	211	0.007	1.9	0.28	0.15	0.35
	Combined	418	0.005	2.2	0.29	0.19	0.35
Auxiliary	Internal	148	0.004	0.81	0.092	0.036	0.14
	External	157	0.002	2.7	0.20	0.082	0.39
	Combined	305	0.002	2.7	0.15	0.056	0.30
Fibre production	Internal	250	0.009	1.9	0.22	0.13	0.28
	External	None					
	Combined	NA					
Finishing	Internal	206	0.028	2.7	0.66	0.50	0.48
	External	375	0.009	30	0.99	0.40	2.3
	Combined	581	0.009	30	0.87	0.46	1.9
Installation	Internal	None					
	External	288	0.003	2.5	0.41	0.24	0.47
	Combined	NA					
Mixing and forming	Internal	231	0.007	1.4	0.30	0.20	0.28
	External	186	0.010	4.1	0.32	0.15	0.52
	Combined	417	0.007	4.1	0.31	0.18	0.40
Other (NEC)	Internal	96	0.008	0.90	0.11	0.048	0.14
	External	244	0.003	6.4	0.24	0.056	0.56
	Combined	340	0.003	6.4	0.20	0.054	0.48

EXPOSURE DATA

Table 28 (contd)

Functional job category	Data set	Fibre concentration (TWA) (fibre/cm^3)					
		Sample size	Minimum	Maximum	Arithmetic mean of sample	Median of sample	Standard deviation of sample
Removal	Internal	5	0.66	3.2	1.6	1.20	0.88
	External	103	0.027	5.4	1.2	0.90	1.0
	Combined	108	0.027	5.4	1.2	0.91	1.0
Combined	Internal	1143	0.004	3.2	0.31	0.18	0.37
	External	1564	0.002	30	0.52	0.20	1.3
	Combined	2707	0.002	30	0.43	0.19	1.0

From Maxim *et al.* (1997) ('internal' samples collected from operations inside a plant manufacturing or processing refractory ceramic fibres; 'external' samples collected from other operations in which refractory ceramic fibres are processed or used)
NA, not applicable; NEC, not elsewhere classified

(ii) *Europe*

In the late 1970s, Head and Wagg (1980) studied airborne concentrations of both total particulate and respirable fibres in 25 manufacturing plants and construction sites in the United Kingdom. Various types of MMVF were evaluated including insulation wools (i.e. glass wool and rock (stone) and slag wool), continuous filament glass fibre, special-purpose fibres: glass microfibres (i.e. fibres with nominal mean diameters ranging from 0.1–3 μm) and refractory ceramic fibres.

Four plants that produced insulation wools were evaluated; one rock (stone) slag wool plant and three glass wool fibre manufacturing plants. The mean concentrations of respirable fibres were in the range 0.12–0.31 fibre/cm^3 in the glass wool fibre plants. A closer examination of these facilities found that the mean concentrations of airborne fibres were lower (i.e. 0.02–0.26 fibre/cm^3) in the basic production areas than in the areas where the fibres were fabricated into finished products, especially when machining operations were involved in the fabrication. The fabrication operations using rock (stone) or slag wool resulted in higher average concentrations of airborne respirable fibres (mean, 0.89; range, 0.03–10.3 fibres/cm^3) (see Table 29) (Head & Wagg, 1980).

For operations using continuous glass filament, principally the manufacture and weaving of textile yarns, the mean total fibre concentrations ranged from 0.006–0.28 fibre/cm^3. However, the overall mean concentrations of airborne respirable fibres were 0.02 fibre/cm^3 or below (Head & Wagg, 1980). The fibres in this group cover user processes only and for convenience have been subdivided into glass microfibres and high-quality aerospace insulation products.

In the manufacture of high efficiency filter paper, including filter production, and in the manufacture of high efficiency fabric type filters, total dust concentrations are characteristically low, the overall range of means being 0.4–1.4 mg/m^3, but with the potential for high respirable fibre counts, the range of overall mean concentrations being 0.8–3.70 fibres/cm^3 and the range of individual values being 0.02–18.83 fibres/cm^3. The maximum fibre count was found at a paper slitting machine (Head & Wagg, 1980).

From 1977–80, scientists from the Institute of Occupational Medicine in Edinburgh, United Kingdom, measured the concentrations of airborne fibres in 13 European plants, of which six produced rock (stone) wool (plants 1, 3, 4, 5, 8 and 9), four, glass wool (plants 2, 6, 7 and 10) and three, continuous glass filaments (plants E, J and N) (i.e. continuous glass filament or textile glass) (Ottery *et al.*, 1984). The data from this study were used as the basis for the exposure assessment component of the study by Saracci *et al.* (1984a,b; IARC, 1988). The workforce at each factory was classified into occupational groups on the basis of measured exposure levels and working areas, and some members of each group were selected at random for personal sampling. A total of 1078 samples were collected for counting of respirable fibres at the rock (stone) wool and glass wool plants, generally over a 7–8-h period. Upon re-analysis, the concentrations of respirable fibres originally reported were determined to be too low by a factor of about two (ranging between 1 and 2.8), and were therefore corrected (Cherrie *et al.*, 1986).

Table 29. Overall mean concentrations of total airborne dust and respirable fibres in insulation wool production plants in the United Kingdom (sampling period not stated)

Fibre type	Plant no.	Mean total dust			Mean respirable fibres[a]		
		No. of samples	Concentration (mg/m^3)	Range	No. of samples	Concentration ($fibre/cm^3$)	Range
Glass wool	Plant 1	32	11.1	0.7–78.2	50	0.31	0.02–1.10
Glass wool	Plant 2	16	4.1	0.5–14.3	35	0.27	0.01–0.79
Glass wool	Plant 3	30	8.9	0.4–51.3	67	0.12	0.003–0.85
Rock (stone)/slag wool	Plant 4	22	6.5	0.7–16.2	55	0.89	0.03–10.3

From Head & Wagg (1980)
[a] Respirable fibres are defined as being < 3 μm diameter and > 5 μm long.
Note: Overall mean values include both breathing-zone and static-sample results.

Tables 30 and 31 present the revised data from the glass wool and rock (stone) wool plants; only unrevised figures from the study by Ottery *et al.* (1984) are available for the three continuous glass filament plants, and these are presented in Table 32.

The range of group arithmetic mean fibre concentrations in the four glass wool plants was 0.01–1.00 fibre/cm^3. The highest concentrations were associated with the manufacture of special fine-fibre ear plugs. In the main production and secondary production groups, the concentrations ranged from 0.01–0.16 fibre/cm^3 (Cherrie *et al.*, 1986). In the rock (stone) wool plants, the combined arithmetic means for the occupational groups were 0.01–0.67 fibre/cm^3. The concentrations of respirable glass fibres measured in the continuous glass filament plants were very low; the occupational group means ranged from 0.001–0.023 fibre/cm^3 (Ottery *et al.*, 1984).

In two glass wool plants in France that were surveyed in the mid-1980s, the mean concentration of airborne fibres < 3 μm in diameter and > 5 μm in length was 0.05–0.18 fibre/cm^3 in different working zones (Kauffer & Vigneron, 1987).

An experimental simulation of a rock (stone) wool production process with conditions similar to those operating in the 1940s was carried out in a Danish pilot plant to determine the effect on concentrations of airborne fibres of the addition of oil to the rock (stone) wools. The addition of oil to the product resulted in a threefold to ninefold reduction in the concentration of airborne fibres in different situations. The TWA concentrations of respirable fibres were about 1.5 fibre/cm^3 when the product was treated with oil and 5.0 fibres/cm^3 in the absence of oil. No substantial differences in concentrations of airborne fibres were observed during the manufacture and handling of batch-produced rock (stone) wool and continuously produced rock (stone) wool (Cherrie *et al.*, 1987).

A survey of the occupational environment of 10 rock (stone) wool and glass wool production plants was conducted by the MMVF industry and the Swedish Work Environment Fund. The products evaluated included insulation rock (stone) and glass wools, continuous glass filament and special-purpose products (earplugs (glass wool) and Inorphil (rock wool)). A pilot study at the factory that produced continuous glass filaments showed that no airborne respirable glass fibres were present, and this factory was therefore excluded from further epidemiological investigation. From the remaining nine factories, 1350 samples were collected. In insulation wool production, the arithmetic mean exposure to total dust and respirable fibres as assessed by gravimetric analysis and PCOM, respectively, are presented in Table 33. The mean concentrations of respirable fibres to which workers were exposed ranged from 0.15–0.27 fibre/cm^3 for rock (stone) wool production workers and 0.14–0.23 fibre/cm^3 for glass wool production workers; maintenance and cleaning workers were found to be exposed to higher concentrations of fibres than process workers. During the production of special-purpose fibres (Table 34), production workers were exposed to higher concentrations of respirable fibres (i.e. 1.4 fibre/cm^3 for rock (stone) wool and 0.62 fibre/cm^3 for glass wool) than in production of insulation wool (0.15 and 0.14 fibre/cm^3, respectively). The

Table 30. Concentrations of respirable fibres (fibre/cm^3)[a] in combined occupational groups and total concentration of airborne dust in mg/m^3 in four European glass wool plants (1977–80)

Combined occupational group	Plant 2			Plant 6			Plant 7			Plant 10		
	No.[b]	Mean[c]	Range	No.[b]	Mean[c]	Range	No.[b]	Mean[c]	Range	No.[b]	Mean[c]	Range
Respirable fibre concentrations												
Preproduction	8	0.01	<0.01–0.01	5	0.01	0.01	5	0.01	0.01–0.02	5	0.01	<0.01–0.03
Production	26	0.01	<0.01–0.03	27	0.03	0.01–0.11	39	0.05	0.01–0.62	61	0.05	<0.01–0.22
Maintenance	4	0.03	0.01–0.06	12	0.04	<0.01–0.17	20	0.07	0.01–0.60	27	0.02	<0.01–0.06
General	10	0.02	0.01–0.04	10	0.02	0.01–0.04	15	0.03	0.01–0.06	12	0.03	<0.01–0.06
Secondary process 1	32	0.05	0.01–0.21	26	0.03	<0.01–0.07	37	0.04	0.01–0.11	36	0.02	<0.01–0.06
Secondary process 2	–			2	0.07	0.05–0.09	23	1.00	0.17–4.02	45	0.16	0.02–1.39
Cleaning	–			4	0.01	0.01–0.02	–					
Airborne dust concentration												
Plant mean and range (mg/m^3)[d,e]	69	[0.6]		79	[1.3]		124	[1.0]		168	[1.3]	

From Cherrie et al. (1986) (personal samples; sampling times 2–7 h; analysed by PCOM)
[a] Including fibres < 3 μm in diameter
[b] Number of measurements
[c] Arithmetic mean, revised in 1986, compared to values of Ottery et al. (1984)
[d] From Ottery et al. (1984)
[e] Overall mean [calculated by the Working Group]

Table 31. Concentrations of respirable fibres (fibre/cm^3)[a] in combined occupational groups and total concentration of airborne dust (mg/m^3) in six European rock (stone) wool plants (1977–80)

Combined occupational group	Plant 1 No.[b]	Plant 1 Mean[c]	Plant 1 Range	Plant 3 No.[b]	Plant 3 Mean[c]	Plant 3 Range	Plant 4 No.[b]	Plant 4 Mean[c]	Plant 4 Range	Plant 5 No.[b]	Plant 5 Mean[c]	Plant 5 Range	Plant 8 No.[b]	Plant 8 Mean[c]	Plant 8 Range	Plant 9 No.[b]	Plant 9 Mean[c]	Plant 9 Range
Respirable dust concentration																		
Preproduction	8	0.08	0.01–0.22	3	0.06	0.03–0.11	7	0.03	0.01–0.07	2	0.01	0.01	1	0.04	0.04	4	0.01	<0.01–0.01
Production	36	0.10	0.02–0.37	28	0.12	0.03–0.32	27	0.06	0.02–0.19	22	0.06	0.02–0.14	19	0.05	0.01–0.13	51	0.05	0.01–0.16
Maintenance	9	0.08	0.05–0.18	8	0.05	0.03–0.10	20	0.05	0.02–0.12	12	0.05	0.01–0.14	9	0.03	0.01–0.07	10	0.04	0.01–0.11
General	16	0.08	0.02–0.37	8	0.07	0.04–0.14	13	0.06	0.02–0.09	7	0.04	0.03–0.07	2	0.04	0.04	23	0.06	0.01–0.36
Secondary process 1	32	0.10	0.03–0.21	11	0.12	0.06–0.23	28	0.08	0.03–0.33	16	0.07	0.01–0.15	24	0.08	0.01–0.20	55	0.06	0.02–0.39
Secondary process 2	11	0.40	0.09–1.40	3	0.34	0.25–0.41	–			–			3	0.25	0.19–0.36	22	0.67	0.06–1.37
Cleaning	–			4	0.13	0.05–0.29	8	0.06	0.02–0.14	5	0.09	0.04–0.11	8	0.09	0.01–0.18	12	0.14	0.02–0.44
Airborne dust concentration																		
Plant mean and range (mg/m^3)[d,e]	101	[2.4]		56	[1.6]		86	[1.1]		53	[1.0]		60	[1.0]		164	[0.7]	

From Cherrie et al. (1986) (personal samples; sampling times 2–7 h; analysed by PCOM)
[a] Including fibres < 3 μm in diameter
[b] Number of measurements
[c] Arithmetic means, revised in 1986, compared to values of Ottery et al. (1984)
[d] From Ottery et al. (1984)
[e] Overall mean [calculated by the Working Group]

Table 32. Concentrations of respirable fibres (fibre/cm^3)[a] in combined occupational groups and total concentrations of airborne dust (mg/m^3) at three European continuous filament glass fibre plants (1977–80)

Combined occupational group	Plant E			Plant J			Plant N		
	No.	Mean	Range	No.	Mean	Range	No.	Mean	Range
Respirable fibre concentrations									
Preproduction	12	0.004	0.001–0.015	—			6	0.009	0.005–0.017
Production I	54	0.002	0.001–0.012	19	0.001	0.001–0.003	44	0.007	0.001–0.039
Production II	—			32	0.001	0.001–0.003	22	0.023	0.005–0.112
Maintenance	16	0.005	0.001–0.022	—			15	0.014	0.006–0.023
General	2	0.005		11	0.001	0.001–0.003	7	0.012	0.008–0.020
Secondary process 1	70	0.002	0.001–0.016	87	0.002	0.001–0.006	27	0.007	0.005–0.017
Secondary process 2	—			—			6	0.022	0.006–0.056
Research and development	10	0.002	0.001–0.003	—					
Total airborne dust concentration									
Plant mean and range (mg/m^3)[a]	145	[1.5]		132	[0.61]		115	[0.9]	

From Ottery et al. (1984)
[a] Overall mean [calculated by the Working Group]

Table 33. Total concentrations of dust and respirable fibres during the production of insulation wool

Occupational group	Mean total dust (mg/m^3)		Mean respirable fibre concentration (fibre/cm^3)	
	Rock (stone) wool	Glass wool	Rock (stone) wool	Glass wool
Process workers	0.8	0.7	0.15	0.14
Maintenance workers	1.1	4.4	0.20	0.23
Cleaning workers	2.9	0.2	0.27	0.15
Mean (all groups)[a]	1.4	1.9	0.20	0.18
Total range	0.1–32	0.1–410	0.01–2.6	0.01–1.8

From Krantz (1988)
[a] Also includes handling of raw material

Table 34. Total concentrations of dust and respirable fibres during the production of special-purpose fibres

Occupational group	Total dust (mg/m^3)		Respirable fibre concentration (fibre/cm^3)	
	Rock (stone) wool	Glass wool	Rock (stone) wool	Glass wool
Process workers	2.0	0.6	1.4	0.62
Maintenance workers	–	0.4	–	0.20
Cleaning workers	–	0.4	–	0.16
Mean (all groups)[a]	2.0	1.0	1.4	0.47
Total range	0.5–2.0	0.1–1.4	0.45–1.9	0.08–2.4

From Krantz (1988)
[a] Also includes handling of raw material

main reason for these higher values was the absence of dust-reducing oil and binder in the special-purpose fibres together with manual handling (Krantz, 1988).

Plato *et al.* (1995b) used the data from this study as well as additional information on employee job titles, work tasks and periods of employment to develop a multiplicative model to assess past exposure to respirable fibres in rock (stone) wool and slag wool production workers in Sweden from 1938–90. This model took into account process changes, as well as the addition of binders and oil, and suggested that the levels of exposure in the MMVF industry had changed considerably during this period. The estimated average concentrations of airborne fibres during the 1940s were in the range 0.78–1.32 fibre/cm^3 which was between 16 and 26 times higher than the concentrations of airborne fibres measured in the 1980s.

Plato et al. (1995a) also evaluated exposure to MMVFs at 11 Swedish plants that manufactured prefabricated wooden houses. All the samples were analysed by PCOM as well as using a modified fibre-counting method specifically developed for this study. The alternative counting method was designed to exclude fibres other than MMVFs. The geometric mean of the concentration of airborne fibres to which insulators were exposed was 0.10 fibre/cm^3 (range, 0.03–0.30 fibre/cm^3) using the standard fibre-counting method, and 0.029 fibre/cm^3 (range, 0.013–0.077 fibre/cm^3) using the modified fibre-counting method. The results of the modified fibre-counting method were validated by SEM.

The concentrations of airborne fibres were also measured in three plants that produced rock (stone) wool insulation materials by personal sampling according to VDI 3492 (Tiesler et al., 1993). The highest fibre concentration measured was 0.44 fibre/cm^3 and concentrations higher than 0.1 fibre/cm^3 were measured in only nine samples of a total of 69 (Draeger & Löffler, 1991). The German Ministry of Labour also published in its regulations for dangerous materials (TRGS 901) a summary of 192 measurements made during the production of glass and rock (stone) wool insulation materials. The 75th percentile of all these measurements of concentration of airborne fibres was 0.095 fibre/cm^3 and the 90th percentile was 0.25 fibre/cm^3. The concentrations of airborne fibres measured during insulation work with glass and rock (stone) wool products (102 samples) were: 75th percentile, 0.11 fibre/cm^3; 90th percentile, 0.40 fibre/cm^3 (TRGS, 1999).

For operations involving the manufacture and fabrication of refractory ceramic fibres, the overall mean concentration of airborne respirable fibres in a production plant was 1.27 fibre/cm^3 (range, 0.06–6.14 fibres/cm^3). The concentrations of airborne fibres during the manufacture and handling of refractory ceramic fibres in another plant were usually lower (mean, 0.44 fibre/cm^3; range, 0.09–0.87 fibre/cm^3). Airborne refractory ceramic fibres were also found to yield a proportion of fibres generally with the same length as glass wool and rock (stone) and slag wool fibres (Head & Wagg, 1980).

The exposure of workers to total dust and airborne fibres during the production and use of products made from refractory ceramic fibres was evaluated in the United Kingdom in the mid-1980s. The products studied included blankets, felts, yarns, papers, vacuum-formed shapes and boards. The data collected were pooled with the data that had been collected during previous surveys of similar operations by the United Kingdom Health and Safety Executive to yield comprehensive results. Fibre concentrations rarely exceeded 1.0 fibre/cm^3 (Friar & Phillips, 1989).

In the mid-1980s, the European manufacturers of refractory ceramic fibres and their products, collectively known as the European Ceramic Fibre Industries Association (ECFIA), developed a product stewardship programme, designated Control and Reduce Exposure (CARE), to monitor and control exposure to airborne fibres associated with their products. Several hundred TWAs were collected from sites of manufacture and end-use and the results were reported at both public forums and in the published literature (Burley et al., 1997; Maxim et al., 1998). The results of this programme are essentially similar to those obtained at plants making and using refractory ceramic fibres in the USA (see Table 29). The TWA concentrations of refractory ceramic fibres varied widely from one functional job category to another, both within and between plants. The analyses indicated that approximately 50% of TWA concentrations measured in the workplace are below 0.25 fibre/cm^3, 71% are below 0.5 fibre/cm^3, 84% are below 1 fibre/cm^3 and 94% are below 2 fibres/cm^3 (Maxim et al., 1998).

Exposure to refractory ceramic fibres and AES wools during manufacturing and oven insulation operations was recently evaluated by Class et al. (2001). Monitoring was performed and fibre concentrations were determined using the WHO criteria. The mean concentrations of refractory ceramic fibres ranged from 0.07–0.39 fibre/cm^3 during manufacturing operations and the mean concentrations of AES wools ranged from 0.06–0.39 fibre/cm^3 during the same operations. The mean concentration of refractory ceramic fibres measured during oven insulation was 0.46 fibre/cm^3, whereas for AES wool, the mean concentration of fibres was 0.36 fibre/cm^3. From these data, the authors concluded that there was essentially no difference between the atmospheric concentrations of fibres produced by refractory ceramic fibres and AES wool during their production and end-use.

Exposure to airborne refractory ceramic fibres and total dust was also evaluated in plants producing aluminosilicate fibres in Poland. The mean concentrations of respirable refractory ceramic fibres ranged from 0.14–1.13 fibre/cm^3 and the mean concentrations of total dust ranged from 0.4–13.6 mg/m^3 (Wojtczak, 1994). In one plant that produced fibre blankets and mats, the mean concentrations of respirable refractory ceramic fibres ranged from 0.07–0.37 fibre/cm^3 while the concentrations of dust ranged from 0.4–2.9 mg/m^3 (Wojtczak et al., 1996). In a plant that manufactured packing cord, insulating tape and paperboard products, the mean concentrations of respirable refractory ceramic fibres ranged from 0.05–0.62 fibre/cm^3, while the concentrations of dust were between 0.6 and 23.2 mg/m^3 (Wojtczak et al., 1997).

(iii) *Other studies*

In 1990, the National Occupational Health and Safety Commission (NOHSC) of Australia specified the National Standard for Synthetic Mineral Fibres and Code of Practice for the Safe Use of Synthetic Mineral Fibres based on the recommendations of an expert working group. A survey of Australian plants manufacturing glass wool, rock (stone) wool and refractory ceramic fibres and user industries was undertaken from 1991–92 to measure the typical levels of exposure to airborne fibres and to monitor adherence to recommended work practices. A total of 1572 samples of both respirable fibres and inspirable dust were collected from more than 250 jobs or processes. The data from this survey are presented in Table 35. Nearly all (i.e. 1225 or 97.7%) of the

Table 35. Exposure to respirable fibres in the MMVF manufacturing industry in Australia

Type of MMVF	Job or process	Typical fibre count before introduction of NOHSC standard (fibre/cm³)	Sample type	Count of respirable fibres after introduction of NOHSC standard (fibre/cm³)			
				No.	Range	Geometric mean	Geometric standard deviation
Glass wool	Furnace operator	< 0.02		4	< 0.01–0.08	0.02	2.39
	Bulk production line	< 00.1–0.2		18	< 0.01–0.10	0.03	1.86
	Forming of specific shape	–		41	< 0.01–0.10	0.02	1.94
	Supervisor	< 0.02		6	< 0.03–0.10	0.04	1.63
	Packer	0.02–0.2		21	< 0.02–0.20	0.04	2.13
	Others	–		4	< 0.02–0.10	0.05	2.00
	All processes		Personal	49	< 0.01–0.20	0.04	1.83
			Static	45	< 0.01–0.10	0.02	2.04
			Total	94	< 0.01–0.20	0.03	2.06
Rock (stone) wool	Furnace operator	< 0.1–0.2	Personal	2	0.04–0.10	0.06	1.91
	Bulk production line	–	Personal	5	0.05–0.07	0.06	1.16
	Forming of specific shape	–	Personal	10	< 0.02–0.08	0.04	1.76
	Packer	0.03–0.4	Personal	7	< 0.02–0.10	0.03	1.89
	Cleaning and maintenance	–	Personal	6	< 0.02–0.20	0.08	2.17
	Supervisor	0.1–0.2		–	–	–	
	All processes		Total	30	< 0.02–0.20	0.05	1.90
Refractory ceramic fibres	Furnace operator	–		4	0.04–0.4	0.15	2.88
	Bulk production line	< 0.05–1.3		4	0.20–0.9	0.42	1.97
	Forming of specific shape	–		31	0.02–2.0	0.34	3.15
	Others	–		3	0.09–2.0	0.38	4.78
	All processes		Personal	27	0.03–2.0	0.38	2.81
			Static	15	0.02–2.0	0.25	3.55
			Total	42	0.02–2.0	0.32	3.09

From Yeung & Rogers (1996)

reported fibre counts were below the established exposure standard of 0.5 fibre/cm^3 (Yeung & Rogers, 1996).

The levels of exposure to airborne refractory ceramic fibres were measured at two factories in Japan. In each factory, one manufacturing and one processing workplace were selected so that a total of four sampling sites were evaluated. Manufacturing site A produced pipes and tubes from refractory ceramic fibres. The workers at manufacturing site B were engaged in the cutting, binding and packaging of fibre blankets. Stationary and personal samples were collected, and analysed using PCOM. The geometric mean concentrations of airborne fibres were 0.10, 0.27, 0.30 and 0.66 fibre/cm^3 for workplace A manufacturing, A processing, B manufacturing and B processing, respectively. The results obtained from personal samples (range, 0.09–3.69 fibres/cm^3) were generally higher than those from stationary samples (range, 0.06–0.86 fibre/cm^3) (Hori *et al.*, 1993).

(b)   Exposure to compounds other than MMVFs in production plants

The assessments of exposure conducted in conjunction with the epidemiological studies of MMVF workers (discussed more fully in section 2) have revealed a number of other substances to which workers in the production environment are exposed. The earlier estimates of these other sources of exposure in production operations were qualitative due to the limited details in the records available. However, methods for the quantitative estimation of exposure to various chemical substances and physical agents present in the workplace, in addition to airborne fibres, have been developed recently using the process of environmental reconstruction.

For example, in several of the plants included in the study of European MMVF workers (Saracci *et al.*, 1984a,b; Simonato *et al.*, 1986a,b, 1987; Boffetta *et al.*, 1992, 1997, 1999), asbestos gloves and aprons were worn by a few individuals for personal protection and as thermal insulation. In four factories, asbestos was used as stitching yarn and cloth. In at least one plant, olivine sand, potentially contaminated with a natural mineral fibre with a composition similar to tremolite asbestos, was also used (Cherrie & Dodgson, 1986).

Exposure to polycyclic aromatic hydrocarbons (PAHs) may have occurred close to the cupola furnaces in three rock (stone) wool plants and one glass wool plant and in one plant in which an electric furnace was used. The possibility of exposure to arsenic from copper slags in one factory was also suggested (Cherrie & Dodgson, 1986). In addition, in one plant included in this study, located in Germany, exposure to coal-tar, bitumen, quartz and asbestos was identified, but not quantified (Grimm, 1983; Grimm *et al.*, 2000). Confirmation that workers in the 1940s were exposed to PAHs, to heavy metals (i.e. zinc and lead) and to carbon monoxide at a Danish slag wool production factory has been obtained from a simulation study (Fallentin & Kamstrup, 1993).

A case–control study of lung cancer nested in the rock (stone) and slag wool component of the European cohort of MMVF production workers presented analyses based upon various indicators of exposure to asbestos, PAHs, crystalline silica, welding fumes, heavy metals (i.e. arsenic, cadmium, chromium and nickel), formaldehyde, diesel exhaust, paints and ionizing radiation (Kjaerheim *et al.*, 2002). These analyses were based on a new method for the subjective assessment of exposure in the 13 plants included in the study over time (Cherrie & Schneider, 1999).

The main epidemiological study of MMVF production workers in the USA (Enterline *et al.*, 1983; Enterline & Marsh, 1984; Enterline *et al.*, 1987; Marsh *et al.*, 1990, 1996, 2001a,b,c), in addition to reporting concentrations of airborne fibres, provided estimates of exposure to chemical substances and physical agents, including several potential carcinogens. Quantitative estimates of exposure to respirable fibres, asbestos, formaldehyde and silica, together with qualitative estimates of exposure to arsenic, asphalts, PAHs, phenolics, radiation and urea were presented in the update of a study on a cohort of 3035 rock (stone) wool and slag wool workers (Marsh *et al.*, 1996). Quantitative estimates of exposure to respirable fibres, formaldehyde and silica, and qualitative estimates for arsenic, asbestos, asphalt, epoxy, PAHs, phenolics, styrene and urea were presented in the update of the US cohort study of 32 110 glass fibre workers (Marsh *et al.*, 2001a). All the exposure estimates were derived from an extensive reconstruction of historical exposure in the 15 plants included in the study (Quinn *et al.*, 2001; Smith *et al.*, 2001).

Quantitative estimates of exposure to several chemical substances, in addition to respirable fibres, were also included in two case–control studies of malignant and non-malignant respiratory disease among employees of the oldest and largest glass fibre manufacturing plant in the USA (Chiazze *et al.*, 1992, 1993), and in a historical cohort mortality study of a continuous glass filament plant in the USA (Chiazze *et al.*, 1997; Watkins *et al.*, 1997). Using a reconstruction of the historical environment of these two plants, quantitative estimates of cumulative exposure to respirable fibres, asbestos, talc, formaldehyde, respirable silica, asphalt fumes and total particulate matter were generated for the glass fibre plant and quantitative estimates of cumulative exposure to respirable fibres, asbestos, refractory ceramic fibres, respirable silica, formaldehyde, total particulate matter, arsenic and total chrome (chromium oxides) were developed for the continuous glass filament plant. These estimates were then incorporated into a conditional logistic regression analysis of lung cancer and non-malignant respiratory disease in the workers at the two plants.

(c) *Exposure of workers during the installation and removal of MMVF insulation products*

Occupational exposure to MMVFs outside production facilities occurs when workers either install materials containing MMVFs or remove (old) MMVFs from their site of installation. Most of the data that have been collected are for MMVF installation although higher levels of exposure have been recorded during removal operations (Jacob *et al.*, 1993; Maxim *et al.*, 1997; Yeung & Rogers, 1996). Moreover, during removal operations, workers are also exposed to many other airborne particles, both fibrous and non-fibrous.

The exposure of insulation workers to airborne MMVFs has been the subject of investigation for some time. Fowler *et al.* (1971) surveyed individuals while they were installing various types of glass fibre insulation in the western USA in the early 1970s (see Table 36). The concentrations of fibres to which the installers were exposed ranged from 0.5–8.0 fibres/cm^3 with a median of 1.26 fibre/cm^3 and a mean of 1.8 fibre/cm^3.

The exposure of workers to MMVFs during the installation of several types of insulation in commercial and residential buildings and at two aircraft installation facilities in the USA was evaluated by Esmen *et al.* (1982). Detailed data from this survey are presented in Table 37. The average exposure of workers for all applications, except the blowing of thermal insulation into attics and operations involved in the fabrication of aircraft insulation, was reported to be in the range of 0.003–0.13 fibre/cm^3. The reported average concentration of fibres during the blowing operations was 1.8–4.2 fibres/cm^3 with measurements in the different operations in the range of 0.50–14.8 fibres/cm^3 and the mean fibre concentration during fabrication of aircraft insulation was in the range of 0.05–1.7 fibre/cm^3 with individual measurements ranging from 0.01–3.78 fibres/cm^3. The fibre concentrations to which installers, with the exception of blowing wool applications, were exposed were deemed to be similar to the concentrations of airborne fibres found in MMVF production operations.

Large surveys of the exposure of installers and end-users of MMVF insulation products to airborne fibres and total dust were conducted in the United Kingdom and Scandinavia from the late 1970s until the mid-1980s (Schneider, 1979a; Head & Wagg, 1980; Hallin, 1981; Schneider, 1984). Full-shift samples were not collected during these surveys (except by Hallin, 1981), but the sampling time was determined to be representative of the particular product or operation being evaluated. Since the equipment used to collect dust samples varied from site to site the results for total dust concentration can only be compared qualitatively between these surveys. The results from the surveys in Sweden and Denmark are presented in Tables 38 and 39, and those from the surveys conducted in the United Kingdom in Table 40.

The distribution of results for the concentrations of respirable fibres had geometric means of 0.22 and 0.14 fibre/cm^3 and geometric standard deviations of 3.3 and 3.8 in the Swedish and Danish surveys, respectively. Very few measurements exceeded 2.0 fibres/cm^3. The geometric mean concentration of respirable fibres was 0.046 fibre/cm^3 (standard deviation, 1.8) in open and well-ventilated spaces and 0.50 fibre/cm^3 (standard deviation, 3.1) in confined and poorly ventilated spaces (Schneider, 1979a; Schneider, 1984). In the survey of exposure during installation of insulation in the United Kingdom, the application of loose fill appeared to generate the highest concentration of respirable fibres (Head & Wagg, 1980).

From 1980–83, the United Kingdom Factories Inspectorate (1987) surveyed factories where MMVFs with a nominal diameter of < 3 μm were used. The concentrations of total dust and airborne fibres measured are shown in Table 41.

The arithmetic mean value of fibre concentrations (as defined by WHO) (see section 1.1.3), determined in nine randomly selected building sites in Europe (where

Table 36. Concentrations and dimensions of airborne fibres measured during various operations using glass fibre insulation

Operation	Parent material (range of mean[a] fibre diameters, μm)	No. of samples	Breathing zone samples	
			Mean fibre concentration (fibres/cm^3)	Range of mean fibre diameters (μm)
Duct wrapping	4.0–7.5	9	1.2	2.3–6.2
Wall and plenum insulation	7.2–10.2	4	4.0	3.5–8.4
Pipe insulation	5.6–8.5	3	0.7	3.1–4.1
Fan-housing insulation	6.9	1	1.6	3.5

From Fowler et al. (1971) (personal samples; sampling times 20–60 min; analysed by PCOM)
[a] Not length-weighted

Table 37. Airborne concentrations of respirable fibres[a] in the final preparation and installation of MMVF insulation, as determined by a combination of phase-contrast and electron microscopy

Product and job classification	No. of samples	Fibre concentration		
		Average (fibre/cm^3)	Range (fibre/cm^3)	Average respirable fraction[b]
Acoustical ceiling installer	12	0.003	0–0.006	0.55
Duct installation				
Pipe covering	31	0.06	0.007–0.38	0.82
Blanket insulation	8	0.05	0.025–0.14	0.71
Wrap around	11	0.06	0.03–0.15	0.77
Attic insulation				
Glass fibre				
Roofer	6	0.31	0.07–0.93	0.91
Blower	16	1.8	0.67–4.8	0.44
Feeder	18	0.70	0.06–1.48	0.92
Rock (stone)/slag wool				
Helper	9	0.53	0.04–2.03	0.71
Blower	23	4.2	0.50–14.8	0.48
Feeder	9	1.4	0.26–4.4	0.74
Building insulation installer	31	0.13	0.013–0.41	0.91

Table 37 (contd)

Product and job classification	No. of samples	Fibre concentration		
		Average (fibre/cm^3)	Range (fibre/cm^3)	Average respirable fraction[b]
Aircraft insulation				
Plant A				
Sewer	16	0.44	0.11–1.05	0.98
Cutter	8	0.25	0.05–0.58	0.98
Cementer	9	0.30	0.18–0.58	0.94
Others	7	0.24	0.03–0.31	0.99
Plant B				
Sewer	8	0.18	0.05–0.26	0.96
Cutter	4	1.7	0.18–3.78	0.99
Cementer	1	0.12	–	0.93
Others	3	0.05	0.012–0.076	0.94
Glass fibre duct				
Duct fabricator	4	0.02	0.006–0.05	0.66
Sheet–metal worker	8	0.02	0.005–0.05	0.65
Duct installer	5	0.01	0.006–0.20	0.87

From Esmen *et al.* (1982) (average concentration of fibres/cm^3 is equal to concentrations derived from both PCOM and electron microscopy; total fibre concentration was not included in the table for calculation of the average respirable fraction (respirable/total); results cannot be directly compared to results obtained by PCOM only; when counting and sizing by electron microscopy, only fibres of diameter ≤ 1 μm were treated.
[a] < 3 μm in diameter
[b] Arithmetic mean of respirable fibre concentration/total fibre concentration

glass and rock (stone) wool insulation materials were used in different manners) was reported to be 0.04 fibre/cm^3. This mean value was based on 22 stationary samples; measurements were made using the VDI 3492 method (electron microscopy with EDXA). The maximum concentration reported was 0.2 fibre/cm^3. When personal sample measurements were analysed, the mean product fibre concentration was 0.07 fibre/cm^3 and the maximum value was 0.45 fibre/cm^3. The glass and rock (stone) wool insulation materials were found to contribute about one-third of the total mineral fibre measured; one-third was attributed to gypsum fibres and the remaining one-third to other inorganic fibres (Tiesler *et al.*, 1990).

In a survey of the installation of insulation materials at industrial sites in the European chemical industry (outdoors) and in shipyards (indoors), 22 personal sampling measurements were made using the VDI 3492 method. The concentrations of

Table 38. Concentrations of total dust and respirable fibres[a] during installation of insulation in Sweden (1979–80)

Operation	Total dust (mg/m^3)		Respirable fibres (fibre/cm^3)	
	Mean[b]	Range	Mean[b]	Range
Attic insulation, existing buildings	11.6	1.7–21.7	1.11	0.1–1.9
Insulation of new buildings	2.63	0.5–11.1	0.57	0.07–1.8
Technical insulation	3.14	0.4–25	0.37	< 0.01–1.39
Acoustical insulation	1.8	1.7–1.9	0.15	0.11–0.18
Spraying	13.5	1.3–43.7	0.51	0.13–1.1
Hanging of fabric	4.18	3.6–5.2	0.60	0.30–0.76

From Hallin (1981)
[a] < 3 μm in diameter
[b] Calculated by the previous IARC Working Group (IARC, 1988)

Table 39. Concentrations of total dust and respirable fibres[a] during installation of insulation in Denmark

Operation	Total dust (mg/m^3)		Respirable fibres (fibre/cm^3)	
	Mean	Range	Mean	Range
Attic insulation, existing buildings	26.8	1.5–134	0.89	0.04–3.5
Insulation of new buildings	12.6	0.22–44	0.10	0.04–0.17
Technical insulation	7.1	1.8–12.8	0.35	0.03–1.6
Application in industrial products	0.88	0.83–0.91	0.05	0.01–0.11
Hot–house substrate	3.00	0.61–3.9	0.06	0.03–0.09

From Schneider (1984)
[a] ≤ 3 μm in diameter

WHO product fibres were between 0.003 and 0.29 fibre/cm^3, and the arithmetic mean was 0.10 fibre/cm^3. The concentrations of WHO product fibres during disassembly work at the same sites were also analysed. The mean value of 12 personal sample measurements was 0.32 fibre/cm^3 (range, 0.05–0.77 fibre/cm^3) (Julier et al., 1993).

Insulation materials made of rock (stone) and glass wool are regularly removed and replaced in European power plants. The concentrations of inorganic fibres, excluding asbestos and gypsum, during these procedures (calculated from personal measurements) were reported to range between 0.21 and 0.99 fibre/cm^3 (Böckler et al., 1995).

Table 40. Concentrations of total dust and respirable fibres[a] in breathing zone and static samples taken during installation and application of MMVFs in the United Kingdom

Product, use or process	Total dust (mg/m^3)			Respirable fibres (fibre/cm^3)		
	No. of samples	Mean	Range	No. of samples	Mean	Range
Construction insulation (glass fibre)						
Blankets	9	35.5	8.2–90	12	0.75	0.24–1.76
Loose fill	4	30.9	5.0–59.7	6	8.19	0.54–20.9
Fire protection (mineral wool)	9	17.2	1.9–51.5	22	0.77	0.16–2.57
Industrial product insulation (one plant)	4	0.8	0.6–1.0	12	0.10	0.02–0.36
Manufacture and use of high–temperature insulation and mouldings (refractory ceramic fibres)	6	1.7	0.7–5.2	11	0.48	0.09–0.87
Manufacture of stack block insulation and engine silencer insulation (refractory ceramic fibres)	11	9.8	1.5–22.9	30	2.2	0.35–5.64

From Head & Wagg (1980) (personal and area samples; sampling times at least 4 h; analysed by PCOM)

[a] ≤ 3 μm in diameter

Table 41. Concentrations of total dust and respirable fibres[a] during the manufacture and use of special-purpose fibres and refractory ceramic fibres

Process	Mean concentration of total dust (mg/m^3)	Total fibres/cm^3 (mean counts)[b]
Manufacture of glass fibre paper	0.47–2.28	2.9–1.3
Conversion of glass fibre paper	0.17–0.49	0.53–15.1
Manufacture of refractory ceramic fibres	0.83–4.0	0.49–9.2
Use of refractory ceramic fibres	–	2.7–17.1

From United Kingdom Factories Inspectorate (1987)
[a] < 3 μm in diameter
[b] Determined by TEM

In an effort to characterize better the concentrations of airborne fibres associated with the installation of glass fibre insulation in homes, and to determine the proportion of airborne fibres that were glass fibres, a survey of batt and loose fill insulation in the USA was undertaken in the early 1990s (Jacob *et al.*, 1992). The results of this survey are presented in Tables 42 and 43. The arithmetic mean concentration of total airborne fibres present during the installation of batt insulation was 0.22 fibre/cm^3. Approximately 60% of this total was glass fibres and approximately 30% of these glass fibres were respirable (< 3 μm in diameter). During the application of blowing wool, the total concentrations of airborne fibres were higher, with arithmetic means of 1.0 and 2.1 fibres/cm^3, depending upon the product type. The mean concentrations of glass fibres were 0.7 and 1.7 fibre/cm^3 and those of respirable glass fibres were 0.3 and 0.8 fibre/cm^3, respectively in batt and loose fill installation. Approximately 70–80% of the total airborne fibres measured were glass fibres.

A survey of MMVF insulation products was also conducted at several industrial construction sites where workers were installing or removing insulation in Montreal, Quebec, Canada in the early 1990s. The samples were analysed using both PCOM and TEM and the results are presented in Table 44 (Perrault *et al.*, 1992).

In the largest survey of the exposure of end-users to airborne MMVF fibres (Lees *et al.*, 1993; Breysse *et al.*, 2001), nearly 1200 samples were collected during the fabrication and installation of MMVF products for residential, commercial and industrial use. The samples were analysed using PCOM or SEM, and the results of the PCOM are presented in Tables 45 and 46. The results indicated that during installation of MMVF insulation in homes, the geometric mean concentrations of airborne fibres were less than 1.0 fibre/cm^3 for all exposure groups, except for installers and feeders of glass fibre loose-fill insulation without binder (7.7 fibres/cm^3 for installers and 1.7 fibre/cm^3 for feeders), and for installers of loose rock (stone) and slag wool (1.9 fibre/cm^3) (Lees

Table 42. Arithmetic mean concentration of airborne fibres (fibre/cm³) associated with installation of batt insulation

Sampling location	Total fibres[a]	Total glass fibres[b]	Respirable fibres[c]	Respirable glass fibres[d]
Before installation				
Arithmetic mean	0.008	0.001	0.002	0.00005
95% confidence limits[e]	0.006–0.012	0.001–0.002	0.001–0.003	0.00001–0.00011
95th percentile[f]	0.026	0.003	0.006	0.0003
Number of samples	26	26	15	15
Installers				
Arithmetic mean	0.22	0.13	0.059	0.042
95% confidence limits	0.18–0.27	0.099–0.16	0.049–0.073	0.032–0.052
95th percentile	0.56	0.34	0.14	0.11
Number of samples	60	60	32	32
After installation				
Arithmetic mean	0.012	0.004	0.001	0.0002
95% confidence limits	0.005–0.028	0.001–0.010	0.001–0.002	0.0001–0.0004
95th percentile	0.021	0.006	0.003	0.0009
Number of samples	26	26	15	15

From Jacob *et al.* (1992)
[a] Total fibres, NIOSH 7400 A rules, both filters and cowls
[b] Total glass fibres, NIOSH 7400 A rules, both filters and cowls, glass fibres only
[c] Respirable fibres, NIOSH 7400 B rules, both filters and cowls
[d] Respirable glass fibres, NIOSH 7400 B rules, both filters and cowls, glass fibres only
[e] 95% confidence intervals of the arithmetic mean
[f] 95th percentile of the data (95% of all measurements were less than this value)

et al., 1993). The geometric mean concentrations of airborne fibres during the fabrication and installation of commercial and industrial MMVF insulation products were all found to be less than 1.0 fibre/cm³ (Breysse *et al.*, 2001).

Koenig and Axten (1995) reviewed the concentrations of airborne fibres associated with the installation of rock (stone) and slag wool ceiling tiles and various other commercial and industrial insulation products manufactured from rock (stone) and slag wool in the USA. The arithmetic mean concentration of airborne fibres measured during the installation of wet-felted and moulded acoustical ceiling tiles was 0.26 fibre/cm³ with a range of 8-h TWA exposure from 0.02–0.82 fibre/cm³. The concentrations of airborne fibres measured for certain other rock (stone) and slag wool products are presented in Table 47. The overall mean exposure during installation operations involving these products was 0.09 fibre/cm³ and all the 8-h TWAs measured were well below 0.5 fibre/cm³.

Table 43. Arithmetic mean concentration of airborne fibres (fibre/cm^3) associated with installation of cubed and milled blowing-wool insulation

Sampling location	Total[a]	Total glass fibres[b]	Respirable fibres[c]	Respirable glass fibres[d]
Before installation				
Arithmetic mean	0.004	0.0008	0.001	0.0001
95% confidence limits[e]	0.003–0.005	0.007–0.001	0.001–0.002	0.00005–0.0002
95th percentile[f]	0.008	0.002	0.002	0.0004
Number of samples	38	38	24	24
Loader, cubed				
Arithmetic mean	0.38	0.20	0.12	0.084
95% confidence limits	0.33–0.42	0.17–0.23	0.10–0.15	0.06–0.10
95th percentile	0.80	0.50	0.30	0.24
Number of samples	99	99	49	49
Loader, milled				
Arithmetic mean	0.54	0.37	0.22	0.18
95% confidence limits	0.42–0.69	0.26–0.50	0.16–0.31	0.11–0.26
95th percentile	1.25	0.99	0.45	0.42
Number of samples	31	31	16	16
Installer, cubed				
Arithmetic mean	1.03	0.68	0.37	0.30
95% confidence limits	0.93–1.13	0.60–0.76	0.31–0.44	0.24–0.35
95th percentile	1.95	1.66	0.79	0.71
Number of samples	100	100	52	52
Installer, milled				
Arithmetic mean	2.1	1.7	0.91	0.82
95% confidence limits	1.5–2.7	1.2–2.5	0.58–1.3	0.53–1.3
95th percentile	6.0	5.2	2.8	2.7
Number of samples	31	31	15	15
After installation				
Arithmetic mean	0.004	0.001	0.001	0.0001
95% confidence limits	0.003–0.004	0.001–0.002	0.0007–0.002	0.00005–0.0002
95th percentile	0.008	0.003	0.002	0.0003
Number of samples	62	62	38	38

From Jacob *et al.* (1992)
[a] Total fibres, NIOSH 7400 A rules, both filters and cowls
[b] Total glass fibres, NIOSH 7400 A rules, both filters and cowls, glass fibres only
[c] Respirable fibres, NIOSH 7400 B rules, both filters and cowls
[d] Respirable glass fibres, NIOSH 7400 B rules, both filters and cowls, glass fibres only
[e] 95% confidence intervals of the arithmetic mean
[f] 95th percentile of the data (95% of all measurements were less than this value)

Table 44. Concentrations of respirable fibres on construction sites as measured by phase-contrast optical microscopy

Sites	No. of samples	Geometric mean (fibre/cm^3)
Glass wool	17	0.01
Rock (stone) wool (blown)	10	0.32
Rock (stone) wool (sprayed on)	16	0.15
Refractory ceramic fibres		
Site A (patching furnace insulation)	40	0.64
Site B (installing insulation panels)	41	0.39
Site C (installing refractory bricks)	46	3.51

From Perrault *et al.* (1992)

Additional information on exposure during the installation of rock (stone) and slag wool products (see Table 26) is included in the aforementioned NAIMA database (Marchant *et al.*, 2002). These results essentially substantiate those of Koenig and Axten (1995). Further data on the concentrations of airborne fibres associated with the installation of glass wool products as presented in Tables 48 and 49 (Marchant *et al.*, 2002) are consistent with the results obtained by Lees *et al.* (1993).

Data on the concentrations of airborne fibres measured in the USA during the installation and removal of refractory ceramic fibre products have been given in Table 29 (Maxim *et al.*, 1997, 2000a). The overall arithmetic mean of TWA values measured from 1993–96 was 0.41 fibre/cm^3. The average concentrations of airborne refractory ceramic fibres in the workplace were higher for insulation removers. The arithmetic mean actual TWA over the period 1993–96 was 1.2 fibre/cm^3 (Maxim *et al.*, 2000a). [The Working Group noted that the composition of refractory ceramic fibre at removal may be different from its composition at assembly.]

Finally, the survey of plants manufacturing glass wool, rock (stone) wool and refractory ceramic fibre and of user industries in Australia which was discussed previously, produced considerable data on exposure to airborne fibres from MMVF products during their installation and removal (Yeung & Rogers, 1996). These data, summarized in Table 50, indicate that, as was the case for manufacturing operations, the geometric mean exposure of individuals installing and/or removing these insulation products was below the established exposure standard of 0.5 fibre/cm^3 (Yeung & Rogers, 1996).

Table 45. Summary of 8-h TWA estimates of exposure to fibres during insulation of dwellings by product/occupational category as measured by phase-contrast optical microscopy

MMVF product	Occupation	No. of TWAs	Fibre concentration (fibre/cm^3)[a]			
			Mean	Standard deviation	Geometric mean	Geometric standard deviation
Fibre glass batts	Installer	19	0.06	0.05	0.05	2.41
Rock (stone) and slag wool batts	Installer	6	0.11	0.04	0.10	1.45
Fibrous glass loose-fill insulation with binder	Installer	8	0.15	0.13	0.09	3.82
	Feeder	7	0.05	0.04	0.03	3.06
Fibrous glass loose-fill insulation without binder	Installer	4	1.96	1.23	1.52	2.56
	Feeder	1	0.85	–	–	–
Rock (stone) and slag wool, loose	Installer	5	0.97	1.06	0.52	3.76
	Feeder	4	0.18	0.15	0.14	2.20

From Lees et al. (1993)
[a] Respirable fibres (≤ 3 μm in diameter) only

Table 46. Summary of task length average exposure estimates by product/occupation as measured by phase-contrast optical microscopy

MMVF product	Occupation	No. of TLAs[b]	Fibre concentration (fibre/cm^3)[a]			
			Mean	Standard deviation	Geometric mean	Geometric standard deviation
Glass fibre duct board	Fabricator	14	0.05	0.10	0.03	2.30
	Installer	4	0.03	0.01	0.03	1.40
Glass fibre duct liner	Fabricator	26	0.04	0.06	0.03	2.31
	Installer	5	0.32	0.06	0.32	1.17
Glass fibre duct wrap	Installer	4	0.68	0.97	0.35	3.33
Glass fibre pipe and vessel insulation	Installer	23	0.04	0.04	0.03	2.54
Rock (stone) and slag wool ceiling tiles	Installer	14	0.24	0.12	0.22	1.64
Glass fibre batts (prefabricated homes)	Installer	3	0.19	0.06	0.19	1.33
Rock (stone) and slag wool, loose fill (prefabricated homes)	Installer	5	0.13	0.06	0.11	1.71
Rock (stone) and slag wool building safing	Installer	5	0.16	0.12	0.12	2.21
General glass fibre products	Fabricator	22	0.16	0.12	0.11	2.84

From Breysse *et al.* (2001)
[a] Respirable fibres (< 3 μm in diameter) only
[b] TLA, task-length average. The geometric mean TLAs for each product/occupational group are generally slightly lower than the arithmetic mean concentrations.

Table 47. Exposure to respirable fibres associated with installation of insulation products manufactured from rock (stone) and slag wool

Product installed	Area type[a]	Tasks performed	Fibre concentrations (fibre/cm^3)		
			No. of samples	8-h TWA	Range
Attic insulation 1	Confined	Blower	5	0.07	0.04–0.62
	Open	Helper	3	0.03	0.05–0.13
	Confined	Blower	4	0.19	0.02–0.30
Attic insulation 2	Open	Helper	4	0.05	0.02–0.09
Sound attenuation blanket	Moder. open	Cutting/place	9	0.16	0.11–0.24
Insulation fabrication shop	Open	Cutting/assembling/packing	5	0.07	0.03–0.11
		Cutting/assembling/packing	4	0.07	0.03–0.19
		Cutting/assembling/packing	4	0.05	0.03–0.07
		Cutting/assembling/packing	4	0.04	0.03–0.07
		Assembling/packing	4	0.06	0.02–0.03
Installation of outdoor pipe-covering	Open	Cutting/installation/cover	1	0.02	0.02
		Cutting/installation/cover	2	0.05	0.04–0.05
		Cutting/installation/cover	1	0.05	0.05
Spray-on fire proofing	Open	Nozzle man	8	0.13	0.03–0.29
		Labourer	2	0.02	0.07–0.14
		Pump operator	2	0.01	0.02
Spray-on fire proofing	Open	Nozzle man	8	0.42	0.03–1.10
		Labourer	6	0.09	0.03–0.29
		Labourer	4	0.04	0.01–0.05
		Pump operator	3	0.03	0.01–0.07

From Koenig & Axten (1995)

[a] Confined, confined area such as an attic; moder. open, moderately open areas such as within a factory with no high bay; open, outdoors or in a very large enclosed area

Table 48. Exposure to glass wool fibres by job description

Job description	Sample size	Exposure (fibre/cm^3)			
		Mean	Standard deviation	Median	Range
Assembly	34	0.04	0.06	0.02	0.01–0.35
Feeder	63	0.36	0.37	0.20	0.01–2.18
Installer	232	0.45	0.85	0.18	0.01–7.49
Other[a]	9	0.16	0.14	0.07	0.03–0.37

From Marchant et al. (2002) (personal samples; sampling times greater than 4 h; analysed by NIOSH 7400 method with B counting rules)
[a] Includes cutting or sawing with power tools and maintenance

Table 49. Exposure during installation of glass wool by product type and job description

Product type	Job description	Sample size	Exposure (fibre/cm^3)			
			Mean	Standard deviation	Median	Range
Blowing wool with binder	Feeder	6	0.09	0.06	0.06	0.04–0.19
	Installer	13	0.39	0.32	0.28	0.09–1.13
Blowing wool without binder	Feeder	49	0.44	0.39	0.35	0.01–2.18
	Installer	84	0.99	1.21	0.62	0.04–7.49

From Marchant et al. (2002) (personal samples; sampling times greater than 4 h; analysed by NIOSH 7400 method using B counting rules)

(d)   Exposure in residential, commercial and public buildings

The exposure of the occupants of a building to airborne MMVFs may occur during installation, maintenance, or removal operations, as a result of physical damage or degradation of insulation or as fibres are released over time. Several studies, using a variety of sampling and analytical techniques, have evaluated and reported on the nature and magnitude of such exposure.

In the late 1960s, concern was expressed over health problems associated with the possible erosion of glass fibre used to line ventilation and heating ducts. Cholak & Schafer (1971) tested six different ventilation-system ducts and found some glass fibres in settled dust, but no evidence of erosion of the fibres. Thirteen air-transmission systems lined with glass fibre were studied to determine their contributions of glass fibres to the air. The average concentrations of glass fibres in ambient air in the air leaving air-supply

Table 50. Exposure to MMVFs during installation and removal

Type of MMVF	Process	Pre-NOHSC standard Typical fibre count (fibre/cm³), location/job	Type of sample	Post-NOHSC standard respirable fibre count (fibre/cm³)			
				No.	Range	Geometric mean	Geometric standard deviation
Glass fibre	Installation of bonded products	0.1–0.3 ceiling space < 0.1–3.5 ceiling space, short-term sample < 0.01–0.1 plant room	Personal Static Total	23 29 52	< 0.01–0.80 < 0.01–0.40 < 0.01–0.80	0.12 0.03 0.06	2.23 2.74 3.14
	Removal of glass fibre products	–	Personal Static Total	21 12 33	< 0.01–0.20 < 0.01–0.03 < 0.01–0.20	0.04 0.02 0.03	2.20 1.67 2.18
Rock (stone) wool	Installation of bonded products	0.1–0.4 ceiling space 0.01–0.1 lagging of boilers	Personal Static Total	42 303 345	< 0.01–1.50 < 0.01–1.0 < 0.01–1.50	0.03 0.02 0.02	2.71 1.87 2.07
	Removal of rock (stone) wool products	–	Personal Static Total	1 17 18	0.5 < 0.01–0.05 < 0.01–0.50	– 0.01 0.04	– 1.73 2.76
Refractory ceramic fibre	Installation of bonded products[a]		Personal	21	< 0.01–1.5	0.04	2.48
	Application of unbonded materials	0.7 firewall 0.03–0.2 fire damper 0.02–1.8 delagging in furnace 0.07–1.5 lagging in furnace 0.01–1.4 lagging of boilers	Personal Static Total	28 99 127	< 0.01–1.4 < 0.01–0.4 < 0.01–1.4	0.04 0.02 0.02	7.12 2.60 3.63
	Removal of refractory ceramic fibres	–	Personal Static Total	13 62 75	< 0.01–0.3 < 0.01–0.1 < 0.01–0.3	0.05 0.02 0.02	2.99 1.86 2.21

From Yeung & Rogers (1996)
[a] Mixed fibres (rock (stone) wool and refractory ceramic fibre)

systems, and in building areas were extremely low (below 0.005 fibre/cm^3). In some cases, there was a decrease in the fibre concentration after fibre-containing outdoor air had passed through the air-transmission system (Balzer et al., 1971).

Five studies were conducted in Europe using PCOM combined with PLM as a primary analytical device, and selected buildings that had acoustical ceiling tiles made of MMVF or sprayed-on MMVF ceilings. Rindel et al. (1987) reported that in a study in 24 kindergartens in Denmark, all had concentrations of respirable fibres of MMVF below 0.001 fibre/cm^3, but the sampling time was only 1 h. Skov et al. (1987) measured the fibre concentrations in 14 Danish town halls and found airborne MMVFs in only one building. Gaudichet et al. (1989) studied 79 buildings containing various MMVF products and found concentrations of respirable MMVFs ranging from none detected to 0.006 fibre/cm^3.

Nielsen (1987) and Schneider et al. (1990) reported on measurements made in 105 rooms containing acoustical ceiling tiles. The average concentrations of airborne respirable MMVFs were less than 0.0001 fibre/cm^3 and ranged from none detected to 0.002 fibre/cm^3.

Schneider (1986) reported on measurements made in 16 schools and one office building in Greater Copenhagen where the concentrations of airborne respirable MMVFs ranged from non-detectable to 0.08 fibre/cm^3.

Two studies used PCOM to measure the concentrations of airborne fibres before, during and after the installation of MMVF products in homes. In the study of Van der Wal et al. (1987) that made measurements in eight homes, the average concentration of fibres just before the installation of insulation began was 0.012 (range 0.001–0.03) fibre/cm^3. During installation, the average peak fibre concentration was 0.03 (range, 0.01–0.65) fibre/cm^3. When measurements were repeated the next day at the same locations, the average concentration had decreased to 0.004 (range, 0.001–0.01) fibre/cm^3. Jacob et al. (1992) reported that the mean concentrations of respirable glass fibres measured in 74 homes before and after the installation of batt or blowing wool insulation were 0.0002 fibre/cm^3 or less, while during the installation the mean concentrations of respirable glass fibre were 0.04 fibre/cm^3 during batt installation and 0.30–0.82 fibre/cm^3 during installation of blowing wool insulation. The authors reported that on the night following the completion of the insulation work, the concentration of airborne fibres had dropped to that measured before the insulation work began.

Jaffrey et al. (1990) reported the concentrations of airborne MMVF in 11 dwellings in the United Kingdom during and after the disturbance of loft insulation. Approximately 250 samples were collected and analysed by TEM–EDXA and a few additional samples were analysed by PCOM. The TWA concentrations measured over a 4-h period during the physical disturbance of the insulation in the lofts ranged up to 0.03 fibre/cm^3. The personal measurements of exposure to MMVF during this period were up to 0.2 fibre/cm^3. However, no contamination of the living space was detected although the access doors to the lofts remained open throughout the disturbance and sampling period.

Tiesler and Draeger (1993) and Tiesler *et al.* (1993) reported more than 130 measurements of indoor concentrations of airborne fibres made in various buildings. Most samples were collected in offices, but some schools, private houses and laboratories were also sampled. The buildings selected contained visible and uncoated glass and rock (stone) slag wool products which represented worst-case conditions. The authors concluded that the total indoor concentrations of respirable inorganic fibres generally averaged 0.0046 fibre/cm^3, ranging from none detected to 0.038 fibre/cm^3. The fibres of MMVF products comprised 12.5% of the mean total respirable inorganic mineral fibres detected, averaging 0.0006 fibre/cm^3 and ranging from none detected to 0.0057 fibre/cm^3. Thirty-nine simultaneously collected outdoor samples were reported. The concentration of inorganic nonasbestos respirable fibre outdoors averaged 0.00499 fibre/cm^3. The authors concluded that concentrations of fibres indoors and outdoors were generally of the same order of magnitude.

Fischer (1993) reported the concentrations of airborne respirable fibres measured in seven offices, one school and one dwelling, and in a group of six buildings (a shop, offices, school, kindergartens) in which the users had complained about possible exposure to fibre dust from ceiling boards. To ensure a conservative estimate, the ceiling boards in the buildings selected were in direct contact with indoor air. The concentrations of respirable inorganic fibres averaged 0.0018 fibre/cm^3; the concentration of MMVF products was 0.0003 fibre/cm^3 or 17% of the respirable inorganic fibres. The concentration of respirable MMVFs outdoors was of the same order of magnitude.

Dodgson *et al.* (1987b) measured the concentrations of airborne fibres in 10 homes using PCOM and SEM before, 24 h after and 7 days after the installation or disturbance of MMVF insulation in their lofts. The mean concentrations of MMVF measured 24 h after the installation of insulation in new homes was not different from the concentrations measured before work began; the fibre concentrations determined by SEM ranged from 0.0001–0.0003 fibre/cm^3.

Carter *et al.* (1999) reported the concentrations of airborne MMVFs measured in 51 residential and commercial buildings throughout the USA. All samples were analysed by PCOM and 50 randomly selected samples were analysed by SEM–EDXA. The mean concentration of all respirable fibres counted using PCOM was 0.008 fibre/cm^3. Ninety-seven per cent of the respirable fibres identified by SEM–EDXA were determined to be organic in nature. The concentrations of inorganic fibres determined by SEM–EDXA, which included MMVFs, were found to be less than 0.0001 fibre/cm^3.

1.4.2 Environmental occurrence

Man-made vitreous fibres may be released into the environment during their production, installation, erosion, removal and/or disposal. However, few studies have attempted to quantify the concentration of MMVFs present in ambient air, and those that have been conducted have often failed to distinguish between MMVFs and mineral fibres or 'other inorganic fibres' or to attempt to define fibre size or chemical composition. Despite these limitations, several studies (described below) have indicated the concentrations of MMVFs which may be present in the general environment.

Spurny and Stöber (1981) identified inorganic fibres in air samples from an urban non-industrial area and in clean rural air in the former Federal Republic of Germany. The total concentrations of fibres ranged from 0.004–0.015 fibre/cm^3. Friedrichs et al. (1983) also published data on the number of fibres present in the ambient air in samples collected in the former Federal Republic of Germany during 1982. The concentrations of glass fibre ranged from 0.00036–0.00249 fibre/cm^3 and were higher on week days than at the weekend.

Lanting and Den Boeft (1983) reported a uniform background concentration of non-asbestiform inorganic fibres of 10^4 fibres/m^3 (i.e. 0.01 fibre/cm^3) at all their rural and urban sampling sites in the Netherlands.

Iburg et al. (1987) updated earlier studies of concentrations of fibres in the ambient air measured in the former Federal Republic of Germany. The concentrations were reported to range from 0.0008–0.0020 fibre/cm^3 at urban crossroads. Noack and Böckler-Klusemann (1993) published similar findings. They reported concentrations of inorganic fibres > 5 μm length excluding asbestos and gypsum at urban locations in Germany to be 0.00070–0.00325 fibre/cm^3 and Goldmann and Kruger (1989) reported concentrations of 0.00496–0.00589 fibre/cm^3 in urban non-occupational situations. Rödelsperger et al. (1989) also published data indicating that the average concentrations of airborne inorganic fibres ≥ 5 μm in length excluding asbestos and gypsum in ambient air exceeded 0.001 fibre/cm^3 in urban and rural areas of Germany, while Schnittger (1993) reported that concentrations of airborne inorganic nonasbestos fibres with diameters below 3 μm in ambient air were of the order of 0.0022 fibre/cm^3 in an urban area of Germany.

Three studies have reported on the concentrations of glass fibres measured in outdoor air in France (Gaudichet et al., 1989), Germany (Höhr, 1985) and the USA (Balzer, 1976). In these studies, the concentrations of glass fibres ranged from 2×10^{-6} fibres/cm^3 in a rural area to 0.0017 fibre/cm^3 near a city. These concentrations were considered to represent a very small percentage of the total fibre and total suspended particulate matter in ambient air.

Most recently, a study of the exposure of individuals to respirable inorganic and organic fibres during everyday activities was conducted (Schneider et al., 1996). Four groups (suburban schoolchildren, rural retired persons, office workers and taxi drivers) containing five persons each were monitored to assess their exposure to fibres for 24 h four times over a 1-year period. The fibres were sized by SEM and the elemental composition of individual fibres was determined by EDXA. For inorganic fibres, excluding asbestos and gypsum, the geometric mean concentration was around 0.005 fibre/cm^3. The proportion of these inorganic fibres that had an elemental composition similar to that of MMVF was less than about one quarter.

1.5 Regulations and guidelines

Many industrialized and developing countries have, or refer to, exposure limit values for inert dusts, 'nuisance' dusts or particles. Several of these countries use the same limit values for insulation wool fibres and dusts (ILO, 2000). Table 51 summarizes the regulations enforced in selected countries. The absence of information for a particular country should not be taken to imply that the country does not have regulations with regard to MMVFs.

In December 1997, the European Union (EU) published the classification of man-made (vitreous) silicate fibres under the Substances Directive (European Commission, 1997). [The Working Group noted that this classification is to be reviewed by the EU by December 2002.] According to the EU system, all MMVF wools are considered to be irritants and are classified for carcinogenicity according to the following rules. Man-made vitreous fibre wools with length-weighted geometric mean diameter, less two standard errors, greater than 6 µm are exempt from carcinogenicity classification. [The Working Group noted that the classification and labelling group of the European Commission has agreed to change 'standard error' to standard deviation.] Untested insulation wool fibres with a diameter ≤ 6 µm and with a content of $Na_2O + K_2O + CaO + MgO + BaO$ that exceeds 18% by weight are classified as category 3, 'possible carcinogen'. If the content of $Na_2O + K_2O + CaO + MgO + BaO$ is less than or equal to 18%, the fibre is classified as category 2, 'probable carcinogen'. A category 3 fibre can be exempted from classification as a possible carcinogen if it fulfils one of the following criteria:

(1) a short-term biopersistence test by inhalation has shown that the fibres longer than 20 µm have a weighted half-life of less than 10 days;
(2) a short-term biopersistence test by intratracheal instillation has shown that the fibres longer than 20 µm have a weighted half-life of less than 40 days;
(3) an appropriate intraperitoneal test has shown no evidence of excess carcinogenicity; or
(4) relevant pathogenicity or neoplastic changes have been found to be absent in a suitable long-term inhalation test.

The protocol by which these criteria are established has been proposed by the European Commission (Bernstein & Riego-Sintes, 1999).

Germany (Bundesgesetzblatt, 2000) has banned the production and use of the following fibre-containing products in building and technical applications for insulation against heat and noise:

– MMVF wools with mass percentage greater than 18% of the oxides of Na, K, Ca, Mg and Ba;
– preparations and articles containing in excess of 0.1% (mass fraction) of MMVF wools.

Table 51. Examples of national regulations and guidelines pertaining to limits for occupational exposure to MMVFs

Country	Reference	Products regulated[a]	Exposure limits	
			Fibre/cm³	Dust (if no specific limit given for fibres)
Australia	NOHSC (2001)	Synthetic mineral fibres	0.5 respirable	2 mg/m³ inspirable dust; secondary standard in situations in which almost all the airborne material is fibrous
Austria	ILO (2000)	Man-made vitreous (silicate) fibres	0.5 WHO fibre	
Canada	National Research Council (NRC) (2000)	MMVF insulation wool products	Each province has it own, e.g. Alberta 1 (0.5 for refractory ceramic fibres)	
Denmark	Arbejdstilsynet (2001)	Glass, (rock) stone and slag wool	1 WHO fibre	
Finland	ILO (2000)	None		10 mg/m³
France	INRS (1999)	Mineral wool fibres	1 for glass, rock (stone) and slag wool, 0.6 for refractory ceramic fibres	
Germany	(see text)			
Italy	ILO (2000)	None	1 (fibre with diameter < 3 μm)	5 mg/m³ total dust
Japan	ILO (2000)	None		2.9 mg/m³ respirable dust
Republic of Korea	Kim et al. (1999)			Permissible level of exposure for glass and rock (stone) wool is 10 mg/m³.
Netherlands	ILO (2000)	None	1	

Table 51 (contd)

Country	Reference	Products regulated[a]	Exposure limits	
			Fibre/cm³	Dust (if no specific limit given for fibres)
New Zealand	NRC (2000)	Man-made mineral fibres	1 (fibre with diameter < 3 μm)	
Norway	ILO (2000)	Synthetic mineral fibres	1	
Poland	NRC (2000)	Man-made mineral fibres	2 (length > 5 μm)	
Sweden	AFS (1996)	Inorganic synthetic fibres	1	
Switzerland	ILO (2000)	Mineral wool	0.5 WHO fibres	
United Kingdom	ILO (2000)	MMVF	2	5 mg/m³ (fibre or mass limit applies, whichever is achieved first)
USA	(see text)			

Swedish National Board of Occupational Safety and Health (AFS) (1996); Arbejdstilsynet (2001); ILO (2000); INRS (1999); Kim *et al.* (1999); Australian National Occupational Health & Safety Commission (NOHSC) (2001); National Research Council (NRC) (2000)
[a] If None, dust is regulated as inorganic dust.

This regulation does not apply to MMVFs that fulfil one of the following exoneration criteria:
(1) a suitable intraperitoneal test showed no evidence of increased carcinogenicity;
(2) the half-life after intratracheal instillation of 2 mg of a fibre suspension containing fibres of length > 5 μm, diameter < 3 μm, and with a length-to-diameter ratio in excess of 3:1 (WHO fibres) does not exceed a maximum of 40 days;
(3) the carcinogenicity index (KI) (mass percentages of oxides of Na, K, Ca, Mg, Ba and B minus twice the mass percentage of aluminum oxide) is at least 40;
(4) glass fibres developed for high-temperature applications with a classification temperature greater than 1000 °C when their half-life, determined by method 2, above, does not exceed 65 days.

The German TRK Wert (Technical limit value) for fibres has been specified by the Deutsche Forschungsgemeinschaft (2001) as follows:
– High-temperature glass fibres: 0.5 fibre/cm^3
– All other applications: 0.25 fibre/cm^3, except in specific areas where refractory ceramic fibres, polycrystalline fibres or special glass fibres are used: existing installations where refractory ceramic fibres and special-purpose glass fibres are processed, processing of refractory ceramic fibres and polycrystalline fibres; for finishing operations, installation, assembly, mixing, forming, packaging at place of fibre production and polycrystalline fibres for which the limit value of 0.5 fibre/cm^3 is applicable until December 31, 2002.

In the USA, the Occupational Safety and Health Administration (OSHA) has not developed a specific exposure limit for fibres and the limits for total dust (15 mg/m^3) and respirable dust (5 mg/m^3) apply. However, certain organizations and the MMVF industry have issued their own recommendations.

- The American Conference of Governmental Industrial Hygienists (2001) has adopted the following Threshold Limit Values for continuous filament glass fibres, 1 fibre/cm^3 and 5 mg/m^3 respirable dust; for glass wool, slag wool, rock (stone) wool and special-purpose glass fibres, 1 fibre/cm^3; and for refractory ceramic fibres, 0.2 fibre/cm^3.
- In 1999, the US Navy reduced their exposure standard for all MMVFs from 2 fibres/cm^3 to 1 fibre/cm^3 (National Research Council, 2000).
- The Health and Safety Partnership Program (between OSHA, NAIMA and the users) established a limit of 1 fibre/cm^3 for respirable MMVF insulation wools (ILO, 2000; Marchant et al., 2002).
- The Refractory Ceramic Fibers Coalition (RCFC), a trade organization of manufacturers of refractory ceramic fibres in the USA, adopted a recommended exposure guideline of 0.5 fibre/cm^3 in 1997 (Maxim et al., 1997; National Research Council, 2000).

- The International Labour Office has issued a Code of Practice on safety in the use of synthetic vitreous fibre insulation wools (ILO, 2000). The Code of Practice covers general and specific measures for prevention and protection, information, education and training and surveillance of exposure and workers' health.

2. Studies of Cancer in Humans

The epidemiological studies of MMVFs published before 1987 were reviewed by IARC (1988). These studies are reviewed here only briefly if no updates have been published since. The epidemiological studies reported since 1988 include updates of previously published studies, notably two large cohort studies from the United States of America (United States University of Pittsburgh study) and Europe, as well as new studies. Studies conducted within the MMVF industry have attempted to separate the different types of MMVF, although this was not always possible, in particular for rock (stone) wool and slag wool, and for glass wool and continuous glass filament. The results of studies of workers exposed to both glass wool and continuous glass filament were considered more relevant to glass wool, which is the predominant source of exposure in these plants. In most studies of installers and studies conducted in the general population, no reliable differentiation between different types of MMVF was possible and they are generally included in section 2.5. The studies described in sections 2.1–2.4 were carried out on production and maintenance workers. The epidemiological studies available have not directly addressed the issue of biopersistence of MMVFs. However, indirect information can be obtained by considering the results according to the types of fibre manufactured in specific plants (e.g. special-purpose glass fibres are produced in plants 6 and 10 of the United States University of Pittsburgh study).

2.1 Glass wool (see Table 52)

Two major mortality studies have been conducted on glass wool and continuous glass filament workers, one in the USA and one in Europe. The results published before 1987 were reviewed by IARC (1988). The results of the updates of these studies and of the new studies have been evaluated below. A study in Canada of a cohort of 2557 male glass wool workers was also reviewed in the 1988 *IARC Monographs*, but has not been updated since (Shannon *et al.*, 1984, 1987). From these studies, there was *inadequate evidence* for the carcinogenicity of glass wool in humans (IARC, 1988).

Table 52. Studies of the health effects of exposure to glass wool

Reference, plants	Description, employment period, follow-up, definition of cases	No. of deaths or cases, type of cancer, controls	Exposure categories	No. of cases	Relative risks (95% CI)	Comments
US University of Pittsburgh						
Cohort studies						
Marsh et al. (1990) 8 plants	11 380 male workers[a] employed 1945–63, follow-up 1946–85	340 deaths from respiratory cancer	'Glass wool'		SMR 1.12 [1.00–1.24]	Local rates
			Time since first employment			
			< 10 years	11	0.92 [0.46–1.64]	
			10–19 years	49	1.08 [0.81–1.44]	
			20–29 years	118	1.11 [0.92–1.33]	
			≥ 30 years	162	1.15 [0.98–1.34]	
			Duration of employment			
			< 10 years	190	1.21 ($p < 0.05$)	
			10–19 years	56	0.98	
			20–29 years	62	1.09	
			≥ 30 years	32	0.97	
Marsh et al. (2001a) 8 plants (same as above)	26 679 male and female workers[a] employed 1945–78, follow-up 1946–92	733 deaths from respiratory cancer	'Glass wool' Special-purpose glass fibres Mostly glass wool[b] Long-term workers (≥ 5 years)	81 243 138	SMR [1.07 (0.99–1.15)] [1.09 (0.87–1.36)] 1.18 (1.04–1.34) 1.06 (0.90–1.26)	Local rates
		63 deaths from buccal cavity and pharynx cancer	'Glass wool'		1.11 (0.85–1.42)	
		64 deaths from cancer of the bladder and other urinary organs	'Glass wool'		1.07 (0.82–1.37)	

Table 52 (contd)

Reference, plants	Description, employment period, follow-up, definition of cases	No. of deaths or cases, type of cancer, controls	Exposure categories	No. of cases	Relative risks (95% CI)	Comments
Nested case–control studies						
Marsh et al. (2001a); Stone et al. (2001) 10 plants (Glass fibres including continuous glass filament plants)	631 deaths from respiratory cancer[a] diagnosed 1970–92 (men)	570 controls	Ever exposed to respirable fibres continuous glass filament glass wool + continuous glass filament[b] mostly glass wool[b]	622 356 183	Odds ratio 1.37 (0.55–3.42) 1.0 1.01 (0.69–1.47) 1.06 (0.71–1.60)	Adjusted for smoking Adjusted for smoking Adjusted for smoking
Chiazze et al. (1992, 1993) plant 9 of Marsh et al. (2001a) glass wool + continuous glass filament plant[b]	166 deaths from lung cancer[a] diagnosed 1940–82 (men)	387 controls	< 100 fibres/cm^3–day 100–299.99 fibres/cm^3–day ≥ 300 fibres/cm^3–day	98 37 27	Odds ratio 1.0 1.72 (0.77–3.87) 0.58 (0.20–1.71)	Adjusted for smoking and other potential confounders

Table 52 (contd)

Reference, plants	Description, employment period, follow-up, definition of cases	No. of deaths or cases, type of cancer, controls	Exposure categories	No. of cases	Relative risks (95% CI)	Comments
European study						
Cohort studies						
Plato et al. (1995c) Sweden; one glass wool plant included in Boffetta et al. (1997)	1970 male and female workers[a] employed before 1978, mortality follow-up 1952–90	14 deaths from lung cancer	Duration of employment with 20-year lag		SMR 0.97 (0.57–1.69)	Local rates
			< 2 years	5	2.24 (0.73–5.23)	
			2–9 years	5	1.14 (0.37–2.66)	
			10–19 years	0		
			≥ 20 years	1	0.94 (0.02–5.21)	
			Total	11	1.21 (0.68–2.30)	
	Incidence follow-up 1958–89	17 cases of lung cancer	Duration of employment		SIR	
			< 2 years	3	0.72 (0.15–2.12)	
			2–9 years	11	1.15 (0.57–2.05)	
			10–19 years	1	0.31 (0.01–1.74)	
			≥ 20 years	2	1.45 (0.18–5.24)	
			Total	17	0.93 (0.54–1.48)	

Table 52 (contd)

Reference, plants	Description, employment period, follow-up, definition of cases	No. of deaths or cases, type of cancer, controls	Exposure categories	No. of cases	Relative risks (95% CI)	Comments
Boffetta et al. (1997) 5 glass wool plants in Finland, Italy, Norway, Sweden and United Kingdom	6936 male and female workers[a] employed 1933–77, follow-up until 1992	140 deaths from lung cancer			SMR 1.27 (1.07–1.50)	National rates
			Technological phase			
			early	19	1.07 (0.64–1.67)	
			intermediate	100	1.40 (1.14–1.70)	
			late	21	1.02 (0.63–1.56)	
		10 deaths from cancer of the buccal cavity and pharynx			1.47 (0.71–2.71)	
Boffetta et al. (1999) 3 glass wool plants in Finland, Norway and Sweden; included in Boffetta et al. (1997)	2611 male and female workers[a], follow-up until 1995	40 cases of lung cancer			SIR 1.28 (0.91–1.74)	National rates
			Time since first employment		Relative risk	Adjusted for age, gender, country and technological phase
			≤ 19 years	10	1.0	
			20–29 years	15	1.9 (0.8–4.8)	
			≥ 30 years	15	2.3 (0.6–9.2)	p for linear trend = 0.2
			Duration of employment			Adjusted for age, gender, country, technological phase and time since first employment
			1–4 years	23	1.0	
			5–9 years	8	0.8 (0.3–2.0)	
			10–19 years	4	0.8 (0.3–2.4)	
			≥ 20 years	1	0.7 (0.08–5.3)	p for linear trend = 0.5
			Technological phase			Adjusted for age, gender, country and time since first employment
			early	20	0.6 (0.2–1.9)	
			late	20	1.0	

Table 52 (contd)

Reference, plants	Description, employment period, follow-up, definition of cases	No. of deaths or cases, type of cancer, controls	Exposure categories	No. of cases	Relative risks (95% CI)	Comments
Nested case–control study						
Gardner et al. (1988) 1 United Kingdom glass wool plant from the European study	73 deaths from lung cancer employed 1946–78, follow-up until 1984	506 controls	Super-fine glass wool Glass wool	2 31	Odds ratio 1.3 (0.3–5.8) 1.1 (0.7–1.9)	Potential asbestos exposure odds ratio, 1.5 (0.8–2.5)
Other studies						
Cohort studies						
Shannon et al. (1984, 1987) Canada	2557 male workers employed 1955–77, ≥ 90 days, follow-up until 1984	Plant-only workers 19 deaths from lung cancer	All cases Exposed ≥ 5 years and ≥ 10 years since first exposure	19 13	SMR 1.99 [1.20–3.11] 1.82 [0.97–3.11]	Local rates
			Duration of employment < 5 years 5–< 10 years 10–< 15 years 15–< 20 years 20–< 25 years 25–< 30 years ≥ 30 years	6 3 3 2 1 2 2	SMR 2.91 2.88 1.79 1.16 0.86 1.59 3.25	Local rates

Table 52 (contd)

Reference, plants	Description, employment period, follow-up, definition of cases	No. of deaths or cases, type of cancer, controls	Exposure categories	No. of cases	Relative risks (95% CI)	Comments
Shannon et al. (1984, 1987) (contd)			Time since first employment		SMR	Provincial rates
			< 5 years	1	4.31	
			5 –< 10 years	1	1.68	
			10 –< 15 years	2	2.11	
			15 –< 20 years	3	1.91	
			20 –< 25 years	2	1.04	
			25 –< 30 years	5	2.09	
			≥ 30 years	5	2.70	
Moulin et al. (1986) France	1374 male workers[a] employed 1975–84	5 cases of lung cancer	Production workers duration of exposure		SIR 0.74 (0.24–1.72)	Regional rates
			1–9 years	2	1.82 (0.22–6.57)	
			10–19 years	1	0.63 (0.02–3.48)	
			≥ 20 years	1	0.56 (0.01–3.10)	
		19 cases of 'upper respiratory and alimentary tract' cancer			2.18 (1.31–3.41)	

SMR, standardized mortality ratio; SIR, standardized incidence ratio; respiratory cancer, ICD8, 160–163

[a] Workers employed for ≥ 1 year

[b] For this review, workers exposed to 'mostly glass wool' or 'glass wool and continuous glass filaments' are considered as being exposed to glass wool, because the pattern of exposure among these groups results predominantly in exposure to glass wool.

2.1.1 United States University of Pittsburgh cohort

(a) Cohort studies

The United States (US) University of Pittsburgh cohort consists of male workers employed for 1 year or more between 1945 and 1963 in production or maintenance at one or more of 11 glass fibre and six rock (stone) wool and slag wool plants (Table 53). Of the 11 glass fibre plants, three produced continuous glass filament, two glass wool and continuous glass filament and six mostly glass wool. The original report and the updates presented the mortality statistics collected until 1982 (Enterline *et al.*, 1983; Enterline & Marsh, 1984; Enterline *et al.*, 1987). The 1985 follow-up study was reported by Marsh *et al.* (1990) and included 16 661 workers. Death certificates were obtained for 96.2% of those identified as deceased. The data were analysed according to three different follow-up periods, time since first employment and duration of

Table 53. Characteristics of the plants participating in the study by the US University of Pittsburgh, USA, on MMVFs

Plant no.	Location	Principal product
Glass fibre plants		
1	Parkersburg, WV	Mostly wool[a]
2	Ashton, RI	Filament
4	Kansas city, KS	Mostly wool[a]
5	Huntington, PA	Filament
6	Santa Clara, CA	Mostly wool[a]
9	Newark, OH	Wool and filament
10	Waterville, OH	Wool and filament[b]
11	Defiance, OH	Mostly wool[a]
14	Shelbyville, IN	Mostly wool[a]
15[c]	Kansas City, KS	Wool and filament[b]
Rock (stone)/slag wool plants		
3	Alexandria, IN	Rock (stone)/slag wool[d]
7	Tacoma, WA	Rock (stone)/slag wool[d]
8	Wabash, IN	Rock (stone)/slag wool[d]
12	Birmingham, AL	Rock (stone)/slag wool[d]
13	S. Plainfield, NJ	Rock (stone)/slag wool[d]
17	Joplin, MO	Rock (stone)/slag wool[e]

Plants 6 and 10 produced special-purpose glass fibres (small-diameter, < 1.5 μm).
[a] Includes some filament operations
[b] Consists of one facility devoted to filament manufacturing and one devoted to wool.
[c] Two of the original glass fibre plants (plants 15 and 16) included in Marsh *et al.* (1990) were combined (as plant 15) in Marsh *et al.* (2001a) because the workers moved freely between the adjacent manufacturing sites.
[d] Extended follow-up: N-cohort
[e] Extended followup: O-cohort

employment. The expected numbers of deaths were based on cause-specific mortality rates for white men in the USA and on the rates from the county where the plant was located. In addition, Poisson regression modelling was used to investigate the dependence of standardized mortality ratios (SMRs) on possible combinations of exposure to fibres with potential confounding variables. The SMR for all malignant tumours for the whole cohort was 1.08 based on local rates.

The US University of Pittsburgh cohort included eight glass fibre plants: two producing glass wool and continuous glass filament (start of production, 1938–50) and six producing mostly glass wool (start of production, 1946–52). The study population consisted of 11 380 male workers employed for 1 year or more (or for six months at two plants that produced small-diameter glass fibres (less than 1.5 μm)) between 1945 and 1963 (1940–63 for plant 9) in these eight plants. Results are presented for 'fibrous glass wool and both' combining 'fibrous glass wool' ('mostly glass wool') and 'fibrous glass both' ('glass wool and continuous glass filament'). For the purposes of this review, workers exposed to 'mostly glass wool' or 'glass wool and continuous glass filament' are considered as being exposed to glass wool, because the pattern of exposure among these groups results predominantly in exposure to glass wool. The SMR for respiratory cancer (ICDs 8 160–163) (including cancer of the larynx) for workers exposed to glass wool was 1.12 [95% confidence interval (CI), 1.00–1.24] (340 cases) based on local reference rates. The average exposure to fibres was 0.047 fibre/cm^3. The average exposure to (respirable fibres) < 3 μm diameter was 0.039 fibre/cm^3 for all glass fibre plants. There was no positive relationship between respiratory cancer and duration of employment (Marsh et al., 1990).

The US University of Pittsburgh cohort study for the glass fibre plants was extended until 1992 and expanded to include a more complete characterization of the work histories and the racial composition of the cohort, a nested case–control study of respiratory cancer, a survey of tobacco smoking habits and a retrospective assessment of exposure (Buchanich et al., 2001; Marsh et al., 2001a,b,c; Quinn et al., 2001; Smith et al., 2001; Stone et al., 2001; Youk et al., 2001). The expanded cohort included female employees, workers employed after the original cohort end-date of 1963 and workers from additional manufacturing sites. Thus the study covered 10 glass fibre plants (two from the original cohort were combined), including eight plants that produced glass wool and two that produced continuous glass filament (see Table 53). This study examined the mortality experience between 1946 and 1992 of 32 110 production or maintenance workers (5431 workers exposed to continuous glass filament, 15 718 to glass wool and continuous glass filament and 10 961 mostly to glass wool) employed for at least one year between 1945 and 1978 with certain exceptions. Firstly, a six-month employment criterion was applied to male workers in the original cohort from two plants (6 and 10) where special-purpose glass fibres, relatively low-solubility glass fibres or quartz (pure silica) fibres (small-diameter (less than 1.5 μm)) were produced, and secondly, a starting date of 1940 was used for one plant (plant 9). The cause of 98.8% of deaths was identified and 0.6% of the study subjects were lost to follow-up. The whole cohort of glass fibre workers, including those who worked with glass wool and continuous glass filament covered 935 581 person–years and was about evenly divided between short-term (< 5 years; 47.9%) and long-term (≥ 5 years; 52.1%) workers. Of the glass-fibre workers, 5675 (17.7%) had been employed for 20 or more years and 15 766 (49.1%) were followed up for 30 or more years (Marsh *et al.*, 2001a).

As described by Smith *et al.* (2001) and Quinn *et al.* (2001), profiles for the historical exposure of individual workers were developed using an approach that integrated epidemiological methods with those used by industrial hygienists. Quantitative estimates of exposure were made for respirable fibres, formaldehyde and crystalline silica, and qualitative estimates for other agents such as arsenic, asbestos, asphalt, polycyclic aromatic hydrocarbons (PAHs) and styrene. The exposure was estimated from the date of plant start-up until closure or until 31 December 1987, which was the latest common work-history end-date. The median average intensity of exposure to respirable fibres computed across all individual workers was 0.035 fibre/cm^3. When calculated by plant, this value ranged from 0.001 fibre/cm^3 for workers in one plant that produced continuous glass filament to 0.167 fibre/cm^3 for workers in a plant that produced mostly glass wool. The median cumulative exposure was 1.441 fibres/cm^3–months for all workers, ranging from 0.086 to 6.382 fibres/cm^3–months. The average exposure of long-term workers (employed for at least 5 years) during their first five years of exposure was similar to that of short-term workers (employed for less than 5 years) (Marsh *et al.*, 2001a).

For the whole glass fibre cohort (10 plants) using local county rates, the SMR for all causes of mortality was reduced (SMR, 0.90; 95% CI, 0.88–0.92) as was that for all cancers (SMR, 0.94; 95% CI, 0.90–0.98) during 1960–92 (8436 deaths from all causes and 2243 deaths from cancer). For all workers, the local county-based SMRs for respiratory cancer increased with calendar time and time since first employment, but not with duration of employment. The short-term workers had an excess of respiratory cancer (SMR, 1.12; 95% CI, 1.01–1.24) (378 cases) compared with an SMR of 1.03 (0.94–1.12) (496 cases) for long-term workers. The SMRs for long-term workers did not increase with calendar time, duration of employment or time since first employment (Marsh *et al.*, 2001a).

An analysis restricted to the eight glass wool plants resulted in an SMR for respiratory cancer of [1.06 (95% CI, 0.99–1.14)] (733 deaths) (local county comparison). The SMR for respiratory cancer in the four plants producing special-purpose glass fibres (local county comparison) was [1.06 (95% CI, 0.97–1.15)] (490 cancer deaths). In general, comparison with national reference rates provided similar results (Marsh *et al.*, 2001a).

Buchanich *et al.* (2001) inferred from a survey of the tobacco-smoking habits of the US University of Pittsburgh cohort that male glass fibre workers had higher estimated point prevalence rates of ever smoking than the corresponding general US populations and than most of the states where the study plants were located. The method of Axelson and Steenland (1988) was used to make an indirect adjustment of the SMRs for respiratory cancer to account for potential confounding by smoking. The adjustments were based on data on smoking prevalence within the cohort estimated from a random sample of cohort members (Marsh et al., 2001b) and the relative risk for ever versus never smoking as estimated from the nested case–control study of respiratory cancer (Marsh et al., 2001a). The adjustment suggested that cigarette smoking may account for the excess in respiratory cancer observed for the cohort of male glass-fibre workers (SMR adjusted for age-adjusted prevalence of ever smoking and based on local county rates, 0.89). The same conclusion was reached regardless of which of several alternatives were used to adjust local county rate-based SMRs for respiratory cancer. All SMRs that were statistically significantly elevated when unadjusted were reduced to non-statistically significant levels when adjusted for smoking (Marsh et al., 2001b).

As part of the on-going mortality surveillance programme for the US MMVF industry, mortality from mesothelioma [mesotheliomas were not identified before the 8th ICD revision in 1968] was investigated from the 1992 follow-up of the US University of Pittsburgh study (Marsh et al., 2001c). A manual search of all death certificates of 9060 glass-fibre workers revealed that seven of the death certificates issued for the glass wool workers mentioned the word mesothelioma. A subsequent review of medical records and pathology specimens for one (plant 9) of the seven workers deemed this one death as having a 50% chance of being due to mesothelioma. Five of the seven workers who had died had potentially been exposed to asbestos while working in the glass fibre industry or in other jobs. No death coded as pleural cancer was observed in the glass wool cohort.

With the exception of respiratory cancer, no statistically significant excesses of mortality were observed among the cancer site categories 'buccal cavity and pharynx' (SMR, 1.11; 95% CI, 0.85–1.42; 63 cancer deaths) or 'bladder and other urinary organs' (SMR, 1.07; 95% CI, 0.82–1.37; 64 cancer deaths) (local county comparison) (Marsh et al., 2001a).

(b) Nested case–control studies

Marsh et al. (2001a), Stone et al. (2001) and Youk et al. (2001) performed a nested case–control study as part of the US University of Pittsburgh cohort study of glass-fibre workers (10 plants) in which workers at the continuous glass filament plants represented the lower exposure groups. The investigators identified 713 men who had died from respiratory cancer during 1970–92 and one control per case. The potential controls were at risk during 1970–92 and alive and at risk at the age at which the case had died; controls were also matched by date of birth. A telephone interview was conducted with the study subject or a knowledgeable informant for [88.6%] of cases and 80.2% of controls (Stone et al., 2001). [The proportion of respondent type was not given.] There were 516 matched sets for which data on smoking (631 cases and 570 controls) were available for analysis.

Marsh et al. (2001a) reported an increased risk for respiratory cancer for combined non-baseline levels of exposure to respirable fibres (odds ratio, 1.37; 95% CI, 0.55–

3.42), adjusted for smoking. The duration of exposure and cumulative exposure to respirable fibres (adjusted for smoking) did not appear to be associated with an increased risk for respiratory cancer and no apparent increase in risk with increasing time since first employment in the plant was noted. There was some evidence of an elevated risk for respiratory cancer associated with non-baseline levels of average intensity of exposure to respirable glass wool, but this was not statistically significant when adjusted for smoking and there was no apparent trend with increasing exposure. The analysis by product group ('continuous glass filament', 'glass wool and continuous glass filament' and 'mostly glass wool') used continuous glass filament as the baseline category. After adjustment for smoking, the odds ratios for the 'glass wool and continuous glass filament' and 'mostly glass wool' categories were close to unity when compared to the baseline for continuous glass filament, and were 1.01 (95% CI, 0.69–1.47) and 1.06 (95% CI, 0.71–1.60), respectively.

Youk et al. (2001) explored the possible exposure–response relationship between respiratory cancer and exposure to respirable fibres or formaldehyde using exposure-weighting. None of the categorized measurements of exposure to respirable fibres using time lags and unlagged/lagged time windows showed a statistically significant association with risk for respiratory cancer ($p > 0.49$ for each). All of the estimated odds ratios for exposure-weighted models were lower than the estimated odds ratio of 1.37 (95% CI, 0.55–3.42) for the unweighted model. No pattern of increasing risk for respiratory cancer with increasing levels of cumulative exposure or average intensity of exposure to respirable fibres was seen.

Stone et al. (2001) extended the exposure–response analysis within the case–control study to include quantitative measures of exposure to respirable fibre for the US University of Pittsburgh cohort. Quantitative measures of formaldehyde and crystalline silica (mainly quartz) were made as these substances were considered as potential confounders and effect modifiers. Neither the average intensity of exposure nor the cumulative exposure to respirable fibres showed a statistically significant association with risk for respiratory cancer in any of the hundreds of fractional polynomial models considered.

Chiazze et al. (1992) reported on a case–control study of male workers employed for one year or more between 1 January 1940 and 31 December 1962, and followed up until 1982, at one glass fibre plant (producing glass wool and continuous glass filament) in the USA. This plant is included in the US University of Pittsburgh cohort (plant 9). The investigators identified 166 deaths due to lung cancer. The controls were cohort members matched on year of birth (within 2 years) and survival to end of the follow-up period or death (within 2 years). The response rate was 88% for cases and 79% for controls. Interviews were completed with proxies for 144 of these cases. Eighty per cent of the interviews were conducted face-to-face and the remaining 20% by telephone. Moste of the interviewees were proxy respondents (88%), of whom 87% reported having been in contact with the subject at least once a month. Data available for analysis included information on work history, demographic information (including smoking habits) and Chiazze et al. (1993) added information on exposure for a profile constructed for the years 1934–87 that included estimates of cumulative exposure to respirable fibres. The odds ratio for smoking (> 6 months versus never) and lung cancer was 26.2 (95% CI, 3.32–207). The odds ratios for lung cancer adjusted for smoking and other potential confounders were 1.72 (95% CI, 0.77–3.87) and 0.58 (95% CI, 0.20–1.71) for cumulative exposure to respirable fibre categories of 100–299.99 fibres/cm^3–days and ≥ 300 fibres/cm^3–days, respectively. The lowest exposure category, < 100 fibres/cm^3–days, served as the reference group.

2.1.2 European glass fibre cohort

(a) Cohort studies

Plato et al. (1995c) investigated the mortality and cancer incidence in Sweden among 3539 male and female workers (1970 from a glass wool plant, 1187 from a large rock (stone) wool/slag wool plant and 382 from a small rock (stone) wool/slag wool plant), employed for at least one year before 1978. These plants were included in the European MMVF cohort study (see Boffetta et al., 1997). Of the 3539 subjects, 245 had emigrated before the study and 41 were lost to follow-up. The mortality analysis was based upon the remaining 3253 subjects, 738 of whom died between 1952 and 1990. Cancer incidence was followed from 1958 to 1989. The SMR and SIR were analysed using regional and national reference rates. When compared with regional reference rates, there was a slightly increased excess for overall mortality (SMR, 1.02; 95% CI, 0.95–1.10) for the total cohort including the rock (stone)/slag wool component. Comparison with regional reference rates showed no increased risk for mortality from all cancers for workers at the glass wool plant (SMR, 1.00; 95% CI, 0.82–1.22; 102 cancer deaths) or from lung cancer (ICD-8 162) (SMR, 0.97; 95% CI, 0.57–1.69; 14 lung cancer deaths). Neither was there an increased risk for lung cancer mortality associated with a longer duration of employment (length of employment < 2 years, SMR, 1.47; 95% CI, 0.48–3.44; 2–9 years, SMR, 0.92; 95% CI, 0.37–1.90; 10–19 years, SMR, 0.39; 95% CI, 0.01–2.19, ≥ 20 years, SMR, 0.94; 95% CI, 0.02–5.21). There was an excess in lung cancer mortality for workers with 30 years of latency (SMR, 1.43; 95% CI, 0.74–3.05). [No data on smoking or co-exposure were available.]

Boffetta et al. (1997) extended the follow-up of cancer mortality for the European cohort study of MMVF production workers in 13 factories from 1982 (except for one continuous glass filament plant where it was until 1983) until 1990 in Denmark, Italy, Norway and Sweden, 1991 in Germany and 1992 in the United Kingdom (see Table 54). Information on work history was available until 1977. The population under study was the workforce (male and female) ever employed (i.e. with at least 1 year of employment in Sweden and in one of the two factories in the United Kingdom) between the year production started (1933–50) and 1977; follow-up was successful for 97.7% of the workers and the cause of 99.5% of deaths was known. Five factories (1 in Finland, 1 in Italy, 1 in Norway, 1 in Sweden and 1 in the United Kingdom) in which glass wool was

Table 54. Plants and populations included in the European MMVF study

Production process, plant no. and country	No. of workers
Rock (stone)/slag wool	
1 Denmark	4 585
3 Norway	473
4 Norway	460
5 Norway	875
8 Sweden	384
9 Sweden	1 194
12 Germany	2 137
Total, rock (stone)/slag wool	10 108
Glass wool	
2 Finland	924
6 Norway	644
7 Sweden	2 022
10 United Kingdom	4 145
14 Italy	600
Total, glass wool	8 335
Continuous glass filament	
11 United Kingdom (Northern Ireland)	1 837
14 Italy	1 722
Total, continuous filament	3 559
Total, cohort	22 002

From Boffetta *et al.* (1997)

produced employed a total of 8335 workers (6936 workers had ≥ 1 year of employment and contributed 167 675 person–years of observation). Exposure to asbestos occurred in one plant in Finland (1946–48) and one plant in the United Kingdom (1946–62) (Cherrie & Dodgson, 1986). No information on other potential sources of workplace co-exposure or on smoking habits were available. Among the cohort of glass-wool workers employed for one year or more, excesses for all causes of death (SMR, 1.05; 95% CI, 1.00–1.10; 1679 deaths) and all malignant neoplasms (SMR, 1.11; 95% CI, 1.01–1.22; 460 cancer deaths) were observed based on national mortality rates. There was an increased risk for lung cancer (SMR, 1.27; 95% CI, 1.07–1.50; 140 lung cancer deaths). An analysis of lung cancer mortality by technological phase in all workers and in workers employed for one year or more did not reveal a trend of higher risks during the earlier technological phases. No trend in lung cancer mortality was associated with duration of employment. The SMR for cancer of the buccal cavity and pharynx was 1.47 (95% CI, 0.71–2.71). One death from mesothelioma was observed in the glass-wool workers.

The cancer incidence of 2611 glass-wool production workers was followed up until 1995. The subjects had been employed for one year or more (68 523 person–years) in Finland, Norway or Sweden (3 plants from the Boffetta *et al.* (1997) study) (Boffetta *et al.*, 1999), an increased incidence of lung cancer (SIR, 1.28; 95% CI, 0.91–1.74; 40 cases) was observed. A trend was suggested for time since first employment ($p = 0.2$), but not with employment during the earlier technological phases. The SIR for cancers of the oral cavity, pharynx and larynx was 1.41 (95% CI, 0.80–2.28; 16 cases).

(b)   Nested case–control study

Gardner *et al.* (1988) reported a study of 73 employees (66 men and seven women) who had died from lung cancer and 506 matched controls through 1984 (as did Simonato *et al.* (1987) in their follow-up of the cohort) from a glass wool plant in the United Kingdom which was included in the European glass fibre cohort. Superfine fibres (diameters, 1–3 µm and 2–5 µm) had been manufactured at this plant for two periods between 1949 and 1968. Up to eight controls were matched for sex and year of birth (within 2.5 years) with each case. The controls were selected at random from all eligible workers who were alive and had been employed for one year or more at the time the case died. [The response rates and proportions of respondent type were not reported.] The odds ratios and CIs were computed by conditional logistic regression for matched case–control sets with a variable matching ratio. The odds ratio for employment in jobs entailing exposure to 'superfine glass wool' was 1.3 (95% CI, 0.3–5.8) and that for employment in jobs in which workers were exposed to glass wool was 1.1 (95% CI, 0.7–1.9). There was no evidence of a relationship between lung cancer and fibre diameter, duration of exposure or time since first exposure. The results by broad occupational group were similar to those from the cohort study. [The Working Group noted that the study did not indicate a differential risk for lung cancer in workers exposed to 'finer-diameter' (superfine) glass fibres; however, the exposure levels were low and the number of cases was small. Data on smoking and co-exposure were not available.]

2.1.3   *Other cohort studies*

Shannon *et al.* (1984, 1987) reported a cohort study of 2557 men who had worked for 90 days or more between 1955 and 1977 in a glass wool plant in Sarnia, Canada. The cohort was followed for deaths to the end of 1984 and 97% of the cohort was traced. No data on historical exposure were available, but samples taken since 1978 suggested that fibre concentrations were rarely > 0.2 fibre/cm^3 and the mean concentrations in most areas of the plant were < 0.1 fibre/cm^3. The cohort was divided into three groups of workers: plant only, office only and 'mixed exposure'. For the plant-only group, the SMR for lung cancer based on 19 deaths and provincial (Ontario) rates was 1.99 [95% CI, 1.20–3.11]. In the office-only and mixed-exposure groups combined, there were two deaths from lung cancer compared to [2.4 expected (SMR, 0.83; 95% CI, 0.10–3.01)]. For plant-only workers who had been exposed for five years or more and for whom ten or more years had elapsed since first exposure, there were 13 deaths from lung cancer (SMR, 1.82 [95% CI, 0.97–3.11]). There was no trend of increasing lung cancer risk with increasing duration of employment or time since first employment.

Moulin et al. (1986) reported a cohort study in France at a glass wool production factory. The cohort included 1374 male workers who were employed at any time during 1975–84 for at least one year. The incidence of cancer during this period was ascertained from the social insurance records of the company, and the diagnoses were obtained from various medical sources. Five workers with lung cancer were identified in the whole cohort (SIR, 0.74; 95% CI, 0.24–1.72). Nineteen cases of cancers of the 'upper respiratory and alimentary tract' were observed (SIR, 2.18; 95% CI, 1.31–3.41). In particular, there was an excess of cancers of the larynx (5 observed; SIR, 2.30), pharynx (5 observed; SIR, 1.40) and buccal cavity (9 observed; SIR, 3.01). The excess was limited to production workers, and among this group the SIR increased with duration of employment. A survey of the cigarette smoking habits of the 1983 workforce indicated slightly lower smoking levels than those reported in a survey conducted in France in 1979. [The Working Group noted that the study was initiated because 'an industrial physician had noted an excess of cancers in the upper respiratory and alimentary tracts' in the factory and that the authors did not report whether any case of 'upper respiratory and alimentary tract' cancer was identified in addition to the index cases. Both the observed and expected numbers of workers with at least 10 years of exposure who developed lung cancer were very small. The Working Group also noted that the reference population may not have been appropriate.]

2.2 Continuous glass filament (see Table 55)

In the studies until 1987 evaluated by the previous Working Group, there was *inadequate evidence* for the carcinogenicity of continuous glass filaments in humans (IARC, 1988) (see section 2.1).

(a) United States University of Pittsburgh cohort

Marsh et al. (1990) reported the 1985 follow-up of the cohort of MMVF workers in the USA (see description of the study in section 2.1). The US University of Pittsburgh cohort included three plants that produced continuous glass filament (start of production, 1941–51). This part of the study included 3435 male workers employed for one year or more between 1945 and 1963. The SMR for respiratory cancer for workers exposed to continuous glass filament was 0.98 (84 cases) based on local reference rates. The average concentration of fibres to which workers were exposed was 0.011 fibre/cm^3.

In the 1992 follow-up, in the study of 10 glass fibre plants in the USA (described in detail in section 2.1), two plants (2 and 5) mainly manufactured continuous glass filaments. The two plants taken together had an SMR for respiratory cancer of 1.04 (95% CI, 0.87–1.22) for all workers and an SMR of 0.96 (95% CI, 0.76–1.19) for long-term workers (≥ 5 years) (local county comparison) (Marsh et al., 2001a). The SMRs

Table 55. Studies of cancer incidence in workers exposed to continuous glass filament

Reference, plants	Description, employment period, follow-up	No. of deaths or cases (controls), type of cancer	Exposure categories	No. of cases	Relative risks (95% CI)	Comments
US University of Pittsburgh						
Cohort studies						
Marsh et al. (1990) 3 plants	3435 male workers[a], employed 1945–63, follow-up 1946–85	84 deaths from respiratory cancer	Time since first employment		SMR	Local rates
			< 10 years	6	1.03	
			10–19 years	8	0.47	
			20–29 years	42	1.22	
			≥ 30 years	28	0.99	
			Duration of employment			
			< 10 years	51	1.13	
			10–19 years	12	0.61	
			20–29 years	18	1.21	
			≥ 30 years	3	0.54	
Marsh et al. (2001a) 2 plants	5431 male and female[a] workers, employed 1945–78, follow-up 1946–92	141 deaths from respiratory cancer	Long-term workers (≥ 5 years employment)	81	SMR 1.04 (0.87–1.22) 0.96 (0.76–1.19)	Local rates
Case–control study						
Marsh et al. (2001a) 1970–92 Male workers	Plant 2	61 deaths from respiratory cancer			Odds ratio 1.60 (0.95–2.69)	Adjusted for smoking
	Plant 5	31 deaths from respiratory cancer			0.54 (0.31–0.94)	Adjusted for smoking

Table 55 (contd)

Reference, plants	Description, employment period, follow-up	No. of deaths or cases (controls), type of cancer	Exposure categories	No. of cases	Relative risks (95% CI)	Comments
European cohort study						
Boffetta et al. (1997) 2 plants United Kingdom (Northern Ireland), Italy	1940 male and female workers[a] employed 1946–61 follow-up until 1990	Deaths from lung cancer Deaths from cancer of the buccal cavity and pharynx		14 2	SMR 1.11 (0.61–1.86) 1.63 (0.20–5.87)	National rates
US Georgetown University						
Chiazze et al. (1997) Cohort study	2933 white male workers[a] employed 1951–91 follow-up until 1991	47 deaths from lung cancer 2 deaths from cancer of the buccal cavity and larynx			SMR 1.17 (0.86–1.55) 0.87 (0.11–3.16)	Local rates
Chiazze et al. (1997) Case–control study	45 deaths from lung cancer[a] white men 1951–91	122 controls	Cumulative exposure to respirable glass fibres 0 > 0.005 fibre/cm^3–days > 0.005 fibre/cm^3–days	35 10 8	Odds ratio 1.0 0.91 (0.36–2.25) 0.78 (0.28–2.20)	Smokers

Table 55 (contd)

Reference, plants	Description, employment period, follow-up	No. of deaths or cases (controls), type of cancer	Exposure categories	No. of cases	Relative risks (95% CI)	Comments
US Georgetown University (contd)						
Watkins et al. (1997) Cohort study	1074 white women[a]	4 deaths from lung cancer			SMR 0.72 (0.20–1.85)	Local rates
	494 black men[a], employed 1951–91, follow-up until 1991	2 deaths from lung cancer			0.30 (0.04–1.07)	
Canadian cohort study						
Shannon et al. (1990)	1465 men and women[a], employed 1951–86	11 deaths from lung cancer			SMR 1.36 [0.68–2.4]	Local Ontario rates
			Cumulative exposure to dust ($\geq$ 15 years since first exposure)			
			< 5 years	1	1.38	
			5–9 years	2	1.56	
			10–24 years	2	1.71	
			$\geq$ 25 years	2	0.67	

[a] Employed for $\geq$ 1 year
SMR, standardized mortality ratio; SIR, standardized incidence ratio; respiratory cancer, ICD-8, 160–163

for respiratory cancer for the two continuous glass filament plants were (male workers only): plant 2, 1.18 [95% CI, 0.96–1.44] and plant 5, 0.85 [95% CI, 0.60–1.10]. The results of the exposure–response analysis for the continuous glass filament plants were included in the study of glass-wool workers, described in section 2.1. Adjustment for smoking had very little effect (Marsh et al., 2001b). One person had died from mesothelioma according to the death certificate, but this diagnosis was not confirmed by a review of medical records and pathology specimens (Marsh et al., 2001c).

In a nested case–control study among male workers of this cohort (see section 2.1 for description of the study), the smoking-adjusted odds ratios for respiratory cancer for the two continuous glass filament plants compared with the baseline plant (plant 9) were: plant 2, 1.60 (95% CI, 0.95–2.69) and plant 5, 0.54 (95% CI, 0.31–0.94) (Marsh et al., 2001a).

(b)   *European cohort*

In the study by Boffetta et al. (1997), described in detail in section 2.1, separation into distinct technological phases was not applicable to the process of continuous filament production. For 1940 continuous filament workers employed for one year or more in one plant in the United Kingdom (Northern Ireland) or one in Italy, contributing 35 293 person–years of observation, the SMR for overall mortality was 1.22 (95% CI, 1.05–1.40; 191 deaths) and that for overall cancer mortality was 1.04 (95% CI, 0.76–1.39; 45 cases). The SMR for lung cancer was 1.11 (95% CI, 0.61–1.86; 14 cases) and a non-statistically significant increase in SMR was seen for cancer of the buccal cavity and pharynx, based upon two deaths (SMR, 1.63; 95% CI, 0.20–5.87).

(c)   *United States Georgetown University cohort*

Chiazze et al. (1997) studied a cohort of 2933 white male production workers employed for one year or more between 1951 and 1991 in a continuous glass filament plant in the USA, which was followed up until 1991. B fibres ('respirable fibres') (average diameter, 3.5 µm) were produced only from 1963–68 and glass fibres of 10–12 µm diameter were produced throughout the study period. Three per cent of the cohort members were lost to follow-up and cause of 96.3% of deaths was known. Information on 'respirable glass fibre' and on potentially confounding exposure to asbestos, refractory ceramic fibres, respirable silica, formaldehyde, 'total chrome' [presumed to be chromium oxides] and arsenic was available. This information was not presented in the SMR analysis (only the results of the case–control study were presented using this information, see below). The SMRs were calculated using national and county mortality rates. For all causes of death and all malignant neoplasms, there were deficits in mortality when compared with local rates (SMR for all causes, 0.92; 95% CI, 0.84–1.01; all malignant neoplasms, SMR, 0.96; 95% CI, 0.78–1.18). The SMR for lung cancer in white males, based on local mortality rates, showed a non-statistically significant increase (SMR, 1.17; 95% CI, 0.86–1.55; 47 deaths) and a non-significant deficit in mortality due to malignant neoplasms of the buccal cavity and pharynx (SMR, 0.87; 95% CI, 0.11–3.16; 2 deaths).

In a nested case–control study of this cohort, Chiazze et al. (1997) reported a study of 47 white men who had died from lung cancer between 1951 and 1991. Controls were matched on year of birth (within 2 years) and survival to end of follow-up or death (within 2 years). Information on demographic factors, including smoking, was obtained from interviews, and a reconstruction of the historical working environment was used to identify the agents to which the workers were potentially exposed (such as asbestos, refractory ceramic fibres, total particulate matter, respirable silica, formaldehyde, total chrome [presumed to be chromium oxides] and resins (binder)). Information on exposure was available for 45 (96%) and information on smoking habits was available for 35 (75%) of the cases. The odds ratio for lung cancer among workers exposed to respirable glass fibres (B fibres) was below unity (odds ratio, 0.91; 95% CI, 0.36–2.25). For smokers, the odds ratio for lung cancer among workers exposed to respirable glass fibres was further reduced (odds ratio, 0.78; 95% CI, 0.28–2.20). None of the other substances to which workers at the plant were potentially exposed was associated with an increase in lung cancer risk for this population.

In the same cohort, Watkins et al. (1997) studied 1074 white women, 130 black women and 494 black men (with the same entrance criteria, follow-up and information on other potential sources of exposure as used by Chiazze et al., 1997). A total of 107 white women died during the period of investigation and relatively few deaths were attributable to any one specific cause. There were no significant excesses or deficits in mortality by cause, including cancer, among the white women, when compared with national mortality rates. For black men, the SMRs for all cancers combined were below unity when calculated using either national (SMR, 0.84; 95% CI, 0.46–1.41) or local county standards (SMR, 0.82; 95% CI, 0.45–1.38; 14 deaths). Based upon local rates, the SMRs for lung cancer were below unity for both white women (SMR, 0.72; 95% CI, 0.20–1.85; 4 cases) and black men (SMR, 0.30; 95% CI, 0.04–1.07; 2 cases). Only four of the black women died during the study period.

(d) Canadian cohort

Shannon et al. (1990) reported the results of a cohort study in Canada. The cohort consisted of 1465 men and women who had worked for a total of at least one year at a continuous glass filament plant in Guelph, Ontario between 1951 (when the operations began) and 1986. Ninety-six per cent of the potential study subjects were traced. Data on the history of exposure to 'dust' were not available for the plant until 1978 and previous dust concentrations were estimated. These estimates were made by two groups of employees who were asked to rank dustiness for individual jobs and departments over time on a scale from 0 (fresh air) to 5 (the dustiest conditions ever experienced). When there was disagreement the estimates were averaged. In dust samples taken between 1979 and 1987, the time-weighted averages were between 0.02 and 0.05 fibre/cm^3. The highest value observed for any sample was 0.91 fibre/cm^3. The SMRs were calculated based upon local (Ontario) mortality rates. The overall mortality risk was decreased for both men (SMR, 0.76 [95% CI, 0.60–0.94]; 82 deaths) and women (SMR, 0.95 [95% CI, 0.52–1.6]; 14 deaths), and there was also a deficit in risk for all cancers for both men (SMR, 0.99 [95% CI, 0.64–1.5]; 25 deaths) and women (SMR, 0.67 [95% CI, 0.18–1.7]; 4 deaths). A non-statistically significant increase in risk for lung cancer, based upon 11 deaths (SMR, 1.36 [95% CI, 0.68–2.4]) was reported for the total cohort. The risk for lung cancer was not associated with increasing estimates of cumulative exposure to dust. [The Working Group considered that the limitations of this study include small size of the cohort and the lack of data on industrial hygiene and smoking habits.]

2.3   Rock (stone) wool and slag wool (see Table 56)

2.3.1   Cohort studies

In their evaluation of the studies of the carcinogenicity of rock (stone) wool and slag wool, the previous Working Group considered that there was *limited evidence* for the carcinogenicity of rock (stone) wool and slag wool in humans (IARC, 1988) (see section 2.1). The studies from the USA indicated statistically significant excess mortality from respiratory cancer; however, there was no relationship with time since first exposure, duration of exposure or time-weighted measurements of fibre exposure. The European study showed an overall excess of lung cancer mortality that was not statistically significant and an increasing risk with time since first exposure. The highest (and statistically significant) excess of lung cancer was found after more than 20 years of follow-up in workers who had first been exposed to rock (stone) and slag wool during the early technological phase. In addition to the studies described in the previously evaluated papers, Boffetta *et al.* (1992) reported the results from a Poisson regression analysis of the European cohort followed up for mortality until 1982–83 which showed no significant association of lung cancer with surrogates for fibre exposure.

(a)   United States University of Pittsburgh cohort

Two follow-up studies of this cohort have been published since 1988 (Marsh *et al.*, 1990, 1996). The 1985 follow-up used (Marsh *et al.*, 1990) (see section 2.1 for description of the study), data on 1846 male workers from six plants producing rock (stone) wool and slag wool. During the follow-up period, 73 cohort members died from respiratory cancer giving an overall SMR of 1.36 [95% CI, 1.06–1.71] when local rates were used. The average concentration of fibres of diameter < 3 μm was 0.351 fibre/cm^3. There was no positive relationship between respiratory cancer and duration of employment; men employed for less than 10 years had an SMR of 1.43 [95% CI, 1.01–1.96], 38 deaths. The Poisson regression analysis showed no statistically significant pattern of increasing risk associated with any of the indicators of exposure.

STUDIES OF CANCER IN HUMANS

Table 56. Studies of cancer in workers exposed to rock (stone) wool and slag wool

Reference, plants	Description, employment, follow-up	No. of deaths, cases (controls), type of cancer	Exposure categories	No. of cases	Relative risks (95% CI)	Comments
US University of Pittsburgh						
Cohort studies						
Marsh et al. (1990) 6 plants	1846 male workers[a] employed 1945–63, follow-up 1946–85	73 deaths from respiratory cancer			SMR 1.36 [1.06–1.71]	Local rates
			Time since first employment			
			< 10 years	2	0.89	
			10–19 years	13	1.56	
			20–29 years	24	1.37	
			≥ 30 years	34	1.32	
			Duration of employment			Local rates
			< 10 years	38	1.43 [1.01–1.96]	
			10–19 years	15	1.46	
			20–29 years	11	1.18	
			≥ 30 years	9	1.18	

Table 56 (contd)

Reference, plants	Description, employment, follow-up	No. of deaths, cases (controls), type of cancer	Exposure categories	No. of cases	Relative risks (95% CI)	Comments

Marsh et al. (1996) 5 plants	N-cohort (cohort participating in the new programme): 3035 male and female workers[a] employed 1945–78	71 deaths from respiratory cancer (68 in men)	Time since first employment
			N-cohort (men only)
			  < 10 years
			  10–19 years
			  20–29 years
			  ≥ 30 years
			O-cohort
				2
				13
				23
				30
					SMR
					0.58
					1.22
					1.35
					1.06
						Local rates

1 plant	O-cohort (cohort from the original plant): 443 male workers[a] employed 1945–63
	
	Follow-up until 1989
		32 deaths from respiratory cancer	< 20 years
			20–29 years
			≥ 30 years
			
			Duration of employment
			N-cohort (men only)
			  < 10 years
			  10–19 years
			  20–29 years
			  ≥ 30 years
			O-cohort
			  < 10 years
			  10–19 years
			  ≥ 20 years
				3
				8
				21
				
				39
				15
				8
				6
				
				15
				7
				10
					0.95 [0.20–2.78]
					1.41 [0.61–2.78]
					1.71 [1.06–2.61]
					
					SMR
					1.14
					1.34
					1.07
					0.89
					
					1.32
					2.02
					1.61
						Asbestos exposure

Table 56 (contd)

Reference, plants	Description, employment, follow-up	No. of deaths, cases (controls), type of cancer	Exposure categories	No. of cases	Relative risks (95% CI)	Comments
Nested case–control study						
Marsh et al. (1996) 5 plants	*N-cohort* 54 deaths[a] from respiratory cancer (men)	107 male controls	Cumulative exposure to respirable fibres <3 fibres/cm³–months 3–14 fibres/cm³–months 15–39 fibres/cm³–months ≥40 fibres/cm³–months		Odds ratio 1.0 0.70 0.59 0.71	Unadjusted for smoking p for linear trend = 0.76
	54 deaths from respiratory cancer (men)	101 male controls	<3 fibres/cm³–months 3–14 fibres/cm³–months 15–39 fibres/cm³–months ≥40 fibres/cm³–months		1.0 0.64 0.55 0.58	Adjusted for smoking p for linear trend = 0.64
1 plant	*O-cohort* 24 deaths[a] from respiratory cancer (men) 1970–89	47 controls	Duration of employment <2 years 2–4 years 5–19 years ≥20 years		Odds ratio 1.0 1.62 0.23 0.85	Unadjusted for smoking p for linear trend = 0.21
	18 deaths from respiratory cancer	31 controls	<2 years 2–4 years 5–19 years ≥20 years		1.0 1.82 0.33 0.73	Smokers only p for linear trend = 0.47
USA						
Nested case–control study						
Wong et al. (1991) 9 plants slag wool workers (4 plants also in Marsh et al., 1990, 1996)	55 men who died from lung cancer[a] 1970–89	98 male controls who had died from other causes	Exposed/unexposed Exposed ≥7 fibres/cm³–months exposed <7 fibres/cm³–months	50 27	Odds ratio 0.90 (0.23–3.49) 0.94 (0.23–3.78) 0.86 (0.42–1.79) 0.98 (0.47–2.04)	NIOSH exposure classification Unadjusted for smoking Adjusted for smoking Unadjusted for smoking Adjusted for smoking

Table 56 (contd)

Reference, plants	Description, employment, follow-up	No. of deaths, cases (controls), type of cancer	Exposure categories	No. of cases	Relative risks (95% CI)	Comments
European study						
Cohort studies						
Plato et al. (1995c) Sweden 2 plants (included in Boffetta et al., 1997)	1569 male and female workers employed[a] before 1978, follow-up 1952–90 for mortality	13 deaths from lung cancer			SMR [1.57 (0.83–2.68)] [1.02 (0.55–1.75)]	Local rates National rates
			Duration of employment with 20-year lag:			Local rates
			< 2 years	1	1.10 (0.28–6.12)	
			2–9 years	5	2.69 (0.87–6.27)	
			10–19 years	1	0.87 (0.02–4.89)	
			≥ 20 years	2	1.43 (0.17–5.16)	
			Plant-specific cumulative fibre exposure (fibres/cm³–years):		SMR	
			< 1	7	2.01 (0.81–4.13)	
			1–2	4	2.45 (0.67–6.21)	
			> 2	2	0.62 (0.08–2.24)	
	Follow-up 1958–89 for incidence	13 cases of lung cancer	Duration of employment:		SIR	Local rates
			< 2 years	1	0.69 (0.02–3.84)	
			2–9 years	7	2.12 (0.85–4.37)	
			10–19 years	3	1.63 (0.34–4.76)	
			≥ 20 years	2	1.61 (0.20–5.83)	
		13 cases of stomach cancer			SIR 1.71 (0.91–2.93)	Local rates

Table 56 (contd)

Reference, plants	Description, employment, follow-up	No. of deaths, cases (controls), type of cancer	Exposure categories	No. of cases	Relative risks (95% CI)	Comments
Boffetta et al. (1997) 7 plants Denmark, Germany, Norway and Sweden Mortality study	4912 male and female workers[a] employed 1933–77, follow-up until 1990–91	97 deaths from lung cancer			SMR 1.34 (1.08–1.63) Relative risk[b]	National rates Adjusted for age, calendar year, country, technological phase and duration of employment
			Time since first employment			
			≤ 9 years	10	1.0	
			10–19 years	26	1.3 (0.6–3.0)	
			20–29 years	29	1.2 (0.5–3.1)	
			≥ 30 years	32	1.4 (0.4–4.6)	p for linear trend = 0.67
			Duration of employment			Adjusted for age, calendar year, country, technological phase and time since first employment
			1–4 years	31	1.0	
			5–9 years	21	1.4 (0.8–2.4)	
			10–19 years	21	1.0 (0.5–1.8)	
			≥ 20 years	24	1.6 (0.8–3.1)	p for linear trend = 0.27
			Technological phase			Adjusted for age, calendar year, country, duration of employment and time since first employment
			Late	76	1.0	
			Intermediate	12	1.0 (0.5–2.3)	
			Early	9	1.1 (0.4–2.8)	
		8 deaths from oral cancer + cancer of the pharynx			SMR 1.33 (0.57–2.61)	National rates
		6 deaths from cancer of the larynx			1.96 (0.72–4.27)	
		8 deaths from cancer of the oesophagus			1.25 (0.54–2.46)	

Table 56 (contd)

Reference, plants	Description, employment, follow-up	No. of deaths, cases (controls), type of cancer	Exposure categories	No. of cases	Relative risks (95% CI)	Comments
Consonni et al. (1998) Denmark, Germany, Norway and Sweden 7 plants	9603 male workers employed until 1977 follow-up until 1990–91	159 deaths from lung cancer	Cumulative exposure		Relative risk[b]	Adjusted for age, calendar period, country, time since first employment and employment status
			≤ 0.007 fibre/cm^3–years	39	1.0	
			0.008–0.136 fibre/cm^3–years	40	1.3 (0.8–2.4)	
			0.137–1.367 fibre/cm^3–years	40	1.2 (0.7–2.1)	
			> 1.368 fibres/cm^3–years	40	1.5 (0.7–3.0)	p for linear trend = 0.4
		97 deaths from lung cancer in workers with ≥ 1 year of employment	≤ 0.139 fibre/cm^3–years	25	1.0	
			0.140–0.729 fibre/cm^3–years	24	0.9 (0.4–2.0)	
			0.730–2.622 fibres/cm^3–years	24	0.8 (0.3–1.9)	
			> 2.622 fibres/cm^3–years	24	1.0 (0.4–2.7)	p for linear trend = 1.0
Boffetta et al. (1999)	3685 male and female workers[a] employed 1933–77 follow-up 1994–95	73 cases of lung cancer			SIR 1.08 (0.85–1.36)	National rates Adjusted for age, gender, country and technological phase
			Time since first employment		Relative risk[b]	
			≤ 9 years	7	1.0	
			10–19 years	21	1.8 (0.7–4.7)	
			20–29 years	25	2.4 (0.9–6.8)	
			≥ 30 years	20	3.0 (0.8–10.5)	p for linear trend = 0.1
			Duration of employment (15-year lag)			Adjusted for age, gender, country, technological phase and time since first employment
			1–4 years	33	1.0	
			5–9 years	11	1.0 (0.5–2.1)	
			10–19 years	10	1.2 (0.5–2.6)	
			≥ 20 years	5	2.0 (0.7–6.2)	p for linear trend = 0.4
			Technological phase			Adjusted for age, gender, country and time since first employment
			Late	50	1.0	
			Intermediate	14	0.8 (0.4–1.7)	
			Early	9	0.8 (0.3–2.0)	p for linear trend = 0.5
		31 cases of cancer of the oral cavity, pharynx or larynx			SIR 1.46 (0.99–2.07)	National rates

Table 56 (contd)

Reference, plants	Description, employment, follow-up	No. of deaths, cases (controls), type of cancer	Exposure categories	No. of cases	Relative risks (95% CI)	Comments
Case–control study						
Kjaerheim et al. (2002) 7 plants	133 cases of lung cancer; rock (stone) wool/slag wool male workers employed 1937–76, follow-up 1971–96	513 male controls	Cumulative fibre exposure in quartiles		Odds ratio	Adjusted for age, country and tobacco smoking
			All workers			
			quartile 1	33	1.0	
			quartile 2	32	0.86 (0.47–1.56)	
			quartile 3	33	0.91 (0.51–1.63)	
			quartile 4	34	0.51 (0.28–0.93)	p for linear trend = 0.04
			Workers employed > 1 year			
			quartile 1	12	1.0	
			quartile 2	3	2.08 (0.36–11.91)	
			quartile 3	26	0.85 (0.34–2.15)	
			quartile 4	34	0.52 (0.21–1.30)	p for linear trend = 0.11
			Cumulative fibre exposure, in quartiles, lagged 15 years			
			All workers			
			quartile 1	36	1.0	
			quartile 2	36	1.25 (0.66–2.34)	
			quartile 3	30	1.02 (0.54–1.93)	
			quartile 4	30	0.67 (0.35–1.27)	p for linear trend = 0.17
			Workers employed > 1 year			
			quartile 1	23	1.0	
			quartile 2	5	2.00 (0.41–9.83)	
			quartile 3	18	0.76 (0.27–2.17)	
			quartile 4	29	0.63 (0.28–1.42)	p for linear trend = 0.19

Table 56 (contd)

Reference, plants	Description, employment, follow-up	No. of deaths, cases (controls), type of cancer	Exposure categories	No. of cases	Relative risks (95% CI)	Comments
Kjaerheim et al. (2002) (contd)			Duration of exposure in rock (stone)/slag wool industry		Odds ratio	
			All workers			
			Unexposed	7	1.0	
			1 year	58	1.24 (0.47–3.26)	
			2–6 years	32	0.86 (0.32–2.31)	
			7–40 years	35	0.85 (0.32–2.26)	*p* for linear trend = 0.23
			Workers employed > 1 year			
			Unexposed	6	1.0	
			1 year	2	0.51 (0.05–5.50)	
			2–6 years	32	0.60 (0.17–2.08)	
			7–40 years	35	0.65 (0.20–2.12)	*p* for linear trend = 0.63
			Duration of exposure in rock (stone)/slag wool industry, lagged 15 years			
			All workers			
			Unexposed	28	1.0	
			1 year	51	1.57 (0.82–2.99)	
			2–5 years	28	1.20 (0.61–2.34)	
			6–39 years	25	0.97 (0.48–2.00)	*p* for linear trend = 0.15
			Workers employed > 1 year			
			Unexposed	18	1.0	
			1 year	4	1.06 (0.21–5.30)	
			2–5 years	28	1.01 (0.44–2.34)	
			6–39 years	25	0.98 (0.39–2.47)	*p* for linear trend = 0.96

[a] Employed for ≥ 1 year
[b] Poisson regression analysis
SMR, standardized mortality ratio; SIR, standardized incidence ratio; respiratory cancer, ICD-8, 160–163

Marsh et al. (1996) extended the follow-up period for workers employed between 1945 and 1978 to 1989. The extended cohort (N-cohort) included 2762 men and 273 women who had been employed for one year or more in production or maintenance. One factory (plant 17) had closed and declined to participate further; this plant had evidence of exposure to asbestos and it was therefore designated to a separate subcohort (O-cohort). Vital status for the members of both cohorts was determined as of 31 December 1989. From the five other plants (the N-cohort), quantitative estimates of exposure to total airborne fibres, respirable fibres, formaldehyde and silica were computed together with qualitative estimates of potential exposure to asbestos, arsenic, asphalt, radiation and PAHs (Marsh et al., 1993). The pattern of findings of respiratory cancer after the extended update was generally consistent with the results of the previous follow-up. Seventy-one workers from the five plants died from respiratory cancer. This increase was non-significant (SMRs, 1.24 [95% CI, 0.97–1.56] and 1.17 [95% CI, 0.91–1.48], based on US and local county comparisons rates, respectively). For the O-cohort (plant 17), however, 32 men died from respiratory cancer, resulting in statistically significant increases of the SMRs (SMRs, 1.91 [95% CI, 1.31–2.70] and 1.52 [95% CI, 1.04–2.15], using national and local county rates, respectively). There was no evidence of a positive association between mortality from respiratory cancer, duration of employment and year since first employment in the five plants (N-cohort), but in the O-cohort a positive association was seen with time since first employment (SMRs, 0.95, 1.41 and 1.71 [95% CI, 0.20–2.78, 0.61–2.78 and 1.06–2.61, respectively] in the groups with < 20 years, 20–29 and ≥ 30 years since first employment, respectively) (Marsh et al., 1996). Indirect adjustments for smoking (see section 2.1) reduced the SMR for respiratory cancer, among male workers in the six plants (N- and O-cohorts) taken together, from a statistically significant SMR of 1.24 [95% CI, 1.0–1.51] to an SMR of 0.96 [95% CI, 0.78–1.17] based on 100 observed deaths and using local county rates (Marsh et al., 2001b).

Relative risk regression modelling of internal cohort rates for respiratory cancer in the N-cohort revealed no consistent evidence of an association with any of the exposure indices for respirable fibre considered, with or without adjustment for potential confounding factors that included year of hire, plant and estimated co-exposure for individual workers (Marsh et al., 1996).

A manual search of death certificates of 1011 rock (stone) and slag wool workers from the N-cohort revealed two death certificates that mentioned the word 'mesothelioma'. A subsequent review of medical records and pathology specimens for one of the two cases deemed this death as definitely not due to pleural cancer. In a mortality analysis, only one of the two cases among rock (stone) and slag wool workers was coded as a pleural cancer (ICD-9 163.9) (this mesothelioma was seen in a worker from plant 17 where there was potential exposure to asbestos (Marsh et al., 1996; Marsh et al., 2001c)). [The Working Group noted the limitations of death certificates for the identification of mesothelioma diagnoses: deaths due to mesothelioma would not have been detected in this study if the word 'mesothelioma' had not appeared on the death certificate.]

Nested case–control studies

A nested case–control study of the six rock (stone) and slag wool plants in the US cohort identified 78 men who had died from respiratory cancer between 1970 and 1989. Controls were matched on age and time period and were randomly selected from men who had died during the period 1970–89. Information on smoking habits was available for all of the 54 cases and 94% of 95 controls in the five-plant N-cohort, and for 92% of the 24 cases and 87% of 37 controls in the asbestos-contaminated O-cohort. The odds ratios for cumulative exposure to respirable fibres in the N-cohort showed a decreasing trend (odds ratios, 1.00, 0.70, 0.59 and 0.71; global test p-value = 0.76) for cumulative exposures of < 3, 3–14, 15–39 and ≥ 40 fibres/cm^3–months, respectively. After adjustment for smoking, the odds ratios were 1.0, 0.64, 0.55 and 0.58 (global test p-value = 0.64) for the same exposure groups. Further adjustment for potential confounding by co-exposure revealed no evidence of an increasing trend in odds ratios for any of the levels of exposure to respirable fibre. In the O-cohort, the odds ratios for the categories for duration of employment (< 2, 2–4, 5–19 and ≥ 20 years) showed no evidence of an increasing trend when no adjustment was made for smoking or when calculated for smokers only (Marsh *et al.*, 1996).

A case–control study nested in a cohort of 4841 men employed for one year or more in the rock (stone) wool and slag wool industry was reported by Wong *et al.* (1991). The workers studied were employed in nine plants, four of which were also included in the University of Pittsburgh study (Marsh *et al.*, 1990, 1996). Most of the plants had started producing slag wool, or using slag wool in their products, in the 1940s. Altogether, 504 workers who had died between 1970 and 1989 were included; 61 of whom had died from lung cancer. It was not possible to locate relatives for three of the workers and three others were excluded because matching controls could not be found. The final analysis included 55 cases and 98 controls who had died from other causes. Data on exposure were obtained from two industrial hygiene surveys, one by the National Institute for Occupational Safety and Health (NIOSH) (Fowler, 1980); and the other by the Pittsburgh group (Esmen *et al.*, 1979a). The estimates of exposure to airborne fibres calculated for each case and control ranged from 0–0.25 fibres/cm^3 according to the NIOSH model and from 0–0.21 fibres/cm^3 according to the Pittsburgh classification. Information on smoking habits was collected by interview. All cases had been cigarette smokers, but only 81% of the controls had been smokers. According to the NIOSH classification, 93% of the cases and controls had been exposed to slag wool. The odds ratio of ever versus never exposure to slag wool, adjusted for smoking, was 0.94 (95% CI, 0.23–3.78). When data were divided into two exposure groups (< or ≥ 7 fibres/cm^3–months), the odds ratio after adjustment for smoking was 0.98

(95% CI, 0.47–2.04). [The results were not adjusted for exposure to asbestos although four plants had used a limited amount of asbestos for a few years.]

(b) *European study*

Plato *et al.* (1995b,c) reported the results of the follow-up study from 1952–90 (for cancer mortality) and from 1958–89 (for cancer incidence) of three Swedish plants included in the European MMVF cohort (see Boffetta *et al.*, 1997 and section 2.1). This cohort comprised 1569 male and female workers employed for at least one year before 1978, in the two plants that produced rock (stone) wool and slag wool. The duration of employment was categorized as < 2, 2–9, 10–19 or ≥ 20 years. The cumulative exposure to fibres, specific for plant and job, was estimated for the period from 1938–90, for 1487 workers (1329 men and 158 women), contributing 34 392 person–years of observation, in rock (stone) and slag wool plants. Exposure to fibres was classified as low, medium or high, the cut-points used were 1.0 fibre/cm^3–years and 2.0 fibres/cm^3–years. Lags of 20 and 30 years were used. Both national and county rates were used in the external comparisons to produce standardized mortality ratios and standardized incidence ratios (SMRs and SIRs). A Poisson regression analysis with internal comparisons was also performed. The SMR for cancer at all sites for the two rock (stone) wool and slag wool plants combined, based on 66 deaths, was very similar when either national or regional rates were used for comparison [SMR, 0.96; 95% CI, 0.74–1.22 using regional rates; and SMR, 0.87; 95% CI, 0.68–1.11 using national rates]. Based on 131 cancer cases and regional rates, the SIR for cancer at all sites for the two plants combined was [1.15 (95% CI, 0.95–1.35)]. In contrast to the finding in the whole European cohort, the SMR for lung cancer was higher when compared with regional than national rates [SMR, 1.57; 95% CI, 0.83–2.68; and 1.02; 95% CI, 0.55–1.75, using regional and national rates, respectively, based on 13 deaths]. In the larger rock (stone) wool and slag wool plant, the SMR for lung cancer was 2.40 (95% CI, 1.24–4.19), based on 12 deaths and regional rates. No trend was seen with duration of employment when 20 years of latency and regional rates were used. Neither the plant-specific nor the job-group-specific estimates of exposure to fibres gave a monotonic increase in SMRs for lung cancer (the lowest SMRs of 0.62 (95% CI, 0.08–2.24) and 0.95 (95% CI, 0.20–2.78), were found in the groups exposed to the highest fibre concentrations using regional reference rates). Internal comparison also revealed no monotonic increase in the relative risks with increasing estimates of exposure. The SIR for lung cancer was 1.65 (95% CI, 0.88–2.83), based on 13 cases and regional rates. It did not increase with increasing duration of employment. Based on 13 cases and comparison with regional rates, an increased incidence of stomach cancer was seen (SIR, 1.71; 95% CI, 0.91–2.93).

In their extended follow-up of the European cohort (until 1990 in Denmark, Norway and Sweden and 1991 in Germany) Boffetta *et al.* (1997) (see section 2.1) included 10 108 rock (stone) wool and slag wool workers from seven plants (1 in Denmark, 3 in Norway, 2 in Sweden (see Plato *et al.*, 1995a,b) and 1 in Germany) contributing 221 871 person–years. The cohort included 4912 workers employed for one year or more who contributed 114 228 person–years; 29% of these person–years took place 20 years or more after the date of first employment. Three hundred and twenty-two of the rock (stone) wool and slag wool workers were first employed during the early technological phase of the industry (when exposure to fibres was estimated to be high), 603 were first employed during the intermediate phase and 9183 during the late phase (corresponding to modern technology). Copper slags were used periodically between 1940 and 1944 in Denmark, and there was potential exposure to asbestos (for a few workers in Denmark between 1962 and 1982, and for a large number of workers in Germany between 1941 and 1970) (Cherrie & Dodgson, 1986). A Poisson regression analysis of all male workers with known duration of employment showed that the SMR for overall mortality, in rock (stone) wool and slag wool workers employed for one year or more, was 1.04 (95% CI, 0.98–1.10; 1281 deaths), the SMR for cancer at all sites was only slightly elevated (SMR, 1.08; 95% CI, 0.97–1.21; 325 deaths) and the SMR for lung cancer was 1.34 (95% CI, 1.08–1.63; 97 deaths), compared with national mortality rates. The subcohorts from Denmark and Germany accounted for 70% of all observed deaths from lung cancer. The lung cancer mortality was highest among workers with the longest time since first employment (p for trend, 0.19). Only small differences in the SMRs were seen in the analysis by technological phase. Among workers employed for one year or more, the SMRs for the early, intermediate and late technological phases, respectively, were 1.51 (95% CI, 0.69–2.87; 9 deaths), 1.49 (95% CI, 0.77–2.60; 12 deaths) and 1.30 (95% CI, 1.02–1.63; 76 deaths) for lung cancer. In the group first employed in the early technological phase and with 30 years or more since first employment, the SMR for lung cancer was 1.46 (95% CI, 0.59–3.02; 7 deaths). Using the Poisson regression analysis the relative risks for lung cancer with increasing time since first employment (adjusted for duration of employment, technological phase, age, calendar year and country) were 1.3 (95% CI, 0.6–3.0), 1.2 (95% CI, 0.5–3.1) and 1.4 (95% CI, 0.4–4.6; p for linear trend, 0.67) (10–19, 20–29 and ≥ 30 years since first employment). After adjustment for time since first employment and duration of employment, no differences in relative risk according to technological phase were seen (relative risk, 1.0 (95% CI, 0.5–2.3) and 1.1 (95% CI, 0.4–2.8), using the late technological phase as the reference). Workers employed for 20 years or more had a relative risk for lung cancer of 1.6 (95% CI, 0.8–3.1; 24 deaths), but the trend with increasing duration of employment was not monotonic (p-value for trend, 0.27).

In this cohort, four deaths from mesothelioma were recorded (two among workers with < 1 year of employment and two in the factory from Germany [expected numbers of deaths were not given]). Statistically non-significant elevations were found for oral and pharyngeal cancer (SMR, 1.33; 95% CI, 0.57–2.61; 8 deaths); laryngeal cancer (SMR, 1.96; 95% CI, 0.72–4.27; 6 deaths) and oesophageal cancer (SMR, 1.25; 95% CI, 0.54–2.46; 8 deaths) (Boffetta et al., 1997). No associations with time since first employment or duration of employment were seen for these cancer sites.

Consonni et al. (1998) analysed mortality from lung cancer by estimated fibre exposure in the cohort described by Boffetta et al. (1997). Individual cumulative exposure to fibres (fibres/cm^3–years) and lifetime maximum annual exposure (fibres/cm^3) were calculated, for year-specific estimates of exposure based on measurements of the concentrations of respirable fibres made in six of the seven plants during 1977–80, in combination with reconstruction of historical work processes, theoretical considerations and results obtained from simulation experiments on fibre emissions (Krantz *et al.*, 1991). These data were used to assign numerical coefficients to the factors that had been identified as exposure determinants (addition of oil as a dust-suppressing agent, production rate, presence of ventilation system and degree of manual handling). These numerical coefficients were tabulated over the years and combined in a multiplicative way with the current fibre concentrations to obtain estimates of past concentrations of fibres for every plant and year of production. These estimates were combined in turn with individual work histories to provide an individual exposure assessment for each year of employment, from which the estimates of the cumulative and maximum exposure were derived. A 15-year lag was applied to the cumulative exposure variable and, in alternative models, a clearance with a half-life of five years and two years, respectively, was taken into account. The overall mean concentration of respirable fibres was measured as 0.06 fibre/cm^3. There was a very high correlation between the estimates of cumulative exposure to fibres and duration of employment ($r = 0.92$), and an even higher correlation when lung clearance was taken into account ($r > 0.99$). Poisson regression models were adjusted for employment status, time since first employment (< 10, 10–19, 20–29, ≥ 30 years), country, age and calendar year (in quinquennia). The analysis was based on 159 deaths from lung cancer (97 in workers with ≥ 1 year of employment). The relative risk for lung cancer was independent of employment status, but increased with increasing time since first employment. [The Working Group noted that relative risks for time since first employment were not given.] No trend in risk for lung cancer was seen with cumulative exposure among workers employed for one year or more. In the analysis that included all workers, a non-significant trend in the relative risks was found. No trend was found associated with maximum annual exposure. The incorporation of pulmonary clearance rates did not change these results.

In the countries with national cancer registries, follow-up for cancer incidence was extended until 1994 (in Denmark) and 1995 (in Norway and Sweden) (Boffetta *et al.*, 1999). The cohort included 3685 male and female workers employed in plants that produced rock (stone) wool and slag wool for one year or more, contributing 92 562 person–years; 32.5% of the person–years took place 20 years or more after the date of first employment. Completeness of follow-up ranged from 89.8% to 98.3% in the different plants. National cancer rates were used in the SIR analyses and when production started before these rates had become available, the rates were linearly extrapolated back to the date of first production. Internal comparison with Poisson regression was also performed. No elevation in risk for cancer at all sites was found in the rock (stone) wool or slag wool workers (SIR, 0.97; 95% CI, 0.88–1.06; 468 cases). The SIR for lung cancer was 1.08 (95% CI, 0.85–1.36; 73 cases). The analysis of risk by time since first employment showed a non-significant trend. No association between risk and duration of employment or technological phase was found. The overall SIR for cancer of the oral cavity, pharynx and larynx was 1.46 (95% CI, 0.99–2.07; 31 cases) for the six plants; at one plant in Norway the SIR was 2.72 (95% CI, 1.09–5.61; 7 cases). There was no clear relationship with time since first employment. [The Working Group noted that no similar increase was seen in other rock (stone) wool or slag wool plants and that 14 of the 31 cancers observed in the whole cohort were lip cancers.]

A case–control study nested in this cohort (Kjaerheim et al., 2002) included 133 of 196 cases of lung cancer in men (134 cases identified in the mortality study, 45 from the incidence study and 17 identified from the extended follow-up in Denmark) that occurred between 1971 and 1996 (in Denmark), 1995 (in Norway and Sweden) and 1991 (in Germany, deaths only). Two control groups (group 1, incidence density-sampled controls; group 2, matched to cases by year of death) were combined in the main analyses, and comprised 513 controls matched by age and plant. Information on residential and general occupational history and on tobacco smoking habits came from personal interviews with the patients and controls or their next of kin. The response rate was 67.9% for cases. The proportion of interviews performed out of the controls contacted was 59.2% out of the number of possible respondents contacted. Exposure assessments within the industries were based on information obtained from panels of experts and outside the industry on the basis of self-reported life-time occupational histories (Cherrie et al., 1996; Cherrie & Schneider, 1999). Variables indicating ever exposure, ever high exposure, duration of exposure and cumulative exposure were calculated for seven agents within the rock (stone) wool and slag wool industry. Similar variables, excluding cumulative exposure, were calculated for six sources of co-exposure occurring both within and outside the industry and for a further 12 sources of exposure that were found only outside the industry. All conditional logistic regression models included adjustment for age group and plant; additional adjustments were made for tobacco smoking and potential occupational confounders to which a significant number of workers were exposed (i.e. asbestos, polycyclic aromatic hydrocarbons and silica). No association was found between exposure to rock (stone) or slag wool fibres and risk for lung cancer for cumulative exposure to fibres of rock (stone) or slag wool assessed with a 15-year lag; the smoking-adjusted odds ratios in the second to fourth quartile of exposure were 1.25 (95% CI, 0.66–2.34), 1.02 (95% CI, 0.54–1.93) and 0.67 (95% CI, 0.35–1.27). Similar results were obtained when only workers employed for more than one year were included, with other indicators (e.g. duration of exposure) of exposure to fibres of rock (stone) wool or slag wool, and after adjustment for co-exposure. [Only four of the 11 workers (Boffetta et al., 1997) who died from lung cancer and were first employed during the early technological phase (among whom the SMR for lung cancer was particularly elevated) were included in the case–control study.] The 1985 age-adjusted prevalence of current smokers among controls was 20% greater than in the general population.

2.4 Refractory ceramic fibres (see Table 57)

2.4.1 Cohort study

A cohort study of workers at two plants in the USA that produced refractory ceramic fibres included 927 male workers employed for one year or more between 1952 and 1997. The mortality data were presented in a conference abstract (Lemasters et al., 2001) and in a paper addressing risk analysis (Walker et al., 2002). The estimated exposure ranged from 10 fibres/cm^3 (8-h TWA) in the 1950s to < 1 fibre/cm^3 in the 1990s. No significant increase in cancer mortality was reported. [The Working Group noted that neither the observed nor the expected numbers of cancers other than lung cancer were given.] Six deaths from lung cancer were observed versus 9.35 expected, SMR, 0.64 (95% CI [0.24]–1.27). No cases of mesothelioma were observed. [The Working Group noted that the details of cohort definition and period of follow up were not clear, and there was no analysis of risk in relation to time since first exposure or exposure surrogates. The small number of study subjects, especially those with adequate latency, limits the informativeness of the study.]

2.4.2 Case–control study

A case–control study including 45 men with lung cancer and 122 controls was nested within a cohort of 2933 white men employed in a plant manufacturing continuous glass filament (described in detail in section 2.2) (Chiazze et al., 1997). Exposure to respirable glass fibres, asbestos, refractory ceramic fibres (used at the plant for high-temperature heat insulation, but not manufactured there), and a number of other sources of exposure was assessed by a procedure of reconstruction of historical exposure conditions. The risk of lung cancer was lower in workers exposed to a cumulative dose of refractory ceramic fibres of 0.01–1 fibre/cm^3–days (odds ratio, 0.36 (95% CI, 0.04–3.64); 1 case), and those exposed to 1–40 fibres/cm^3–days (odds ratio, 0.30 (95% CI, 0.11–0.77); 7 cases), than in workers not exposed to fibres. The odds ratios were not adjusted for exposure in the workplace to other fibres or for tobacco smoking, but the trends in odds ratios were similar when the analysis was restricted to smokers. [The Working Group noted that exposure to refractory ceramic fibres may have been difficult to separate from other sources of exposure in the workplace in view of the small number of cases and the large number of sources of exposure.]

2.5 MMVF (not otherwise specified) (see Table 58)

A number of epidemiological studies in MMVF user industries and in the general population have provided information on cancer risk following exposure to MMVFs, but the ability of these studies to distinguish between exposure to the different types of MMVF was limited.

Table 57. Studies of lung cancer in workers exposed to refractory ceramic fibres

Reference	Description, employment, follow-up	Type of cancer, no. of deaths/cases (controls)	Exposure categories	No. of cases	Relative risks (95% CI)
USA studies					
Cohort study					
Lemasters et al. (2001); Walker et al. (2002)	927 male workers[a] employed 1952–97	6 deaths from lung cancer			0.64 ([0.24]–1.27)
Case–control study					
Chiazze et al. (1997) Co-exposure in continuous glass filament plant	45 men with lung cancer[a]	122 controls (white men)	Cumulative exposure < 0.01 fibre/cm^3–days 0.01–0.999 fibre/cm^3–days 1.0–39.24 fibres/cm^3–days	37 1 7	1 0.36 (0.04–3.64) 0.30 (0.11–0.77)

[a] Employed for ≥ 1 year

Table 58. Studies of cancer in workers exposed to MMVF (fibre type not otherwise specified)

Reference, study population	Description, employment, follow-up	Type of cancer, no. of deaths/ cases (controls)	Exposure categories	No. of cases	Relative risks (95% CI)	Comments
Cohort study						
Gustavsson et al. (1992) Sweden Exposure to MMVFs 11 factories	1465 men employed[a] before 1972, follow-up 1969–88 (mortality)	14 deaths from lung cancer	Time since first exposure		SMR	Information on smoking habits was obtained for 73% of workers in 8 of 13 factories 1 lung cancer death with exposure level unknown
			< 10 years	1	0.59 [0.02–3.28]	
			10–19 years	4	0.70 [0.19–1.79]	
			≥ 20 years	9	0.75 [0.34–1.42]	
			Exposure to fibres during 8-h shift by occupational group			
			< 0.02 fibre/cm^3–years	9	0.78 [0.35–1.49]	
			0.02–0.08 fibre/cm^3–years	1	0.58 [0.02–3.28]	
			0.05–0.13 fibre/cm^3–years	1	0.33 [0.01–1.86]	
			0.05–0.17 fibre/cm^3–years[b]	2	0.85 [0.10–3.01]	
		22 deaths from stomach cancer	Time since first exposure			
			< 10 years	3	2.52	
			10–19 years	7	1.90	
			≥ 20 years	10	1.24	
			Duration of employment			
			< 10 years	11	2.16	
			10–19 years	3	0.98	
			≥ 20 years	6	1.25	
Nested case–control study						
Martin et al. (2000) France Incidence	310 cases of lung cancer, 1978–89	1225 controls	33 cases exposed to MMVFs		Odds ratio 0.73 (0.32–1.70)	Adjusted for socioeconomic status and exposure to asbestos

Table 58 (contd)

Reference, study population	Description, employment, follow-up	Type of cancer, no. of deaths/cases (controls)	Exposure categories	No. of cases	Relative risks (95% CI)	Comments
Case–control studies						
Lung cancer						
Kjuus et al. (1986) Norway Incidence	176 cases of lung cancer, 1979–83	176 controls	Rock (stone) wool/glass fibre	13	Odds ratio 1.0 (0.4–2.5)	Adjusted for smoking
Siemiatycki (1991) Canada Population-based case–control study Incidence	857 cases of lung cancer, 1979–85	1360 cancer controls, 533 population controls	> 5 years of exposure glass wool rock (stone) wool and slag wool	11 10	Odds ratio 1.2 (0.5–2.5) 1.2 (0.5–2.7)	90% CI, adjusted for age, smoking, other demographic factors and occupational exposure
Brüske-Hohlfeld et al. (2000); Pohlabeln et al. (2000) Germany Incidence	3498 men with lung cancer, workers in several types of industry 1988–96	3541 controls	Ever versus never exposed to MMVF (rock (stone) wool/slag wool and glass wool) Duration of exposure	304	1.48 (1.17–1.88)	Adjusted for smoking and exposure to asbestos
			0–3 years	51	1.68 (0.98–2.88)	
			3–10 years	69	1.38 (0.86–2.20)	
			10–20 years	76	1.17 (0.77–1.77)	
			20–30 years	61	1.69 (1.01–2.81)	
			≥ 30 years	47	2.03 (1.04–3.95)	
			Ever exposed to MMVF and never exposed to asbestos Cumulative exposure to MMVF (fibre·year)			Adjusted for smoking
			0.1–0.4	30	1.41 (0.73–2.72)	
			> 0.4	29	1.20 (0.63–2.30)	

Table 58 (contd)

Reference, study population	Description, employment, follow-up	Type of cancer, no. of deaths/cases (controls)	Exposure categories	No. of cases	Relative risks (95% CI)	Comments
Mesothelioma						
Rödelsperger et al. (2001) Germany	125 men with mesothelioma 1988–91	125 male controls	Ever exposed to MMVFs Geometric mean (fibres–year)	55	3.08 (1.17–8.07)	Adjusted for exposure to asbestos
			≤ 0.015	10	0.78 (0.16–3.77)	
			> 0.015–0.15	11	3.11 (0.56–17.2)	
			> 0.15–1.5	20	7.95 (0.88–72.3)	
			> 1.5	14	5.43 (0.72–41.0)	
			Exposed to MMVFs without asbestos	2	15.1 (1.05–218)	Matched for age and region of residence
Larynx, hypopharynx						
Marchand et al. (2000) France Incidence	201 men with hypopharyngeal cancer 296 men with laryngeal cancer 1989–91	295 controls	Ever exposed *Rock (stone) wool/slag wool*	130	1.23 (0.79–1.91)	Odds ratios adjusted for exposure to asbestos, age, smoking and alcohol habits
			Larynx cancer	51	1.61 (0.85–3.04)	
			Epilarynx cancer	99	1.51 (0.90–2.52)	
			Hypopharynx cancer			
			Continuous glass filaments			
			Larynx cancer	8	0.44 (0.15–1.31)	
			Hypopharynx cancer	8	0.91 (0.30–2.76)	
			Refractory ceramic fibres			Adjusted for age, smoking and alcohol habits
			Larynx cancer	16	1.28 (0.51–3.22)	
			Hypopharynx cancer	7	0.78 (0.26–2.38)	
			Microfibres			
			Larynx cancer	16	1.28 (0.51–3.22)	
			Hypopharynx cancer	7	0.78 (0.26–2.38)	

Table 58 (contd)

Reference, study population	Description, employment, follow-up	Type of cancer, no. of deaths/ cases (controls)	Exposure categories	No. of cases	Relative risks (95% CI)	Comments
Colon						
Goldberg et al. (2001) Canada Incidence (same study population as Siemiatycki, 1991)	497 men with colon cancer 1979–85	1514 cancer controls 533 population controls	> 5 years of exposure Rock (stone) wool/slag wool Glass fibres	8 6	1.9 (0.8–4.6) 1.9 (0.7–5.4)	Adjusted for age, smoking, occupational and non-occupational exposure
Registry-based studies						
Breast, ovary						
Weiderpass et al. (1999) Finland Incidence	23 638 women with breast cancer 1971–95		MMVF Postmenopausal women Medium-to-high exposure Low exposure		SIR 1.32 (1.05–1.66) 1.01 (0.90–1.12)	Adjusted for birth cohort, follow-up period, socioeconomic status, mean no. of children, mean age at delivery of first child and turnover rate for each job title
Vasama-Neuvonen et al. (1999) Finland Incidence	5072 women with ovarian cancer 1971–95		MMVF		1.3 (0.9–1.8)	

SMR, standardized mortality ratio; SIR, standardized incidence ratio
[a] Employed for > 1 year
[b] For this exposure category, the highest exposure that occurred in 1975–80 was estimated as 0.20–0.25 fibre cm^{-3}–years.

2.5.1 Cohort studies

Gustavsson *et al.* (1992) studied a cohort of 2807 male workers in Sweden (employed for at least 1 year before 1972) in 11 factories that produced prefabricated wooden houses. A total of 1465 of the workers had been exposed to MMVFs. The workers were assigned to one of four exposure categories: category 0, no exposure; category 1, background level (truck drivers, repairmen, transportation workers); category 2, full-time employment in locations where MMVF was handled, but without direct handling of MMVF (e.g. wood cutters); category 3, direct handling of MMVF during most of the working day (e.g. insulators). Current exposure to MMVF was measured and past exposure was estimated. Exposure varied from 0.02–0.25 fibre/cm^3 per 8-h shift. The follow-up period was from 1969 to 1988 for mortality and from 1969 to 1985 for incidence and was more than 99% complete. A deficit in mortality for lung cancer (SMR, 0.68; 95% CI, 0.37–1.13; 14 deaths) and incidence (SIR, 0.47; 95% CI, 0.24–0.85; 11 cases) was reported. An analysis of mortality from stomach cancer showed an excess (SMR, 1.59; 95% CI, 1.00–2.41; 22 deaths), based on local rates. An analysis of stomach cancer incidence based on national rates was concordant (SIR, 1.78; 95% CI, 1.15–2.63; 25 cases). Analysis by duration of employment showed that this excess of mortality from stomach cancer was seen mainly among workers who had been employed for less than 10 years.

A study on a cohort of 135 037 male construction workers in Sweden reported in 1987 and evaluated by the previous Working Group (IARC, 1988) has not been updated (Engholm *et al.*, 1987). In a nested case–control study, the relationship between lung cancer in 424 workers and their exposure to MMVFs and to asbestos was examined. After adjustment for smoking habits and population density in the area of residence, an odds ratio of 1.21 (95% CI, 0.60–2.47) was reported for exposure to MMVFs (adjusted for exposure to asbestos). Twenty-three cases of pleural cancer were observed among the cohort [SIR, 2.13; 95% CI, 1.35–3.20]. [The Working Group noted that misclassification of exposure to asbestos and the difficulty in differentiating exposure to asbestos from exposure to MMVFs may have resulted in residual confounding.]

Nested case–control studies

In a study of workers in French national electricity and gas companies, 310 workers with lung cancer and 1225 controls were included in a nested case–control study during 1978–89. Thirty-three of the cases were exposed to MMVF as opposed to eight of the controls. The odds ratio after adjustment for socioeconomic status and exposure to asbestos was 0.73 (95% CI, 0.32–1.70) (Martin *et al.*, 2000).

2.5.2 Population-based case–control studies

Population-based case–control studies can be used to investigate exposure to MMVF in various circumstances. However, assessment of exposure in such studies is usually less valid and precise than in industry-based cohort studies.

(a) Lung cancer

Kjuus *et al.* (1986) conducted a case–control study of lung cancer (176 incident cases in men and 176 controls selected from one county in Norway) during 1979–83 and its association with occupational exposure and smoking habits. No association was seen between lung cancer and exposure to MMVF (rock (stone) wool or glass fibre) (odds ratio, 1.0; 95% CI, 0.4–2.5; 13 cases) after adjustment for smoking.

Siemiatycki (1991) conducted a population-based case–control study among male residents of Montreal, Canada, to explore associations between hundreds of occupational circumstances and cancer at several sites which had been diagnosed between 1979 and 1985. The study included interviews with 3730 patients with cancer at 11 major sites (response rate, 82%), including 857 patients with a pathologically confirmed diagnosis of lung cancer (response rate, 79.2%) from 19 hospitals in Montreal. Two sets of controls were selected: 1360 cancer controls (excluding patients with cancers at several sites that were anatomically contiguous), and 553 out of 740 potential population controls (response rate, 72%). Detailed job histories and information on relevant potential confounding variables were obtained during the interviews. The job histories were converted by a team of chemists and industrial hygienists into a history of occupational exposure. After adjustment for age, smoking, demographic factors and occupational exposure, the odds ratio compared with both control groups combined, for the group with substantial exposure to MMVF (i.e. > 5 years at medium or high frequency), for exposure to glass wool was 1.2 (90% CI, 0.5–2.5; 11 cases). The odds ratio for exposure to rock (stone) and slag wool was 1.2 (90% CI, 0.5–2.7; 10 cases).

Two population-based case–control studies were conducted at the Bremen Institute for Prevention Research and Social Medicine and the National Research Centre for Environmental Health in Germany during 1988 and 1996. Pooled analyses of exposure to fibres (mainly outside the MMVF production industry) were reported (Brüske-Hohlfeld *et al.*, 2000; Pohlabeln *et al.*, 2000). The final analysis included 3498 cases of lung cancer in men and 3541 controls. The response rates for cases were 69% and 77% and those for controls were 68% and 41%, for the Bremen Institute for Prevention Research and Social Medicine (1988–1993) and the National Research Centre for Environmental Health (1990–1996) studies, respectively. Both studies included pathologically confirmed cases of incident primary lung cancer. Information on smoking habits was obtained by standardized interview. Industries and job titles were grouped according to 21 industry codes and 33 job categories, respectively. The assessment of exposure was based on self-reported occupational histories obtained from personal interviews using a standardized questionnaire and two supplementary questionnaires to address exposure to MMVF. This information was used by two industrial hygiene experts to estimate cumulative exposure to MMVFs for a subset of the study group. A total of 304 cases and 170 controls were classified as ever exposed to MMVFs. The odds ratio after adjustment for smoking and exposure to asbestos was 1.48 (95% CI, 1.17–1.88) and a twofold risk was calculated for those who had been exposed for more than 30 years (odds ratio, 2.03; 95% CI, 1.04–3.95). The odds ratio for ever versus never self-reported exposure to MMVFs was 1.30 (95% CI, 0.82–2.07), after adjustment for smoking, in a subset of subjects who reported no exposure to asbestos (59 cases and 39 controls). No clear trend of increasing lung cancer risk with increasing cumulative exposure to MMVF was observed in this subset after adjusment for smoking. [The Working Group noted that the response rate of controls in the National Research Centre for Environmental Health study was low. The numbers of cases and controls reported in the two publications were inconsistent.]

(b) Mesothelioma

In a study designed to evaluate the association between occupational factors and mesothelioma, Rödelsperger et al. (2001) selected 137 German men with pathologically confirmed malignant mesothelioma recruited between 1988 and 1991 from clinics in Hamburg, Germany. Cases were matched on region of residence, sex, and year of birth (within 5 years) to controls selected randomly from population registries. The study team interviewed 125 patients almost all of whom were undergoing treatment at one of two specialized hospitals in Hamburg, and who were willing to grant a personal interview. The 125 controls were interviewed by trained interviewers using a similar protocol. [The Working Group noted that the response rates appear to have been higher for the cases (91%) than the matched controls (63%), but the basis for calculation of these participation rates was not provided.] The mean duration of lifetime employment was 42 years for cases and 43 years for controls. The self-reported job histories obtained from both cases and controls were used to estimate exposure to MMVFs and asbestos and to classify individuals as either exposed or not exposed to these fibres. The investigators did not differentiate between the types of MMVF to which the men were exposed. A statistically significant increase in risk for mesothelioma (odds ratio, 3.08; 95% CI, 1.17–8.07; 55 cases) was reported for individuals classified as ever exposed to MMVF (versus never exposed) and adjusted for exposure to asbestos. In addition, the odds ratios for MMVF exposure, adjusted for level of asbestos exposure, defined as geometric mean × 5 (fibre–year) showed an increased risk with increasing level of exposure to MMVFs. Of the 125 cases and 125 controls, the investigators estimated that only two cases and two controls were exposed to MMVFs alone (i.e. in the absence of exposure to asbestos) (odds ratio, 15.1; 95% CI, 1.05–218) compared with nine cases and 65 controls who reported no exposure to either asbestos or MMVFs.

[The Working Group noted the likely misclassification of exposure due to information bias and errors in exposure assessment in these analyses leading to possible imprecision in the adjustment for exposure to asbestos when examining an independent effect of MMVFs, and the potential imprecision in the risk estimates resulting from the small number of cases exposed to MMVFs only. Concerns about the catchment areas for cases and controls were also expressed.]

(c) *Cancer at sites other than the lung and pleura*

Marchand *et al.* (2000) studied hospital-based cases and controls in France to assess the relation between laryngeal and hypopharyngeal cancer and exposure to asbestos and MMVFs. The study subjects included 296 men with incident cancers of the larynx and 201 with incident cancers of the hypopharynx out of 664 patients newly diagnosed with squamous-cell carcinoma between 1989 and 1991. The 295 controls had been diagnosed with a non-respiratory cancer. A detailed job history and information on smoking and alcohol consumption were recorded. A job–exposure matrix was used to estimate exposure to asbestos and to four groups of MMVF (rock (stone) wool/slag wool/glass wool, refractory ceramic fibres, continuous glass filament and microfibres). In the group ever exposed to rock (stone) wool/slag wool and glass wool, the odds ratio for laryngeal cancer, after adjustment for smoking and alcohol consumption, was 1.33 (95% CI, 0.91–1.95; 130 cases) and the odds ratio for hypopharyngeal cancer was 1.55 (95% CI, 0.99–2.41; 99 cases). When the odds ratios were adjusted for exposure to asbestos, the highest odds ratio for rock (stone) wool and slag wool was for cancer of the epilarynx (odds ratio, 1.61; 95% CI, 0.85–3.04). For continuous glass filament, the odds ratios were 0.44 (95% CI, 0.15–1.31; 8 cases) for laryngeal cancer and 0.91 (95% CI, 0.30–2.76; 8 cases) for hypopharyngeal cancer. The odds ratios for exposure to refractory ceramic fibres were 1.28 (95% CI, 0.51–3.22; 16 cases) for laryngeal cancer and 0.78 (95% CI, 0.26–2.38; 7 cases) for hypopharyngeal cancer.

In the same study population as described for the Canadian case–control study of lung cancer by Siemiatycki (1991), Goldberg *et al.* (2001) investigated the association between exposure to MMVFs and colon cancer. The study included 497 male patients with a pathologically confirmed diagnosis of colon cancer, 1514 controls with cancers at other sites (excluding several sites including the lung and peritoneum) and 533 population-based controls, frequency matched on age during 1979–85. There was a non-significant increase in the risk for colon cancer (odds ratio, 1.9; 95% CI, 0.8–4.6; 8 cases), adjusted for age, smoking, demographic factors and occupational exposure and compared with both control groups combined, for the group with substantial exposure (i.e. > 5 years at medium or high levels) to rock (stone) wool or slag wool. For the same group, a non-significant increase in the risk of colon cancer was reported (odds ratio, 1.9; 95% CI, 0.7–5.4; 6 cases).

2.5.3 *Registry-based studies*

Weiderpass *et al.* (1999) and Vasama-Neuvonen *et al.* (1999) reported results from a study that linked incidence of breast cancer (23 638 cases) (Weiderpass *et al.*, 1999) and ovarian cancer (5072 cases) (Vasama-Neuvonen *et al.*, 1999) with 324 job titles from the 1970 census (for 892 591 Finnish women) during 1971–95. The job titles were classified with respect to 33 agents by means of a measurement-based, period-specific, national job–exposure matrix. The data were analysed by fitting Poisson regression models and adjusted for birth cohort, follow-up period, socioeconomic status, mean number of children, mean age at delivery of first child and turnover rate for each job title. For postmenopausal breast cancer, Weiderpass *et al.* (1999) reported an SIR of 1.32 (95% CI, 1.05–1.66) for medium to high levels of exposure to MMVFs. Vasama-Neuvonen *et al.* (1999) reported an elevated risk for ovarian cancer for women exposed to MMVFs (SIR, 1.3; 95% CI, 0.9–1.8). [The Working Group noted that the usefulness of the results was limited by potential misclassification of exposure and that the job–exposure matrix was not gender-specific and may not have reflected the relevant routes of exposure for the cancers under study.]

3. Studies of Cancer in Experimental Animals

3.1 Continuous glass filament

3.1.1 *Intraperitoneal injection* (see Table 59)

Rat: Groups of 28–50 female Wistar rats, 12–15 weeks old, were treated with one of three test fibres (ES3, ES5 or ES7) [chemical composition presumed to be E-glass] derived from continuous filaments and administered as a single (10 mg) or two weekly (2 × 20 mg) intraperitoneal injections in 2 mL saline or by intraperitoneal laparotomy (50-mg and 250-mg doses). The median dimensions and doses of the glass filament test fibres were as follows: ES3, 16.5 μm × 3.7 μm, 50 and 250 mg/rat; ES5, 39 μm × 5.5 μm, 10, 40 and 250 mg/rat; ES7, 46 μm × 7.4 μm, 40 mg/rat. Groups of control animals were similarly treated with saline only, ground glass (40, 50 or 250 mg) or UICC/A chrysotile asbestos (6 or 25 mg). Animals were observed for life. Median survival times were 111, 107, 121 and 119 weeks for the groups given 10 mg ES5, 40 mg ES5, 40 mg ES7 and 40 mg ground glass, respectively. The corresponding mean survival times of animals with tumours were 106, 119, 126 and 129 weeks, respectively. No statistically significant increase in the incidence of abdominal tumours was observed in the groups treated intraperitoneally with ES3, ES5 or ES7 compared with controls treated with saline or ground glass. The tumour incidences were (from low to high dose): for ES3, 6% and 9%; for ES5, 4%, 11% and 7%; for ES7, 2%; for ground glass, 4–8%; for saline, 4–6%; and for chrysotile, 77–81% (Pott *et al.*, 1987). [The Working Group noted that administration by laparotomy shortened the survival times and that the number of fibres injected was much smaller in these studies (as these fibres were relatively coarse (i.e. > 3 μm in diameter)) than in studies on glass wool carried out in the same laboratory.]

3.2 Glass wool

3.2.1 *Inhalation exposure* (see Table 60)

(a) Rat

Groups of 46 young adult male Sprague-Dawley rats [age at outset not stated] were exposed by whole-body inhalation to ball-milled glass fibre [chemical composition not

Table 59. Study of carcinogenicity of glass filaments administered by intraperitoneal injection in female Wistar rats

Fibre type	Median diameter (μm)	Median length (μm)	Mass (mg)	No. of tumours	No. of rats	Percentage of tumours	Age in weeks
E-glass, continuous filament (ES3)	3.7	16.5	50	3	48	6	15
	3.7	16.5	250	4	46	9	
E-glass, continuous filament (ES5)	5.5	39	10	2	50	4	12
	5.5	39	40	5	46	11	
	5.5	39	250	2	28	7	
E-glass, continuous filament (ES7)	7.4	46	40	1	47	2	
Saline (control)			1 mL	2	32	6	5
			4 mL	2	45	4	15
Ground glass (control)			40	2	45	4	12
			50	4	48	8	15
			250	4	48	8	
Chrysotile (positive control)	0.15	9	6	26	34	77	12
	0.15	9	25	25	31	81	

From Pott et al. (1987)

given] (700 fibres > 5 μm/cm^3; 24% with diameters < 3 μm) or to amosite asbestos (3100 fibres > 5 μm/cm^3; 38% with diameters < 3 μm) for 5 h per day on five days per week for three months followed by observation for 21 months. [The aerosol concentrations estimated as WHO fibres/cm^3 were approximately 168 for glass fibre and 1178 for amosite.] A group of 46 untreated rats served as negative controls. Four to 10 rats per exposure group were killed at 20 and 50 days; 3, 6, 12 and 18 months; and the remainder at 24 months. No pulmonary tumours were observed in any of the rats that died or were killed prior to 24 months. The incidences of tumours observed in rats killed at 24 months were not significantly above the level typical for this species: glass fibre, 2/11 (adenomas); amosite asbestos, 3/11 (two adenomas and one carcinoma); controls, 0/13 (Lee et al., 1981). [The Working Group noted the short period of exposure and the small number of animals involved. No fibres tested were longer than 10 μm and the study was therefore inconclusive.]

Groups of 24 male and 24 female Wistar IOPS AF/Han rats, 8–9 weeks of age, were exposed by whole-body inhalation to dust at concentrations of 5 mg/m^3 (respirable particles) from French (Saint Gobain) glass fibre [type and composition not stated] (42% of fibres < 10 μm in length, 69% < 1 μm in diameter; 48 WHO fibres/cm^3) or a Canadian chrysotile fibre (6% respirable fibres > 5 μm in length; 5901 WHO fibres/cm^3) for 5 h per day on five days per week for 12 or 24 months. An unspecified number of rats were

Table 60. Chronic inhalation studies of insulation glass wools and special-purpose glass fibres in rodents (rats and hamsters)

Test substance	Aerosol fibres (numbers and dimensions)	Positive control	Test system (no. at risk); observation time	Exposure	Lung dose	No. of thoracic tumours/no. of animals	Comments (positive control tumour incidence)	Reference
Insulation glass wools								
Rat								
Glass fibre	~168 WHO f/cm^3	Amosite, ~1178 WHO f/cm^3	Sprague-Dawley rats, male; 24 mo	Whole-body, 5 h/d, 5 d/wk, for 3 mo		2/11 (adenomas)	Short exposure period, small number (11) of animals at risk (amosite: 3/12 tumours)	Lee et al. (1981)
Glass fibre + resin	240 WHO f/cm^3 (10 f L > 20 μm, 10 mg/m^3 respirable dust; D, 52% < 1 μm; L, 72% 5–20 μm	Chrysotile, 10 mg/m^3, 3800 WHO f/cm^3	56 (48) SPF Fischer rats, male and female; 24 mo; lifetime	Whole-body, 7 h/d, 5 d/wk, for 12 mo	mg f/lung: 1.9 at 12 mo 0.5 at 21 mo	1/48 (adenocarcinoma); fibrosis, 0	Type of glass fibre not specified (chrysotile: 12/48 tumours; fibrosis)	Wagner et al. (1984)
Glass fibre − resin	323 WHO f/cm^3 (19 f L > 20 μm, 10 mg/m^3 respirable dust; D, 47% < 1 μm; L, 58% 5–20 μm	Chrysotile, 10 mg/m^3, 3800 WHO f/cm^3	56 (47) SPF Fischer rats, male and female; 24 mo; lifetime	Whole-body, 7 h/d, 5 d/wk, for 12 mo	mg f/lung: 0.9 at 12 mo 0.2 at 21 mo	1/47 (adenoma); fibrosis, 0	Type of glass fibre not specified (chrysotile: 12/48 tumours; fibrosis)	Wagner et al. (1984)
Glass fibre; French (Saint Gobain)	48 WHO f/cm^3, 5 mg/m^3, respirable dust; D, 69% < 1 μm; L, 42% > 10 μm	Chrysotile, 5 mg/m^3; L, 6% > 5 μm	45 Wistar rats, male/female; 28 mo	Whole-body for 24 mo	ND	1/45	Type of glass fibre not specified (chrysotile: 9/47 tumours; fibrosis)	Le Bouffant et al. (1984)
Insulsafe II building insulation	30 f/cm^3; L, > 10 μm; D, < 1 μm; 100 total f/cm^3; 10 mg/m^3; D, 1.4 μm mean; GMD, 1.2 μm; L, 37 μm mean; GML, 24 μm	Crocidolite, 3000 total f/cm^3, 90 f/cm^3 (L, > 10 μm)	52 Osborne-Mendel rats, female; lifetime	Nose-only, 6 h/d, 5 d/wk, for 24 mo	3 × 10^4 f/mg dry lung at 3 mo	Tumours: 0/52; fibrosis, 0	Aerosolized fibres were short. Asbestos: low tumour incidence (3/57; ~5%) + fibrosis (some)	Smith et al. (1987)
Manville 901 building insulation	232 WHO f/cm^3 (73 f L > 20 μm), 30 mg/m^3; D, 1.4 μm mean; GMD, 1.3 μm; L, 16.8 μm mean; GML, 13.1 μm	Chrysotile, 10 mg/m^3, 10 600 WHO f/cm^3	140 (119) Fischer rats, male; lifetime	Nose-only, 6 h/d, 5 d/wk, for 24 mo	f L > 5 μm/ lung: 42 × 10^6 at 12 mo; 82 × 10^6 at 24 mo f L > 20 μm/ lung: 3 × 10^6 at 12 mo; 5 × 10^6 at 24 mo	Lung tumours: 7/119 (1 carcinoma); fibrosis, 0	Chrysotile: 13/69 lung tumours; 1/69 mesothelioma	Hesterberg et al. (1993); Hesterberg & Hart (2001)

Table 60 (contd)

Test substance	Aerosol fibres (numbers and dimensions)	Positive control	Test system (no. at risk); observation time	Exposure	Lung dose	No. of thoracic tumours/no. of animals	Comments (positive control tumour incidence)	Reference
Insulsafe II building insulation	246 WHO f/cm^3 (90 f L > 20 μm), 30 mg/m^3; D, 0.9 μm mean; GMD, 0.7 μm; L, 18.3 μm mean; GML, 13.7 μm	Chrysotile, 10 mg/m^3, 10 600 WHO f/cm^3	140 (112) Fischer rats, male; lifetime	Nose-only, 6 h/d, 5 d/wk, for 24 mo	f L > 5 μm/lung: 69 × 10^6 at 12 mo; 182 × 10^6 at 24 mo; f L > 20 μm/lung: 7 × 10^6 at 12 mo; 6 × 10^6 at 24 mo	Lung tumours: 3/112; fibrosis, 0	Chrysotile: 13/69 lung tumours; 1/69 mesothelioma	Hesterberg et al. (1993); Hesterberg & Hart (2001)
Manville building insulation	25 f/cm^3; L, > 10 μm; D, < 1 μm; 100 total f/cm^3; 12 mg/m^3; D, 1.4 μm mean; GMD, 1.1 μm; L, 31 μm mean; GML, 20 μm	Crocidolite, 3000 f/cm^3, total f/cm^3, 90 f/cm^3; L, > 10 μm	57 Osborne-Mendel rats, female; lifetime	Nose-only, 6 h/d, 5 d/wk, for 24 mo	2000 f/mg dry lung at 3 mo	Tumours: 0/57; fibrosis, 0	Aerosolized fibre concentration was very low. Asbestos: few tumours (5%) + fibrosis (some)	Smith et al. (1987)
Owens Corning building insulation	5 f/cm^3; L, > 10 μm; D, < 1 μm; 25 total f/cm^3; 9 mg/m^3; D, 3 μm mean; GMD, 3 μm; L, 114 μm mean; GML, 83 μm	Crocidolite, 3000 f/cm^3 total f/cm^3; 90 f/cm^3; L, > 10 μm	58 Osborne-Mendel rats, female; lifetime	Nose-only, 6 h/d, 5 d/wk, for 24 mo	600 f/mg dry lung at 3 mo	Tumours: 0/58; fibrosis, 0	Aerosolized fibre concentration was very low and most fibres were very coarse and thick. Asbestos: few tumours (5%) + fibrosis (some)	Smith et al. (1987)
Owens Corning building insulation	5 and 15 mg/m^3, no fibre dimensions; no aerosol concentrations specified	None	500 Fischer 344 rats; lifetime	Whole-body, 7 h/d, 5 d/wk for 86 wks	ND	Tumour, 0/500; fibrosis, 0	No data on concentrations of fibres in aerosol or in the lung. No positive asbestos control	Moorman et al. (1988)

Hamster

Test substance	Aerosol fibres (numbers and dimensions)	Positive control	Test system (no. at risk); observation time	Exposure	Lung dose	No. of thoracic tumours/no. of animals	Comments (positive control tumour incidence)	Reference
Insulsafe II building insulation	30 f/cm^3; L, > 10 μm; D, < 1 μm; 100 total f/cm^3; 10 mg/m^3; D, 1.4 μm mean; GMD, 1.2 μm; L, 37 μm mean; GML, 24 μm	Crocidolite, 3000 total f/cm^3, 90 f/cm^3 (L, > 10 μm)	60 Syrian golden hamsters, male; lifetime	Nose-only, 6 h/d, 5 d/wk, for 24 mo	1 × 10^4 f/mg dry lung at 3 mo	Tumours: 0/60; fibrosis, 0	Aerosolized fibres were short. Asbestos-exposed animals had no tumours (some did have fibrosis)	Smith et al. (1987)
Glass fibre	~168 WHO f/cm^3	Amosite ~1178 WHO f/cm^3	Hamsters; 24 months	Whole-body, 5 h/d, 5 d/wk, for 3 mo	ND	0/9	Short exposure period, small number (9) of animals at risk (amosite: 0/5 tumours)	Lee et al. (1981)

Table 60 (contd)

Test substance	Aerosol fibres (numbers and dimensions)	Positive control	Test system (no. at risk); observation time	Exposure	Lung dose	No. of thoracic tumours/no. of animals	Comments (positive control tumour incidence)	Reference
Manville 901 building insulation	339 WHO f/cm³ (134 fL > 20 µm), 30 mg/m³; GMD, 0.84 µm; GML, 12.4 µm	Amosite (mid), 165 WHO f/cm³ (38 fL > 20 µm); amosite (high), 263 WHO f/cm³ (69 fL > 20 µm)	125 (81) Syrian golden hamsters, male; lifetime	Nose-only, 6 h/d, 5 d/wk, for 18 mo	fL > 5 µm/lung: 32 × 10⁶ at 12 mo; 77 × 10⁶ at 18 mo fL > 20 µm/lung: 1 × 10⁶ at 12 mo; 5 × 10⁶ at 18 mo	Lung tumours or mesotheliomas, 0/81; fibrosis, 0	Amosite (mid): 22/85 mesotheliomas, no lung tumours, fibrosis Amosite (high) 17/87 mesotheliomas, no lung tumours, fibrosis	Hesterberg et al. (1999); McConnell et al. (1999)
Manville building insulation	25 f/cm³; L, > 10 µm; D, < 1 µm; 100 total f/cm³; 12 mg/m³; D, 1.4 µm mean; GMD, 1.1 µm; L, 31 µm mean; GML, 20 µm	Crocidolite, 3000 f/cm³, total f/cm³; 90 f/cm³; L, > 10 µm	66 Syrian golden hamsters, male; lifetime	Nose-only, 6 h/d, 5 d/wk, for 24 mo	1000 f/mg dry lung at 3 mo	Tumours; 0/66; fibrosis, 0	Aerosolized fibre concentration was very low. Asbestos-exposed animals had no tumours (some did have fibrosis)	Smith et al. (1987)
Owens Corning building insulation	5 f/cm³; L, > 10 µm; D, < 1 µm; 25 total f/cm³; 9 mg/m³; D, 3 µm mean; GMD, 3 µm; L, 114 µm mean; GML, 83 µm	Crocidolite, 3000 f/cm³, total f/cm³; 90 f/cm³; L, > 10 µm	61 Syrian golden hamsters, male; lifetime	Nose-only, 6 h/d, 5 d/wk, for 24 mo	500 f/mg dry lung at 3 mo	Tumours: 0/61; fibrosis, 0	Aerosolized fibre concentration was very low and most fibres were very coarse and thick. Asbestos-exposed animals had no tumours (some did have fibrosis)	Smith et al. (1987)
Special-purpose glass fibres								
Rat								
JM 100	10 mg/m³ respirable dust; D, 0.3 µm; L, 71% < 10 µm	Chrysotile, 10 mg/m³	100 Fischer rats, 50 male, 50 female; lifetime	Whole-body, 7 h/d, 5 d/wk, for 12 mo	ND	0/55; fibrosis, 0	JM 100 fibres were short (chrysotile: 11/56 tumours; fibrosis)	McConnell et al. (1984)
JM 100	1436 WHO f/cm³ (108 fL > 20 µm), 10 mg/m³ respirable dust; D, 97% < 1 µm; L, 93% 5–20 µm	Chrysotile, 10 mg/m³, 3800 WHO f/cm³	56 (48) SPF Fischer rats, male and female; 24 mo; lifetime	Whole-body, 7 h/d, 5 d/wk, for 12 mo	mg f/lung: 4.5 at 12 mo 2.1 at 21 mo	1/48 (adenocarcinoma); fibrosis, 0	JM 100 fibres were short (chrysotile: 12/48 tumours; fibrosis)	Wagner et al. (1984)

Table 60 (contd)

Test substance	Aerosol fibres (numbers and dimensions)	Positive control	Test system (no. at risk); observation time	Exposure	Lung dose	No. of thoracic tumours/no. of animals	Comments (positive control tumour incidence)	Reference
JM 100	332 WHO f/cm^3, 5 mg/m^3 respirable dust; D, 95% < 1 μm; L, 60% > 10 μm, 25% > 20 μm	Chrysotile, 5 mg/m^3, 6000 WHO f/cm^3	48 Wistar rats, male and female; 28 mo	Whole-body, 5 h/d, 5 d/wk, for 24 mo	ND	0/48	JM 100 fibres were short (chrysotile: 9/47 tumours; fibrosis)	Le Bouffant et al. (1984; 1987)
JM 104/475	252 WHO f/cm^3, 3 mg/m^3; D, 0.42 μm median; L, 4.8 μm median	Crocidolite, 162 WHO f/cm^3, chrysotile, 131 WHO f/cm^3	108 Wistar rats, female; lifetime	Nose-only, for 12 mo	WHO f × 10^6: 70 at 12 mo 25 at 24 mo	1/107	JM 104/475 fibres were short (crocidolite: 1/50 tumours; chrysotile: 0/50 tumours)	Muhle et al. (1987)
Manville Code 100	530 f/cm^3; L, > 10 μm; D, ≤ 1 μm; 3000 total f/cm^3; 3 mg/m^3; D, 0.4 μm mean; GMD, 0.4 μm; L, 7.5 μm mean; GML, 4.7 μm	Crocidolite, 3000 f/cm^3, total f/cm^3, 90 f/cm^3; L, > 10 μm	57 Osborne-Mendel rats, female; lifetime	Nose-only, 6 h/d, 5 d/wk, for 24 mo	2 × 10^6 f/mg dry lung at 2 mo	Tumours: 0/57; fibrosis, 0	Low survival rates: < 50% of rats survived to 24 mo (including controls). Asbestos: few tumours (3/57; 5%) + fibrosis	Smith et al. (1987)
JM 475	5 and 15 mg/m^3; no fibre dimensions; no aerosol concentrations specified	None	500 Fischer 344 rats; 18 mo; lifetime	Whole-body, 7 h/d, 5 d/wk, for 86 wks	ND	Tumours, 0; fibrosis, 0	No data on fibre concentrations in aerosol or in the lung. No positive asbestos control	Moorman et al. (1988)
JM 475	1066 WHO f/cm^3 (38 f > 20 μm/cm^3); 0.2 μm < D (more than 60% of fibres) < 0.4 μm	Amosite, 981 WHO f/cm^3 (89 f > 20 μm/ cm^3)	24 mo, 83 (38) Wistar rats	Whole body, 7 h/d, 5 d/wk, for 12 mo	2241 × 10^6 WHO f/lung; 11 × 10^6 f > 20 μm/lung	4/38 (adenomas); fibrosis: negligible	Authors concluded tumours/fibrosis similar to that of controls (5% tumours in air control).	Davis et al. (1996a); Cullen et al. (2000)
104 E	975 WHO f/cm^3 (72 f > 20 μm/cm^3); 0.2 μm < D (more than 60% of fibres) < 0.4 μm	Amosite, 981 WHO f/cm^3 (89 f > 20 μm/ cm^3)	24 mo, 83 (43) Wistar rats	Whole body, 7 h/d, 5 d/wk, for 12 mo	2356 × 10^6 WHO f/lung; 83 × 10^6 f > 20 μm/lung	12/43 (7 carcinomas, 3 adenomas and 2 mesotheliomas); fibrosis: +	Authors concluded that amosite and 104 E were fibrogenic and tumorigenic.	Davis et al. (1996a); Cullen et al. (2000)

Table 60 (contd)

Test substance	Aerosol fibres (numbers and dimensions)	Positive control	Test system (no. at risk); observation time	Exposure	Lung dose	No. of thoracic tumours/no. of animals	Comments (positive control tumour incidence)	Reference
Hamster								
Manville Code 100	530 f/cm^3; L, > 10 μm; D, ≤ 1 μm; 3000 total f/cm^3; 3 mg/m^3; D, 0.4 μm mean; GMD, 0.4 μm; L, 7.5 μm mean; GML, 4.7 μm	Crocidolite, 3000 f/cm^3, total f/cm^3; 90 f/cm^3; L, > 10 μm	69 Syrian golden hamsters, male; lifetime	Nose-only, 6 h/d, 5 d/wk, for 24 mo	1 × 10^6 f/mg dry lung at 2 mo	Tumours: 0/69; fibrosis, 0	Low survival rates: < 25% of hamsters survived to 24 mo (including controls). Asbestos-exposed animals had no tumours (some did have fibrosis).	Smith et al. (1987)
JM 475	310 WHO f/cm^3 (109 f L > 20 μm), 37 mg/m^3; GMD, 0.70 μm; GML, 11.8 μm	Amosite (mid): 165 WHO f/cm^3 (38 f L > 20 μm), 3.5 mg/m^3; amosite (high) 263 WHO f/cm^3 (69 f L > 20 μm), 7 mg/m^3	125 (83) Syrian golden hamsters, male; lifetime	Nose-only, 6 h/d, 5 d/wk, for 18 mo	f L > 5 μm/lung: 49 × 10^6 at 12 mo 234 × 10^6 at 18 mo f L > 20 μm/lung: 6 × 10^6 at 12 mo 30 × 10^6 at 18 mo	Lung tumours: 0/83; mesothelioma, 1/83; fibrosis, 0	Amosite (mid): 22/85 mesotheliomas, no lung tumours, fibrosis Amosite (high) 17/87 mesotheliomas, no lung tumours, fibrosis	Hesterberg et al. (1999); McConnell et al. (1999)

f, fibre; L, length; D, diameter; total f, any particle having L:D ≥ 3; f > 20, fibres with length > 20 μm; f/cm^3, no. of fibres per cm^3 of air. Some concentrations were expressed as fibres > 5 μm length, others as total fibres; authors did not always specify. GMD, geometric mean diameter; GML, geometric mean length; lifetime: until survival rate of air controls is ≤ 20%; typically ~30 months. Thoracic tumours: lung tumours, including adenomas and carcinomas; WHO, respirable fibres as defined by World Health Organization, L > 5 μm, D < 3 μm, L:D ≥ 3; h, hour; d, day; wk, week; mo, month; ND, not determined killed either immediately after treatment or after various periods of observation (4, 7, 12 and 16 months after exposure for rats exposed for 12 months and 4 months after exposure for rats exposed for 24 months). The incidences of pulmonary carcinoma in the various treatment groups were 1/45 (French glass fibre), 9/47 (chrysotile) and 0/47 (control rats) (Le Bouffant et al., 1984). [The Working Group noted that, because data on survival were not reported, the exact incidence of tumours could not be ascertained.]

Groups of 56 SPF Fischer rats (equal numbers of males and females) [age unspecified] were exposed by inhalation to dust at concentrations of approximately 10 mg/m^3 glass fibre or chrysotile for 7 h per day on five days per week for 12 months (cumulative exposure, 17 500 mg × h/m^3 for each group). The samples of fibrous dust used (and the size distributions of airborne fibres > 5 μm in length) were: glass fibre with resin coating [chemical composition not given] (72% fibres < 20 μm in length; 52% < 1 μm in diameter; 240 WHO fibres/cm^3), glass fibre without resin coating [source unspecified] (58% < 20 μm in length; 47% < 1 μm in diameter; 323 WHO fibres/cm^3) and UICC Canadian chrysotile (39% > 10 μm in length; 29% > 0.5 μm in diameter). To study dust retention, groups of six rats per exposure group were killed at the end of the exposure period or one year after the end of exposure. The remainder were kept until natural death [survival times not reported]. During the period 500–1000 days after the start of exposure, the incidences of pulmonary adenomas and carcinomas were: 1/48 (adenocarcinoma; glass fibre with resin), 1/47 (adenoma; glass fibre without resin), 12/48 (1 adenoma and 11 adenocarcinomas; chrysotile) and 0/48 (untreated controls) (Wagner et al., 1984). [The Working Group noted that because data on survival were inadequate, the exact incidence of tumours could not be ascertained.]

Groups of 52–61 female Osborne-Mendel rats, 100 days old, were exposed by inhalation (nose-only) to dusts from various types of glass wool and special purpose fibres, UICC crocidolite asbestos or clean air (negative controls) for 6 h per day on five days per week for two years and then observed for life. A group of 125 untreated rats served as additional negative controls. The three glass wools tested were:

— CT loose 'blowing wool' building insulation Insulsafe II [chemical composition similar to MMVF11] (mean dimensions of aerosol fibres, 37 μm × 1.4 μm; 4.4 mg/m^3; 100 total fibres/cm^3; 30 fibres/cm^3 with length > 10 μm and diameter < 1 μm);

— Manville building insulation with binder [chemical composition similar to MMVF10] (mean dimensions of aerosol fibres, 31 μm × 1.4 μm; 9.9 mg/m^3; 100 total fibres/cm^3; 25 fibres/cm^3 with length > 10 μm and diameter < 1 μm); and

— Owens Corning binder-coated appliance insulation [chemical composition not given] (mean dimensions of aerosol fibres, 3.0 μm × 114 μm; mass concentration, 7.0 mg/cm^3; 19% respirable; 25 total fibres/cm^3; 5 fibres/cm^3 > 10 μm in length and ≤ 1.0 μm in diameter).

The mass of human respirable fibre was estimated for each aerosol from data obtained from nylon cyclone air samplers. [The Working Group noted that no information was available on respirability in the rat and that for the Manville building insulation and the Owens Corning appliance insulation fibres, the retained lung burden was relatively low.] No pulmonary tumours were observed in any of the groups exposed to any of the types of glass fibre or in the controls. There was no effect of glass fibre on survival and little pulmonary cellular change. The tumour incidences in rats exposed to UICC crocidolite asbestos were: 3/57 rats (one mesothelioma and two carcinomas) (Smith et al., 1987). [The Working Group noted that the counting criteria for the retained fibres were not stated.]

A total of 500 Fischer 344 rats, five weeks of age, were exposed in whole-body chambers to Owens Corning insulation glass wool [chemical composition not given] or air filtration fibre glass fibres for 7 h per day on five days per week for 86 weeks. The target exposure concentration was 15 mg/m^3 for both treatments. [The Working Group noted that the physical dimensions (length and diameter) of the fibres and the numbers of fibres in the exposure aerosol and the lungs were not characterized and the lung burden was not given.] No lung fibrosis, lung cancer or mesotheliomas were observed in the rats exposed to fibres. The incidence of mononuclear-cell leukaemia was statistically elevated in the groups exposed to fibres when compared with the air control group. However, this tumour type is observed at relatively high incidences (22% in the current study) in ageing Fischer 344 rats (Moorman et al., 1988).

Groups of 140 male Fischer 344/N rats, eight weeks of age, were exposed by nose-only inhalation to glass wool building insulation fibres (Manville 901 (MMVF10) or CertainTeed® Insulsafe II (MMVF11)) for 6 h per day on five days per week for up to two years and observed for life (until 20% survival). The target aerosol concentrations were 3, 16 or 30 mg/m^3. At a concentration of 30 mg/m^3, MMVF10 contained an average of 232 WHO fibres/cm^3 and MMVF11, 246 WHO fibres/cm^3 (including 73 and 90 fibres/cm^3 longer than 20 μm, respectively). The average fibre dimensions were 1.4 μm × 16.8 μm (MMVF10) and 0.9 μm × 18.3 μm (MMVF11). The negative control rats were exposed under similar conditions to filtered air. Another group of rats was exposed to size-selected NIEHS medium-chrysotile asbestos at 10 mg/m^3 (10.6 × 10^3 WHO fibres/cm^3, no detectable fibre > 20 μm). Lung fibre burdens and lung pathology and histopathology were evaluated after 3, 6, 12, 18 and 24 months of exposure and after various combinations of exposure and recovery periods (3 + 21, 6 + 18, 12 + 12, 18 + 6, 24 + 6 months; no exposure to fibres occurred during the recovery period). Lung fibre burdens were dose- and time-dependent, increasing with both level of exposure and duration. After 24 months of exposure to ~250 WHO fibres/cm^3, the lung fibre burdens, expressed as fibres > 20 μm/lung × 10^6 were 5 and 6 for MMVF10 and MMVF11, respectively. During exposure of rats to either of the two glass wools, dose-related and time-dependent lung inflammation was observed; however, after a post-exposure recovery period of several months, the lungs showed no evidence of inflammation. In the rats exposed to glass wool, no lung fibrosis or mesothelial tumours were seen and the incidences of lung tumours were not significantly elevated over background levels for this species (range, 0.8–7.5% in 112–120 rats at risk; 3.3% in concurrent air controls). The incidences of tumours in the rats exposed to chrysotile were 13/69 (18.9%) lung tumours and 1/69 (1.4%) pleural mesotheliomas (Hesterberg et al., 1993; Hesterberg & Hart, 2001).

(b) Hamster

Groups of 60–70 male Syrian golden hamsters, 100 days old, were examined after exposure by nose-only inhalation to dusts of various types of glass wool and special-purpose fibres, UICC crocidolite asbestos or clean air (negative controls) for 6 h per day on five days per week for two years, and were then observed for life. A group of 125 untreated hamsters served as additional negative controls. The three glass wools tested were:
— CT loose 'blowing wool' building insulation Insulsafe II [chemical composition similar to MMVF11] (37 µm × 1.4 µm; 4.4 mg/m^3; 100 total fibres/cm^3);
— Manville building insulation with binder [chemical composition similar to MMVF10] (31 µm × 1.4 µm; 9.9 mg/m^3; 100 total fibres/cm^3); and
— Owens Corning binder-coated appliance insulation [chemical composition not given] (3.0 µm diameter; mass concentration, 7.0 mg/cm^3; 19% respirable, 25 fibres/cm^3 with 5 fibres/cm^3 > 10 µm in length and ≤ 1.0 µm in diameter).

A second group of 38 animals was also exposed to Owens Corning binder-coated appliance insulation because of a high death rate in the first group that was unrelated to exposure to fibres. No pulmonary tumours were observed in the group exposed to glass fibre or the room-control group. The incidences of pulmonary tumours in the other groups were: 1/58 (chamber controls; carcinoma) and 0/58 (crocidolite). None of the types of glass fibre affected survival or caused pulmonary lesion (Smith et al., 1987). [The Working Group noted that the counting criteria for the retained fibres were not stated and that the lifetime of the hamsters was surprisingly long.]

Groups of 30–35 hamsters [sex, strain and age unspecified] were exposed by whole-body inhalation to glass fibre (700 fibres/cm^3 > 5 µm in length; 24% with length < 3 µm) or amosite asbestos (3100 fibres/cm^3 > 5 µm in length; 38% with length < 3 µm) [no information on diameters was given] for 5 h per day on five days per week for three months and were then observed for 21 months. [Aerosol concentrations estimated as WHO fibre/cm^3 (length > 5 µm, diameter < 3 µm, length/diameter ≥ 3) were approximately 168 for glass fibre and 1178 for amosite.] One group of 30 unexposed animals served as controls. Groups of 1–8 animals per exposure group were killed at 50 days and 3, 6, 12 and 18 months, and the remainder at 24 months. No pulmonary tumour was observed in any group (0/9 in the exposed group killed at 24 months) (Lee et al., 1981). [The Working Group noted the short exposure period and the small number of animals evaluated.]

A total of 125 male Syrian golden hamsters, 13–15 weeks of age (~145 g bw), were exposed by nose-only inhalation to building insulation glass fibre (Manville 901, coded MMVF10a) at a concentration of 30 mg/m^3 (339 WHO fibres/cm^3, including 134 fibres > 20 µm/cm^3) for 6 h per day on five days per week for up to 78 weeks (18 months) and observed for life (i.e. until 20% survival; ~88 weeks). The MMVF10a sample was a thinner version of the original MMVF10 used in the chronic inhalation studies in rats summarized above. In order to compare results obtained using MMVF10a with those of a known human carcinogen, three additional groups of 125 hamsters were concurrently exposed to amosite asbestos at the following aerosol concentrations: 36 (10 fibres > 20 µm), 165 (38 fibres > 20 µm) and 263 (69 fibres > 20 µm) WHO fibres/cm^3. The geometric mean dimensions of the aerosol fibres were 0.84 µm × 12.4 µm for MMVF10a and 0.55 µm × 8.4 µm for amosite. Another group of 125 hamsters was exposed under similar conditions to filtered air only (negative controls). After 13, 26, 52 and 78 weeks of exposure and after various combinations of exposure and recovery periods, 9–10 hamsters were randomly selected from each exposure group and killed; five of the hamsters were evaluated for lung burden and for histopathology and the remaining four or five were evaluated for lung and pleural cell proliferation. Lung depositions after 6 h of exposure were fairly similar in animals exposed to MMVF10a and high dose amosite: 8.4×10^5 WHO fibres/lung (1.6 fibres > 20 µm) for MMVF10a and 20.8×10^5 WHO fibres/lung (2 fibres > 20 µm) for high-dose amosite. However, after 78 weeks of exposure, lung burdens for the two fibres were no longer similar: numbers of fibres > 20 µm/lung were 4.6×10^6 for MMVF10a but 144×10^6 for high-dose amosite; numbers of WHO fibres per lung were ninefold greater for amosite than for MMVF10a; and the number of fibres > 20 µm/lung were fivefold greater for amosite. After 78 weeks of exposure followed by six weeks of recovery, the lung burdens of MMVF10a were reduced by 66% (WHO fibres/lung) and 95% (fibres > 20 µm/lung); in contrast, the lung burdens of amosite were reduced by only 21% and 38%, respectively. Exposure to MMVF10a induced lung inflammation, which was no longer evident after a recovery period of six weeks; no pulmonary fibrosis or tumours were observed in these hamsters (81 hamsters at risk). In contrast, all three doses of amosite induced widespread lung fibrosis and 3%, 22% and 17% pleural mesotheliomas in the 36, 165 and 263 WHO fibres/cm^3 dose groups, respectively (87 hamsters at risk) (Hesterberg *et al.*, 1997, 1999; McConnell *et al.*, 1999).

3.2.2 *Intraperitoneal injection*

(*a*) Rat

Groups of female Wistar rats, 8–12 weeks of age, received either single intraperitoneal injections of 2 or 10 mg, or four weekly injections of 25 mg, German glass wool [chemical composition not given] (59% fibres < 3 µm in length), different doses of UICC chrysotile A or 100 mg of one of seven suspensions of granular dust in 2 mL saline. The German glass wool was administered at doses of 24, 120 and 1200×10^6 fibres > 5 µm in length. The animals were kept until natural death. In the groups treated with glass fibre, lesions reported by the authors as mesotheliomas and spindle-cell sarcomas were observed at incidences of 1/34, 4/36 and 23/32 for the doses of 24, 120 and 1200×10^6 fibres > 5 µm in length, respectively, with corresponding average survival times of 518, 514 and 301 days. Tumour incidences ranged from 6/37 (2 mg dose) to 25/31 (25 mg dose) in the groups treated with chrysotile, with average survival times of 468–407 days. No abdominal tumours were observed in the control animals treated with saline (Pott et al., 1976).

Groups of female Wistar rats, 8–10 weeks old, were given intraperitoneal injections containing a total dose of either 70 mg (two weekly injections) or 180 mg (six weekly injections) of MMVF11 (manufactured by CertainTeed®). The number of fibres was 0.4×10^9 or 1.0×10^9 WHO fibres/rat for the 70 mg and 180 mg doses, respectively, and the approximate median fibre dimensions were 14.6 μm × 0.77 μm. Rats were kept for life and then examined for tumours. Forty rats served as concurrent negative controls and received intraperitoneal injections of saline. The incidences of peritoneal mesotheliomas for the 70 mg and 180 mg dose groups were 12/40 (30%) and 16/23 (70%), respectively. Negative control rats did not develop any mesotheliomas (Pott, 1995; Roller et al., 1996; Roller & Pott, 1998).

Five groups of 40–53 male or female Wistar rats, 8–10 weeks old, were given 5–40 weekly intraperitoneal injections of 25 mg B-01-09 glass wool to a total dose of 125–1000 mg glass wool. The doses expressed as fibres > 5 μm/rat for each of the five groups were 2.5, 5, 10 (females), 10 (males) and 20×10^9, respectively, the highest dose being 20 times more than the highest dose recommended by the European Union guidelines (see section 1.5 for details). Approximate median fibre dimensions were 9 μm × 0.7 μm. Rats were kept for life (~30 months) and then examined for tumours. Two hundred and eight male and female rats were kept as concurrent negative controls and treated with 1–20 weekly intraperitoneal injections of saline. The incidences of peritoneal mesotheliomas for the five groups treated with glass wool were 3/40 (8%), 4/40 (11%), 3/40 (8%), 10/40 (21%) and 33/40 (66%), respectively. The incidence of abdominal tumours in negative control rats was 1/208 (Pott, 1995; Roller et al., 1996).

Four groups of 40 female Wistar rats, 8–10 weeks old, were given a series of weekly intraperitoneal injections containing either a thin or thicker version of an experimental glass wool (B-09-0.6 or B-09-2.0). The median dimensions of the two fibres were 3.3 μm × 0.49 μm and 10.5 μm × 1.2 μm, respectively. The total dose per rat (in saline) was: 100 or 300 mg (B-09-0.6) and 150 or 450 mg (B-09-2.0) (injections ranging from 2 × 50 mg to 9 × 50 mg/animal). The numbers of fibres for the four doses were: 2 and 6×10^9 WHO fibres per rat for B-09-0.6 and 1 and 3.2×10^9 WHO fibres per rat for B-09-2.0. Forty rats were kept as concurrent negative controls and were given three weekly intraperitoneal injections of 2 mL physiological saline solution. Rats were kept for life (30 months) and then examined for tumours. The incidences of peritoneal mesotheliomas for the four dose groups were 3% (100 mg B-09-0.6), 10% (300 mg B-09-0.6), 23% (150 mg B-09-2.0) and 53% (450 mg B-09-2.0). No tumours were observed in the negative control rats. Exposure to longer and thicker fibres resulted in more tumours (Roller et al., 1996, 1997).

Groups of 22–24 male Wistar rats [age unspecified] received an intraperitoneal injection of 144 mg of a glass fibre standard building insulation wool (MMVF10) or

6.1 mg of amosite asbestos and were observed for life to monitor the development of peritoneal mesotheliomas. The doses of the two fibres were 973×10^6 WHO fibres/rat (MMVF10) and 410×10^6 WHO fibres/rat (amosite). The numbers of fibres > 20 μm for each of the two fibres were 555×10^6 (including 119 with a diameter < 0.95 μm) and 71×10^6 (including 63 with a diameter < 0.95 μm), respectively. Tumours were diagnosed by the macroscopic presence of peritoneal mesotheliomas; when the diagnosis was in doubt, microscopy was performed. The incidence of tumours for MMVF10 was 13/22 (59%) and for amosite 21/24 (88%). [Estimated survival fractions for deaths from mesothelioma were reported.] In their consideration of these fibres and others tested, the authors suggested a link between the incidence of intraperitoneal tumours and two other factors:
— the number of fibres > 20 μm injected; and
— the biopersistence of fibres > 5 μm in rat lungs following intratracheal instillation (biopersistence is described more fully in section 4 of this monograph) (Miller et al., 1999).

(b) Hamster

Groups of 40 female Syrian golden hamsters, 8–12 weeks old, received single intraperitoneal injections of 2 or 10 mg glass wool (59% of fibres shorter than 3 μm) [chemical composition not given] or UICC chrysotile A in 1 mL saline. Animals were observed for life. No tumour of the abdominal cavity was found (Pott et al., 1976).

3.3 Special-purpose glass fibres

3.3.1 *Inhalation exposure*

(a) Rat

Groups of 24 male and 24 female Wistar IOPS AF/Han rats, 8–9 weeks of age, were exposed by whole-body inhalation to dust of US JM 100 glass fibre [chemical composition not given] at concentrations of 5 mg/m^3 (respirable particles) (97% respirable fibres < 5 μm in length; 43% total fibres < 0.1 μm in diameter; 332 respirable fibres > 5 μm/cm^3) or a Canadian chrysotile fibre (6% respirable fibres > 5 μm in length; 5901 WHO fibre/cm^3) for 5 h per day on five days per week for 12 or 24 months. An unspecified number of rats were killed either immediately after treatment or after a period of observation (4, 7, 12 and 16 months after exposure for rats exposed for 12 months; 4 months after exposure for rats exposed for 24 months). The incidences of pulmonary carcinoma were 9/47 (chrysotile), 0/48 (JM 100 glass fibre) and 0/47 (control rats) (Le Bouffant et al., 1984). [The Working Group noted that, because of the lack of data on survival, the exact incidence of tumours could not be ascertained.]

Two studies of similar design, A and B, using animals from the same source were conducted concurrently in different laboratories (study B was part of the study by Wagner et al. (1984), reviewed in detail below). Groups of 50 male and 50 female SPF

Fischer 344 rats, 7–8 weeks of age, were exposed by whole-body inhalation to approximately 10 mg/m^3 respirable dust [size unspecified] of JM 100 [chemical composition not given] or UICC Canadian chrysotile for 7 h per day on five days per week for 12 months and were observed for life. Fifty rats of each sex served as chamber controls. Groups of 3–5 rats per group were killed at 3, 12 and 24 months. No pulmonary tumours were observed after 3 or 12 months of exposure. The incidences of pulmonary tumours in the chrysotile-exposed male rats killed after 24 months of exposure were 2/4 (one adenoma, one adenocarcinoma; study A) and 0/4 (study B). The incidences of pulmonary tumours (adenomas and adenocarcinomas combined) in rats from study A observed for life were: chrysotile, 9/29 males and 2/27 females; JM 100 glass fibre, 0/28 males and 0/27 females; control, 3/27 males and 0/26 females. The incidences in study B were: chrysotile, 7/24 males and 5/24 females; JM 100 glass fibre, 1/24 males and 0/24 females; control, 0/24 males and 0/24 females. Thus, chrysotile asbestos was positive for carcinogenicity, while JM 100 glass was negative (McConnell *et al.*, 1984). [The Working Group noted that the dimensions of the fibres used in study A were not reported.]

Groups of 56 SPF Fischer rats (equal numbers of males and females) [age unspecified] were exposed by inhalation to dust at concentrations of approximately 10 mg/m^3 glass fibre [chemical composition not given] or chrysotile for 7 h per day on five days per week for 12 months (cumulative exposure, 17 500 mg × h/m^3 for each group). The fibrous dust samples used (and the size distributions of those airborne fibres > 5 μm in length) were: JM 100 glass fibre (93% < 20 μm in length; 97% < 1 μm in diameter; 1436 WHO fibres/cm^3) and UICC Canadian chrysotile (39% > 10 μm in length; 29% > 0.5 μm in diameter). To study dust retention, groups of six rats per exposure group were killed at the end of the exposure period and one year after the end of exposure. The remainder were kept until natural death [survival times not reported]. During the period 500–1000 days after the start of exposure, the incidence of pulmonary adenomas and carcinomas was: 1/48 (adenocarcinoma; JM 100), 12/48 (one adenoma and 11 adenocarcinomas; chrysotile) and 0/48 (untreated controls) (Wagner *et al.*, 1984). [The Working Group noted that, because of inadequate data on survival, the exact tumour incidence could not be ascertained.]

In the study by Smith *et al.* (1987), described in detail in section 3.2.1, exposure of female Osborne-Mendel rats by nose-only inhalation to Manville code 100 fibres without binder (mean dimensions of aerosol fibres, 7.5 μm × 0.4 μm; highest concentration, 2.4 mg/m^3; 3000 total fibres/cm^3; 530 fibres/cm^3, length > 10 μm, and diameter < 1 μm) for 6 h per day on five days per week for two years did not produce pulmonary tumours.

Female Wistar rats, 12 weeks old, were exposed by nose-only tubes to fibre aerosols for 5 h per day on four days per week for up to one year (total exposure time, 1000 h). One group of 108 rats was exposed to JM 104/475 glass microfibre (median fibre dimensions, 4.8 μm × 0.42 μm; 90% < 12.4 μm long) and two groups of 50 rats were exposed either to South African crocidolite (median dimensions, 1.5 μm × 0.27 μm) or to Calidria chrysotile (from California, USA; median fibre dimensions, 6.0 μm × 0.67 μm). Aerosol concentrations (expressed as mg/m^3 and as WHO fibres/cm^3) for the three fibres were as follows: JM 104/475, 3.0 mg/m^3 (252 fibres/cm^3), crocidolite, 2.2 mg/m^3 (162 fibres/cm^3) and chrysotile, 6.0 mg/m^3 (131 fibres/cm^3). Two groups of negative controls were also kept. One group comprised 55 rats exposed to clean air and the other 50 untreated rats. The lung burdens were: JM 104/475, 0.4 mg fibre/lung; crocidolite, 0.6 mg fibre/lung and chrysotile, 0.3 mg fibre/lung after six months of exposure; 0.6, 0.7 and 0.3 after 12 months of exposure; and 0.2, 0.4 and 0.03 after 12 months of exposure and a 12-month post-exposure recovery period. The proportions for pulmonary tumours were 1/107 (glass fibre; carcinoma), 1/50 rats (crocidolite; adenocarcinoma), 0/50 (chrysotile) and 0/105 (negative controls). The authors suggested that the low incidence of tumours seen after exposure to crocidolite might have been because the lung burden of < 1 mg dust was relatively low, and the absence of tumours after exposure to Calidria chrysotile might be because these fibres were less persistent than those of UICC chrysotile (Muhle et al., 1987).

In a study by Moorman et al. (1988), described in detail in section 3.2.1, whole-body exposure of 500 Fischer 344 rats to JM 475 special-purpose fibres for 7 h per day on five days per week for 86 weeks, at target exposure concentrations of 5 and 15 mg/m^3, did not produce lung fibrosis, lung cancer or mesotheliomas. The incidence of mononuclear-cell leukaemia was statistically elevated in the groups exposed to fibres when compared to the air control group, but this tumour type is typically observed at relatively high incidences (22%) in ageing Fischer 344 rats. [The Working Group noted that the physical dimensions and the number of fibres in the exposure aerosol and the lungs were not characterized and that the lung burden was not stated.]

In a series of inhalation studies (Davis et al., 1996b; Searl et al., 1999; Cullen et al., 2000), 475 and 104E glass fibres were compared with amosite asbestos. Groups of male Wistar rats were exposed to aerosols containing approximately 1000 fibres longer than 5 μm/cm^3, more than 60% of which had diameters between 0.2 and 0.4 μm. Lung burdens and some clearance data for each fibre type (by length category) were presented. The changes in chemistry of fibres during their residence in the lungs were monitored by SEM-EDXA. The numbers of thoracic tumours observed in animals at risk were: 2/38 for controls, 12/43 for 104E, 18/42 for amosite and only 4/38 for 475 glass. Mesotheliomas occurred only in the groups treated with 104E and amosite. All the tumours observed in animals treated with 475 glass were adenomas. There was little correlation between final lung burdens and severity of effect and the authors suggested a role for fibre leaching that was seen only with the relatively inert 475 glass.

(b) *Hamster*

In the study by Smith et al. (1987), described in detail in section 3.2.1, exposure of male Syrian golden hamsters by nose-only inhalation to Manville code 100 fibres without binder (mean dimensions of aerosol fibres, 7.5 μm × 0.4 μm; highest concentration, 2.4 mg/m³; 3000 total fibres/cm³; 530 fibres/cm³ with length > 10 µm and diameter < 1 µm) for 6 h per day on five days per week for two years did not affect survival and caused no pulmonary lesions or tumours.

In a study described in section 3.2.1, 125 Syrian golden hamsters were exposed by nose-only inhalation to JM 475 glass fibre (coded MMVF33) at a concentration of 37 mg/m³ (310 WHO fibres/cm³, including 109 fibres > 20 µm/cm³). The geometric mean dimensions of fibres in the aerosol were 0.7 µm × 11.8 µm. Time-points, evaluations and numbers of hamsters tested were similar to those described above. The lung deposition of MMVF33 (11.5×10^5 WHO fibres, including 2.2×10^5 fibres > 20 µm) after 6 h of exposure, and of MMVF10a and high-dose amosite was similar. However, after 78 weeks of exposure, the lung burden for MMVF33 was between those of MMVF10a and amosite asbestos, i.e. 5, 30 and 144×10^6 fibres > 20 µm/lung for MMVF10a, MMVF33 and high-dose amosite, respectively; or, expressed as WHO fibres × 10^6/lung, 77, 234 and 612, respectively. After 78 weeks of exposure plus six weeks of recovery, the lung burdens in animals treated with MMVF33 were reduced by 40–62% (compared with reductions of 66–95% in animals treated with MMVF10a and little or no significant clearance after treatment with the high dose of amosite). All hamsters exposed to MMVF33 had lung fibrosis after six months of exposure and one pleural mesothelioma was seen in 83 animals at risk (1.2%) (Hesterberg *et al.*, 1997, 1999; McConnell *et al.*, 1999).

(c) *Guinea-pig*

Groups of 31 male albino guinea-pigs [age unspecified] were exposed by whole-body inhalation to fibres > 5 µm in length, at concentrations of 700 fibres/cm³ of ball-milled glass fibre (24.2% fibres with diameter < 3 µm) or 3100 fibres/cm³ UICC amosite asbestos (38% with diameter < 3 µm), for 6 h per day on five days per week for three months and were then observed for 21 months. One group of 31 unexposed animals served as controls. Groups of 1–10 animals per exposure group were killed at 50 days and at 3 months of exposure, and at 6, 12 and 18 months post-exposure; the remainder were killed at 24 months post-exposure. No pulmonary tumour was observed in animals that were killed or died before the end of the study. Bronchoalveolar adenomas were observed in 2/7 animals treated with glass fibre, 0/5 animals treated with amosite and 0/5 controls killed at the end of the study (Lee *et al.*, 1981). [The Working Group noted the short exposure period and the small number of animals evaluated.]

3.3.2 *Intraperitoneal injection*

Rat: Groups of female Wistar rats, 8-12 weeks old, received single intraperitoneal injections of 2, 10 or 50 mg (the latter in two doses) [JM 104] glass fibre (code104; approximate mean fibre dimensions, 10 µm × 0.2 µm) [source of production and chemical composition not given], 20 mg JM 112 glass fibre (code 112; approximate mean dimensions, 30 μm × 1 μm) [source of production and chemical composition not given], 2 mg UICC crocidolite or 50 mg corundum. Average survival times were 673, 611 and 361 days for the groups treated with 2, 10 and 50 mg JM 104, respectively, and 610 and 682 days for the groups treated with JM 112 and crocidolite, respectively. Abdominal tumours were diagnosed macroscopically and some were also identified by microscopy. Dose-related increases in the incidences of abdominal tumours (mesotheliomas, sarcomas and, rarely, carcinomas) were observed in the groups treated with the finer JM 104 glass fibre: 20/73 (2 mg), 41/77 (10 mg) and 55/77 (50 mg). The incidences in the groups treated with JM 112 and with crocidolite were 14/37 and 15/39, respectively. Of the rats that received injections of granular corundum, 3/37 had tumours in the abdominal cavity; mean survival time was 746 days (Pott et al., 1976).

To test the effect of fibre durability on tumorigenicity, eight groups of 46–48 female Wistar rats (range of body weights, 158–166 g) were injected intraperitoneally with either a low-durability glass wool (Bayer B1 or B2 [chemical composition not given]) developed for filters, or with two durable glass fibres (Bayer B3 and JM 475 glass microfibre) of very similar chemical compositions, but different size distributions. Several sizes of fibres of the low durability glass wool were administered in one or two doses as follows: thick (~1.5 μm median diameter) fibres in short, medium and long fibre ranges, designated B1K, B1M and B1L (median lengths, 7.4, 10.7 and 17.8 μm, respectively), with a maximum dose of 0.24×10^9 fibres per rat; and thin (0.5 μm mean diameter) fibres, in two length ranges, B2S and B2L (median lengths, 4.2 and 6 μm, respectively), to a maximum dose of 10^9 fibres > 5 μm per rat. The two more-durable glass fibre compositions, B3 (two length ranges, B3K and B3L, 3.3 and 5.6 μm, respectively) and JM 475, were administered at maximum doses of 0.45 and 0.33×10^9 fibres > 5 μm per rat, respectively. Negative control rats were given five intraperitoneal injections of 2 mL physiological saline solution at weekly intervals: 2/50 (4%) developed abdominal tumours. Exposure to low-durability glass fibre (B1 or B2) at any fibre number, dose or dimension tested did not significantly elevate the number of abdominal tumours. The incidence of tumours in the rats exposed to B1 or B2 ranged from 0–11% (11% in the group exposed to the thickest and longest fibres). Both of the durable glass fibres induced elevated incidences of tumours: B3K and B3L, 64–66%; JM 475, 17%. When the effects of two doses of the same fibre were compared, some dose–response relationships became apparent, but when two length populations of the same fibre were compared, no consistent relationship between tumorigenicity and fibre number or fibre length was apparent (Pott et al., 1991).

In a study using fibres similar to those used by Pott et al. (1976), three groups of 44 female Wistar rats, four weeks of age, were given intraperitoneal injections of 2 or 10 mg [JM 104E] glass fibre (code 104; milled for 2 h [size not given]) [source of production and chemical composition not given] or 2 mg of [JM 475] glass fibre (code 100; median fibre dimensions, 2.4 μm × 0.33 μm) [source of production and chemical composition not given]. Abdominal tumours were observed in 14/44 rats that received 2 mg JM 104E, in 29/44 rats that received 10 mg JM 104E and in 2/44 rats that received

2 mg JM 475. The first tumour was observed 350 days (50 weeks), 252 days (36 weeks) or 664 days (95 weeks) after the start of treatment with 2 mg JM 104E, 10 mg JM 104E and 2 mg JM 475, respectively. In three positive-control groups that received intraperitoneal injections of 0.4, 2 or 10 mg UICC chrysotile B, tumours developed in 9/44, 26/44 and 35/44 rats, respectively; the first tumour-bearing rat was found at 522 days (75 weeks), 300 days (43 weeks) and 255 days (36 weeks) after start of treatment in the groups treated with 0.4, 2 and 10 mg UICC chrysotile B, respectively. A negative-control group treated with 2 mg granular corundum dust had a tumour incidence of 1/45; the first tumour was found 297 days (42 weeks) after injection. The tumours observed in both the test and control groups were mesotheliomas or sarcomas. The groups treated with 0.4 mg chrysotile B or with JM 475 contracted an infection during month 21 which might have reduced the tumour incidence (Pott et al., 1984a).

Groups of female Sprague Dawley rats, 8 weeks of age, received single intraperitoneal injections of 2 mg or 10 mg [JM 475] glass fibre (JM 100; median fibre dimensions, 2.4 µm × 0.33 µm) [chemical composition not given] in 2 mL saline. The median survival times were 90 weeks and 79 weeks for the groups treated with 2 mg and 20 mg, respectively. Sarcomas, mesotheliomas and (rarely) carcinomas were seen in 21/54 animals treated with the 2 mg dose and 24/53 animals treated with the 10 mg dose; the first tumour was seen after 53 weeks in each group. Three tumours were found in both groups of 54 rats that received two injections each of either 20 mg Mount St Helen's volcanic ash or 20 mL saline alone (median survival time, 93 and 94 weeks, respectively; first tumour after 79 weeks of administration of volcanic ash and 94 weeks of administration of saline) (Pott et al., 1987).

Groups of 32 female Wistar rats, five weeks of age, received single intraperitoneal injections of 0.5 mg or 2.0 mg [JM 475] glass fibre (code 104; median fibre dimensions, 3.2 µm × 0.18 µm), 2.0 mg of [JM 475] glass fibre treated with 1.4 M hydrochloric acid for 24 h; or 0.5 mg or 2.0 mg South African crocidolite (median fibre dimensions, 2.1 µm × 0.20 µm) in 1 mL saline or saline alone. A group of 32 animals that received three intraperitoneal injections of titanium dioxide (total dose, 10 mg) served as another control. The animals were observed for life. Median survival times were 116, 110, 107, 109, 71, 130 and 120 weeks for rats that received 0.5 mg and 2.0 mg glass fibre, acid-treated glass fibre, 0.5 and 2.0 mg crocidolite, titanium dioxide and saline only, respectively. After exclusion of tumours of the uterus, the observed incidences of sarcomas, mesotheliomas and (rarely) carcinomas of the abdominal cavity were 5/30 (first tumour after 88 weeks) in animals treated with 0.5 mg glass fibre, 8/31 (first tumour after 84 weeks) with 2.0 mg glass fibre and 16/32 (first tumour after 56 weeks) with acid-treated glass fibre. The incidences of tumours in the groups treated with crocidolite were 18/32 (low-dose crocidolite; first tumour after 79 weeks) and 28/32 (high-dose crocidolite; first tumour after 52 weeks); the incidence of tumours in the saline-treated control group was 2/32 (first tumour after 113 weeks). None of the above-mentioned tumours was found in the group treated with titanium dioxide (Muhle et al., 1987; Pott et al., 1987).

Groups of female Sprague-Dawley rats [initial numbers not specified], eight weeks of age, received one injection of 5 mg [JM 104E] glass fibre (code 104; cut and ground for 1 h in an agate mill; median fibre dimensions, 4.8 µm × 0.29 µm). A negative-control group received 5 mg granular titanium dioxide. In the group treated with JM 104E glass fibre, 44/54 rats developed abdominal tumours; the median survival time was 64 weeks, and the average survival time of the tumour-bearing animals was 67 weeks. In the group treated with titanium dioxide, 2/52 rats had abdominal tumours; the median survival time was 99 weeks, and the average survival time of the tumour-bearing animals was 97 weeks (Pott et al., 1987).

Groups of Wistar rats [initial number and sex unspecified], four weeks of age, received one injection of 5 mg [JM 104E] glass microfibre (code 104, cut and ground for 1 h in an agate mill: median fibre dimensions, 4.8 µm × 0.29 µm). A negative control group received 5 mg granular titanium dioxide. Treatment with the JM 104E glass fibre induced abdominal tumours in 20/45 rats; the median survival time of the group was 34 weeks, and average survival time of the tumour-bearing rats was 49 weeks. None of the 47 rats treated with titanium dioxide developed abdominal tumours; median survival time of the group was 102 weeks (Pott et al., 1987).

A group of 25 female Osborne-Mendel rats, 100 days of age, received a single intraperitoneal injection of 25 mg Manville code 100 fibres (geometric mean fibre dimensions, 4.7 µm × 0.4 µm; 19% of fibres > 10 µm in length and 0.2–0.6 µm in diameter) in 0.5 mL saline. A group of 25 rats was injected with saline only and another group of 125 rats was untreated. All animals were observed for life. The median average life span was significantly shorter in rats treated with fibres (593 days) than in animals treated with saline (744 days) or untreated controls (724 days). Mesotheliomas were found in 8/25 of the rats treated with glass fibres and in 20/25 rats injected with 25 mg UICC crocidolite (5% ≥ 5 µm in length; mean, 3.1 µm), but in neither of the control groups (Smith et al., 1987).

Groups of 54 female Sprague-Dawley rats, eight weeks of age, were injected intraperitoneally with 5 mg of an untreated or treated fibre derived from JM code 104 glass microfibre that had been chopped and ground; 50% were < 4.8 µm in length. Treated fibres were soaked in either 1.4 mol/L hydrochloric acid (HCl) or 1.4 mol/L sodium hydroxide (NaOH) at room temperature for either 2 or 24 h. The weight loss of the fibres treated with HCl was 25% (2 h) or 33% (24 h) [weight loss with NaOH not reported]. Imaging by SEM revealed no modifications of the fibres. [The Working Group noted that presumably this applied to all of the treatments, although this was not stated. The incidence of abdominal tumours was reported in the form of a graph showing tumour incidence over time, but the actual numbers and types of tumour were not reported.] The incidences of tumours (estimated from the graph) in animals that received untreated JM 104 were the same as those that received the NaOH-treated JM 104 (approximately 80%). Administration of JM 104 treated with HCl, however, markedly reduced the tumour incidence: ~60% tumour incidence for 2-h HCl treated fibres and < 10% tumour incidence for 24-h HCl treated fibres. To test their biodurability 2 mg of the treated

(24 h-HCl) and untreated JM 104 fibres were injected intratracheally into rats, and their lungs were examined nine months later for fibres. The number of fibres × 10^6/lung for HCl-treated, NaOH-treated and untreated JM 104 were 31, 76 and 295, respectively. These data suggest greater biopersistence of untreated JM 104 than of NaOH- or HCl-treated JM 104, which does not completely parallel tumour incidences for the three test fibres (Pott et al., 1988).

Female Wistar rats [initial number not specified], eight weeks of age, were given intraperitoneal injections of saline solution containing suspensions of JM 475 glass microfibre (5 mg; 680 × 10^6 WHO fibres) or UICC Canadian chrysotile asbestos. The rats treated with chrysotile were given one of three gravimetric doses (0.05, 0.25 or 1.00 mg containing 40 × 10^6, 202 × 10^6 or 808 × 10^6 WHO fibres, respectively). The median dimensions for JM 475 test fibres were: length, 2.6 µm; diameter, 0.15 µm; those for chrysotile fibres were: length, 0.67 µm; diameter, 0.05 µm. The lengths of the JM 475 and chrysotile fibres were 90% < 9.6 µm and 90% < 2.1 µm, respectively. [A number of other fibres were also included in this study.] A group of 102 control rats was injected with saline. Rats were kept until ~130 weeks post-injection, at which time all surviving rats were killed, necropsied and examined for abdominal tumours. When tumours were present, they were examined histologically, and rats with uterine tumours were eliminated from the study. [These tumours metastasize quickly to other organs and therefore could be mistaken for sarcomas or mesotheliomas (the incidence of uterine tumours in controls treated with saline was ~15%).] The numbers of rats at risk for abdominal sarcoma/mesothelioma examined were 53 for the JM 475 and 34 for the chrysotile experiments, respectively. The incidences of tumours (excluding rats with uterine tumours) for JM 475 and the three increasing doses of chrysotile were 34/53, 12/36, 23/34 and 30/36, respectively; 2/102 control rats developed mesotheliomas (Pott, 1989; Pott et al., 1989).

Groups of 40 female Wistar rats, 8–10 weeks of age, were given intraperitoneal injections containing a total dose of either 17 mg or 50 mg JM 753 glass fibre. The numbers of fibres for the two doses were 1 × 10^9 and 3 × 10^9 WHO fibres per rat, respectively. The approximate median dimensions of fibres were 3.3 µm × 0.22 µm. Rats were maintained for life (until 30 months) and examined for tumours. A group of 40 concurrent negative control rats were given intraperitoneal injections of saline. The incidences of peritoneal mesotheliomas for the two doses of JM 753 were 30/36 (83%) and 36/39 (92%), respectively. Negative control rats did not develop mesotheliomas (Roller et al., 1996).

In two series of intraperitoneal injection studies, groups of 24 male Wistar rats, 12 weeks of age, were each injected with approximately 10^9 fibres longer than 5 µm. The fibres were counted by phase contrast optical microscopy (PCOM). The fibres tested were: JM 475 glass, amosite (Davis et al., 1996a; Miller et al., 1999) and JM 104E glass (Cullen et al., 2000). After injection of JM 475 glass, 8/24 of the rats developed abdominal tumours compared with 21/24 after treatment with amosite or 104E glass.

No control groups were included in these studies, but sizes of fibres and data on animal survival were fully reported.

3.3.3 Intratracheal instillation

(a) Rat

Groups of female Wistar rats aged 11 weeks at the start of the study received 20 weekly intratracheal instillations of a high dose of 0.5 mg/dose (total dose, 10 mg) JM 104/475 [composition not specified] glass fibre (median dimensions of fibres, 3.2 µm × 0.18 µm) or South African crocidolite (median dimensions of fibres, 2.1 µm × 0.2 µm) in 0.3 mL saline or saline alone. The median lifespans were 107, 126 and 115 weeks for the groups receiving glass fibres, crocidolite and saline, respectively. In the group treated with glass fibres, 5/34 rats developed pulmonary tumours (one adenoma, four carcinomas); the mean lifespan of rats with tumours was 113 weeks. In the group treated with crocidolite, 15/35 rats developed lung tumours (11 carcinomas and four mixed tumours); the mean lifespan of tumour-bearing animals was 121 weeks and the first tumour was observed after 89 weeks). Pulmonary tumours did not occur in the 40 control animals or in historical controls of this strain (Pott *et al.*, 1987).

A group of 22 female Osborne-Mendel rats, 100 days old, received five weekly intratracheal instillations of a high dose of 2 mg [JM 475] glass fibres (10 mg total dose; geometric mean dimensions of fibres, 4.7 µm × 0.4 µm; 19% of fibres > 10 µm in length and 0.2–0.6 µm in diameter) in 0.2 mL saline. A group of 25 rats was injected with saline only and another group of 125 animals was untreated. All animals were observed for life. The median lifespan was longer in rats treated with glass fibres (783 days) than in the rats treated with saline (688 days) or untreated controls (724 days). No tumour of the respiratory tract was observed in any group. Of 25 rats treated under similar conditions with UICC crocidolite (5% fibres > 5 µm in length; mean, 3.1 ± 10.2 µm), two developed bronchoalveolar tumours (Smith *et al.*, 1987). [The Working Group noted the relatively small number of animals in the study and the low tumour response in positive controls, which made interpretation of the results difficult.]

(b) Hamster

Two groups of 136 or 138 male Syrian golden hamsters [age unspecified] were given eight weekly intratracheal instillations in 0.15 mL saline of 1 mg of two different samples prepared from JM 104 glass fibre [chemical composition not given]. The two samples were wet-milled in a ball mill for 2 or 4 h, respectively, resulting in different length distributions: fibres milled for 2 h; length, 50% < 7.0 µm; diameter, 50% < 0.3 µm; fibres milled for 4 h: length, 50% < 4.2 µm; diameter, 50% < 0.3 µm). Two control groups received eight intratracheal instillations of 1 mg of either UICC crocidolite (length, 50% > 2.1 µm; diameter, 50% > 0.2 µm) as a positive control, or granular titanium dioxide as a negative control (total doses, 8 mg). The incidences of thoracic tumours in the group treated with the longer glass fibres were: 48/136 (5 lung carcinomas, 37 mesotheliomas, 6 sarcomas); in the group treated with shorter glass fibres, 38/138 (6 lung carcinomas, 26 mesotheliomas, 6 sarcomas); in the group treated with crocidolite, 18/142 (9 lung carcinomas, 8 mesotheliomas, 1 sarcoma); and in the group treated with titanium dioxide, 2/135 (sarcomas). The total duration of the experiment was 113 weeks. Nearly all of the tumour-bearing animals survived for up to 18 months after the first instillation (Pott *et al.*, 1984b).

Six groups of 35 male and 35 female Syrian golden hamsters, 16 weeks of age, received intratracheal instillations in 0.2 mL 0.005% gelatine in saline of 1 mg JM 104 glass fibres (58% < 5 μm in length; 88% < 1.0 μm in diameter) [chemical composition not given], 1 mg glass fibre plus 1 mg benzo[*a*]pyrene, 1 mg crocidolite (UICC standard reference sample; 58% > 5 μm in length; 63% > 0.25 μm in diameter), 1 mg crocidolite plus 1 mg benzo[*a*]pyrene, 1 mg benzo[*a*]pyrene in gelatine solution in saline or vehicle alone, once every two weeks for 52 weeks. The experiment was terminated at 85 weeks, at which time 53, 43, 43, 50, 48 and 46 animals were still alive in the six groups, respectively. Tumours of the respiratory tract were found only in hamsters treated with benzo[*a*]pyrene alone. The incidences of respiratory tract tumours were as follows: in the group given benzo[*a*]pyrene alone, 3/63 (2 carcinomas and 1 sarcoma, plus 4 papillomas); in the group given crocidolite plus benzo[*a*]pyrene, 3/52 (2 carcinomas and 1 sarcoma, plus 1 papilloma); and in the group given glass fibre plus benzo[*a*]pyrene, 2/66 (2 sarcomas, plus 2 papillomas) (Feron *et al.*, 1985). [The Working Group noted the relatively short observation time and the absence of tumours in the positive, crocidolite-treated groups.]

3.3.4 Intrapleural injection

(a) Mouse

Four groups of 25 BALB/c mice [sex and age unspecified] received single intrapleural injections of a high dose of 10 mg of one of four different samples of borosilicate glass fibres [chemical composition not given] in 0.5 mL distilled water. The material for injection was obtained by separating each of two original samples with average diameters of 0.05 μm and 3.5 μm into two samples, one with lengths of several hundred microns and the other with lengths of < 20 μm. Animals were killed at intervals of two weeks until 18 months (a total of 37 mice survived at this time). No pleural tumour was found in any of the treated animals, whereas mesotheliomas were observed in 2/150 mice given intrapleural injections of chrysotile or crocidolite [dose not stated] in a parallel experiment. The author concluded that the pleural cavity of mice might be very resistant to tumour induction by any type of mineral fibre (Davis, 1976). [The Working Group noted the small number of animals used, the relatively short observation time and the low response in the positive controls.]

(b) *Rat*

Groups of 32–36 SPF Wistar rats (twice as many males as females), 13 weeks of age, received single intrapleural injections in 0.4 mL saline of 20 mg glass fibre (a borosilicate; 30% of fibres 1.5–2.5 μm in diameter; maximum diameter, 7 μm; 60% > 20 μm in length) [chemical composition not given], 20 mg glass powder (a borosilicate; diameter < 8 μm) or 20 mg of one of two different samples of Canadian SFA chrysotile. Rats were kept until natural death; the average survival times were 774, 751, 568 and 639 days for the groups treated with glass fibre, glass powder and the two chrysotile samples, respectively. No injection-site tumour was observed in the group treated with glass fibre; a single mesothelioma occurred in the group treated with glass powder (after 516 days). The incidences of tumours in the two groups treated with chrysotile were 23/36 and 21/32; the first deaths of animals with tumours occurred after 325 and 382 days (Wagner *et al.*, 1973).

Three groups of 16 male and 16 female Wistar rats, 10 weeks of age, received single intrapleural injections of 20 mg of a fine US JM 100 glass fibre (99% of fibres < 0.5 μm in diameter; median diameter, 0.12 μm; 2% > 20 μm in length; median length, 1.7 μm) [chemical composition not given] or a coarser US JM 110 glass fibre (17% of fibres < 1 μm in diameter; median diameter, 1.8 μm; 10% > 50 μm in length; median length, 22 μm) [chemical composition not given] in 0.4 mL saline or saline alone. Animals were kept until natural death; mean survival times were 716, 718 and 697 days, for the mice treated with fine fibres, coarse fibres and saline, respectively. Between 663 and 744 days after inoculation, 4/32 animals given the finer glass fibre had mesotheliomas. No pleural tumour occurred in animals treated with the coarser glass fibre or in controls that received saline (Wagner *et al.*, 1976).

Groups of 30–130 female Osborne-Mendel rats, 12–20 weeks old, received a single intrathoracic implantation of one of 72 different types of natural and man-made mineral fibres [chemical compositions not given], 19 of which were uncoated or resin-coated glass fibres. The test substances were mixed in 10% gelatine, and 40 mg of each type of glass in 1.5 mL gelatine was smeared on a coarse fibrous glass pledget which was implanted into the left thoracic cavity. The rats were observed for 24 months after treatment and were compared with untreated controls and controls implanted with the pledget alone. The incidence of pleural mesothelioma in animals that survived for more than 52 weeks varied from 0/28 to 20/29 depending on fibre size. The most carcinogenic fibres were those < 1.5 μm in diameter and > 8 μm in length. When two of the glass fibre preparations (diameter, > 0.25 μm) were leached to remove all elements except silicon dioxide, the incidences of pleural mesotheliomas were 2/28 and 4/25 (Stanton *et al.*, 1977, 1981).

Groups of 32–45 male SPF Sprague-Dawley rats, three months of age, received single intrapleural injections of 20 mg JM 104 glass fibre (mean length, 5.89 μm; mean diameter, 0.229 μm) [chemical composition not given], 20 mg UICC chrysotile A (mean dimensions, 3.21 μm × 0.063 μm), 20 mg UICC crocidolite (mean dimensions,

3.14 μm × 0.148 μm) in 2 mL saline, or received 2 mL saline alone. Animals were kept until natural death; the mean survival times for whole groups (and for animals with tumours) were 513 (499), 388 (383), 452 (470) and 469 days, respectively. The incidences of thoracic tumours were as follows: group that received glass fibre, 6/45 (mesotheliomas); groups treated with chrysotile and crocidolite, 15/33 (1 carcinoma and 14 mesotheliomas) and 21/39 (mesotheliomas), respectively. No thoracic tumours occurred in the 32 control animals (Monchaux et al., 1981).

Groups of 48 SPF Sprague-Dawley rats [sex and age unspecified] received single intrapleural injections of 20 mg fibrous glass dusts in 0.5 mL saline or chrysotile in 0.5 mL saline. The dust samples used (and the size distributions of those fibres longer than 1 μm) were: English glass fibre with resin coating (70% fibres < 5 μm in length; 85% < 1 μm in diameter), English glass fibre after removal of resin (57% < 5 μm in length; 85% < 1 μm in diameter), US JM 100 glass fibre (88% < 5 μm in length; 98.5% ≤ 1 μm in diameter) [chemical composition not given] and UICC African chrysotile [fibre sizes unspecified]. The animals were kept until natural death [survival times unspecified]. One mesothelioma occurred in the group treated with English glass fibre [whether coated or uncoated was not specified], four in the group treated with JM 100 glass fibre and six in the group treated with chrysotile. No mesothelioma was observed in a group of 24 controls treated with saline (Wagner et al., 1984).

3.4 Rock (stone) wool

3.4.1 *Inhalation exposure* (see Table 61)

Rat: Groups of 24 male and 24 female Wistar IOPS AF/Han rats, eight to nine weeks old, were exposed by inhalation to dust at concentrations of 5 mg/m^3 (respirable particles) of French (Saint-Gobain) resin-free rock (stone) wool [type of rock unspecified] (40% of fibres < 10 μm in length, 23% < 1 μm in diameter) or a Canadian chrysotile fibre (6% respirable fibres > 5 μm in length) for 5 h per day on five days per week for 12 or 24 months. An unspecified number of rats was killed either immediately after treatment or after a period of observation (of 7, 12 or 16 months after exposure for animals exposed for 12 months; 4 months after exposure for those exposed for 24 months). No pulmonary tumour was observed in 47 rats treated with rock (stone) wool or in 47 untreated controls; nine pulmonary tumours were seen in the 47 rats treated with chrysotile (Le Bouffant et al., 1984). [The Working Group noted that, because of the lack of survival data, the exact incidences of tumours could not be ascertained.]

Groups of 48 SPF Fischer rats [sex and age unspecified] were exposed by inhalation to dust concentrations of approximately 10 mg/m^3 resin-free rock (stone) wool [type of rock unspecified] or UICC Canadian chrysotile for 7 h per day on five days per week for 12 months. The size distribution of those airborne fibres longer than 5 μm was: 71% of rock (stone) wool fibres ≤ 20 μm in length, 58% ≤ 1 μm in diameter; 16% of chrysotile fibres ≥ 20 μm in length, 29% ≥ 0.5 μm in diameter. Six rats were

Table 61. Inhalation studies on the carcinogenicity of rock (stone) wool and slag wool in rats

Test substance	Fibre dimensions: length (L), diameter (D)	Dosing schedule/ cumulative exposure (mg/m³ × h)	Duration of exposure	Period of observation	No. of animals examined	No. of animals with tumours[a]	Histo- logical type[b]	Median or average survival time (weeks)
Exposure by inhalation to respirable dust at a concentration of 5 mg/m³ rock (stone) wool (male and female Wistar rats, 8–9 weeks old) (Le Bouffant et al., 1984)								
French resin-free rock (stone) wool	40% fibres < 10 µm length, 23% ≤ 1 µm diameter	ND	5 h/day, 5 days/week, for 52 or 104 weeks	12–28 months	47	0	–	NG
Canadian chrysotile	6% fibres > 5 µm length	ND	5 h/day, 5 days/week, for 52 or 104 weeks	12–28 months	47	9	9A	NG
Control	–	–	5 h/day, 5 days/week, for 52 or 104 weeks	12–28 months	47	0	–	NG
Exposure by inhalation to respirable dust at a concentration of ~10 mg/m³ rock (stone) wool (SPF Fischer rats) (Wagner et al., 1984)								
Resin-free rock (stone) wool	71% fibres ≤ 20 µm length, 58% ≤ 1 µm diameter	17495	7 h/day, 5 days/week, for 52 weeks	12 months– lifetime[c]	48	2	2A	NG
UICC Canadian chrysotile	16% fibres ≥ 20 µm length, 29% ≥ 0.5 µm diameter	17499	7 h/day, 5 days/week, for 52 weeks	12 months– lifetime[c]	48	12	11AdCa 1A	NG
Control	–	–	7 h/day, 5 days/week, for 52 weeks	12 months– lifetime[c]	48	0	–	NG

Table 61 (contd)

Test substance	Fibre dimensions: length (L), diameter (D)	Dosing schedule/ cumulative exposure (mg/m³ × h)	Duration of exposure	Period of observation	No. of animals examined	No. of animals with tumours[a]	Histo- logical type[b]	Median or average survival time (weeks)
Exposure by nose-only inhalation to dust clouds (7.8 mg/m³) of slag wool (female Osborne-Mendel rats, 100 days old) (Smith et al., 1987)								
Slag wool	GML, 22 µm GMD, 0.9 µm	200 fibres/cm³	6 h/day, 5 days/week, for 104 weeks	Lifetime[c]	55	0	–	97
Crocidolite	L, 5% > 5 µm	3000 fibres/cm³	6 h/day, 5 days/week, for 104 weeks	Lifetime[c]	57	3	1M, 2 BT	109
Chamber controls	–	–	6 h/day, 5 days/week, for 104 weeks	Lifetime[c]	59	0	–	108
Room controls	–	–	6 h/day, 5 days/week, for 104 weeks	Lifetime[c]	125	0	–	103
Exposure by nose-only inhalation to respirable dust at concentrations of 3, 16 and 30 mg/m³ rock (stone) wool (MMVF21) (male Fischer 344 rats, 8 weeks old) (McConnell et al., 1994)								
Rock (stone) wool (MMVF21)	GML, 13.0 µm GMD, 0.94 µm	9 672	6 h/day, 5 days/week, for 104 weeks	28 months	114	5	4A 1Ca	~104
	GML, 15.4 µm GMD, 0.90 µm	50 232	6 h/day, 5 days/week, for 104 weeks	28 months	115	5	4A 1Ca	~104
	GML, 14 µm GMD, 0.98 µm	94 848	6 h/day, 5 days/week, for 104 weeks	28 months	114	5	4A 1Ca	~104
Crocidolite	GML, 4.1 µm GMD, 0.28 µm	13 000	6 h/day, 5 days/week, for 44 weeks	28 months	106	14	10A 5Ca 1M	~100

Table 61 (contd)

Test substance	Fibre dimensions: length (L), diameter (D)	Dosing schedule/ cumulative exposure (mg/m³ × h)	Duration of exposure	Period of observation	No. of animals examined	No. of animals with tumours[a]	Histo-logical type[b]	Median or average survival time (weeks)
Control	—	—		28 months	126	2	2A	~104

Exposure by nose-only inhalation to respirable dust at concentrations of 3, 16 and 30 mg/m³ slag wool (MMVF22) (male Fischer 344 rats, 8 weeks old) (McConnell et al., 1994)

Test substance	Fibre dimensions	Dosing schedule	Duration of exposure	Period of observation	No. of animals examined	No. of animals with tumours[a]	Histological type[b]	Median or average survival time (weeks)
Slag wool (MMVF22)	GML, 12.3 µm GMD, 0.84 µm	9 672	6 h/day, 5 days/week, for 104 weeks	28 months	116	2	1A 1Ca	~104
	GML, 13.2 µm GMD, 0.84 µm	50 232	6 h/day, 5 days/week, for 104 weeks	28 months	115	0		~104
	GML, 15.2 µm GMD, 0.87 µm	93 288	6 h/day, 5 days/week, for 104 weeks	28 months	115	3	2A 1Ca	~104
Crocidolite	GML, 4.1 µm GMD, 0.28 µm	13 000	6 h/day, 5 days/week, for 44 weeks	28 months	106	14	10A 5Ca 1M	~100
Control	—	—	6 h/day, 5 days/week, for 104 weeks	28 months	126	2	2A	~104

NG, not given; GML, geometric mean length; GMD, geometric mean diameter

[a] Tumours of the lung, pleura, thorax or abdominal cavity
[b] A, adenoma; AdCa, adenocarcinoma; BT, bronchoalveolar tumour; Ca, relatively undifferentiated epidermoid carcinoma; M, mesothelioma
[c] Lifetime, until survival rate was ≤ 20%

removed from each group at the end of the exposure period to study dust retention, and a similar number of animals was killed one year later for the same purpose. The remainder were kept until natural death [survival times not reported]. During the period 500–1000 days after the start of exposure, lung adenomas (1 with some malignant features) occurred in 2/48 rats in the group treated with rock (stone) wool; 11 adenocarcinomas and one adenoma (with some malignant features) were observed in 48 rats treated with chrysotile. No lung tumour was observed in a group of 48 untreated controls (Wagner et al., 1984). [The Working Group noted that, because of inadequate data on survival, the exact tumour incidences could not be established.]

Groups of 140 male Fischer 344 rats, eight weeks of age, were exposed in nose-only inhalation chambers to rock (stone) wool (MMVF21) at concentrations of 3, 16 or 30 mg/m^3 for 6 h per day on five days per week for 104 weeks. These concentrations corresponded to average numbers of WHO fibres/cm^3 of 34, 150 and 243 and numbers of fibres > 20 μm long of 13, 74 and 114, respectively. The geometric mean diameter of the test fibres was about 0.95 μm and the geometric mean of the fibre length was about 14 μm. The retained lung burdens/mg of dry lung tissue after 24 months were: 242×10^3 WHO fibres and 39.5×10^3 fibres > 20 μm for the 30 mg/m^3 dose, 217×10^3 and 25.6×10^3, respectively, for the 16 mg/m^3 dose and 55.7×10^3 and 9.0×10^3, respectively, for the 3 mg/m^3 dose. A subsequent post-exposure period lasted until approximately 20% of the animals in the air-control group survived which occurred at approximately 28 months. Mortality of the treated animals was similar to that of unexposed control animals. Four lung adenomas and one lung carcinoma each were observed among 114, 115 and 115 rats in the 3 mg/m^3, 16 mg/m^3 and 30 mg/m^3 dose groups, respectively. In the control group, two adenomas were observed in 126 rats. A positive control group was treated with crocidolite asbestos (10 mg/m^3) with a geometric mean diameter of 0.28 μm, determined by SEM, and a geometric mean length of 4.1 μm, determined by light microscopy. [The Working Group noted that exposure to crocidolite asbestos was terminated after 10 months because of increased morbidity and mortality.] The lung burden/mg dry lung tissue retained after 104 weeks was 759×10^3 WHO fibres and 41.1 fibres $\times 10^3$ > 20 μm. Ten pulmonary adenomas, five pulmonary carcinomas and one mesothelioma were reported among 106 crocidolite-treated rats (McConnell et al., 1994).

3.4.2 *Intratracheal instillation* (see Table 62)

(a) Rat

Groups of 40–59 female Wistar rats, 15 weeks of age, were treated by intratracheal instillation with 5 mg (10 weekly doses of 0.5 mg each) and 10 mg (20 weekly doses of 0.5 mg each) of a rock (stone) wool suspension and were then observed for 131 weeks. The low-dose group was exposed to a total of 4×10^6 fibres and the high-dose group, to 8×10^6 fibres (length, > 5 μm; diameter, < 2 μm; aspect ratio > 5:1) [chemical composition and size distribution not given]. At the end of the experiment, the lungs were

Table 62. Carcinogenicity studies on the intratracheal instillation of rock (stone) wool in rats and hamsters

Substance	Instillation schedule (if > 1 injection/week)	Total number of fibres	Observation period (weeks)	No. of animals with tumours/no. of animals examined[a,b]	Histological types[c]	Comments	Reference
Rock (stone) wool	5 mg (10 × 0.5 mg)	[d]4 × 10^6	131	0/59 rats 5/59 rats	PLT OLT		Pott et al. (1994)
Rock (stone) wool	10 mg (20 × 0.5 mg)	[d]8 × 10^6	131	0/40 rats 4/40 rats	PLT OLT		
Control	Saline solution (20 × 0.4 mL)		131	0/40 rats 2/40 rats	PLT OLT		
Tremolite (positive control)	2.5 mg (5 × 0.5 mg)	[d]70 × 10^6	131	3/38 rats 6/38 rats	PLT OLT	[$p = 0.023$][e]	
Tremolite (positive control)	7.5 mg (15 × 0.5 mg)	[d]300 × 10^6	131	8/37 rats 4/37 rats	PLT OLT	[$p = 0.0025$][e]	
Rock (stone) wool	10 mg (5 × 2 mg)		104	0/20 hamsters	–	GML, 296 µm GMD, 6.1 µm	Adachi et al. (1991)
Control	Saline solution		104	0/20 hamsters	–		

GML, geometric mean length; GMD, geometric mean diameter
[a] Female Syrian hamsters, ~ 80 g
[b] Female Wistar rats, 15 weeks of age
[c] PLT, primary lung tumours: adenomas, adenocarcinomas, squamous cell tumours and cystic keratinizing squamous-cell tumours; OLT, other lung tumours: fibrosarcomas, lymphosarcomas, mesotheliomas or lung metastases from tumours at other sites
[d] $D < 2$ µm; $L > 5$ µm; aspect ratio > 5:1
[e] Versus control group treated with saline solution

perfused with formalin and examined histopathologically. None of the 59 rats treated with 5 mg of the suspension developed a primary lung tumour, but other lung tumours (fibrosarcomas, lymphosarcomas, mesotheliomas and lung metastases from tumours at other sites) were found in 5/59 rats. In the group that received 10 mg of the suspension, no primary lung tumours were observed in 40 rats; 4/40 animals had other lung tumours. A control group (40 rats) was treated with 20 weekly injections of 0.4 mL saline solution. No primary lung tumours were observed; other lung tumours were found in 2/40 rats. In a positive-control group, 38 and 37 female Wistar rats were injected intratracheally with tremolite: 2.5 mg (5 doses of 0.5 mg each) or 7.5 mg (15 doses of 0.5 mg each). [No information was given on the fibre dimensions.] The total number of fibres in the low dose was 70×10^6 and in the high dose, it was 300×10^6 fibres (length, > 5 μm; diameter, < 2 μm; aspect ratio, > 5:1). In the group that received the low dose, three primary tumours of the lung and six other tumours of the lung were observed in the 38 rats examined. In the group that received the high dose, the numbers were eight primary tumours of the lung and four other tumours of the lung in 37 rats (Pott *et al.*, 1994).

(b)   Hamster

A group of 20 female Syrian hamsters (weight, 80 g) was injected intratracheally with 2 mg rock (stone) wool suspended in 0.2 ml saline once a week for five weeks. The average diameter of the fibres was 6.1 μm and the average length was 296 μm. Two years after the first administration, all hamsters were killed and routine autopsies were performed. No tumours were reported in the treated group or in a control group of 20 hamsters (Adachi *et al.*, 1991). [The Working Group noted that the relevance of this study is limited because of the very long and thick fibres used in this experiment.]

3.4.3   *Intraperitoneal injection* (see Table 63)

Rat: Groups of female Sprague-Dawley rats [initial numbers unspecified], eight weeks of age, received three weekly intraperitoneal injections of 75 mg Swedish rock (stone) wool [type of rock unspecified] (median fibre length, 23 μm; median diameter, 1.9 μm), a single injection of 10 mg of a fine fraction prepared from the rock (stone) wool sample (median fibre length, 4.1 μm; diameter, 0.64 μm) or 40 mg (two injections) of granular volcanic ash from Mount St Helen's in 2 mL saline solution. The median survival times were 77, 97 and 93 weeks for the animals given the 75 mg and 10 mg doses of rock (stone) wool and volcanic ash, respectively; the median lifespan of a control group that received two injections of 2 mL saline was 94 weeks. A high incidence of tumours in the abdominal cavity was observed following treatment with 75 mg of the original rock (stone) wool sample: (45/63; lifespan of first animals with tumour, 39 weeks) and a slight increase in tumour incidence in animals treated with 10 mg of the fine fraction (6/45; first tumour after 88 weeks) compared with a tumour incidence of 3/54 in the groups of controls treated with volcanic ash and saline (Pott *et al.*, 1987). [The thicker and longer fibres induced more tumours than the thinner and shorter ones.]

Table 63. Carcinogenicity studies on the intraperitoneal injection of fibres of rock (stone) and slag wool in rats

Test substance	Injection schedule (weekly intervals)	No. of fibres L > 5 μm and differences from WHO definition[a]	L (μm); D (μm); L/D > 3/1 (median of all lengths, all diameters); different fibre definitions stated[b]	Observation period (weeks), strain, gender[c]	Median survival time (weeks)	No. of rats with tumours[d]/ no. of rats examined	Histo-logical tumour types[e]	Comments	Reference
Rock (stone) wool									
Rock (stone) wool (Sweden)	3 × 25 mg	NG	L, 23; D, 1.9	134, SpD, F	77	45/63	meso/sarc	[$p < 0.001$]	Pott et al. (1987)
Rock (stone) wool (Sweden), fine	1 × 10 mg	NG	L, 4.1; D, 0.64	134, SpD, F	97	6/45	meso/sarc		
Volcanic ash	2 × 20 mg	(granular dust *control*)[j]		134, SpD, F	93	3/54	meso/sarc		
Saline control	5 × 2 mL	(saline control)		134, SpD, F	94	3/54	meso/sarc		
Rock (stone) wool	1 × 25 mg	NG	NG	104, SpD, M and F	NG	3/40	meso	duration of study 104 weeks [results not conclusive]	Maltoni & Minardi (1989)
Water control	NG		(water control)	111 (death of last rat), W, F		0/40			
Basalt wool, G & H	5 × 15 mg	59 × 10[6] WHO with L/D > 5/1	L, 17; D, 1.1 L/D > 5/1		79	30/53	meso/sarc	strong adhesions of the abdominal organs [$p < 0.0001$]	Pott et al. (1989)
Titanium dioxide	5 × 20 mg	(granular dust *control*)[j]		130, W, F	109	2/53	meso/sarc		
Saline control	5 × 2 mL	(saline control)		130, W, F	111	2/102	meso/sarc		

Table 63 (contd)

Test substance	Injection schedule (weekly intervals)	No. of fibres L > 5 μm and differences from WHO definition[a]	L (μm); D (μm); L/D > 3/1 (median of all lengths, all diameters); different fibre definitions stated[b]	Observation period (weeks), strain, gender[c]	Median survival time (weeks)	No. of rats with tumours[d]/ no. of rats examined	Histological tumour types[e]	Comments	Reference
Basalt wool, G & H	1 × 25 mg	5 × 10⁶; L/D > 5/1; L > 5; D < 2 μm	L, 13.8; D, 1.08 L/D > 5/1	130, W, F	110	1/38	meso/sarc	lung infection months 12–13; mortality 10 of 48 at start	Pott et al. (1991)
Basalt wool, G & H	5 × 30 mg	30 × 10⁶; L/D > 5/1; L > 5; D < 2 μm	L, 13.8; D, 1.08 L/D > 5/1	130, W, F	84	15/21	meso/sarc	lung infection months 12–13; mortality 15 of 36 at start [$p < 0.001$]	
Carbon, activated	5 × 50 mg	(granular dust control)[j]		131, W, F	122	1/25	meso/sarc	lung infection months 12–13; mortality 11 of 36 at start	
Saline control	5 × 2 mL	(saline control)		130, W, F	106	2/50	meso/sarc	lung infection months 12–13; mortality 22 of 72 at start	
Basalt 'superthin'	2 × 25 mg	NG	17% with L > 5 and D < 3	Lifetime		5/40	meso	proportion of fibres 12% of nonfibres particles 88% [$p < 0.006$]	Nikitina et al. (1989)

Table 63 (contd)

Test substance	Injection schedule (weekly intervals)	No. of fibres L > 5 μm and differences from WHO definition[a]	L (μm); D (μm); L/D > 3/1 (median of all lengths, all diameters); different fibre definitions stated[b]	Observation period (weeks), strain, gender[c]	Median survival time (weeks)	No. of rats with tumours[d]/ no. of rats examined	Histo-logical tumour types[e]	Comments	Reference
Basalt 'ultrathin'	2 × 25 MG	NG	41% with L > 5 and D < 3	Lifetime		7/50	meso	[$p < 0.01$]	Davis et al. (1996b); Roller et al. (1996)
Chrysotile	2 × 25 MG	NG	76% with L > 5 and D < 3	Lifetime		27/60	meso	[$p < 0.001$]	
Saline control			(saline control)[j]			0/110			
M-stone[f]	1 × 8.5 mg	100 × 10^6	L, 10.1; D, 0.84; L/D > 5/1	130, W, F	104	2/32	meso	[$p = 0.030$]	
M-stone[f]	1 × 8.5 mg	100 × 10^6	L, 10.1; D, 0.84; L/D > 5/1	130, W, M	90	2/36	meso	[$p = 0.037$]	
M-stone[f]	1 × 25.5 mg	300 × 10^6	L, 10.1; D, 0.84; L/D > 5/1	130, W, F	103	9/32	meso	[$p = 0$]	
M-stone[f]	1 × 25.5 mg	300 × 10^6	L, 10.1; D, 0.84; L/D > 5/1	130, W, M	93	8/36	meso	[$p = 0$]	
M-stone[f]	2 × 42.5 mg	1000 × 10^6	L, 10.1; D, 0.84; L/D > 5/1	130, W, M	90	22/35	meso	[$p = 0$]	
B-20-2.0[f,g]	1 × 6 mg	80 × 10^6	L, 7.8; D, 0.77; L/D > 5/1	130, W, F	105	2/32	meso	[$p = 0.030$]	

Table 63 (contd)

Test substance	Injection schedule (weekly intervals)	No. of fibres L > 5 μm and differences from WHO definition[a]	L (μm); D (μm); L/D > 3/1 (median of all lengths, all diameters); different fibre definitions stated[b]	Observation period (weeks), strain, gender[c]	Median survival time (weeks)	No. of rats with tumours[d]/ no. of rats examined	Histological tumour types[e]	Comments	Reference
B-20-2.0[f,g]	1 × 6 mg	80 × 10⁶	L, 7.8; D, 0.77; L/D > 5/1	130, W, M	119	15/36	meso	[p = 0]	
B-20-2.0[f,g]	1 × 18 mg	240 × 10⁶	L, 7.8; D, 0.77; L/D > 5/1	130, W, F	96	7/32	meso	[p = 0]	
B-20-2.0[f,g]	1 × 18 mg	240 × 10⁶	L, 7.8; D, 0.77; L/D > 5/1	130, W, M	107	12/34	meso	[p = 0]	
B-20-2.0[f,g]	2 × 30 mg	800 × 10⁶	L, 7.8; D, 0.77; L/D > 5/1	130, W, M	88	21/35	meso	[p = 0]	
Crocidolite	5 × 0.1 mg	42 × 10⁶	L, 1.8; D, 0.19; L/D > 5/1	130, W, F	85	25/32	meso		
Crocidolite	5 × 0.1 mg	42 × 10⁶	L, 1.8; D, 0.19; L/D > 5/1	130, W, M	89	32/48	meso		
Silicon carbide	5 × 50 mg	(granular dust *control*)[j]		130, W, F	105	1/47	meso		
Silicon carbide	5 × 50 mg	(granular dust *control*)[j]		130, W, M	109	0/71			
Silicon carbide	20 × 50 mg	(granular dust *control*)[j]		130, W, F	107	0/45			
Silicon carbide	20 × 50 mg	(granular dust *control*)[j]		130, W, M	104	0/70			
Saline control	20 × 2 mL	(saline *control*)[j]		130, W, F	110	0/93			
Saline control	20 × 2 mL	(saline *control*)[j]		130, W, M	103	1/69	meso		

Table 63 (contd)

Test substance	Injection schedule (weekly intervals)	No. of fibres L > 5 μm and differences from WHO definition[a]	L (μm); D (μm); L/D > 3/1 (median of all lengths, all diameters); different fibre definitions stated[b]	Observation period (weeks), strain, gender[c]	Median survival time (weeks)	No. of rats with tumours[d]/ no. of rats examined	Histological tumour types[e]	Comments	Reference
B-20-06[f,l]	1 × 3.5 mg	400 × 10^6	L, 3.6; D, 0.30; L/D > 5/1	130, W, F	103	12/40 (6 macr.[h])	meso	final histol.: 11 rats with meso[k] [p = 0]	Davis et al. (1996b); Roller et al. (1996)
B-20-06[f,l]	1 × 8.5 mg	1000 × 10^6	L, 3.6; D, 0.30; L/D > 5/1	130, W, F	96	17/40 (5 macr.[h])	meso	final histol.: 19 rats with meso[k] [p = 0]	
B-20-06[f,l]	1 × 25 mg	3000 × 10^6	L, 3.6; D, 0.30; L/D > 5/1	130, W, F	79	30/40	meso	final histol.: 31 rats with meso[k] [p = 0]	
B-20-06[f,l]	3 × 25 mg	9000 × 10^6	L, 3.6; D, 0.30; L/D > 5/1	130, W, F	38	27/31	meso	final histol.: 27 rats with meso[k] [p = 0]	
Saline control	3 × 2 mL	(saline control)		130, W, F	121	0/38		final histol.: 0 rat with meso[k]	
Tremolite	1 × 3.3 mg	57 × 10^6	L, 3.4; D, 0.29	130, W, F	98	10/39	meso		
Tremolite	1 × 15 mg	260 × 10^6	L, 3.4; D, 0.29	130, W, F	74	30/40	meso		
MMVF21[f,g]	2 × 30 mg	400 × 10^6	L, 16.9; D, 1.02; L/D > 5/1	130, W, F	54	37/38	meso	final histol.: 37 rats with meso[k] [p = 0]	Davis et al. (1996b); Roller et al. (1996)

Table 63 (contd)

Test substance	Injection schedule (weekly intervals)	No. of fibres L > 5 μm and differences from WHO definition[a]	L (μm); D (μm); L/D > 3/1 (median of all lengths, all diameters); different fibre definitions stated[b]	Observation period (weeks), strain, gender[c]	Median survival time (weeks)	No. of rats with tumours[d]/ no. of rats evaluated	Histological tumour types[e]	Comments	Reference
MMVF21[f,g]	5 × 30 mg	1000 × 10⁶	L, 16.9; D, 1.02; L/D > 5/1	130, W, F	51	33/38	meso	final histol.: 33 rats with meso[k] [$p = 0$]	
R-stone-E3[f,i]	4 × 28.5 mg	400 × 10⁶	L, 16.9; D, 1.03; L/D > 5/1	130, W, F	120	0/30		final histol.: 0 rat with meso[k] [$p = 1$]	
R-stone-E3[f,i]	9 × 28.5 mg	900 × 10⁶	L, 16.9; D, 1.03; L/D > 5/1	130, W, F	120	4/35 (4 macr.[h])	meso	final histol.: 1 rat with meso[k] [$p = 0.0005$]	
Crocidolite	5 × 0.1 mg	42 × 10⁶	L, 1.8; D, 0.19; L/D > 5.1	130, W, F	100	20/39	meso		
Untreated control	(untreated control)			115		0/37		final histol.: 0 rat with meso.[k]	
MMVF21	183.1 mg	10⁹		Lifetime or until signs of debilitation, W, M	40	19/20	meso		Miller et al. (1999)
Amosite	6.1 mg	10⁹		Lifetime or until signs of debilitation, W, M	73	21/24	meso		

Table 63 (contd)

Test substance	Injection schedule (weekly intervals)	No. of fibres L > 5 μm and differences from WHO definition[a]	L (μm); D (μm); L/D > 3/1 (median of all lengths, all diameters); different fibre definitions stated[b]	Observation period (weeks), strain, gender[c]	Median survival time (weeks)	No. of rats with tumours[d]/ no. of rats examined	Histo-logical tumour types[e]	Comments	Reference
Slag wool									
Slag wool Rheinstahl	2 × 20 mg	NG	L, 26; D, 2.6; L/D > 5/1	158, W, F	111	6/99	meso/sarc		Pott et al. (1987)
Slag wool Zimmermann	2 × 20 mg	NG	L, 14; D, 1.5; L/D > 5/1	155, W, F	107	2/96	meso/sarc		
Saline control	2 × 2 mL		(saline control)	150, W, F	101	0/48			
Slag wool	5 × 30 mg	250 × 10^6; L/D > 5/1; L > 5 μm; D < 2 μm	L, 9.0; D, 1.21; L/D > 5/1	131, W, F	106	2/28	meso/sarc	lung infection months 12–13; mortality 8 of 36 at start	Pott et al. (1991)
Saline control	5 × 2 mL		(saline control)	130, W, F	106	2/50	meso/sarc	lung infection months 12–13; mortality 22 of 72 at start	
MMVF22[f, m]	1 × 20 mg	400 × 10^6	L, 8.7; D, 0.77; L/D > 5/1	130, W, F	99	4/40 (3 macr.[h])	meso	final histol.: 5 rats with meso[k]	Davis et al. (1996b); Roller et al. (1996)
MMVF22[f, m]	1 × 50 mg	1000 × 10^6	L, 8.7; D, 0.77; L/D > 5/1	130, W, F		8/40	meso	final histol.: 8 rats with meso[k]	

Table 63 (contd)

Test substance	Injection schedule (weekly intervals)	No. of fibres L > 5 µm and differences from WHO definition[a]	L (µm); D (µm); L/D > 3/1 (median of all lengths, all diameters); different fibre definitions stated[b]	Observation period (weeks), strain, gender[c]	Median survival time (weeks)	No. of rats with tumours[d]/ no. of rats examined	Histological tumour types[e]	Comments	Reference
MMVF22[f, m]	3 × 50 mg	2900 × 10⁶	L, 8.7; D, 0.77; L/D > 5/1	130, W, F	101	18/38	meso	final histol.: 18 rats with meso[k]	
Saline control	3 × 2 mL	(saline control)		130, W, F	121	0/38		Final histol.: 0 rat with meso[k]	
Tremolite	1 × 3.3 mg	57 × 10⁶	L, 3.4; D, 0.29	130, W, F	98	10/39	meso		
Tremolite	1 × 15 mg	260 × 10⁶	L, 3.4; D, 0.29	130, W, F	74	30/40	meso		
MMVF22	129.6 mg	10⁹		Lifetime or until signs of debilitation, W, M	94	13/24	meso		Miller et al. (1999)
Amosite	6.1 mg	10⁹		Lifetime or until signs of debilitation, W, M	73	21/24	meso		

Table 63 (contd)

G & H, Grünszweig & Hartmann; NG, not given; lifetime, until survival rate is ≤ 20%.

[a] Aspect ratio > 5/1 in Pott et al. (1989, 1991) and Roller et al. (1996); diameter < 2 μm in Pott et al. (1991) and Roller et al. (1996)

[b] Aspect ratio > 5/1 in Pott et al. (1989, 1991) and Roller et al. (1996)

[c] SpD, Sprague Dawley; W, Wistar; M, male; F, female; period of observation after first injection

[d] Calculated from the stated percentage for some groups

[e] Correct histopathological diagnosis of tumours of the abdominal cavity requires considerable experience. Differential diagnoses of sarcomatous mesothelioma, sarcoma (sarc) and metastases of uterus carcinoma have to be considered in addition to other rare tumour types. Such experience was available for the large study by Roller et al. (1996) at which time it was clear that only mesotheliomas (meso) could be attributed to fibrous dusts. In this study, when mesotheliomas were excluded, 49 tumours of the uterus and 18 other tumours were found in the abdominal cavity of 406 female rats (4.4%) and 32 tumours in 661 male rats (4.8%). Three histopathologists with experience in the pathology of rodent neoplasia reviewed a large series of slides from this study of which a few presented some diagnostic difficulty. This most commonly involved uterine adenocarcinoma with widespread peritoneal metastases. There was good agreement among the histopathologists on the percentage or absence of malignant mesothelioma. In most cases, these neoplasms appeared similar to other malignant mesotheliomas that have been induced in this animal model. The initial diagnoses were unanimous in 85% of the cases. Following discussion, final agreement was reached in 99% of the cases. A comparison of the final diagnosis made by the panel with the original diagnosis made at the Fraunhofer Institute revealed agreement in 98% of the cases (Davis et al. 1996b). A summary of former results of Pott et al. showed tumours in the abdominal cavity in 23 of 886 female Wistar rats (tumours of the uterus were excluded as far as possible) after intraperitoneal administration of non-fibrous dusts or fibres thicker than 3 μm (2.6 %). After injection of saline, tumours were reported in the abdominal cavity in 13 of 491 rats (2.6%) (Pott et al., 1993).

[f] The chemical composition was published in Roller et al. (1996) and a description of the experimental method in Pott et al. (1993).

[g] This experimental vitreous fibre was produced with a chemical composition analogous to early commercial rock (stone) wool fibres.

[h] Rats with abdominal tumours were diagnosed only macroscopically at the time of publication.

[i] This experimental vitreous fibre had a chemical composition expected to have a low biodurability and, consequently, a low carcinogenic potential.

[j] Statistical test performed versus the control group(s) in italics or their sum

[k] The final data on the histopathological findings were used for calculation of the dose–response relationships and published subsequently (Roller et al., 1997; Roller & Pott, 1998).

[l] Same chemical composition as B-20-2.0 (similar to rock (stone) fibres, see footnote [g]), but with shorter lengths and smaller diameters

[m] This experimental vitreous fibre was produced with a chemical composition analogous to early commercial slag wool fibres.

Groups of 20 male and 20 female Sprague-Dawley rats, 6–8 weeks of age, received an intraperitoneal injection of 25 mg rock (stone) wool fibres [type not specified; no information on size was given for the test material] and were killed after 104 weeks. A complete autopsy was performed on all animals together with a histopathological examination of the peritoneum. Mesotheliomas were reported in 3/40 rats. The average latency time was 80 weeks. The control animals were injected with water and no tumours were reported (Maltoni & Minardi, 1989). [The Working Group noted that the results are not conclusive as the number of fibres administered and other important details were not presented.]

Female Wistar rats [initial numbers not specified], eight weeks of age, received five weekly intraperitoneal injections of 15 mg of a basalt wool (total dose, 75 mg) [produced by Grünzweig and Hartmann, Germany, chemical composition not given]. The median length of the fibres was 17 µm and the median diameter was 1.1 µm. The total number of WHO fibres (with L/D > 5/1) injected was 59×10^6. The median life-span was 79 weeks after first treatment. A post-mortem examination of the abdominal cavity was made. Parts of tumours or organs in which macroscopic tumour tissue was found were investigated by histopathological examination. Mesotheliomas were diagnosed in 30/53 treated animals (lifespan of first animal with tumour, 54 weeks). In a control group that received five weekly injections of saline, mesothelioma was observed in 2/102 animals and the median lifespan was 111 weeks after first treatment, showing a clear increase in the mortality of the fibre-treated group. In a further group of rats, treated with 100 mg titania, the median survival time was 109 weeks and 2/53 rats had mesotheliomas or sarcomas (Pott et al., 1989)

In a chronic study, three groups of rats [age, sex and strain not specified] received two monthly intraperitoneal injections of 25 mg of basalt or chrysotile asbestos dusts suspended in saline [chemical composition of dust was not given]. Forty animals in the first group received basalt dust (17% of the dust fibres had length > 5 µm and diameter < 3 µm). Fifty animals of the second group received basalt dust (41% of the dust fibres had length > 5 µm and diameter < 3 µm). Sixty rats in the third group were injected with chrysotile asbestos dust. Fibres with length > 5 µm and diameter < 3 µm comprised 76% of the fibre fraction of the asbestos dust. The granular fractions of all dusts varied from 87.9%–88.9%. Control animals (110 rats) were injected intraperitoneally with the saline solution alone. Animals were observed for life. All the materials tested induced peritoneal mesotheliomas. The numbers were 5/40, 7/50 and 27/60 rats in the first basalt dust-, second basalt dust- and chrysotile asbestos-treated groups, respectively. No mesotheliomas were found in control animals (Nikitina et al., 1989).

Female Wistar rats [initial numbers not specified], weighing approximately 190 g, received a single intraperitoneal injection of 25 mg or five weekly injections of 30 mg (total dose, 150 mg) basalt wool (produced by Grünzweig & Hartmann, Germany) [chemical composition not given]. The median length of the fibres was 13.8 µm and the median diameter was 1.08 µm. The numbers of fibres injected (length, > 5 µm; diameter, < 2 µm; aspect ratio, > 5:1) were 5×10^6 and 30×10^6, respectively. The survival time was much reduced by an infectious disease of the lungs during months 12 and 13 of the study, the cause of which could not be unequivocally determined. In the low-dose group, 10/48 rats died from the infection and 15/36 in the high-dose group. The mean lifespan was 110 weeks in the low-dose group and 84 weeks in the high-dose group. For those rats that developed tumours, the mean lifespan was 128 weeks in the low-dose group and 89 weeks in the high-dose group. Macroscopic tumours were observed in 1/38 rats in the low-dose group and in 15/21 rats in the high-dose group. A control group of 72 rats was treated with five weekly injections of 2 mL saline solution. Of these animals, 22 died from the infection and the median lifespan was 106 weeks. Two of the remaining 50 rats developed an abdominal tumour (Pott et al., 1991). [The Working Group found it difficult to interpret the study since the impact of the infection was unknown.]

Four groups of 40 female Wistar rats, 8–10 weeks of age, received intraperitoneal injections of one of two rock (stone) wools (MMVF21 and R-stone E3) suspended in 2 mL saline administered as two or five weekly injections of 30 mg MMVF21, or four or nine weekly injections of 28.5 mg of R-stone E3. A negative-control group of 40 female Wistar rats received no treatment and a positive-control group of 40 female rats received five intraperitoneal injections of 0.1 mg of a UICC crocidolite sample. The animals were observed up to 130 weeks after first injection. The doses, dose schedules and characteristics of all the fibres tested are given in Table 63, together with data on median survival and incidence of mesotheliomas. The incidences of mesotheliomas were: 37/38 and 33/38 in the rats treated with MMVF21; 0/30 and 4/35 in the rats treated with R-stonewool E3; 20/39 in the rats treated with crocidolite; and 0/37 in untreated control rats (Pott et al., 1993; Davis et al., 1996b; Roller et al., 1996).

A further experiment was reported as part of this study. Five groups of 32–36 male or 32–36 female Wistar rats, 8–10 weeks of age, were administered M-stone (a typical rock (stone) wool; Manville Technical Center, Denver, CO, USA; median length of the fibres was 10.1 µm and the median diameter was 0.84 µm) in 2 mL saline. The rats received either single intraperitoneal injections of 8.5 or 25.5 mg (males and females) or two injections (at a two-week interval) of 42.5 mg (males only). Five other groups of 32–36 male or 32–36 female Wistar rats, 8–10 weeks of age, received intraperitoneal injections of experimental rock (stone) wool B-20-2.0 (Bayer, Germany; median length of the fibres was 6.6 µm and the median diameter was 0.83 µm) suspended in 2 mL saline and administered as a single injection of 6 or 18 mg (males and females) or as two weekly injections of 30 mg (males only). Animals were observed for up to 130 weeks. A control group of 96 female and 72 male Wistar rats received 20 weekly intraperitoneal injections of 2 mL saline alone. Macroscopic tumours were investigated histopathologically. The incidence of peritoneal mesotheliomas in the five groups treated with M-stone was 2/32, 2/36, 9/32, 8/36 and 22/35. In the groups treated with B-20-2.0, mesotheliomas were observed in 2/32 females and 15/36 males in the low-dose group, 7/32 females and 12/34 males in the mid-dose group and in 21/35 animals that received the high dose. In the control group treated with saline, no tumour was observed in any of the 93 females and a mesothelioma was observed in one of the 69 males. Two groups of positive control animals (one female and one male) received five weekly intraperitoneal injections of 0.1 mg UICC crocidolite. The median length of the fibres was 1.8 μm and the median diameter 0.19 μm. Mesotheliomas were observed in 25/32 females and 32/48 males examined. Two groups of negative-control animals (48 female and 72 male Wistar rats) received 20 weekly intraperitoneal injections of 50 mg granular silicon carbide. No mesotheliomas were observed in any of the 45 female or 70 male rats examined (Pott *et al.*, 1993; Davis *et al.*, 1996b; Roller *et al.*, 1996).

An additional experiment performed as part of the same study was also reported. Four groups of 40 female Wistar rats, 8–10 weeks of age, received intraperitoneal injections of experimental rock (stone) wool B-20-0.6 (Bayer, Germany) suspended in 2 mL saline either as a single injection of 3.5, 8.5 or 25 mg or as three weekly injections of 25 mg. A negative-control group consisted of 40 female Wistar rats that received three intraperitoneal injections of 2 mL saline alone and two groups of positive controls received injections of either 3.3 or 15 mg tremolite (Libby, Montana). The incidences of mesotheliomas were: 12/40, 17/40, 30/40 and 27/31 in the rats treated with B-20-0.6; 10/39 and 30/40 in the rats treated with tremolite compared to 0/38 in the control rats treated with saline (Pott *et al.*, 1993; Davis *et al.*, 1996b; Roller *et al.*, 1996).

Groups of about 24 male Wistar rats, approximately 12 weeks of age, received two injections of MMVF21 of a mass dose of 183.1 mg suspended in buffered saline. The target dose was 10^9 WHO fibres; the number of fibres with length > 5 μm was 0.8×10^9 per rat and the diameters were < 0.95 μm and > 0.95 μm. Animals were kept for life or until they showed signs of debilitation. At autopsy, the peritoneal contents were examined macroscopically for the presence of peritoneal mesotheliomas. In addition, specimens from the first six animals to develop mesotheliomas were taken for histopathological examination. The median survival time was 281 days. The median survival time of rats that developed mesotheliomas was 284 days. An abdominal mesothelioma was diagnosed in 19/20 animals examined. Tumours were diagnosed by the macroscopic presence of peritoneal mesotheliomas and by microscopy when the diagnosis was in doubt. In a group of positive controls, male Wistar rats were treated with 6.1 mg amosite fibres with the same target concentration of 10^9 WHO fibres. About 99% of the fibres had diameters below 0.95 μm. Mesotheliomas were observed in 21/24 of the rats in this group (Miller *et al.*, 1999). [The Working Group noted that no vehicle control group was included in this study.]

3.4.4 *Intrapleural injection*

Rat: Groups of 48 SPF Sprague-Dawley rats [sex and age unspecified] received single intrapleural injections of 20 mg Swedish rock (stone) wool [type of rock unspecified] or chrysotile in 0.5 mL saline. The dust samples used (and the size distributions of fibres) were: Swedish rock (stone) wool with resin coating (70% fibres < 5 μm in length; 52% < 0.6 μm in diameter), Swedish rock (stone) wool after removal of resin (70% < 5 μm in length; 58% < 0.6 μm in diameter) and UICC African chrysotile [fibre sizes unspecified]. The animals were kept until natural death [survival times unspecified]. Three mesotheliomas occurred in the group treated with rock (stone) wool with resin and two in the group treated with rock (stone) wool without resin; six mesotheliomas occurred in the group treated with chrysotile. No tumour was observed in a group of 24 controls treated with saline (Wagner et al., 1984).

3.5   Slag wool

3.5.1   *Inhalation exposure* (see Table 61)

Rat: A group of 55 female Osborne-Mendel rats, 100 days of age, was exposed by inhalation (nose only) to slag wool dust [type of slag unspecified] (mass concentration, 7.8 mg/m^3; 15.2% respirable fibres — geometric mean diameter, 0.9 µm; geometric mean length, 22 µm; chamber concentration, 200 fibres/cm^3 with 76 fibres > 10 µm in length and ≤ 1.0 µm in diameter) for 6 h per day on five days per week for two years and then observed for life. Groups of 59 chamber and 125 room controls were also kept. No tumour of the respiratory tract was observed in any group. Average survival time in the group treated with slag wool was shorter (677 days) than that of chamber (754 days) and room (724 days) controls. Of 57 rats exposed to UICC crocidolite (3000 fibres/cm^3; 5% fibres ≥ 5 µm in length; mean, 3.1 ± 10.2 µm), two developed bronchoalveolar tumours and one, a mesothelioma (Smith et al., 1987). [No information was given on chemical composition of the slag wool or on animal respirability. The number of slag wool fibres retained was relatively low.]

Groups of 140 male Fischer 344 rats, eight weeks of age, were exposed in nose-only inhalation chambers to three concentrations (3, 16 and 30 mg/m^3) of slag wool (MMVF22) for 6 h per day on five days per week for 104 weeks. These exposure conditions correspond to average numbers of WHO fibres/cm^3 of 30, 131 and 213. The geometric mean diameter of the test fibres was approximately 0.85 µm and the geometric mean length approximately 13 µm. The retained lung burden/mg dry lung tissue after 24 months was: 44.4 × 10^3 WHO fibres and 1.8 × 10^3 fibres > 20 µm for the 3 mg/m^3 dose, 96.7 × 10^3 and 4.5 × 10^3, respectively, for the 16 mg/m^3 dose and 177 × 10^3 and 11.0 × 10^3, respectively, for the 30 mg/m^3 dose. The post-exposure period was continued until approximately 20% of the animals in the air-control group survived; this occurred four months after the end of exposure. Mortality was similar to that observed in the unexposed controls. In the group that received the 3 mg/m^3 dose, one adenoma and one carcinoma were observed in 116 rats. No tumours were observed in the group that received the 16 mg/m^3 dose. In the group treated with 30 mg/m^3 slag wool dust, two pulmonary adenomas and one lung carcinoma were observed in 115 rats. In the control group, two lung adenomas were found in 126 rats. A positive control group was exposed to crocidolite asbestos (10 mg/m^3), with a geometric mean diameter of 0.28 µm determined by SEM and a geometric mean length of 4.1 µm determined by light microscopy. Exposure to crocidolite asbestos was terminated after 10 months because of increased morbidity and mortality. The retained lung burden/mg dry lung tissue after 104 weeks was 759×10^3 WHO fibres and 41.1×10^3 fibres > 20 μm. Ten pulmonary adenomas, five pulmonary carcinomas and one mesothelioma were reported in 106 rats (McConnell *et al.*, 1994).

3.5.2 *Intraperitoneal injection* (see Table 63)

(a) Rat

Groups of female Wistar rats, 15 weeks old, received two weekly intraperitoneal injections of 20 mg of one of two samples of Rheinstahl and Zimmermann (Germany) slag wool [chemical composition not given] in 2 mL saline. The Rheinstahl sample had a median fibre length of 26 μm and a median fibre diameter of 2.6 μm; the Zimmermann sample had a median fibre length of 14 μm and a median fibre diameter of 1.5 μm. The animals were observed for life; median survival times were 111, 107 and 101 weeks for the groups given coarser (Rheinstahl) and finer (Zimmermann) slag wool and for a control group treated with saline alone, respectively. Slight increases in the incidence of sarcomas, mesotheliomas and (rarely) carcinomas of the abdominal cavity were observed with the slag wool samples: 6/99 with the coarser sample (first tumour after 88 weeks) and 2/96 with the finer sample (first tumour after 67 weeks). No tumour occurred in any of the 48 control animals (Pott *et al.*, 1987). [The Working Group noted that, in other studies in this laboratory, the historical incidence of abdominal tumours in animals treated with saline ranged from 0%–6.3%.]

Female Wistar rats [initial numbers not specified], weighing approximately 190 g, received either five weekly intraperitoneal injections of 30 mg slag wool [chemical composition not given] suspended in 2 mL saline or 2 mL saline alone (control group of 72 females). The median length of the fibres was 9.0 μm, the median diameter was 1.21 μm and the number of fibres injected (length > 5 μm, diameter < 2 μm, aspect ratio > 5/1) was 250×10^6. The lifespan of the animals was much reduced by an infectious disease of the lung in months 12 and 13 (the cause could not be diagnosed) which killed 8/36 rats in the treated group. The mean lifespan in both groups was 106 weeks. The mean lifespan for rats that developed tumours was 77 weeks. Macroscopic tumours were observed in 2/28 rats. In the control group, 22/72 animals died from the infection and two of the remaining 50 rats developed an abdominal tumour (Pott *et al.*, 1991). [The Working Group found it difficult to interpret this study since the impact of the infection was unknown.]

Three groups of 40 female Wistar rats, 8–10 weeks old, received a single intraperitoneal injection of 20 or 50 mg or three weekly injections of 50 mg MMVF22 suspended in 2 mL saline. The median length of the fibres was 8.7 μm and the median diameter was 0.77 μm. A control group of 40 female Wistar rats received three intraperitoneal injections of 2 mL saline only. The length of the observation period was 130 weeks. The median survival time was 99 weeks in the group treated with the 20-mg dose, 102 weeks in the group treated with the single 50-mg dose and 95 weeks in the group that received the three 50-mg doses. Necropsy was followed by a macroscopic investigation of the thoracic and abdominal areas for tumours and the organs were transferred into buffered formalin. Macroscopic tumours were investigated histopathologically. The incidence of mesotheliomas was 4/40 in the group given the 20-mg dose, 8/40 after treatment with the single 50-mg dose and 18/38 in the group that received the three 50-mg doses. In the control group treated with saline, no tumour was observed in 38 rats. Two groups of positive controls received an intraperitoneal injection of either 3.3 mg or 15 mg tremolite (Libby, Montana). The median fibre length was 3.4 µm and the median diameter was 0.29 µm; the numbers of fibres injected were 57×10^6 and 260×10^6, respectively. In the group that received 3.3 mg tremolite, 10/39 rats examined had developed mesotheliomas, and mesotheliomas were seen in 30/40 rats in the group treated with 15 mg tremolite (Pott et al., 1993; Davis et al., 1996b; Roller et al., 1996)

Groups of approximately 24 male Wistar rats, approximately 12 weeks of age, received two injections of a slag wool fibre (MMVF22) of a mass dose of 129.6 mg suspended in buffered saline. The target dose was 10^9 WHO fibres. Animals were kept for life or until they showed signs of debilitation. At autopsy, the peritoneal contents were examined macroscopically for the presence of peritoneal mesotheliomas. Tumours were diagnosed by the macroscopic presence of peritoneal mesotheliomas and by microscopy when the diagnosis was in doubt. In addition, specimens from the first six animals to develop mesotheliomas were taken for a histopathological examination. The median survival time was 658 days, and the median survival time of animals with mesothelioma was 695 days. An abdominal mesothelioma was diagnosed in 13/24 rats examined. In a group of positive controls, male Wistar rats were treated with 6.1 mg amosite fibres with the same target concentration of 10^9 WHO fibres. About 99% of the fibres had diameters below 0.95 µm. Mesotheliomas were observed in 22/24 of the rats in this group [no vehicle control group was included in this study] (Miller et al., 1999).

(b) Hamster

A group of 69 male Syrian golden hamsters, 100 days of age, was exposed by inhalation (nose-only) to slag wool dust [type of slag unspecified] (mass concentration, 7.8 mg/m^3; 15.2% respirable fibres; geometric mean diameter, 0.9 µm; geometric mean length, 22 µm; chamber concentration, 200 fibres/cm^3 with 76 fibres/cm^3 > 10 µm in length and ≤ 1.0 µm in diameter) for 6 h per day on five days per week for two years and then observed for life. Groups of 58 chamber and 112 room controls were included in the study. No tumour of the respiratory tract was observed in the treated animals or in room controls; one of 58 chamber controls had a bronchoalveolar tumour. There was no decrease in lifespan (about 660 days). In a group of 58 hamsters exposed to UICC crocidolite (3000 fibres/cm^3; 5% fibres ≥ 5 µm in length; mean, 3.1 ± 10.2 µm), no pulmonary tumour was seen (Smith et al., 1987). [The Working Group noted that the counting criteria for retained fibres were not stated and that the number of slag wool fibres retained was relatively low.]

3.5.3 Intrapleural injection

Rat: Groups of 48 SPF Sprague-Dawley rats [sex and age unspecified] received either a single intrapleural injection of 20 mg German slag wool [type of slag unspecified] or chrysotile in 0.5 mL saline. The dust samples used (and the size distributions of the fibres) were: German slag wool (67% < 5 μm in length; 42% < 0.6 μm in diameter), German slag wool after removal of resin (80% < 5 μm in length; 62% < 0.6 μm in diameter and UICC African chrysotile [sizes of fibres unspecified]. The animals were kept until natural death [survival times unspecified]. No tumour was observed in the group treated with slag wool or in a group of 24 controls treated with saline. Six mesotheliomas occurred in the group treated with chrysotile (Wagner *et al.*, 1984).

3.6 Refractory ceramic fibres

3.6.1 Inhalation exposure (see Table 64)

(a) Rat

A group of 48 SPF Wistar AF/HAN rats [sex and source unspecified], 12 weeks old, was exposed by whole-body inhalation [exposure chamber parameters (size, flow rate, temperature, humidity) and animal husbandry (single or group caging, environmental controls, light/dark cycle, basal diet and water supply) unspecified] to a target concentration of 10 mg/m^3 respirable dust from bulk fibrous ceramic aluminium silicate glass [chemical composition not given] for 7 h per day on five days per week for 12 months (cumulative exposure, 224 days). A group of 40 unexposed rats housed within the same laboratory unit served as controls [age, sex, source and treatment of controls (whether exposed to room or chamber air) unspecified]. Following cessation of exposure to the dust, all the remaining animals were kept until either their natural death or the termination of the experiment at 32 months. The mean concentration of respirable dust was 10.0 ± 4.8 mg/m^3 and that of total dust, 9.6 ± 8.4 mg/m^3. The analysis of respirable dust, by phase contrast optical microscopy (PCOM), showed that the animals were exposed to 95 WHO fibres/cm^3 (> 5 μm in length and < 3 μm in diameter; aspect ratio, > 3:1). Approximately 90% of the fibres were < 3 μm in length and < 0.3 μm in diameter and the ratio of particles (> 1 μm in diameter) to fibres (> 5 μm in length) was approximately 4:1. Survival times did not differ significantly between control and treated animals. Eight animals treated with dust (8/48, 17%) developed pulmonary neoplasms (one adenoma, three carcinomas and four malignant histiocytomas). Pulmonary tumours were not observed in any of the control animals. In addition to the tumours associated with the lung, eight benign and eight malignant tumours [unspecified], including one peritoneal mesothelioma were also found in the group treated with dust. The dust burden of ceramic aluminium silicate in the left lung was converted to whole lung values and ranged from 2.8–6.8 mg (Davis *et al.*, 1984).

Table 64. Studies of the carcinogenicity of refractory ceramic fibres in rats and hamsters

Fibre type	Route of administration	Strain	Sex	Exposure conc. (mg/m³)	Exposure conc. (f/cm³)	Exposure duration (h/d, d/wk), total exposure (weeks), observation (weeks)	No. of animals/ group	Lung tumours (adenoma and carcinoma)	Tumours (pleural or peritoneal)	Lung burden	Reference
Rat											
RCF (NS)	Whole-body inhalation	SPF Wistar AF/HAN	NS	10	95 WHO f/cm³, 90% with D, < 3 µm; prt/fib, 4:1	(7, 5), 52, lifetime or 128	48	8/48 (17%); 0 in controls	0	–	Davis et al. (1984)
Fibrefrax®	Nose-only inhalation	Osborne-Mendel	F	10.8	88 f/cm³ with L, > 10 µm; GMD, 0.9 µm; GML, 25 µm; prt/fib, 33:1	(6, 5), 104, lifetime	55	0/55	–	2.18 × 10⁴ f/mg dry lung	Smith et al. (1987)
Kaolin-based RCF1	Nose-only inhalation	Fischer 344	M	30	187 WHO f/cm³; GMD, ~0.8 µm; GML, 12.8–17.4 µm	(6, 5), 104, lifetime	140	16/123 (13%); controls, 2/130 (1.5%)	2/123 (1.6%); controls, 0	2–7 × 10⁵ WHO f/mg dry lung	Mast et al. (1995a)
Alumina-zirconia silica RCF2	Nose-only inhalation	Fischer 344	M	30	220 WHO f/cm³; GMD, ~0.8 µm; GML, 12.8–17.4 µm	(6, 5), 104, lifetime	140	9/121 (7.4%); controls, 2/130 (1.5%)	3/121 (2.5%); controls, 0	2–7 × 10⁵ WHO f/mg dry lung	

Table 64 (contd)

Fibre type	Route of administration	Strain	Sex	Exposure conc. (mg/m^3)	Exposure conc. (f/cm^3)	Exposure duration (h/d, d/wk), total exposure (weeks), observation (weeks)	No. of animals/group	Lung tumours (adenoma and carcinoma)	Tumours (pleural or peritoneal)	Lung burden	Reference
High-purity RCF3	Nose-only inhalation	Fischer 344	M	30	182 WHO f/cm^3; GMD, ~0.8 µm; GML, 12.8–17.4 µm	(6, 5), 104, lifetime	140	19/121 (10.7%); controls, 2/130 (1.5%)	2/121 (1.7%); controls, 0	$2-7 \times 10^5$ WHO f/mg dry lung	Mast et al. (1995b)
After-service RCF4	Nose-only inhalation	Fischer 344	M	30	206 WHO f/cm^3; GMD, 1.22 µm; GML, 9.8 µm	(6, 5), 104, lifetime	140	4/118 (3.4%); controls, 2/130 (1.5%)	1/118 (0.8%); controls, 0	$2-7 \times 10^5$ WHO f/mg dry lung	
Kaolin-based RCF1	Nose-only inhalation	Fischer 344	M	3 9 16	26 75 120 WHO f/cm^3; GMD, 0.8 µm; GML, 14 µm	(6, 5), 104, lifetime	140	2 (1.6%) 5 (3.9%) 2 (1.6%); controls, 1 (0.8%)	0 1 (0.8%) 0; controls, 0	4.3×10^4 NS 22.1×10^4 WHO f/mg dry lung	
Fibrefrax®	Intra-tracheal instillation	Osborne-Mendel	F	5 × 2 mg (weekly)	(Elutriated from inhalation chamber)	lifetime	22	0	–		Smith et al. (1987)

Table 64 (contd)

Fibre type	Route of administration	Strain	Sex	Exposure conc. (mg/m³)	Exposure conc. (f/cm³)	Exposure duration (h/d, d/wk), total exposure (weeks), observation (weeks)	No. of animals/ group	Lung tumours (adenoma and carcinoma)	Tumours (pleural or peritoneal)	Lung burden	Reference
Refractory ceramic fibre (NS)	Intrapleural injection	SPF Wistar	M/F	20 mg in 0.4 mL	(Ball mill grinding)	lifetime	31	–	3/31 (9.7%), pleural meso	–	Wagner et al. (1973)
Kaolin (fibre A)/ alumina and silica (fibre B)	Intrapleural injection	Alpk:AP (Wistar-derived)	M/F	20 mg in 0.2 mL	(Ball mill grinding and sieving)	lifetime	24 M/ 24 F	–	Fibre A, 0 Fibre B: 1/48, (2.1%, pleural meso); 2/48 (4.1%, peritoneal meso)	–	Pigott & Ishmael (1992)
High Duty™ grade alumino-silicate, vitreous or devitrified	Intrapleural injection	Wistar-Porton	M	20 mg in 0.4 mL	NS	lifetime	19	–	0	–	Carthew et al. (1995)
Refractory ceramic fibre (NS)	Intraperi-toneal injection	Wistar AF/HAN	NS	25 mg in 2 mL	NS; L, 90% < 3 μm; D, < 0.3 μm	NS	32	–	3/32 (9.4%) peritoneal tumours	–	Davis et al. (1984)

Table 64 (contd)

Fibre type	Route of administration	Strain	Sex	Exposure conc. (mg/m³)	Exposure conc. (f/cm³)	Exposure duration (h/d, d/wk), total exposure (weeks), observation (weeks)	No. of animals/group	Lung tumours (adenoma and carcinoma)	Tumours (pleural or peritoneal)	Lung burden	Reference
Fibrefrax®	Intraperitoneal injection	Wistar WU/ Kißlegg	F	45 mg in 2 mL	L, 8.3 μm; D, 0.91 μm	28 months after injection	~50	–	32/47 (68%)	–	Pott et al. (1987, 1989)
Manville				75 mg in 2 mL	L, 6.9 μm; D, 1.1 μm				12/54 (22%) TiO₂ control, 5/53 (9.4%)		
Fibrefrax®	Intraperitoneal injection	Osborne-Mendel	F	25 mg in 0.5 mL	(Elutriated from inhalation chamber)	lifetime	25	–	19/23 (83%); control, 0/25	–	Smith et al. (1987)
RCF1	Intraperitoneal injection	Charles River Wistar	M	110 mg in 2 mL	228 × 10⁶ f > 10 μm	lifetime	18–24	–	21/24 (88%)	–	Miller et al. (1999)
RCF2				188 mg in 2 mL	320 × 10⁶ f > 10 μm				13/18 (72%)		
RCF4				90 mg in 2 mL	81 × 10⁶ f > 10 μm or 10⁹ f > 5 μm				0/22		

Table 64 (contd)

Fibre type	Route of administration	Strain	Sex	Exposure conc. (mg/m³)	Exposure conc. (f/cm³)	Exposure duration (h/d, d/wk), total exposure (weeks), observation (weeks)	No. of animals/ group	Lung tumours (adenoma and carcinoma)	Tumours (pleural or peritoneal)	Lung burden	Reference
Hamster											
Fibrefrax®	Nose-only inhalation	Syrian golden	M	10.8	200 f/cm³, GMD, 0.9 μm; GML, 25 μm; prt/fib, 33:1	(6, 5), 104, lifetime	70	0; control, 1/58	1/70; control, 0	0.86 × 10⁴ f/mg dry lung	Smith et al. (1987)
RCF1	Nose-only inhalation	Syrian golden	M	30	215 WHO f/cm³; GMD, 0.78 μm; GML, 15.9 μm	(6, 5), 78, lifetime	140	0	42/102 (41%); control, 0	1.59 × 10⁵ f/mg dry lung	McConnell et al. (1995)
Fibrefrax®	Intratracheal instillation	Syrian golden	M	5 × 2 mg (weekly)	(Elutriated from inhalation chamber)	lifetime	25	0	–		Smith et al. (1987)
Fibrefrax®	Intraperitoneal injection	Syrian golden	F	25 mg in 0.5 mL	(Elutriated from inhalation chamber)	lifetime	56	0	7/36; control, 0		Smith et al. (1987)

NS, not specified; prt/fib, ratio of particles to fibres; f., fibre; RCF, refractory ceramic fibres; GMD, geometric mean diameter; GML, geometric mean length; lifetime, until survival rate is ≤ 20%; D, diameter; L, length; M, male; F, female; meso, mesothelioma

A group of 55 female Osborne-Mendel rats, 100 days of age, was exposed by nose-only inhalation to an aerosol of a refractory ceramic fibre (Fibrefrax®) [chemical composition not given] at a mass concentration of 10.8 mg/m^3 for 6 h per day on five days per week for two years and then observed for life. A group of 60 controls was exposed to filtered air by nose-only inhalation and 125 room cage controls were kept. The aerosol fibres had a geometric mean diameter of 0.9 µm and a geometric mean length of 25 µm; it contained approximately 88 fibres/cm^3 > 10 µm in length having diameters ≤ 1.0 µm and a particulate to fibre ratio of 33:1. [No information was available on animal respirability.] The fibre content of the lungs was assessed in only a small sample of animals [number unspecified]. Exposure to refractory ceramic fibres had no effect on overall health or survival of the animals. The lung burden of fibres was 2.18 ± 0.99 × 10^4 fibres/mg dry lung weight [particulates and size distribution of fibres unspecified]. [The lung burden of retained fibres was relatively low.] No pulmonary tumours were observed in any of the treated (0/55) or control (0/60) rats (Smith et al., 1987).

Six groups of 140 male weanling Fischer 344 rats were exposed to refractory ceramic fibres by nose-only inhalation. A control group exposed to high-efficiency particulate air (HEPA)-filtered air was included. Four groups of animals were exposed to 30 mg/m^3 of one of three types (kaolin-based RCF1; alumina zirconia silica — AZS RCF2, or high-purity RCF 3) of size-separated refractory ceramic fibres or to an after-service heat-treated (1316 °C for 24 h to simulate after-service material) kaolin-based fibre containing approximately 27% crystalline silica in the form of cristobalite) (RCF4) for 6 h per day on five days per week for 24 months. An additional group of 80 animals was exposed to 10 mg/m^3 chrysotile asbestos. Following the 24-month exposure period, the rats were kept for lifetime observation (until approximately 20% survival at 30 months), at which time they were killed. Aerosol exposure concentrations of 30 mg/m^3 for the refractory ceramic fibres and 10 mg/m^3 for chrysotile were achieved and maintained during the study. The counts of WHO fibres at these concentrations corresponded to 187 ± 53, 220 ± 52, 182 ± 66, 206 ± 48 and 1.06 × 10^4 fibres/cm^3 for RCF1, RCF2, RCF3, RCF4 and chrysotile, respectively. The GMD of RCF1, RCF2 and RCF3 in the aerosol ranged from 0.82–0.88 µm and the GML ranged from 12.8–17.4 µm. The RCF4 was somewhat thicker (GMD, 1.22 µm) and shorter (GML, 9.8 µm). The corresponding dimensions for chrysotile were 0.08 µm and 1.2 µm, respectively. [In view of the known association between fibre length and carcinogenic potential, the differences in mean length between the refractory ceramic fibres and chrysotile make direct comparison difficult.] The particulate (< 3 µm diameter) to fibre ratio was reported to range between 1.02:1 and 1.88:1 for the refractory ceramic fibres. [This estimate has been revised in Maxim et al. (1997), Mast et al. (2000a,b) and Bellmann et al. (2001).] Lung burdens at 24 months ranged from approximately 2 × 10^5 to 7 × 10^5 WHO fibres/mg dry lung tissue. The numbers of exposure-related pulmonary neoplasms (bronchoalveolar adenomas and carcinomas) were significantly increased in all the groups exposed to fibres (except that exposed to RCF4) (RCF1, 16/123 (13%);

RCF2, 9/121 (7.4%); RCF3, 19/121 (15.7%); RCF4, 4/118 (3.4%); the number in the group that received chrysotile was 13/69 (19%) compared with 2/130 (1.5%) in the untreated chamber controls. A few mesotheliomas were observed in each of the groups exposed to fibres (RCF1, 2/123 (1.6%); RCF2, 3/121 (2.5%); RCF3, 2/121 (1.7%); RCF4, 1/118 (0.8%); chrysotile, 1/69 (1.4%)), compared with 0/130 in the untreated controls (Mast et al., 1995a).

The study by Mast et al. (1995a) was followed up with a multiple-dose study of the kaolin-based refractory ceramic fibre (RCF1) conducted in the same laboratory using the identical lot of RCF1, animal source and experimental design. Four groups of 140 weanling Fischer 344 rats were exposed by nose-only inhalation to HEPA-filtered air (chamber controls) or to 3, 9 or 16 mg/m^3 of RCF1 (corresponding to 26, 75 and 120 WHO fibres/cm^3 and a reported particulate to fibre ratio of 0.9–1.5:1 [as noted above, this has since been recalculated]; GMD, 0.8 μm; GML, 14 μm) for 6 h per day on five days per week for 24 months. Following the 24-month period of exposure, the rats were kept for lifetime observation (until approximately 20% survival), at which time they were killed (30 months). The lung fibre burden (WHO fibres/mg dry lung weight) at 24 months ranged from 4.3×10^4 in the animals treated with 3 mg/m^3 to 22.1×10^4 in the group that received 16 mg/m^3. The numbers of pulmonary neoplasms (adenoma and carcinoma combined) showed no statistically significant increase at any of the concentrations of RCF1 tested when compared with control animals and were within the range reported as typical in the male Fischer 344 rat. At 30 months, when the animals were killed, a single very small mesothelioma was seen in one rat that had been exposed to 9 mg/m^3 RCF1 (Mast et al., 1995b).

[The Working Group noted that the greater particulate fraction of RCF1 (than other RCFs) could have influenced the development of inflammation and subsequent carcinogenic response in the chronic inhalation studies of RCF1. The extent of this influence is difficult to assess quantitatively (Yu et al., 1994; Hesterberg et al., 1995a,b; Mast et al., 1995a; Gelzleichter et al., 1996a; Bernstein et al., 1997; Creutzenberg et al., 1997; Maxim et al., 1997; Brown, 2000; Brown et al., 2000a; Mast et al., 2000a,b; Bellmann et al., 2001).]

(b) *Hamster*

A group of 70 male Syrian golden hamsters, 100 days old, was exposed by nose-only inhalation to refractory ceramic fibres (Fibrefrax®) [chemical composition not given] at a mass concentration of 10.8 mg/m^3 for 6 h per day on five days per week for 24 months. The control groups consisted of 58 chamber controls exposed to air and 112 room cage controls. A group of 58 hamsters was exposed to UICC crocidolite (7 mg/m^3; 3000 fibres/cm^3; 95% ≤ 5 μm in length; 90 fibres/cm^3 > 10 μm long). The exposure aerosol fibres had a GMD of 0.9 μm and a GML of 22 μm. Assuming a concentration of 200 fibres/cm^3, the aerosol was calculated to contain approximately 88 fibres/cm^3 > 10 μm in length and ≤ 1.0 μm in diameter. A respirable mass fraction of 35 ± 7% (percentage weight) was estimated. The particle-to-fibre ratio was reported to be 33:1.

The fibre content of the lungs (0.86 × 10⁴ fibres/mg dry lung) was assessed in a small sample of animals [actual number of animals and time of sampling unspecified]. At 10 months of treatment, one hamster exposed to the refractory ceramic fibre (1/70) developed a mesothelioma. A chamber control hamster (1/58) had a secretory bronchoalveolar tumour. No pulmonary tumours were seen in the hamsters exposed to crocidolite although there was a significant incidence of bronchoalveolar hyperplasia (11/58) and interstitial fibrosis (14/58) (Smith *et al.*, 1987).

Two groups of 140 weanling male Syrian golden hamsters were exposed by nose-only inhalation to either HEPA-filtered air (chamber controls) or to 30 mg/m³ (215 WHO fibres/cm³; GMD, 0.78 μm; GML, 15.9 μm) size-selected refractory ceramic fibres (RCF1) for 6 h per day on five days per week for 18 months. (This study was conducted with the identical lot of refractory ceramic fibres and in the same laboratory as the studies reported by Mast *et al.*, 1995a,b.) Once the period of exposure had ended the hamsters were kept until approximately 20% survival and then killed (20 months). A group of 80 hamsters exposed to 10 mg/m³ (3.0 × 10³ WHO fibres/cm³; GMD, 0.08 μm; GML, 0.98 μm) NIEHS chrysotile acted as the positive control. Survival was unaffected by exposure to RCF1 but was significantly reduced by chrysotile. Lung burden of fibres (WHO fibres/mg dry lung) at 18 months was 1.59 × 10⁵ and 16.5 × 10⁵ in the groups treated with RCF1 and chrysotile, respectively. No pulmonary neoplasms occurred in any experimental group. Mesotheliomas developed in 42/102 (41%) of the animals exposed to RCF1, but not in any other group. While the first mesothelioma was found at 10 months of treatment, most (24/42) were not seen until after 18 months. Many were readily visible at necropsy (57%) and involved both the visceral and parietal pleura (McConnell *et al.*, 1995). [The Working Group noted that all the hamsters had been treated with tetracycline for an intestinal infection that they developed during the study.]

3.6.2 Intratracheal instillation

(a) Rat

A group of 22 female Osborne-Mendel rats, 100 days old, received five weekly intratracheal instillations of 2 mg refractory ceramic fibres (Fibrefrax®) [chemical composition not given], which had been elutriated from the chambers of a nose-only inhalation study. The fibres in the inhalation aerosol had a GMD of 0.9 μm, a GML of 25 μm, contained approximately 88 fibres/cm³ with lengths > 10 μm and diameters ≤ 1.0 μm, and a particulate-to-fibre ratio of 33:1. [The length and diameter distributions and particulate count of the instilled material were unspecified.] A group of 25 control females was treated with saline. The treated animals and controls were then kept for their natural lifespan. Treatment with the refractory ceramic fibre had no effect on median average lifespan (698 days) compared with controls treated with saline (688 days). No pulmonary tumours developed in any of the rats treated with refractory ceramic fibres or with saline. Of 25 rats treated similarly with UICC crocidolite [size distribution of elutriated material unspecified], 2/25 (8%) developed bronchoalveolar tumours (Smith *et al.*, 1987).

(b)   Hamster

A group of 25 male Syrian golden hamsters, 100 days old, received five weekly intratracheal instillations of 2 mg refractory ceramic fibres (Fibrefrax®) [chemical composition not given], which had been elutriated from the chambers of a nose-only inhalation study. The fibres in the inhalation aerosol had a GMD of 0.9 μm, a GML of 25 μm, contained approximately 88 fibres/cm^3 with lengths > 10 μm and diameters ≤ 1.0 μm, and a particulate-to-fibre ratio of 33:1. [The length and diameter distributions and particulate count of the instilled material were unspecified.] A group of 24 males treated with 0.2 mL saline acted as controls. The treated animals and controls were then kept for their natural lifespan. The median average lifespan of hamsters treated with refractory ceramic fibres (446 days) was significantly shorter than that of the controls treated with saline (567 days). No pulmonary tumours were seen either in any hamster treated with refractory ceramic fibres or in any control animal treated with saline. Of 27 hamsters treated similarly with UICC crocidolite [size distribution of elutriated material unspecified], 20/27 (74%) developed bronchoalveolar tumours. Of these primary tumours, 13 were benign and seven were malignant (Smith *et al.*, 1987).

3.6.3   *Intrapleural injection*

Rat: Groups of 31–36 SPF Wistar rats (twice as many males as females) [number of each sex unspecified], 13 weeks old, received a single intrapleural injection, in 0.4 mL sterile saline, containing 20 mg suspended solids of numerous different fibres and dusts including ceramic aluminium silicate fibres [chemical composition not given] prepared by grinding. The mean survival time was 736 days, which was similar to that in controls (728 days) being used at the same time in other experiments. Thirty-one of the animals injected with refractory ceramic fibres were examined by microscopy and three were found to have pleural mesotheliomas (9.7%). The first appeared 743 days after treatment. In this experiment, pleural mesotheliomas were observed in 23/36 (64%) of the rats injected with SFA chrysotile (20 mg); the first tumour appeared 325 days after injection [non-neoplastic disease unspecified] (Wagner *et al.*, 1973).

Groups of 24 male and 24 female Alpk:AP (Wistar-derived) rats, eight weeks old, received a single intrapleural injection of 20 mg suspended solids of aluminosilicate fibres in 0.2 mL saline (fibre A made from kaolin and fibre B from alumina and silica, prepared by grinding and sieving) [source and purity unspecified] (treatment was arranged in replicates and took place over a period of five weeks) and animals were observed for life or until 85% mortality when the survivors were killed [months of total exposure at termination unspecified]. The fibre dimensions were as follows: fibre A:

diameter ≤ 3 μm, 66%; length ≥ 10 μm, 80%; fibre B: diameter ≤ 3 μm, 92%; length ≥ 10 μm, 46%). Control animals received an equivalent injection of 0.2 mL sterile saline. A similar group of animals was injected with UICC chrysotile A [length and diameter distribution unspecified] at the same concentration of suspended solids. At 24 months, survival was not affected by the treatment (control, 79%; fibre A, 96%; fibre B, 83%; chrysotile, 100%) and was not significantly different between groups at termination (control, 29%; fibre A, 41%; fibre B, 29%; chrysotile, 29%). No mesotheliomas were observed in animals treated with fibre A. Treatment-related neoplastic disease was limited to the development of pleural (1/48, 2.1%) or peritoneal mesotheliomas in (2/48, 4.1%) of the rats treated with fibre B (the peritoneal mesotheliomas were the result of a partial deposition of the dose in the peritoneum) and 7/48 (14.5%) pleural mesotheliomas in rats treated with chrysotile A (Pigott & Ishmael, 1992).

Three groups of 19, 21 and 24 male Wistar-Porton rats, weighing 200 g, received a single intrapleural injection of 0.4 mL saline containing 20 mg suspended solids of High Duty™ grade aluminosilicate fibres prepared from commercially available fibre blanket [length, diameter, chemical composition and quantity of non-fibrous particulates unspecified] or one of two samples of fibres that had previously been devitrified at 1400 °C or 1200 °C for 14 days. A control group of 30 rats received intrapleural saline only. Following treatment, the rats were observed for life. There was no statistically significant difference in mean survival between control and treated animals and cumulative mortality curves for control and treated groups were very similar. The study was apparently [read from a graph] terminated 900 days after treatment [actual time of termination unspecified]. No pleural mesotheliomas were observed in any of the three groups exposed to fibres or in the control group (Carthew *et al.*, 1995).

3.6.4 *Intraperitoneal injection*

(*a*) *Rat*

A group of 32 Wistar AF/HAN rats [sex and age unspecified] received a single 2 mL intraperitoneal injection of 25 mg refractory ceramic fibres [chemical composition not given] suspended in buffered saline [treatment of the control group unspecified]. The refractory ceramic fibres were collected from inhalation exposure chambers in use at that time for an ongoing animal study and the material collected was as similar as possible [length, diameter, particle mass and size of refractory ceramic fibres used for injection unspecified] to that entering the inhalation chamber (approximately 90% of the fibres were < 3 μm in length and < 0.3 μm in diameter and the ratio of particles (> 1 μm in diameter and > 5 μm in length) to fibres was approximately 4:1). At the end of the study [duration unspecified], 3/32 (9.4%) of the treated and 2/29 of the control rats had developed peritoneal tumours [unspecified] (Davis *et al.*, 1984).

In two preliminary studies, groups of approximately 50 female Wistar WU/Kiβlegg rats, 8 weeks old, received intraperitoneal injections (once a week for five weeks) of one of the following: Fibrefrax® refractory ceramic fibre wool (total dose, 45 mg; fibre length, 8.3 µm; fibre diameter, 0.91 µm) [chemical composition not given], Manville refractory ceramic fibre wool (total dose, 75 mg; fibre length, 6.9 µm; fibre diameter, 1.1 µm) [chemical composition not given] or titanium dioxide (total dose, 100 mg) [P25 from Degussa, Germany] suspended in 2 mL saline. A group of 102 animals served as controls and received 2 mL of saline once a week for five weeks. Necropsy at 28 months after injection showed the incidence of macroscopic abdominal tumours was 32/47 (68%) in the first group treated with Fibrefrax® (first tumour death at 30 weeks), 12/54 (22%) in the second group treated with Manville (first tumour death at 60 weeks), 5/53 (9.4%) in the group treated with titanium dioxide (first tumour death at 38 weeks) and 2/102 (2%) in the control group treated with saline (first tumour death at 93 weeks). The median survival time in both the groups of animals treated with refractory ceramic fibres was reduced (51 and 91 weeks, respectively) compared with that in the controls treated with saline (111 weeks) (Pott *et al.*, 1987, 1989). [The Working Group noted that these two publications presented essentially the same data, with minor discrepancies. It is unclear whether any histopathological examination was performed.]

A group of 25 female Osborne-Mendel rats, 100 days old, received a single intraperitoneal injection in 0.5 mL saline of 25 mg refractory ceramic fibre (Fibrefrax®) [chemical composition not given], which had been elutriated from the chambers of a nose-only inhalation study. The fibres in the inhalation aerosol had a GMD of 0.9 µm and a GML of 25 µm. [The length and diameter distributions and particulate counts for the elutriated material injected were unspecified.] A group of 25 females treated with 0.5 mL saline alone served as controls. Treated and control animals were then kept for their natural lifespan. The mean lifespan in rats treated with refractory ceramic fibres was significantly shorter (480 ± 32 days) that that in controls (744 ± 28 days). Abdominal mesotheliomas [histological description and type unspecified] developed in 19/23 (included a single fibrosarcoma) of the animals injected with refractory ceramic fibres and in none of the controls (0/25) (Smith *et al.*, 1987).

Groups of 18–24 male Charles River Wistar rats, approximately 12 weeks old, were injected intraperitoneally on two consecutive days with either RCF1 (total dose, 110 mg; 228 × 10^6 fibres > 10 µm length and < 0.95 µm diameter), RCF2 (total dose, 188 mg; 320 × 10^6 fibres > 10 µm length and < 0.95 µm diameter) or RCF4 (total dose, 90 mg; 81 × 10^6 fibres > 10 µm length and < 0.95 µm diameter) suspended in 2 mL sterile saline. The refractory ceramic fibres were obtained from the same original stock as those used in the earlier inhalation studies by Mast *et al.* (1995a,b) and McConnell *et al.* (1995). The mass of refractory ceramic fibres required to deliver a target dose of 10^9 fibres > 5 µm in length was calculated from optical fibre-sizing data. No saline-injected controls were used. Following injection, animals were kept for life or until they became moribund and were killed. By macroscopic examination, the presence of mesotheliomas (responsible for the death of the animal) was found in 21/24 of the rats treated with RCF1, 13/18 of those treated with RCF2 and 0/22 rats injected with RCF4. The median survival time was lower after treatment with the RCF1 (337 days) and RCF2 (376 days) compared with RCF 4 (725 days) (Miller *et al.*, 1999).

(b) *Hamster*

Two groups of female Syrian golden hamsters (56 in total), 100 days of age, received a single intraperitoneal injection, in 0.5 mL saline of 25 mg refractory ceramic fibres (Fibrefrax®) [chemical composition not given], which had been elutriated from the chambers of a nose-only inhalation study. The fibres in the inhalation aerosol had a GMD of 0.9 µm and a GML of 25 µm. The length and diameter distributions and particulate count of the elutriated material injected were unspecified.] A group of 25 females treated with 0.5 mL saline served as controls. Treated animals and controls were kept for life. After intraperitoneal injection with refractory ceramic fibres, 21/36 hamsters in the first group and 15/36 in the second group died within 30 days from acute haemorrhagic peritonitis and vascular collapse. The mean lifespans of the surviving hamsters were significantly reduced (462 and 489 days, in the first and second group, respectively) compared with that of control animals injected with saline (560 days). Abdominal mesotheliomas were found in 2/15 hamsters in the first group and in 5/21 hamsters in the second group [histological description and type unspecified]. No tumours were found in the controls (Smith *et al.*, 1987).

3.7 Newly developed fibres

3.7.1 *Inhalation exposure*

X-607 (alkaline earth silicate) wool

Rat: In a chronic inhalation study, 140 male Fischer 344 rats, 9–11 weeks of age, were exposed to Johns Manville X-607 and 80 male Fischer 344 rats were exposed to NIEHS medium-length chrysotile asbestos. The X-607 wool contains a relatively high proportion of calcium oxide (38%). Rats were exposed by nose-only inhalation for 6 h per day on five days per week for periods of up to two years. The concentration of X-607 fibres in the aerosol was 30 mg/m^3; there were 174 ± 72 × 10^6 WHO fibres/m^3 and 47 ± 23 × 10^6 fibres > 20 µm/m^3, with geometric mean fibre dimensions of 11 µm × 0.9 µm. For chrysotile the concentration was 10 mg/m^3, with 10 600 × 10^6 WHO fibres/m^3 and no detectable fibres > 20 µm; the geometric mean fibre dimensions were 1.2 µm × 0.3 µm. After 104 weeks of exposure to X 607, the lung burden was 58 × 10^6 WHO fibres/lung and 1 × 10^6 fibres > 20 µm/lung. After 104 weeks of exposure to chrysotile, there were 1600 × 10^6 WHO fibres/lung and no fibres > 20 µm were detected. Animals exposed to X-607 did not develop any lung fibrosis. The incidence of lung tumours in rats treated with X-607 was one adenoma and one carcinoma in 121 animals at risk; this was not significantly different from the incidence in controls kept in air which was two adenomas in 130 rats. The tumour incidence after exposure to chrysotile asbestos was six lung adenomas, six lung carcinomas and one mesothelioma in 69 animals at risk (Hesterberg *et al.*, 1998a).

High-alumina, low-silica wool (HT) fibre (MMVF34)

A group of 140 male Fischer 344 rats, 9–10 weeks of age, was exposed by nose-only inhalation for 6 h per day on five days per week for 104 weeks to 30 mg/m^3 high-alumina, low-silica wool (HT) fibre (MMVF34). This fibre, commercialized in 1995, is characterized by a relatively high content of aluminium and a relatively low content of silica compared with the older MMVF21. The gravimetric concentration was selected to obtain a fibre concentration of at least 250×10^6 WHO fibres/m^3 throughout the exposure period to enable comparison with previous studies carried out on MMVFs in which 30 mg/m^3 was the highest dose. The actual values of the fibre concentration were 291×10^6 WHO fibres/m^3 and 85×10^6 fibres > 20 μm/m^3. The geometric mean of the fibre diameter was 0.87 μm and the geometric mean length was 10.8 μm. The retained lung burden after 24 months was 60.2×10^6 WHO fibres and 3.1×10^6 fibres > 20 μm. A subsequent period of no exposure continued until approximately 20% survival was reached in the control group kept in air at approximately 28 months. Mortality after treatment with HT fibres was comparable with mortality of unexposed controls. The results of the comparative study showed a marked difference in the pathogenicity of the MMVF21 and MMVF34 in terms of their fibrogenic potential: MMVF21 caused pulmonary fibrosis, but MMVF34 did not. In 107 rats exposed to MMVF34, no carcinoma and five adenomas were observed. In the 107 rats in the control group, one carcinoma and three adenomas were found (Kamstrup *et al.*, 1998, 2001).

3.7.2 Intraperitoneal injection

High-alumina, low-silica wool (HT) fibre

Rat: A group of 50 female Wistar rats (body weight approximately 200–230 g), aged 10–12 weeks, received a dose of 2.1×10^9 WHO HT fibres (0.6×10^9 fibres > 20 μm) suspended in 2 mL saline, by intraperitoneal injection. The geometric mean diameter of fibres was 0.65 μm and the geometric mean length was 10.7 μm. Fifty female Wistar rats treated with 2 mL saline served as negative controls. After treatment, animals were kept until survival in one group fell below 20%. At this time, all animals were killed. All animals were necropsied; any gross abnormalities observed during necropsy were examined histopathologically. No induction of mesothelioma was observed in either the group treated with HT fibres or in the control group. These results were compared with those of a study in which the fibre D6 (MMVF21; median diameter, 0.8 μm; median length, 8.5 μm; 1×10^9 WHO fibres; 0.2×10^9 fibres > 20 μm) had been tested by the same laboratory. In that study, mesotheliomas were observed in 32 of 57 treated rats and no mesothelioma in the corresponding control group of 91 rats (Kamstrup *et al.*, 2002). [The Working Group noted that these two injection studies were not contemporaneous.]

A and C wools

Rat: Groups of 51 female Wistar rats, 8–9 weeks of age, were given intraperitoneal injections containing 0.7–35 mg of either glass wool A or glass wool C. The doses, expressed as WHO fibres/rat × 10^6, ranged from 9.2–630. Other groups of 51 rats were injected with either crocidolite asbestos (maximum dose, 0.5 mg or 110 × 10^6 WHO fibres) or saline. Rats were kept for 130 weeks, then killed and examined for abdominal tumours. None of the groups exposed to either fibre A or fibre C had a tumour incidence greater than 10%. The tumour incidences were (from lowest dose to highest dose): groups treated with fibre A, 3/51, 1/51, 1/51 and 3/51; groups treated with fibre C, 5/51, 4/51, 1/51 and 1/51; controls treated with saline, 0/51. The tumour incidence in the group exposed to 0.5 mg crocidolite was 25/51 (Lambré *et al.*, 1998).

F, G and H wools

[F, G and H wools have not been commercialized, but they are included here because data on biopersistence and chronic effects are available for these fibre types. For many of the newly developed fibres which are commercially available in the European Union, only biopersistence data are available.]

Rat: Groups of 51 female Wistar rats, 8–9 weeks old, received intraperitoneal injections containing 1.1–55 mg of three types of rock (stone) wool fibres (F, G and H) [the Working Group noted that, according to the definition given in section 1, these fibres are not a rock (stone) wool] (all calcium-modified silicates) suspended in saline. The doses administered, the corresponding numbers of fibres, and geometric mean lengths and diameters were reported, together with tumour incidence (all fibres were defined as having length > 5 μm, diameter < 2 μm, length:diameter > 5 μm). Doses expressed as WHO fibres × 10^6/rat ranged 5.2–550. Other groups of 51 rats were injected with either crocidolite asbestos (maximum dose, 0.5 mg or 110 × 10^6 WHO fibres) or saline. Rats were kept for 130 weeks then killed and examined for abdominal tumours. The incidences of peritoneal tumours (from lowest dose to highest dose) were: 3/51, 1/51 and 3/51 in rats treated with F wool; 2/51, 1/51 and 2/51 in rats treated with G wool; 3/51, 1/51 and 9/51 (dose of 55 mg or 260 × 10^6 WHO fibres for the latter incidence) in rats treated with H wool; and 4/51, 10/51 and 25/51 in rats treated with crocidolite. In a control group of 102 females treated with saline, no mesothelioma was reported (Lambré *et al.*, 1998).

4. Other Data Relevant to an Evaluation of Carcinogenicity and its Mechanisms

4.1 Deposition, retention and clearance

The deposition of airborne fibrous and non-fibrous particles is defined as the active loss of these particles from the air during respiration, as a result of inelastic encounter of the airborne particles with the respiratory epithelium. Clearance from the site of deposition pertains to the removal of these deposited particles by various processes over time, whereas retention is the temporal persistence of particles within the respiratory system (Morrow, 1984). Thus, the amount retained (R) is defined by the amount deposited (D) minus the amount cleared (C) ($R = D - C$).

The deposition of inhaled fibres in the respiratory tract is a function of their physical characteristics (size, shape and density) and of the anatomical and physiological parameters of the upper and lower airways. Chemical composition has no role in deposition of airborne fibres in the respiratory tract, and, therefore, there is no need to differentiate between fibre types when discussing deposition phenomena. The clearance of deposited fibres from the respiratory tract, however, is dependent on both physical and chemical characteristics of fibres, and therefore the clearance and resulting retention behaviour within the respiratory system can vary widely between different fibre types; the different fibres deposited in the lung are subject to the same clearance processes, which together determine the biopersistence of the fibres. Thus, since a general understanding of deposition, retention and clearance is necessary to appreciate the importance of biopersistence, these topics are discussed briefly below for all fibres together rather than considering specific categories of vitreous fibres separately.

The main processes leading to the deposition of inhaled fibres operate throughout the respiratory tract, whereas the mechanisms that clear deposited fibres from different regions of the respiratory system vary considerably and, therefore, contribute to a different degree to the retained fibre burden at sensitive sites in the lung. When discussing deposition and retention it is, therefore, convenient to divide the respiratory system into three compartments as follows: the extrathoracic region (ET, consisting of anterior and posterior nose, larynx, pharynx and mouth); the tracheobronchial region (TB, consisting of trachea, bronchi and bronchioles down to the terminal bronchioles) and the alveolar–interstitial region (AI, including respiratory bronchioles, alveolar ducts and sacs with alveoli and pulmonary interstitium). A somewhat more detailed classification in which the respiratory system was divided into four compartments was presented by the International Committee on Radiological Protection (ICRP) Task Group (ICRP, 1994). In this scheme the TB region was separated into large bronchi (BB) and small bronchioli (bb) compartments. Since the main clearance process in these two regions is the same (ciliary movement) and differs only in rate, the TB region is considered as one compartment here. The following sections summarize those concepts of deposition, clearance and retention that are known for both spherical and fibrous particles and that are helpful for the evaluation of fibre toxicity and carcinogenicity. Such an evaluation will, to a large degree, be based on a definition of fibre biopersistence as discussed in section 4.1.3.

4.1.1 *Deposition*

Several publications have described in detail the deposition behaviour of inhaled particles throughout the respiratory tract in humans and rats in general, as well as pointing out specific differences that pertain to fibres (Heyder, 1982; Morrow & Yu, 1985; Stöber *et al.*, 1993; ICRP, 1994; Schlesinger *et al.*, 1997; Asgharian & Anjilvel, 1998; Asgharian & Yu, 1988, 1989; Dai & Yu, 1998). Some of the main points are summarized below. Airborne particles and fibres are often described as being respirable or inhalable. The terms 'respirability' and 'inhalability' have very different meanings (defined by the American Conference of Governmental Industrial Hygienists (ACGIH), 2001).

Inhalability refers to the ratio of the particle (fibre) concentration in the inhaled air to that in the ambient air. Inhalability decreases with increasing particle (fibre) size due to increasingly higher settling velocities in air and inertia for increasing particle sizes (> 5–10 μm) before entering the airways; it is also dependent on wind velocity.

Respirability is the ratio of the concentration of airborne particles (fibres) penetrating to the alveolar region of the lung to that in the ambient air. Respirability generally increases with decreasing size of the particles (fibres), unless they become extremely small (i.e. fibrils, of nanometer size).

Using these definitions, the European Committee of Standardization (CEN, 1993) and ACGIH (2001) defined an inhaled particulate mass for those materials that are hazardous when deposited anywhere in the respiratory tract; a respirable particulate mass for those materials that are hazardous when deposited in the alveolar region; and a thoracic particulate mass for those materials that are hazardous when deposited anywhere in the lower respiratory tract (TB and AI regions). Thus, restricting the evaluation of fibres to 'respirable' fibres would be to ignore those fibres depositing in the TB region of the respiratory tract. The TB region is important when considering the pathogenicity of fibres in humans, since it is known from research on workers exposed to asbestos that this region is a potential target area for adverse health effects induced by fibres, in particular, chronic inflammation and bronchogenic carcinoma (Churg & Green, 1998). Acute nasal effects induced by fibres have also been observed; therefore, the inhalable fibre fraction must be considered as well (INSERM, 1999; see also section 4.2).

Two important parameters that affect the deposition of airborne particles (fibres) are their aerodynamic and thermodynamic properties. The equivalent aerodynamic diameter is defined as the diameter of a spherical particle of unit density which has the same terminal settling velocity in still air as the particle (fibre) in question. Deposition due to aerodynamic behaviour becomes less important for particles with sizes < 1 µm. Below a particle size of 0.5 µm, thermodynamic properties prevail, and deposition of these particles is governed mainly by the diffusional movement induced by Brownian motion of gas molecules.

For a spherical particle, the geometric diameter multiplied by the square root of the specific density of the material gives the aerodynamic diameter. For a non-spherical particle, a shape factor also needs to be considered; although, for fibres in particular, their aerodynamic diameter is mostly governed by their geometric diameter, their elongated shape (fibre length) and their specific density. Figure 9 illustrates this for fibres of different aspect ratios (length:diameter) and with a density of 2.7 g/cm^3 (see Oberdörster, 1996).

The main mechanisms by which inhaled fibres deposit in the respiratory tract are impaction, sedimentation, diffusion and interception (Asgharian & Yu, 1988, 1989; Dai & Yu, 1998). Electric charges on fibres can also significantly enhance deposition due to

Figure 9. Correlation between aerodynamic and geometric diameter of fibres of different aspect ratios (β)

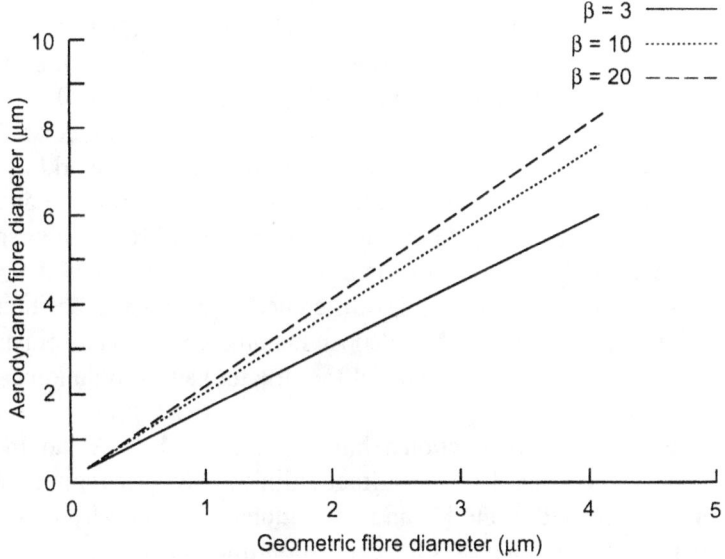

From Oberdörster (1996)
Specific density of fibre material is 2.7 g/cm^3.

generation of image charges in the airway walls; however, in contrast to asbestos fibres, no data on electric charges on vitreous fibres have been published.

Deposition of fibres due to impaction occurs when the airflow encounters rapid changes in direction (e.g. in the nose and conducting airways) and the momentum of the fibre carries it along in a straight line to deposit on the airway wall. The larger the aerodynamic diameter, the greater the deposition efficiency due to impaction, so this mechanism is most effective for aerodynamic diameters > 0.5–1.0 µm.

Sedimentation refers to the settling of fibres due to gravitational forces, which eventually results in the fibres touching the airway wall and depositing on the epithelium. This mechanism also operates mainly on fibres with equivalent aerodynamic diameters of > 0.5–1.0 µm.

Interception is a particularly important mechanism for fibre deposition. Whereas few spherical particles are deposited by interception, significant deposition of fibrous particles occurs by this mechanism (Asgharian & Yu, 1989). Deposition by interception occurs when an airborne particle (fibre) in the airways gets close enough to the airway wall to allow one end to touch the wall. Obviously, for an elongated object such as a fibre, this occurs more readily than for a spherical particle. In particular, fibres are carried along in the airflow while rotating at a variable rate at random orientation, even in a laminar flow (Jeffery, 1922; Asgharian & Yu, 1988), which makes interception an efficient deposition mechanism, especially for longer fibres. Only when fibres enter a laminar airflow perfectly aligned with the flow axis will they not rotate, and even then they become unaligned as soon as a bifurcation is reached.

Although these four mechanisms apply to fibre deposition in humans exposed environmentally or occupationally as well as in rodents exposed experimentally, there are important interspecies differences that need to be considered when interpreting and extrapolating results from rodent inhalation studies to humans. Figure 10 shows the differences for the alveolar region in humans and rats as modelled by Dai and Yu (1998). [The Working Group noted that there are uncertainties associated with these theoretical results.] These authors also calculated the effect of workload (increased minute ventilation) on the efficiency of deposition of fibres in humans. From the results of their model on alveolar deposition of inhaled fibres in humans, rats and hamsters, they reached the following conclusions:

- There is a significant interspecies difference in alveolar deposition of inhaled fibres, i.e. more and larger fibres deposit in humans than in rats or hamsters. This is caused by difference in the size of the structure and ventilation parameters of the airway.
- The alveolar deposition fraction in humans varies with workload. Increasing the workload reduces the deposition fraction in the alveolar region because more fibres are deposited in the ET and TB regions; switching from nose-breathing to mouth-breathing increases the deposition fraction.
- For all species, a peak in deposition occurs with particles or fibres with an aerodynamic diameter between 1 and 2 µm. Increasing the aspect ratio of the fibre

Figure 10. Predicted deposition of inhaled spherical and fibrous particles of different aspect ratios in the alveolar region of humans and rats (β)

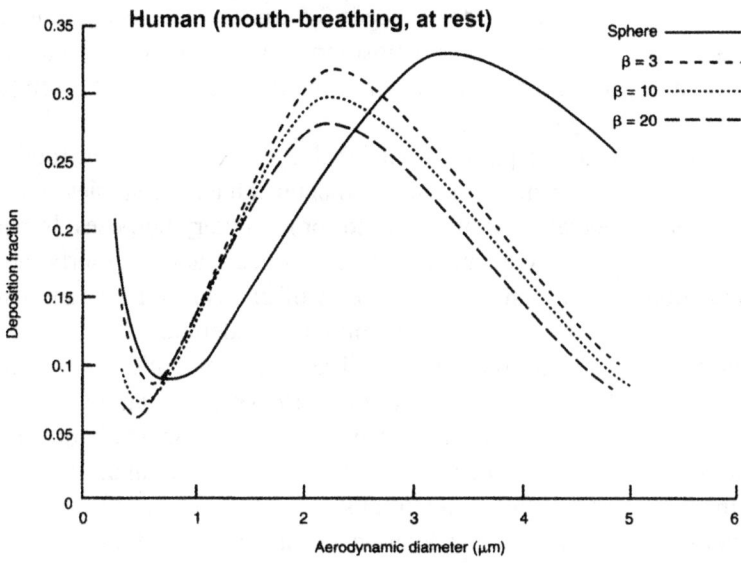

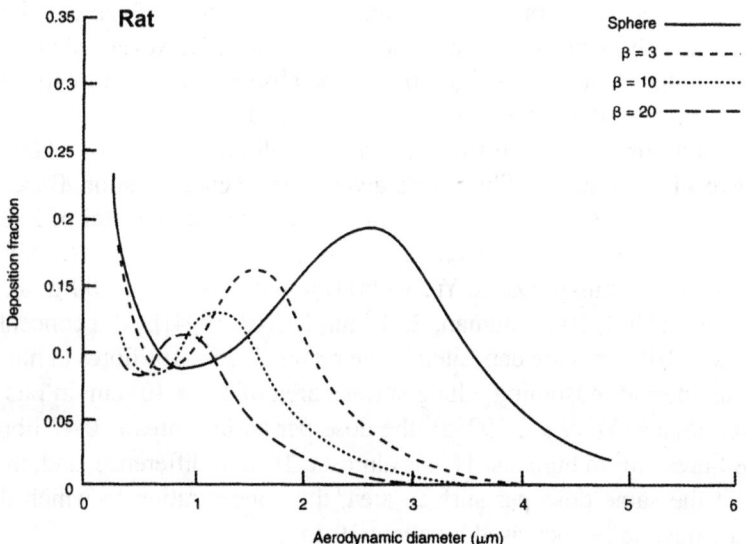

From Dai and Yu (1998)
Specific density is 2.7 g/cm^3. Below 0.5 μm the pattern of deposition is dominated by thermodynamic properties of the particles.

results in a decrease in the peak deposition and the corresponding aerodynamic diameter.
- For rats and hamsters, alveolar deposition is essentially zero when the aerodynamic diameter of the fibres exceeds 3.5 μm and the aspect ratio is > 10. In contrast, considerable alveolar deposition occurs in humans breathing at rest, even when the aerodynamic diameter of the fibres approaches 5 μm (Dai & Yu, 1998).

The conclusion that the respirability of inhaled fibres is lower in rodents than in humans is significant in the design of experimental inhalation studies, the results of which are used for extrapolation to humans and for regulatory purposes. In particular, it would imply that it is not very meaningful to use concentrations of airborne fibres — especially long fibres — to compare the effects of exposure in humans with that in rodents. The dose in the lung is the most important consideration.

The initial deposition patterns of inhaled fibres in rats and mice have been reported by Brody and Roe (1983) and by Warheit *et al.* (1988). These authors found that the preferred site of fibre deposition in the alveolar region of rodents is on first alveolar duct bifurcations. This may also be the site of the initial inflammatory processes and perhaps for the entry of fibres into interstitial sites (see below under clearance).

The discussion in this section has focused so far on the fractions of inhaled fibres deposited in a specific region of the respiratory tract. As shown in Figure 10, deposited fractions of fibres in the alveolar region — according to the model of Dai and Yu (1998) — are significantly lower in rats than in humans so that, for a given inhaled concentration, rats apparently deposit a lower dose than humans. However, given the greater minute ventilation per unit body weight in rats, the absolute amount of deposited fibres may be higher in rats. It is, therefore, of interest to determine the amount of fibres deposited per unit surface area of the alveolar epithelium to discover whether rats or humans receive a higher dose of fibres for a given inhaled concentration. Based on these models, the fraction of fibres with an equivalent aerodynamic diameter of 2 μm and an aspect ratio of 20 deposited in the alveolar region is 2.2% in rats and 22.9% in humans (light work, mouth-breathing, Dai & Yu, 1998) (minute ventilation: 300-g rat, 245 mL/min [Hsieh *et al.*, 1999]; 70-kg human, 25 L/min [ICRP, 1994]). At a concentration of 1 fibre/cm^3, 2.6×10^3 fibres are deposited in the rat and 2.75×10^6 fibres in humans over an 8-h exposure period. Assuming a lung surface area of 5.5×10^3 cm^2 in rats and 6.27×10^5 cm^2 in humans (Yu *et al.*, 1995a), the dose per surface area is 0.47 fibres/cm^2 in rats and 4.38 fibres/cm^2 in humans. This is almost a 10-fold difference, and, in order for rats to deposit the same dose per surface area, the concentration to which they were exposed would have to be increased by about 10-fold.

4.1.2 *Clearance and retention*

The retention kinetics of fibres in the lung are usually influenced by the following variables: chemical composition, fibre size distribution, amount of fibres in the lung and time since exposure (Ellouk & Jaurand, 1994; Muhle *et al.*, 1994; Muhle & Bellman, 1995).

Following deposition of fibres, their retention in different regions of the respiratory tract over time is determined by their clearance rate ($R = D - C$). The general mechanisms for particle clearance have been reviewed (Morrow & Yu, 1985; Oberdörster, 1988; Snipes, 1989; ICRP, 1994; Schlesinger *et al.*, 1997), and models that specifically describe clearance of fibres have been published (Yu *et al.*, 1996, 1998). For the nasal region, ciliary motion-mediated movement of deposited particles has been observed to occur mostly towards the pharynx and to a limited degree towards the nostrils. This is generally a fast process, but it may take up to 24 h (ICRP, 1994). Nose-blowing is a most effective means to clear the anterior region of the nose. Oropharyngeal clearance of particles or fibres deposited there occurs via the gastro-intestinal tract through swallowing.

For the TB region, the main mechanism for fibre clearance is through ciliary motion along the mucociliary escalator, either after phagocytosis by airway macrophages (see below for discussion of fibre length and phagocytosis), or as free fibres. Mucociliary clearance operates throughout the TB region, where cilia are present from the trachea down to the terminal bronchioles. It is generally a fast process, which takes less than 24 h. However, the thin surfactant layer in the conducting airways of the TB region has also been found to promote the embedding of deposited spherical particles into the underlying fluid layer, essentially forcing these particles onto the epithelial cells (Schürch *et al.*, 1990; Gehr *et al.*, 1993). This process together with phagocytosis by airway macrophages may contribute to a long-term retention phase for spherical particles in the TB region lasting for a number of days or weeks, as observed by Stahlhofen *et al.* (1995) in humans. These authors found that the slowly-cleared fraction of the TB deposit decreased with increasing geometric diameter of the particles.

Fibrous particles may be subject to the same mechanisms, and interactions between fibres and epithelial cells would be enhanced by this process, so that translocation of persistent fibres across the TB epithelium could occur. Cigarette smoking has been found to increase the number of short asbestos fibres retained in cells of the bronchial epithelium (Churg *et al.*, 1992; Churg & Stevens, 1995). Smoking-induced impairment of mucociliary clearance, an early functional abnormality in smokers, significantly reduced the clearance of deposited particles (Vastag *et al.*, 1986). This could be one mechanism for the increase in risk for lung cancer in smokers exposed to asbestos (Oberdörster, 1989).

The most important mechanism for the mechanical clearance of particles deposited in the alveolar region is through phagocytosis by alveolar macrophages and subsequent translocation towards the mucociliary escalator. Fibre length is an important parameter as it can limit the ability of alveolar macrophages to completely phagocytose a fibre. Incompletely phagocytosed fibres are likely to come into contact with alveolar cells and translocate to interstitial sites. Further interstitial transport along lymphatic channels can distribute fibres to regional lymph nodes as well as to pleural sites (Oberdörster et al., 1988), provided that fibres are not dissolved along this clearance pathway. The importance of mechanical and chemical processes for the overall retention of these particles in the lung is discussed in the section on biopersistence (section 4.1.3).

Clearance mechanisms in the TB region are generally the same for fibres as for spherical particles; however, there are important differences due to the elongated shape of fibres. Long vitreous fibres can break so that fibres longer than 20 μm — which are incompletely phagocytosable — become smaller and will be taken up and cleared by the alveolar macrophages.

Pleural translocation of deposited refractory ceramic fibres during and after 12 weeks of exposure by inhalation has been demonstrated in rats and hamsters. At the end of exposure and at 12 weeks after exposure, two to three times more fibres longer than 5 μm were counted in the pleural compartment of hamsters than in rats (expressed per cm^2 of pleural surface). Total pleural fibre burdens were more than three orders of magnitude lower than in the lung. This finding may partially explain the greater sensitivity of hamsters to fibre-induced pleural mesothelioma (Gelzleichter et al., 1999).

Important differences in the size of the macrophages between rodents and humans affect phagocytosis of fibres. In general, fibres that are too long to be phagocytosed will remain in the alveolar compartment and be subject to other clearance mechanisms, including dissolution, breakage and translocation to interstitial sites and subsequently to pleural sites. The diameters of alveolar macrophages differ between rats and humans (rat: 10.5–13 μm; human: 14–21 μm) (Crapo et al., 1983; Lum et al., 1983; Sebring & Lehnert, 1992; Stone et al., 1992; Krombach et al., 1997). Alveolar macrophages are very elastic and can spread significantly so that fibres longer than the normal diameter of an alveolar macrophage can be phagocytosed. However, phagocytosis of fibres longer than 20 μm by alveolar macrophages is less likely to occur (Dörger et al., 2000, 2001). Multinucleated giant cells that may have formed by fusion of macrophages upon attempted phagocytosis of longer fibres have been observed (Davis, 1970). If persistent, these longer fibres are thought to be the most potent fibre category associated with tumour induction (Kane et al., 1996).

In addition to their alveolar macrophage-mediated mechanical clearance, dissolution of deposited vitreous fibres in the lung can be an important means of elimination. Short fibres that are ingested by alveolar macrophages encounter an acidic pH (4.5–5) inside the phagolysosome (Lundborg et al., 1995), whereas longer fibres that are incompletely phagocytosed by alveolar macrophages are exposed to extracellular fluid with a pH of ~7.4. In-vitro tests at these two pH levels (simulating the intracellular and extracellular conditions) have been developed to determine the rates of fibre dissolution (Potter &

Mattson, 1991; see section 4.1.4(a)(iii)). In-vitro dissolution and mechanical removal by alveolar macrophages are independent mechanisms: this may explain why there is not always a good correlation between in-vivo retention half-times and in-vitro dissolution rates (Oberdörster, 2000). In-vitro dissolution tests determine only the durability of fibres in a cell-free system, whereas in-vivo retention studies will determine the durability *in vivo* plus other clearance processes which together provide a measure of biopersistence of fibres (see section 4.1.3).

Poorly soluble, low-toxicity, phagocytosable particles and fibres are cleared much more slowly from the Al region of the respiratory tract than from the ET and TB regions when clearance is mediated mainly by alveolar macrophages. For rats, the normal undisturbed overall retention half-time of particles or fibres subject to macrophage-mediated clearance is 60–80 days (Snipes, 1989); for humans, the average overall retention half-time in the alveolar region is several hundred days (Bailey *et al.*, 1982). In both species, the clearance rate is initially fast followed by a slow phase. In a study on human volunteers, a fast-phase retention half-time of tens of days and a slow-phase half-time of hundreds of days were reported for poorly soluble particles (Bailey *et al.*, 1982, 1985). The International Committee on Radiological Protection (ICRP, 1994) discerned three phases of particle retention in the alveolar compartment to account for the experimentally observed retention curves; however, the underlying mechanisms for each phase are not clear. It has yet to be determined whether this pattern is due to the clearance of particles deposited in the central lung regions (closer to the mucociliary escalator which starts at the level of the terminal bronchioles) being somewhat faster than that of particles deposited in the peripheral lung regions (which would have a longer clearance pathway to the terminal bronchioles) or whether it is due to the action of the different alveolar clearance mechanisms for translocation and dissolution described above. However, if a significant number of the retained particles or fibres are removed slowly, there may be serious effects on health, e.g. if the fibres are longer than 20 µm (incompletely phagocytosed, no dissolution, no breakage) or if the material is cytotoxic. In this case, prolongation of the slow phase of clearance (long retention half-time) should be considered as an indicator of potentially increased toxicity or carcinogenicity.

If the dissolution of a retained particle or fibre occurs *in vivo*, the overall retention half-time ($T_{\frac{1}{2}\,total}$) can be significantly shortened, since the clearance rates from mechanical macrophage-mediated clearance (r_{mech}) and from dissolution (r_{dissol}) are combined:

$$r_{total} = r_{mech} + r_{dissol}$$

Or, since the clearance rate and $T_{\frac{1}{2}}$ are correlated by

$$r = \frac{\ln 2}{T_{\frac{1}{2}}},$$

$T_{\frac{1}{2}\,total}$ is determined by

$$\frac{1}{T_{\frac{1}{2}\,total}} = \frac{1}{T_{\frac{1}{2}\,mech}} + \frac{1}{T_{\frac{1}{2}\,dissol}}$$

Whereas the rates of macrophage-mediated mechanical clearance are very different between rats and humans (see above), it is assumed that the rates of in-vivo dissolution of fibres are the same or very similar between species. Since $T_{\frac{1}{2}\,mech}$ is much longer in humans than in rats, a given in-vivo dissolution rate lowers the T½ total more in humans than in rodents. For example, a fibre dissolution rate corresponding to a $T_{\frac{1}{2}\,dissol}$ of 70 days will result in a $T_{\frac{1}{2}\,total}$ of 35 days for rats (half of the normal rat $T_{\frac{1}{2}\,mech}$ of 70 days), and in a $T_{\frac{1}{2}\,total}$ of 59 days (about 14% of the normal $T_{\frac{1}{2}\,mech}$) for humans (calculated using a human $T_{\frac{1}{2}\,mech}$ of 400 days) (Berry, 1999).

The dissolution of vitreous fibres occurs in extracellular and intracellular compartments, so that alveolar clearance of both long and short fibres deposited by inhalation can most often be described by a faster and a slower phase (Bernstein et al., 1996). Yu et al. (1996) developed a comprehensive model for clearance of refractory ceramic fibres from the rat lung, including macrophage-mediated translocation, dissolution and breakage, in order to describe the experimental findings on fibre accumulation from long-term inhalation studies. The inclusion of dissolution rate and breakage rate in the model provided a better interpretation of the experimental data than could be achieved with previously proposed models of fibre clearance.

Macrophage-mediated clearance can be prolonged significantly by several factors that are important for both spherical particles and fibres. These include events associated with particle overload and increased cytotoxicity of the particulate material (ILSI, 2000). In addition, the length of fibrous particles is important when it inhibits complete phagocytosis, as discussed above.

Particle overload occurs when high doses of poorly-soluble particles of low cytotoxicity are chronically deposited in the lung so that their daily rate of deposition exceeds the normal rate of macrophage-mediated clearance (ILSI, 2000). It has been shown in rats that these circumstances occurring together with retarded clearance cause persistent alveolar inflammation, fibrosis and lung tumours (Donaldson, 2000; ILSI, 2000). It has been suggested that the retarded clearance occurs when an average 6% of the volume of the alveolar macrophage is filled with phagocytosed particles, and complete cessation of clearance occurs when 60% of this volume is occupied by particles. Expressed in terms of particle mass, overload occurs when 1–3 mg of particles per gram of rat lung have been deposited (Morrow, 1988).

Other studies have suggested that the overload phenomenon correlates well with the surface area of the retained particles, which may, indeed, be a better dose parameter than particle volume or weight. It should be noted that overload-associated tumorigenic effects have been observed for poorly soluble particles of low toxicity, and only in rats. In contrast, cytotoxic materials such as crystalline silica particles show reduced alveolar macrophage-mediated clearance at lung burdens that are two orders of magnitude lower and presumably through mechanisms unrelated to overload. Impaired clearance of long fibres also involves mechanisms that do not meet the definition of lung overload (Oberdörster et al., 1994; Tran et al., 2000; Oberdörster, 2002).

The concept of lung overload is important with respect to fibres because experimental studies of inhaled fibres in rodents always include non-fibrous material and non-phagocytosable short fibres. Indeed, the non-fibrous particles, if they are biopersistent, can enhance the effects of the fibres or may contribute to chronic effects, such as fibrosis, lung tumours and mesotheliomas. For example, Davis *et al.* (1991) showed that in rats chronically exposed to asbestos fibres in combination with titania or crystalline silica particles, the incidence of lung tumours and mesothelioma was significantly increased compared with that seen with exposure to asbestos alone. Bellmann *et al.* (2001) demonstrated that RCF1 containing a large fraction of non-fibrous material induced a severe retardation of macrophage-mediated clearance of tracer particles, whereas a similar number of RCF1 fibres without the non-fibrous particles did not. The authors suggested that this was due to overload of the alveolar macrophages by the non-fibrous RCF1 particles, which could be the mechanism of lung tumour induction that had been reported in earlier long-term inhalation studies with RCF1 in rats (Mast *et al.*, 1995a,b), a suggestion reiterated by Mast *et al.* (2000b). [The Working Group noted that an amount of low-toxicity non-fibrous particles equivalent to that determined to be retained in the lung by Bellmann *et al.* (2001) would not induce complete cessation of clearance of tracer particles, suggesting that particles or fibres of RCF1 have a greater cytotoxic potential than a low-toxicity particle such as titania, or that an unknown particle or fibre synergism exists. The Working Group also noted that combined exposure to particles and fibres is likely to occur at the workplace, but the ratio of particles to fibres was higher in RCF1 than that ever seen anywhere else, although not usually at levels that would induce overload (Maxim *et al.*, 1997).]

The fact that lower doses of cytotoxic non-fibrous particles can cause the same chronic effects in rats as high doses of low-toxicity fibres underlines the importance of considering not only the fate of the longer, incompletely phagocytosable fibres, but also that of the shorter fibres. The latter should not induce impairment of the alveolar macrophage-mediated clearance function when the alveolar macrophage burdens are below overload. However, if the material under study were cytotoxic, similar effects to those with crystalline silica would be seen.

(*a*) *Studies in animals*

Mast *et al.* (1994) conducted inhalation studies to explore clearance of different types of refractory ceramic fibres. Rats were exposed to samples of RCF1, RCF2, RCF3 and RCF4, 1 μm in diameter and 22–26 μm in length (high purity kaolin, zirconia) and hamsters were exposed to kaolin (RCF1). For comparison, other groups of animals were exposed to chrysotile fibres. Exposure to airborne concentrations ranging between 3 and 30 mg/m^3 took place for 6 h per day on five days per week for 24 months. The lung burden of refractory ceramic fibres was found to be related to exposure: at high doses, it exceeded 10^8 fibres/g dry lung. During the various recovery periods there was a clear reduction in fibre burden. Despite the smaller airway size in hamsters compared with rats, the alveolar deposition of refractory ceramic fibres per breathing cycle in hamsters was found to be higher than that in rats, and the calculated mean size of deposited fibres was larger in hamsters than in rats. Also, alveolar clearance of refractory ceramic fibres was faster in hamsters. The clearance rate in hamsters appeared to be independent of fibre length, but varied with the lung burden. The faster clearance rate in hamsters resulted in a lower accumulation of fibres per unit weight of the lung after a long period of exposure (Yu et al., 1995b).

To study the deposition and clearance of aluminium silicate ceramic fibres in the lung, Yamato et al. (1994a) exposed male Wistar rats to ceramic fibres with a mass median aerodynamic diameter (MMAD) of 3.7 μm for 6 h per day on five days per week for two weeks. The average concentration of fibres was 27.2 mg/m^3 (standard deviation, 9.0). The rats were examined on day 1, and at 1 month, 3 months and 6 months after the end of the exposure period, and the numbers and dimensions of the fibres were analysed by scanning electron microscopy (SEM). No significant differences in the lengths of residual refractory ceramic fibres in the lungs was found between the groups. The geometric mean diameter and number of refractory ceramic fibres, however, decreased according to the clearance period suggesting that the fibres were dissolved at their surface (see also Yamato et al., 1994b). Furthermore, the findings of Yu et al. (1994) who modelled the results of Mast et al. (1994) with RCF1 suggested that the clearance rate of refractory ceramic fibres from the lungs did not depend significantly upon fibre length, but that there was a clear dependence on lung burden — as lung burden increased, the clearance rate was found to decrease. Yu et al. (1996; 1997) used data on refractory ceramic fibres to develop a model for predicting the retention of refractory ceramic fibres in the lung of rats and humans. The model predicted that clearance of refractory ceramic fibres would not be significantly reduced in humans until the fibre concentration approaches 10 fibres/cm^3 during occupational exposure.

Rats were exposed to crocidolite or amosite asbestos or to a variety of man-made vitreous fibres (MMVFs) for five days by nose-only inhalation (6 h per day). In animals exposed to either of the asbestos types or to the more durable MMVFs (as determined by in-vitro assay; see section 4.1.4(a)), shorter fibres tended to clear from the lung more rapidly than longer fibres. However, the reverse was true for the less durable MMVFs. One year after termination of exposure, the percentages of crocidolite fibres (for which in-vivo dissolution is likely to be negligible or nil) retained in the lung were 35% for fibres < 5 μm but 83% for fibres > 20 μm in length. In contrast, the percentage retained after one year for MMVF22 slag wool (a non-biopersistent fibre) was 1% for fibres < 5 μm and < 1% for fibres > 20 μm (Hesterberg et al., 1996a).

In a second similar study, the percentages retained after a 30-day recovery period were 14%, 19% and 63% after treatment with amosite asbestos, RCF1a and high-alumina, low-silica (HT) (rock) stone wool fibres < 5 μm, respectively, and 60%, 50% and 3%, respectively, for amosite asbestos, RCF1a and high-alumina, low silica fibres with average lengths > 20 μm (Hesterberg et al., 1998b). Thus, in all cases, for the more biopersistent and slowly-dissolving fibres the long fibres disappeared more slowly than the shorter fibres. For the less biopersistent fibres the long fibres disappeared considerably faster than the shorter fibres.

In another study in which rats were exposed to fibres by nose-only inhalation (5 days of exposure, for 6 h per day), the weighted lung-retention half-times ($WT_{1/2}$; discussed below) for eight non-biopersistent MMVFs were longer for shorter fibres than for longer fibres. For a commercial rock (stone) wool that was more biopersistent, the $WT_{1/2}$ for long fibres was the same as for short fibres (Bernstein et al., 1996).

The same trend was observed in a fourth study in which rats were exposed to six MMVFs by either intratracheal instillation or inhalation. The investigators observed that, for the more soluble fibres, the long fibre (> 20 μm) fractions appeared to clear faster than the shorter fibre fractions following both inhalation and intratracheal exposure (Morgan, 1994; Bernstein et al., 1996).

On the basis of the experimental findings, a mechanism has been proposed to explain the various length-related patterns of fibre retention. Short fibres are probably phagocytosed and transported fairly quickly by the alveolar macrophages from the lower lung up to the ciliary escalator of the upper airways and then cleared to the oral pharynx, whereas fibres > 20 μm in length are unlikely to be phagocytosed completely and cleared from the airways through similar mechanisms. The clearance of longer fibres from the lower lung would then depend upon dissolution or transverse breakage. Thus, for fibres for which dissolution is negligible (i.e. crocidolite asbestos) or very slow (traditional refractory ceramic fibres) the longer fibres disappear more slowly. The breakage of long fibres would increase the population of short fibres, while at the same time macrophage-mediated clearance would be reducing this population. Thus, the number of short fibres in the lung at any given time would be the combined result of addition by fragmentation and subtraction by translocation (Hesterberg et al., 1998b).

To explain the differences between clearance rates of short and long fibres, other investigators have postulated a different mechanism based on intracellular versus extracellular lung compartments. They suggested that inhaled fibres > 20 μm, which are too long for complete engulfment by macrophages, would tend to be in the extracellular compartment of the lung at near-neutral pH. The shorter fibres would tend to be in the intracellular compartment, i.e. the phagolysosomes of macrophages, where they would be subject to an acidic pH and various digestive factors (Morgan et al., 1982; Christensen et al., 1994; Luoto et al., 1995, 1998). This mechanism is dealt with in more detail below and in section 4.1.4(a)(iv) on dissolution in in-vitro cell culture.

(b) *Studies in humans*

The experimental data on deposition, retention and clearance suggest that some of the parameters found to be important in rodent studies should also be taken into account in human studies.

The bivariate length:diameter distribution of fibres and the proportion of non-fibrous particles are determined by the conventional WHO (or NIOSH) phase contrast optical microscopy (PCOM) methods used in industrial hygiene (NIOSH, 1994; WHO, 1996).

Few data are available on the assessment of fibre retention in human lung tissue.

A case–control study of autopsies from the US MMVF production cohort used analytical transmission electron microscopy (TEM) to determine retention of fibres in the lung. The cases were production workers (101 in glass wool, 11 in rock (stone) and slag wool); the controls were 112 consecutive autopsies from the same hospital. The mean duration of exposure was 11 years and the mean time elapsed since the end of exposure was 12 years. No significant difference was observed in retention of MMVFs in the lung: MMVFs were detected in only 29 of the 112 production workers, of whom 14 had > 200 000 fibres/g dry lung, and 28 of the 112 controls, of whom six had > 200 000 fibres/g dry lung. Moreover, 10 of the 112 cases and two of the 112 controls had more than 1 million asbestos fibres/g dry lung ($p < 0.05$). The authors concluded that either MMVFs disappeared from the lung in less than 12 years (the numbers of fibres detected at autopsy in cases and in controls corresponded to those found after environmental exposure), or workers involved in the production of MMVFs did not inhale enough respirable MMVFs to result in a difference when compared with controls 12 years after the end of exposure, or the fixative fluids of the lung could have altered some retained fibres. They also noted that some of the MMVF production workers had also been exposed to asbestos (McDonald et al., 1990).

Two case reports on refractory ceramic fibre production workers, one in Europe (Sébastien et al., 1994) and one in the USA (INSERM, 1999), found that refractory ceramic fibres were recovered in bronchoalveolar lavage (7 cases in Europe) or in lung tissue (3 cases in the USA). All workers were still exposed to refractory ceramic fibres a few days before analysis. A significant number of refractory ceramic fibres were counted (63–764 fibres/mL in the European cases): most of the fibres had been morphologically and chemically modified (i.e. by ferruginous coating or loss of silicon and aluminium compatible with some leaching). In a review of 1800 bronchoalveolar lavage fluids from consecutive clinical series in the same chest clinic from 1992–97, pseudo-asbestos bodies on refractory ceramic fibres were detected by analytical electron microscopy in samples from nine subjects (0.5%) engaged in metal industries (foundry workers, steel workers, welders). No ferruginous bodies were detected on glass, rock (stone) or slag fibres in this series. This study also demonstrated that refractory ceramic fibres can form ferruginous bodies on durable fibres similar to asbestos. The authors concluded that refractory ceramic fibres have a residence time in the lung of at least several months, and that they interact with alveolar macrophages. Moreover, the reported presence of the typical ferruginous bodies detected by PCOM in lung, bronchoalveolar lavage fluids or induced sputum from end-users of refractory ceramic fibres should be interpreted with caution to ensure that they are not confused with asbestos bodies (Dumortier et al., 2001). Coated fibres have also been observed after exposure of workers to silicon carbide fibres (Kilburn & Warshaw, 1991; Dufresne et al., 1995).

4.1.3 Fibre biopersistence, concepts and definition

The term biopersistence as it applies to the presence of fibres in the lung was adopted to refer to the capacity of fibres to persist and to conserve their chemical and physical features over time in the lung (Hammad, 1984; Bernstein et al., 1996; Hesterberg et al., 1998a,b; Searl et al., 1999; Oberdörster, 2000; Hesterberg & Hart, 2001).

Increasing emphasis has been placed on clearance and retention of MMVFs in discussing the role of mechanistic data in risk assessment for adverse health effects induced by fibres (McClellan, 1997). This is partly because the biopersistence of fibres has been shown to play an important role in the effects on health of man-made and other mineral fibres. The retention kinetics of fibres are usually influenced by their chemical composition, size distribution and amount in the lung and may vary with time after exposure (Ellouk & Jaurand, 1994; Muhle et al., 1994; Muhle & Bellman, 1995, 1997). Several studies have suggested that the oncogenic potential of long MMVFs is determined by their biopersistence (Mast et al., 2000b; Bernstein et al., 2001a; Eastes & Hadley, 1996; Moolgavkar et al., 2001a). Pott et al. (1987) and Bernstein et al. (2001a,b) suggested that a certain minimum persistence of long fibres is necessary before early changes start to appear in the lung. Furthermore, Moolgavkar et al. (2001b) have suggested, on the basis of their model, that fibre-induced cancer risk, in addition to being a linear function of exposure concentration, is also a linear function of weighted half-time of the fibres. There are also significant interspecies differences in cancer susceptibility. Although comparisons between species may be difficult, rats are considered to be the preferable model for assays of toxicity and oncogenicity for particulate materials (ILSI, 2000). In view of the increasing body of experimental evidence for the role of biopersistence in fibre toxicity, it is not surprising that several attempts have been made to reduce the biopersistence of fibres by increasing their biosolubility through changes in the chemical composition of the raw material of the fibres (Guldberg et al., 2000).

The influence of fibre length on fibre toxicity was first shown for asbestos fibres. Vorwald et al. (1951) showed that long fibres (20–50 µm) were associated with both lung and peritoneal disease whereas shorter, ball-milled fibres (3 µm or less) were not. The importance of the in-vivo durability of a fibre and of fibre dimension in the pleural and intraperitoneal cavities was determined in the early studies of Stanton & Wrench (1972) and Pott et al. (1974). Long, thin and durable fibres have the greatest potential to cause tumours. In an early study on the biopersistence of synthetic mineral fibres, Hammad (1984) found that retention of refractory ceramic fibres < 5 µm in length was relatively low following short-term inhalation. With longer fibres, retention increased sharply, reached a peak at a fibre length of 11 µm, and then decreased again; fibres with lengths > 30 µm cleared most slowly. A less durable fibre tested in the same study cleared more rapidly. This knowledge provides the basis for the concept of fibre biopersistence: that is, the longer a fibre persists in the lower respiratory tract, the greater its likelihood to cause effects, especially if it is longer than 20 µm (Bernstein *et al.*, 1996; Hesterberg *et al.*, 1998a,b).

The scheme in Figure 11 summarizes the different processes that contribute to the biopersistence of a fibre (Oberdörster, 2000). The pulmonary retention half-time of a fibre, determined in an in-vivo study, reflects its biopersistence, which is determined by elimination due to physiological clearance processes such as translocation to the larynx by alveolar macrophages, into the interstitium, via the lymphatic system and the pleura, and to physicochemical processes that affect biodurability, such as dissolution, leaching and breaking. As discussed in section 4.1.2, physiological clearance rates such as those for macrophage-mediated removal are very different between humans and rodents; in contrast, it is assumed that physicochemical processes occur at similar rates. The breakage of incompletely phagocytosed long fibres into shorter fragments can decrease their biopersistence significantly; at the same time, the shorter fragments enter the pool of the short fibres, which can result in an apparent increase in their biopersistence.

Figure 11. Factors contributing to the biopersistence of fibres in the lung

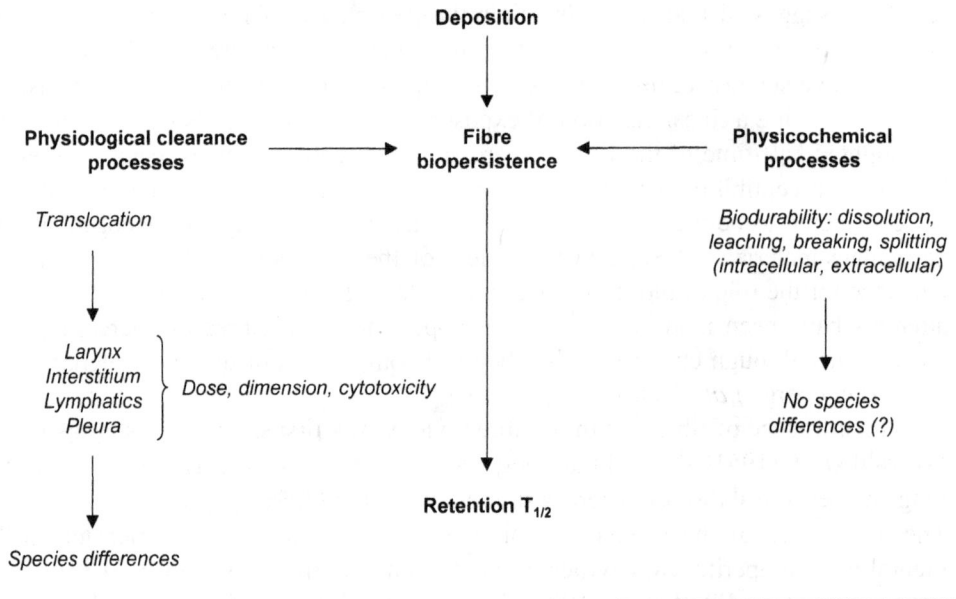

Modified from Oberdörster (2000)

The goal of a biopersistence assay is to assess the potential for accumulation of fibres after long-term exposure. However, it should be noted that the retention time for a fibre derived from a short-term (e.g. 5-day) inhalation assay may be lower than the value determined from a subchronic or chronic inhalation study. Therefore, a retention time determined from a short-term assay should not be used to predict accumulation of fibres in the lung during a long-term study. Various methods are available to determine the retention of fibres in the respiratory tract. Short-term (5-day) inhalation studies in rats have been designed for the determination of fibre retention (Bernstein *et al.*, 1994; Bernstein & Riego-Sintes, 1999). In order to exclude the influence of mucociliary clearance, measurements of retention are generally started some time after exposure. A two-phase pattern of retention is generally observed for the three categories of fibre length, i.e. < 5 µm, 5–20 µm and > 20 µm. For the shorter, phagocytosable fibres, the underlying mechanisms for these phases should be similar to those described for alveolar clearance of non-fibrous particles (see section 4.1.2), i.e. involving clearance pathways of different lengths from the site of deposition by translocation to the mucociliary escalator as well as dissolution in alveolar macrophages; for long fibres, dissolution and breakage can be assumed to be the main mechanisms to explain the two-phase clearance.

(*a*)   *Fibre biopersistence studies* (see Table 65)

In the absence of highly toxic elements or surface structures, a more biopersistent fibre will have a greater potential to induce biological effects in the lung. In addition, biopersistent fibres are more likely than rapidly-clearing fibres to be taken up by the lung epithelium and translocated into the interstitium and on to the pleura and thoracic cavity, where they would continue to cause irritation and inflammation. Thus, a more biopersistent fibre may have an impact on a broader range of target tissues. For example, the overall retention of glass fibres *in vivo* depends not only on their chemical composition, but also on their length. This effect has been attributed to differences in the microenvironment to which long and short fibres are exposed. Although this phenomenon appears to operate with all the glass fibres examined, it does not apply to the MMVFs that dissolve more readily in environments with low pH (Morgan, 1994; Baier *et al.*, 2000).

Overall, recent chronic studies of vitreous and asbestos fibres conducted in rodents indicate a relationship between persistence of fibres in the lung and the severity of their biological effects (for reviews, see Bernstein *et al.*, 2001a,b; Hesterberg & Hart, 2001).

After deposition in the lung, biopersistent fibres such as crocidolite or amosite asbestos exhibited little or no change in chemical composition, surface morphology or dimension (suggesting no significant leaching, dissolution or transverse fragmentation), whereas fibres with lower biopersistence such as MMVF10 and MMVF11 and the newly developed high-alumina, low-silica (HT) stone wool did show changes as follows:

— compositions changed and surfaces often showed signs of erosion;

Table 65. Lung biopersistence, in-vitro dissolution and pathogenicity of selected fibres from inhalation studies in rats

Fibre		Biopersistence: fibres > 20 μm in length; lung clearance rates				In-vitro dissolution (k_{dis}) at pH 7.4 (pH 4.5)*	Pathogenicity (chronic inhalation)		
		Slower pool ($T_{1/2}$)	$WT_{1/2}$ (days)	90% clearance (T_{90}, days)	Reference		Lung fibrosis	Thoracic tumours	Reference
Amosite	Asbestos	1160	418	2095	Hesterberg et al. (1998a)	<1	+	+	McConnell et al. (1999)
Crocidolite	Asbestos	0	817	2770	Hesterberg et al. (1996a)	<1	+	+	McConnell et al. (1994)
MMVF32	E Glass wool	179	79	371	Hesterberg et al. (1998a)	9 (7)	+	+	Davis et al. (1996a)
RCF1a[a]	Refractory ceramic fibre	88	55	227	Hesterberg et al. (1998a)	3	+	+	Mast et al. (1995a,b)
MMVF33	475 Glass wool	155	49	240	Hesterberg et al. (1998a)	12 (13)	+	+/−[b]	Davis et al. (1996a); McConnell et al. (1999)
MMVF21	Rock (stone) wool (96)	613	91	206	Hesterberg et al. (1996a)	20 (72)	+	−	McConnell et al. (1994)
MMVF21	Rock (stone) wool (98)	95	67	264	Hesterberg et al. (1998a)				
MMVF10	901 Glass wool (96)	0	37	123	Hesterberg et al. (1996a)	300 (329)	−	−	Hesterberg et al. (1993)
MMVF10.1[c]	901 Glass wool	30	14.5	69	Hesterberg et al. (2002)		−	−	McConnell et al. (1999)

Table 65 (contd)

Fibre		Biopersistence: fibres > 20 μm in length; lung clearance rates				In-vitro dissolution (k_{dis}) at pH 7.4 (pH 4.5)*	Pathogenicity (chronic inhalation)		
		Slower pool (T½)	WT½ (days)	90% clearance (T_{90}, days)	Reference		Lung fibrosis	Thoracic tumours	Reference
X607	Hybrid fibre	94	10	18	Bernstein et al. (1996)	990	–	–	Hesterberg et al. (1998a)
MMVF11	Glass wool	31	9	38	Hesterberg et al. (1996a)	100 (25)	–	–	Hesterberg et al. (1993)
MMVF22	Slag wool	35	9	37	Hesterberg et al. (1996a)	400 (459)	–	–	McConnell et al. (1994)
MMVF10B	901F Glass wool	20	8	38	Hesterberg et al. (2002)	500	ND	ND	
MMVF35	902 Glass wool	18	7	33	Hesterberg et al. (2002)	150	ND	ND	
MMVF34	HT Stone wool	24	6	19	Hesterberg et al. (1998a)	59 (620)	–	–	Kamstrup et al. (1998)

From Hesterberg et al. (2002); biopersistence data based on 5-day inhalation studies; pathogenicity based on chronic rodent inhalation studies ND, no data (chronic toxicity not evaluated); $T_{½}$, retention half-time of slow clearance phase (if biexponential) or of one clearance phase (if monoexponential); $WT_{½}$, weighted retention half-time; T_{90}, number of days to clear 90% of fibres, calculated from model using estimated fibres/lung directly after exposure (rather than on day 1 of recovery); k_{dis}, ng/cm^2/h; minus sign (–) indicates no fibrosis or tumour incidence not significantly different from background incidence in the test species.

* From Zoitos et al. (1997); Hesterberg et al. (1998a,b); reviewed in Hesterberg & Hart (2001)
[a] RCF1a (biopersistence studies) was modified from RCF1 (chronic studies) to contain fewer non-fibrous particles.
[b] +/– indicates tumorigenicity in hamsters (one mesothelioma in 83 animals), but not in rats; Hesterberg et al., 1998a; Davis et al., 1996a.
[c] MMVF10.1 was size-selected from MMVF10 to have longer and thinner average dimensions. The original MMVF10 was used for previously published chronic and biopersistence studies (Hesterberg et al., 1993; 1998b).

— fibre dimensions decreased; and
— the number of long fibres per lung decreased more rapidly than the number of short fibres per lung (suggesting transverse fragmentation)

(Bernstein et al., 1984; Morgan & Holmes, 1986; Musselmann et al., 1994a,b; Hesterberg et al., 1996b, 1998b).

In two studies, rats were exposed by nose-only inhalation to high concentrations of one of eight different MMVFs or to one of two types of asbestos for five days (6 h per day) and were then kept without further exposure. After post-exposure intervals of up to one year, five to seven animals per exposure group were killed and their lung fibre burdens were evaluated. (Hesterberg et al., 1996c, 1998b).

In the first of these studies, rats were exposed to two standard glass wools used in building insulation (MMVF10 and MMVF11), slag wool (MMVF22) or traditional rock (stone) wool (MMVF21) at approximately 250–350 WHO fibres/cm^3 > 5 µm (including ~100 fibres > 20 µm/cm^3) or to crocidolite asbestos at approximately 2600 WHO fibres/cm^3 > 5 µm (including ~290 fibres > 20 µm/cm^3) (Hesterberg et al., 1996c). In the second study, rats were exposed to a refractory ceramic fibre (RCF1a), a traditional rock (stone) wool (MMVF21), JM E glass wool (a special application microfibre) (MMVF32), JM 475 glass wool (a special application microfibre) (MMVF33) or the newly developed high-alumina, low-silica (HT) stone wool (MMVF34) at ~400 WHO fibres/cm^3 (including ~150 fibres > 20 µm/cm^3) or to amosite asbestos at ~800 WHO fibres/cm^3 > 5 µm (including 235 fibres > 20 µm/cm^3) (Hesterberg et al., 1998b). The MMVFs tested had approximate arithmetic mean dimensions of 1 µm × 20 µm and geometric mean dimensions (GMD) of 0.7–0.9 µm diameter and 12–16 µm length. The crocidolite fibres were smaller than the MMVFs (GMD, 0.3 µm × 4.2 µm), but the amosite fibres were more comparable in size (GMD, 0.5 µm × 7.7 µm).

The rat lungs were analysed at intervals during a one-year post-exposure recovery period for numbers of fibres in the lung, dimensions, morphology and chemical composition. In both of these biopersistence studies, the deposition in the lung of long fibres (> 20 µm) was roughly comparable for the eight MMVFs and the two types of asbestos (crocidolite and amosite) as measured by the lung burden one day after cessation of exposure. However, the deposition of shorter fibres was greater for asbestos than for the MMVFs. The lung burdens of the MMVFs were more similar to those of amosite than to those of crocidolite (both in numbers and dimensions).

In both studies, asbestos fibres cleared much more slowly than the MMVFs: $WT_{1/2}$ retention times for fibres > 20 µm were 817 days for crocidolite and 418 days for amosite, compared with 6–67 days for MMVFs. After one year of recovery, retention of fibres > 20 µm in the lung was 83% for crocidolite and 30% for amosite, compared with 0–10% for the MMVFs.

The different MMVFs had a broad range of retention times: $WT_{1/2}$ values were 49–79 days for the slower-clearing MMVFs (rock (stone) wool, the two special-application glass wools and refractory ceramic fibre), and only 6–9 days for the more rapidly-clearing MMVFs (the two standard building-insulation glass wools, slag wool and high-alumina, low-silica (HT) stone wool). Of all fibre compositions and length categories tested in both of these biopersistence studies, the number of HT rock (stone) wool fibres > 20 μm per lung underwent the most rapid reduction. By 30 days post-exposure, the lung burdens of HT fibres > 20 μm were 3% of those measured on day 1 and by 180 days, they had been reduced to background levels (Hesterberg *et al.*, 1998b).

For asbestos and the more biopersistent MMVFs, the number of short fibres tended to decrease more rapidly than the number of long fibres. The investigators suggested that this was a likely result of macrophage-mediated lung clearance, which is more efficient for shorter fibres. However, for the more biosoluble MMVFs, the numbers of long fibres decreased more rapidly than the number of short fibres. The investigators suggested that this could be a result of a rapid rate of transverse breakage of long, biosoluble fibres into short fibres. Transverse fragmentation of fibres has been demonstrated *in vitro* for rapidly leaching compositions of MMVF (Bauer, 1998a).

During the year of post-exposure recovery, changes in the dimensions of the fibres in the lung were noted. For the less biopersistent MMVFs, mean lengths decreased, suggesting transverse breakage and/or dissolution. The mean lengths of fibres of crocidolite and amosite increased, suggesting selective clearance of shorter fibres by macrophages; their diameters remained unchanged.

Analysis of HT rock (stone) wool fibres deposited in the lung by energy dispersive spectroscopy showed a decrease in the percentage of calcium oxide after the first 30 days of post-exposure recovery (Hesterberg *et al.*, 1998b). After 91 and 365 days of residence in the lung, the percentages of alkali oxides (Na_2O) and alkaline earth oxides (magnesium oxide and/or calcium oxide) in MMVF10, MMVF11 and MMVF22 showed significant decreases, while MMVF21 and crocidolite showed no change in chemical composition (Hesterberg *et al.*, 1996c).

Scanning electron microscopy (SEM; using magnification up to × 50 000) was used to examine the morphology of fibres in the lung at several time points up to three months post-exposure. At one extreme was amosite, in which no morphological changes were observed in any of the fibres in the lung during the three months of monitoring. Most of the refractory ceramic, rock (stone) wool and JM E glass fibres (MMVF32) also underwent no visible changes during the three months. The JM 475 glass fibres (MMVF33) showed a range of morphological changes in the lung, from negligible to fairly severe surface etching. At the other extreme were the slag wool and HT stone wool fibres, most of which developed severe surface etching and other visible indications of deterioration starting as early as one week post-exposure. Common findings in HT stone wool fibres were crater-like surface pits, rounded or blunt ends, dissolved cores and segmented fibre fragments lying end-to-end on the SEM stub filter, the latter indicating transverse breakage during the final processing of samples.

Changes in the numbers of fibres and in their chemistry, dimensions and/or morphology during residence in the lung were more pronounced for the eight MMVFs than for the two types of asbestos. These changes were especially striking in the four MMVFs that cleared most rapidly in the studies of biopersistence, did not produce lung fibrosis or tumours and were non-pathogenic in the chronic studies (the two insulation glass wools, slag wool and HT stone wool). The induction of fibrosis and tumours by a specific fibre type was found to be closely associated with its biopersistence and to a lesser extent with its in-vitro rate of dissolution at pH 7.4 (discussed below). The investigators concluded that these observations supported the hypothesis that the biopersistence of inorganic fibres in the lung is a major determinant of fibre toxicity and is determined in part by the rate of fibre dissolution (Hesterberg et al., 1998b).

The biopersistence of nine MMVFs (four glass wools, four rock (stone) wools and X-607) was evaluated in rats exposed to the fibres for five days (6 h per day) by nose-only inhalation. The fibres included in the study were both commercial and experimental compositions. After various post-exposure time periods, the whole lung was removed and ashed and the fibres from the lung were analysed. At the end of the period of exposure, the lung burdens for the nine fibres ranged from $5.7-33 \times 10^6$ fibres/lung and fibres had geometric mean diameters of ~0.5 μm. The $WT_{1/2}$ of the nine fibres was 11–54 days for WHO fibres and 2.4–45 days for fibres > 20 μm. For one of the fibres, a commercial rock (stone) wool, the number of longer fibres (> 20 μm) per lung declined more rapidly than that of shorter fibres. The $WT_{1/2}$ of the nine fibres correlated quite well with their rates of in-vitro dissolution (Bernstein et al., 1996).

Groups of male Fischer 344 rats were exposed to six MMVFs (three glass wools, two rock (stone) wools and RCF1a) by either nose-only inhalation (5 days, 5 h per day; target aerosol concentration, 30 mg/m^3) or by intratracheal instillation (0.5 mg fibres in 0.3 mL saline per day for 4 days). Subgroups of rats were killed after various post-exposure intervals, the whole lung was ashed and fibres recovered from the lung were analysed. For all the glass wools and rock (stone) wools, longer fibres (> 20 μm) cleared more rapidly than WHO fibres (> 5 μm); for RCF1a, the rate of clearance of both size fractions was about the same. Following intratracheal instillation, the $WT_{1/2}$ of fibres > 20 μm for the glass and rock (stone) wools varied between 3 and 12 days, but RCF1a had a $WT_{1/2}$ of > 1000 days. Following inhalation exposure, the $WT_{1/2}$ of fibres > 20 μm was 4–6 days for the glass and rock (stone) wools, but 64 days for RCF1a. The authors suggested that the very long $WT_{1/2}$ seen for RCF1a seen after intratracheal instillation may have been due to agglomeration in the bronchi (Bernstein et al., 1997).

Rats received eight different types of mineral fibre by intratracheal injection, and the relative biopersistence of the fibres in different size categories was assessed from the changes in mean lung burden, as determined by SEM, at three days and one, six and 12 months after instillation. The samples tested included the commercially available glass fibre 100/475, six size-selected fibre types specifically prepared for research purposes (MMVF10, MMVF21, MMVF22, RCF1, RCF2 and RCF4), and amosite. Short fibres were cleared by cellular processes and long fibres by dissolution and disintegration. The differences in persistence of long fibres (> 20 μm) of these fibre types were correlated with the rates of dissolution measured in vitro. The differences in persistence noted for those fibre types that had also been studied by other research groups were consistent with their findings, both after administration of fibres by inhalation and intratracheal injection. The biopersistence of the different fibres was influenced by their dimensions and solubility. For the six more durable compositions tested (amosite, 100/475, MMVF21, RCF1, RCF2 and RCF4), long fibres tended to clear from the lung more slowly than short fibres. For the two less durable fibres (MMVF10 and MMVF22), in contrast, longer fibres cleared more rapidly than shorter ones (Searl et al., 1999).

(b)   *Fibre biopersistence and pathogenicity*

Several short-term cellular tests for predicting the pathological effects of particulates and fibres are reported in sections 4.2–4.5, but the results often do not correlate with the results of studies on biopersistence *in vivo*. This indicates that, as documented in a recent workshop report, several chemical and biological properties in addition to biopersistence have to be considered in order to screen for pathogenicity of fibres (Fubini et al., 1998).

In an early study, male Fischer 344/N rats were exposed to size-separated respirable fractions of glass fibre with compositions representative of common building-insulation wools. The animals were exposed in nose-only inhalation chambers for 6 h per day on five days per week for 24 months to either 3, 16 or 30 mg/m^3 of two different compositions of glass fibre (MMVF10 and MMVF11). Filtered air was used as a negative control treatment. The findings for glass fibre were compared with those from a concurrent inhalation study of chrysotile asbestos and refractory ceramic fibre. Fibres were recovered from digested lung tissue for the determination of changes in fibre number and morphology. In animals exposed to 30 mg/m^3 MMVF10 or MMVF11, 4.2×10^8 and 6.4×10^8 fibres/g dry lung tissue, respectively, were recovered after 24 months of exposure. For rats exposed to refractory ceramic fibres and chrysotile these numbers were 3.7×10^8 and 1.9×10^{10} fibres/g dry lung, respectively. Exposure to chrysotile asbestos (10 mg/m^3) and to a lesser extent to refractory ceramic fibres (30 mg/m^3) resulted in pulmonary fibrosis as well as mesothelioma and significant increases in the incidence of lung tumours. Exposure to glass fibre was associated with an inflammatory macrophage response in the lungs that did not appear to progress after 6–12 months of exposure. These cellular changes were reversible. No lung fibrosis was observed in the animals exposed to glass fibre, and neither mesotheliomas nor statistically significant increases in the incidence of lung tumours were observed, when compared with the negative-control group (Hesterberg et al., 1993).

Inhalation studies (McConnell, 1994; McConnell et al., 1994; Mast et al., 1995a,b; McConnell et al., 1995) were also conducted in rodents to determine the chronic biological effects of rodent-respirable fractions of MMVFs including refractory ceramic fibres, glass fibre, rock (stone) wool (MMVF21) and slag wool (MMVF22). Animals were exposed by nose-only inhalation for 6 h per day on five days per week for 18 months (Syrian golden hamsters) or 24 months (Fischer 344/N rats). Exposure to 10 mg/m^3 of crocidolite or chrysotile asbestos induced pulmonary fibrosis, lung tumours and mesotheliomas in rats and hamsters, in these chronic inhalation studies.

Exposure of rats to 30 mg/m^3 refractory ceramic fibres also resulted in pulmonary fibrosis and pleural fibrosis as well as a significant incidence of lung tumours and mesotheliomas. In hamsters, 30 mg/m^3 refractory ceramic fibres induced a 41% incidence of mesotheliomas. Exposure of rats to 30 mg/m^3 glass fibre (MMVF10 or MMVF11) or slag wool (MMVF22) was associated with an inflammatory response, but no mesotheliomas or lung tumours were observed after 12 or 24 months of exposure. The same dose of rock (stone) wool (MMVF21) resulted in minimal lung fibrosis and no malignant tumours.

In a chronic inhalation study, male Wistar rats were exposed to an E-glass microfibre (104E; a special-purpose fibre) at a concentration of 1000 fibres > 5 µm/mL of air for 7 h per day on five days per week for 12 months. After a 12-month post-treatment recovery period, the retained lung burden (of fibres of all lengths) was about 30% compared to the burden immediately after treatment. Amosite asbestos and 100/475 microfibres, which were used as controls, were retained at levels of 44% and 28%, respectively. The 104E fibres were thus slightly less persistent in the lungs than amosite. The chemical composition of the 104E fibres did not change during their 24-month residence time in the lungs, but that of the 100/475 microfibres did. Exposure of rats to 104E microfibres and amosite asbestos by inhalation markedly increased the incidence of lung tumours (carcinomas and adenomas) and mesotheliomas and induced pulmonary fibrosis. In contrast, in the animals treated with 100/475, little fibrosis, a few lung adenomas, no carcinomas, and no mesotheliomas were observed (Cullen *et al.*, 2000). [The Working Group noted that the low number of fibres counted may explain some anomalies in the lung fibre burdens.]

Fischer 344 rats were exposed by nose-only inhalation to MMVF21 (traditional rock (stone) wool) and MMVF34 (HT stone wool), for 6 h per day for five days, and then followed post-exposure for up to 12 months. In a second study, rats were exposed for 6 h per day for two years with a post-exposure follow-up until survival in the exposed group was approximately 20%. The short-term study used concentrations of 150 fibres (> 20 µm)/cm^3 and the long-term study 30 mg/m^3. In both these studies, the biopersistence pattern for MMVF34 was similar. The biopersistence of MMVF34 was less than that of MMVF21 when elimination half-lives after short-term inhalation were compared. [The Working Group noted that the weighted half-times ($WT_{1/2}$) reported in the studies by Hesterberg *et al.* (1998b) and Kamstrup *et al.* (1998) increased from six to 17 and 27 days following inhalation for five days, three months and 12 months, respectively.] Only minor histopathological changes were observed in the lungs of rats exposed to MMVF34 compared with those exposed to MMVF21 after up to 18 months. At this time, MMVF21, but not MMVF34, had induced lung fibrosis. From the preliminary histopathological results the authors considered it unlikely that animals exposed to MMVF34 would develop pulmonary tumours in excess of control incidences (Kamstrup *et al.*, 1998).

4.1.4 *Fibre dissolution*

(a) *In-vitro dissolution* (see Table 65)

(i) *Cell-free systems*

The physical and chemical mechanisms whereby fibres may degrade in the lung have been studied extensively *in vitro* using balanced salt solutions ('simulated lung fluids') (Law *et al.*, 1990, 1991; Christensen *et al.*, 1994; Luoto *et al.*, 1994a,b; de Meringo *et al.*, 1994; Hesterberg *et al.*, 1996a; Knudsen *et al.*, 1996; Bauer, 1998a,b). Because fibre dissolution and breakdown in the lung are important determinants of lung clearance, in-vitro dissolution tests have been suggested for screening for toxicity of fibres (Scholze & Conradt, 1987; Eastes & Hadley, 1996).

[The Working Group noted that dissolution of solids may be strongly influenced by the presence of surfactant and molecules that selectively bind to surface ions. Ascorbic acid, which is one of the major antioxidant defences in the lung lining layer, makes crystalline silica more soluble than amorphous silica (Fenoglio *et al.*, 2000).]

The dissolution of silica from a range of industrial MMVFs (including glass wool, rock (stone) wool, slag wool and refractory ceramic fibres) was compared *in vitro* with that of crocidolite, using a solution similar to that of Gamble (1967). The calculated reductions in diameters (based on chemical analysis of the effluent) for the MMVFs ranged from 0.2–3.5 nm/day. The corresponding value for crocidolite was < 0.01 nm/day. The rates of dissolution for glass fibres showed a 15-fold variation; the solubility of samples of rock (stone) wool and slag wool (composition specified) was intermediate among the fibres tested; and the solubility of the refractory ceramic fibres was generally at the lower end of the range (Scholze & Conradt, 1987). Leineweber (1984) also found great variability in the solubility of glass fibres (a 30-fold range in the dissolution rate constants (k_{dis}), from 3.0 to 0.1 $ng/cm^2/h$); a refractory ceramic fibre was found to be fairly insoluble (0.4 $ng/cm^2/h$). These k_{dis} values are much lower (by about one order of magnitude) than those measured by Zoitos *et al.* (1997) and other recent investigators. The dissolution rates (in $ng/cm^2/h$ ± SD) were reported for the following fibres: MMVF21 (23 ± 11), MMVF22 (119 ± 41), MMVF11 (142 ± 39), MMVF10 (259 ± 75) and RCF1 (8 $ng/cm^2/h$; no SD given). Crocidolite has a very low dissolution rate (0.3 ± 0.1). It is assumed that the experimental methods or the sensitivity of the analytical method — comparing weights of residual fibre samples to weights of initial fibre samples — may account for these discrepancies.

Recent studies of in-vitro dissolution and degradation have typically employed a flow-through dissolution system based on the work of Klingholz & Steinkopf (1984) and Leineweber (1984), in which physiological fluid (similar to that described by Gamble, 1967) is passed through a fibre sample. At various time-points, the effluent from the flow-through system is analysed for dissolved fibre components. A dissolution constant (k_{dis}, $ng/cm^2/h$) is then estimated from the effluent data (Potter & Mattson, 1991). The flow rate must be rapid enough to clear the dissolution products from the fibre surface, but not so rapid as to dilute the dissolution products so much that they cannot be analysed accurately. It is also important to note that silica and aluminium compounds vary widely in the extent to which they are involved in the leaching process. This makes analysis of a single component in the effluent unsuitable as a means of comparing the dissolution rates of a wide range of compositions (Mattson, 1994).

A broad range of in-vitro dissolution rate constants (k_{dis}, ng/cm^2/h) were observed for a variety of inorganic fibres tested in a flow-through system at near-neutral pH (i.e. pH 7.4–7.6 which simulates the pH of extracellular fluid). The rates were as follows: < 1 for crocidolite asbestos; 8–12 for biodurable MMVFs, such as refractory ceramic fibres, E-glass microfibre and 475 glass microfibre, and 100–300 for soluble MMVFs, such as building-insulation glass wools and slag wool (Zoitos et al., 1997; Hesterberg et al., 1996a, 1998a). Some fibres have much higher k_{dis} values; for example, JM 909 building-insulation glass fibre which has a k_{dis} > 1000 (Hesterberg & Hart, 2001) and various new or experimental compositions of rock (stone) wool and glass wool which have a k_{dis} > 500 (Hesterberg et al., 1998b). As noted by Hesterberg et al. (1998b), with rare exceptions, the values for k_{dis} for MMVFs at pH 7.4 have shown a good correlation with in-vivo lung clearance and with pathogenicity.

Dissolution rates have also been studied using fluid at pH 4.5, which simulates the acidic pH of the phagolysosome of macrophages (Etherington et al., 1981). Sébastien (1994) found a good correlation between fibre clearance (from the rat lung following intratracheal instillation) and the in-vitro dissolution rate, but only if k_{dis} values at both pH 7.7 and pH 4.5 were taken into consideration. Other researchers have stressed the importance of using dissolution rates at both pHs to predict the in-vivo biopersistence of a fibre (Guldberg et al., 1998, 2000). The HT stone wool (MMVF34) is a case in point: it was observed to be non-persistent in the rat lung (weighted retention half-time of fibres > 20 µm, 6 days; see above), but it did not dissolve rapidly in vitro at near-neutral pH (k_{dis} at pH 7.4, 59; Hesterberg et al., 1998b). However, HT did dissolve rapidly at acidic pH (k_{dis} at pH 4.5, 1100) (Knudsen et al., 1996; Guldberg et al., 1998).

The in-vitro dissolution of seven traditional MMVF compositions was tested. The fibres studied were JM 475/100 glass wool microfibre, building-insulation glass wool (MMVF10), rock (stone) wool (MMVF21), slag wool (MMVF22), three refractory ceramic fibres (RCF1, RCF2 and RCF4)) and amosite asbestos. Fibres were subjected to either a static system (sodium oxalate at initial pH 4.6 or pH 7.0 for 56 days) or a flow-through system (Kanapilly's solution, similar to Gamble's, at pH 7.4). Higher in-vitro dissolution rates of these fibres corresponded with lower biopersistence in rat lung following either intratracheal injection or inhalation of fibres (see above); MMVF10 and MMVF22 were the least durable. After 56 days of treatment with sodium oxalate, the percentage of silica dissolved at pH 4.6 was 31% for MMVF10, 51% for MMVF22, but < 1–6% for all other fibres; dissolution of silica at pH 7.0 was 15% for MMVF10 and 53% for MMVF22, but < 1–6% for all others. Flow-through k_{dis} values (ng/cm^2/h) at pH 7.4 were: 122 and 53 for MMVF10 and MMVF22, respectively, but 0.5–29 for the other fibre types (k_{dis} for amosite not reported) (Searl et al., 1999).

The effect of different chemical compositions of MMVF on their dissolution by alveolar macrophages in culture and in cell-free Gamble's solution was studied. The fibres were exposed to cultured rat alveolar macrophages, culture medium alone, or in Gamble's saline solution for 2, 4 or 8 days. The dissolution of the fibres was studied by measuring the amount of silicon (Si), iron (Fe), and aluminium (Al) in each medium. The macrophages in culture dissolved Fe and Al from the fibres, but the dissolution of Si was more marked in the culture medium without cells and in the Gamble's solution. The dissolution of Si, Fe, and Al was different for different fibres, and increased as a function of time. The Fe and Al content of the fibres correlated negatively with the dissolution of Si from the MMVFs by macrophages, i.e. the higher the content of Fe and Al of the fibres the lower the dissolution rate of Si. The results suggest that the chemical composition of the MMVF has a marked effect on its dissolution. Alveolar macrophages seem to affect the dissolution of Fe and Al from the fibres. This suggests that in-vitro models with cells in the media rather than cell-free culture media or saline solutions would be preferable in dissolution studies of MMVFs (Luoto et al., 1994a).

Glass fibres with a mean diameter of 1 μm of different chemical composition and obtained by different processes were characterized with respect to their solubility under various test conditions and flow rates. The surface morphology was examined using SEM and energy dispersive spectroscopy. The SEM revealed the typical formation of various corrosion patterns: porous, gel-like outer layers, precipation zones, but, in some cases, no modification of the surface aspect was seen. The results show that surface changes depend strongly on the initial composition of the glass and on the test conditions, particularly the flow rate (Lehuédé & de Meringo, 1994).

To evaluate in-vitro dissolution of fibres, JM 475/100 microfibre glass wool and amosite asbestos were suspended and incubated in sodium oxalate (pH 4.6 or 7.0) or in 0.1 M oxalic acid (pH 0.6), and the silicon content of the supernatants was determined at intervals up to 56 days. Changes in pH were found to be minimal over this period. After 56 days of incubation in sodium oxalate at either pH 4.6 or 7.0, JM 475/100 lost about 6% of its silicon content, while amosite lost only 1–2%. After 28 days of incubation in oxalic acid, JM 475/100 lost 29% of its silicon, while amosite lost only 9% (Davis et al., 1996a).

Several researchers have stressed the importance of establishing standardized methods for estimating in-vitro dissolution rates (Mattson, 1994; Zoitos et al., 1997; Guldberg et al., 1998). There is general consensus that a flow-through system similar to that described by Potter and Mattson (1991), with a modified Gamble's solution at 37 °C, would be an appropriate system (Searl et al., 1999). [The Working Group noted that there is still some inter-laboratory variability in methodologies that affects the consistency of the dissolution rates reported.]

(ii) *Mechanisms of in-vitro fibre degradation*

The fibre dissolution model using the rate constant, k_{dis}, assumes simple (congruent) dissolution (all fibre components dissolve at approximately the same rate, which is constant over time), with fibre length and diameter decreasing at constant rates. In this model, the fibre composition does not change during the dissolution process. However, many MMVFs exhibit a high degree of leaching, i.e. incongruent dissolution, in which certain components dissolve more rapidly than others. Leaching of the fibres changes their composition over time and their morphologies show expanding zones of leached-out, lower-density material. This material has been reported to react with other components and form deposits on the fibres (Klingholz & Steinkopf, 1984; Leineweber, 1984; Potter & Mattson, 1991; Christensen *et al.*, 1994; Hesterberg *et al.*, 1996c; Bauer, 1998a,b; Hart *et al.*, 1999). Differences in dissolution behaviour have been related to the chemical composition of the fibres, i.e. high contents of alumina and silica favour congruent dissolution, and to the manufacturing process (Potter & Mattson, 1991; Bauer, 1998a,b). In the manufacture of certain types of glass fibre, heating over a flame is used in the fiberization process, and has been associated with the formation of a resistant surface layer (Bauer, 1998b). Traditionally, fiberization in air at room temperature has been used in the manufacture of most standard thermal and acoustic insulation wools, while heating in a flame has been used for most special-purpose fibres (Hesterberg & Hart, 2001).

Durability is determined not only by dissolution of fibres in extracellular or intracellular fluid, but also by the physical condition of the fibre residuum; the fibre is no longer resident in the lung once it has disintegrated into non-fibrous particles (Scholze & Conradt, 1987). Furthermore, breakage of long fibres in the lung would increase the probability of their clearance by macrophages. A positive correlation between the leaching index (a measure of the change in fibre composition) and breakage was demonstrated in a study of the in-vitro dissolution of 18 compositions of glass fibre in Gamble's solution at pH 7.4. Fibre breakage *in vitro* was apparently caused by leaching-induced weakness followed by mild stress generated by the interwoven nature of the fibres in the test system (Hesterberg *et al.*, 1996a; Bauer, 1998b). Striking fragmentation following in-vitro leaching was also exhibited by another rapidly-leaching MMVF, X-607 fibre (Hesterberg & Hart, 2001).

(iii) *Relationship between in-vivo retention of fibres, in-vitro dissolution rates and fibre composition*

A relationship between in-vivo retention of long fibres and in-vitro dissolution rates (k_{dis}) at near-neutral pH has been reported by Searl *et al.* (1999) and by Maxim *et al.* (1999a,b). In addition, Eastes *et al.* used linear regression techniques to develop equations for the estimation of the dissolution rate of a wide variety of MMVFs. The data used to develop these equations were derived from either data on in-vitro dissolution rates (Eastes *et al.*, 2000a), or from data obtained from in-vivo studies (Eastes *et al.*, 2000b,c). A correlation was shown between the estimate using these equations and the actual values measured *in vitro* for a wide range of compositions of long fibres.

[The Working Group noted that the kinetic and thermodynamic approach used by Eastes *et al.* (2000a,b,c) derived from homogeneous physicochemical systems, may not fully apply to the complex multiphase system of a multicomponent fibre in a biological fluid, where leaching and redeposition processes also occur (Hesterberg *et al.*, 1996a).]

(iv) *Cellular in-vitro systems*

The effect of macrophages on the dissolution of long and short glass wool and rock (stone) wool fibres was studied in a flow-through cell-culture system. In this system, culture medium flowed through a membrane on which the macrophages and fibres were placed. Dissolution was estimated by measuring the concentrations of silicon, iron and aluminium in the culture medium after various time periods. Compared with the traditional static culture system, the flow-through system enables longer testing periods and avoids pH changes resulting in the build-up of dissolution products of cells and fibres (Luoto *et al.*, 1994a).

The dissolution of glass wool was faster in cell-free culture medium than in the presence of macrophages in cell culture. The dissolution of rock (stone) wool was slower than that of glass wool in cell culture. However, in a cell-free culture medium, the dissolution of experimental rock (stone) wool was more pronounced than that of commercial glass wool (Luoto *et al.*, 1995, 1998). Glass wool released silicon faster than rock (stone) wool, but rock (stone) wool released iron and aluminium faster than glass wool. Fibres in the flow-through culture exhibited surface changes, such as fractures, peeling and pitting. In the presence of macrophages, short fibres, < 20 µm in length, were easily phagocytosed whereas longer fibres were incompletely phagocytosed by a large number of alveolar macrophages. Macrophages that had engulfed fibres contained measurable amounts of silicon (Luoto *et al.*, 1998). Man-made vitreous fibres also induced ruffling and blebbing on the surface of rat alveolar macrophages: the exposed cells produced extensions which fastened them to the fibres or to other cells to form clumps or clusters of cells and fibres, each cell engulfing part of a fibre (Luoto *et al.*, 1994b).

Long, partly phagocytosed or non-phagocytosed glass wool fibres seemed to disappear more rapidly than short, phagocytosed glass wool fibres, in agreement with the observations of other investigators (Jaurand, 1994). However, it is possible that the number of long fibres decreases because they break into short fibres and not because they dissolve more rapidly than the short fibres. In general, the dissolution profiles obtained in the flow-through cell culture system were rather similar to those seen in the traditional static system, but the dissolution rate of individual fibres was more rapid in the former. The overall conclusion of these investigators was that cell-culture systems should be chosen rather than cell-free in-vitro systems when assessing fibre durability and dissolution (Luoto *et al.*, 1994a,b, 1995, 1996, 1998).

The dissolution in synthetic simulated lung fluid (Gamble's solution) of two different fibres, biosoluble HT (rock) stone wool and the traditional MMVF21 stone wool was investigated. Both fibres dissolved readily at pH 4.5, possibly due to the presence of organic acids that are able to form complexes with aluminium. This suggestion is supported by the observation that organic acids without this complex-forming ability were ineffective in stimulating dissolution. The HT (rock) stone wool dissolved faster than MMVF21, most probably because of the different ratios of Al/(Al+Si) in the two types of fibre (0.40 and 0.25, respectively) (Steenberg *et al.*, 2001). This ratio has been shown to correlate well with the dissolution rate at pH 4.5 of a range of fibres (Guldberg *et al.*, 2000).

(b)   In-vivo solubility

Christensen *et al.* (1994) suggested that the length of an inhaled fibre may determine the pH environment in the lung to which that fibre is exposed — fibres longer than about 20 μm would be in the extracellular compartment where the pH is 7.4, while shorter fibres would tend to be in the intracellular compartment of macrophages at pH 4.5. Lehuédé *et al.* (1997) found that the changes in fibre composition observed *in vivo* were similar to those observed *in vitro* at pH 7.4, but not at pH 4.5, although only one type of fibre was tested at pH 4.5. See section 4.1.4(*a*)(iv) for a further discussion on fibre dissolution in in-vitro cell cultures of macrophages (Luoto *et al.*, 1994a,b, 1995, 1998).

The retention of fibres in the lung was studied following the exposure of Fischer 344/N rats by inhalation and compared with dissolution in a cell-free flow-through system *in vitro*. The 10 samples of MMVF studied included compositions similar to commercial insulation materials as well as new glass fibres and rock (stone) wools developed to enhance solubility. The compositions of many of these fibres changed with increasing time in the lungs, with a depletion in sodium oxide, calcium oxide and magnesium oxide, and relative enrichment in silica and alumina (Lehuédé *et al.*, 1997). The in-vivo findings agree with those reported by other investigators (Hesterberg *et al.*, 1996a, 1998a,b).

Thus, if fibres are too long for macrophage-mediated clearance and if their chemical composition renders them resistant to breakdown in lung fluids (both intracellular and extracellular), they can be expected to accumulate during chronic exposure and remain in the lung and related tissues for considerable lengths of time. A recent study supports this hypothesis: Syrian golden hamsters were exposed for 18 months (6 h/day, 5 days/week) to comparable dusts of either standard glass fibre insulation or amosite asbestos. After one day of exposure, the lung burdens of longer fibres (> 20 μm in length) were similar for the two fibres (~0.2 million/lung), demonstrating comparable lung deposition for the two types of fibre. However, after 12 months of exposure, the lung burdens of the two types of fibre were different, i.e. 1×10^6 fibres (> 20 μm)/lung for glass wool, but 37×10^6 fibres (> 20 μm)/lung for amosite. Assuming little or no transport by macrophages of fibres > 20 μm in length, the almost 40-fold higher lung burden for amosite suggests that the fibres of the insulation glass wool dissolved or otherwise broke down in lung fluid much more rapidly than those of amosite (Hesterberg et al., 1999; McConnell et al., 1999).

The chronic inhalation effects in rats of X-607 (a rapidly dissolving MMVF that does not fit into any of the traditional categories) were compared with those of RCF1. Fischer 344 rats were exposed to a fibre aerosol by nose-only inhalation for 6 h per day on five days per week for two years. The concentrations of the aerosols of X-607 and RCF1 were similar (~200 WHO fibres/cm^3) and the fibres had similar average dimensions (approximately 20 μm × 1 μm). The deposition of fibres in the lung after 6 h of inhalation was greater for X-607 than for RCF1. However, at all later times, the numbers of fibres per lung (especially fibres > 20 μm) were lower for X-607 than for RCF1, suggesting that X-607 was less biopersistent. Chemical analysis of fibres from the lung revealed more rapid leaching of X-607 than of RCF1; after 78 weeks in the lung, the calcium content of X-607 had decreased from an initial 40% (by mass) to zero, leaving a fibre of essentially pure silica. The chemical composition of RCF1 did not change. Fibres of X-607 also had a much higher in-vitro dissolution rate (k_{dis}, 990 ng/cm^2/h) than RCF1 (k_{dis}, 6 ng/cm^2/h; see below) (Hesterberg et al., 1998a).

Groups of Wistar rats were given a 50-mg dose of one of three MMVFs (glass wool, rock (stone) wool, glass microfibres (JM 100) by intraperitoneal injection. The chemical compositions of the fibres were analysed by inductively coupled plasma spectrometry carried out on the urine of the rats, which was collected at fixed times between day 1 and day 204, and analysed for elements known to be present in the original fibres. At day 204, a piece of omentum was removed at necropsy, ashed and analysed by energy dispersive X-ray diffraction analysis (EDXA) to identify the elements remaining in the fibres. Silicon and aluminium were found to be retained in the fibres from all samples at day 204. Titanium, present at 0.3% (w/w) in the original sample of rock (stone) wool, was not detectable by EDXA at day 204, but small quantitites were detected in urine collected during the first two weeks of exposure. The barium content of the retained glass microfibres (JM 100) had decreased by day 204 and barium was detectable in the urine. The authors concluded that titanium and barium could be suitable biomarkers of exposure to rock (stone) wool and glass microfibres and could also reflect the in-vivo dissolution of these fibres (Wastiaux et al., 1994). [The Working Group noted the preliminary nature of this study and the high dose used and also noted that not all rock (stone) wool contains titanium and barium.]

Baier et al. (2000) used instillation of MMVF10, RCF1a and HT fibre in rats as an in-vivo model to evaluate local biological responses to fibres of different composition. Specimens taken at 2, 7, 30 and 90 days post-instillation were compared by means of SEM and energy-dispersive X-ray spectroscopy (EDS). Fibres of MMVF10 dissolved the most quickly both in extracellular fluids and after phagocytosis; HT fibres were thinned by phagocytes and fragmented, and RCF1a resisted both external dissolution and uptake by macrophages, becoming embedded in granulomatous nodules. In accordance with these observations, the infrared spectra of sequentially-harvested slices of lung lobe showed increasing intensities of absorption bands associated with granuloma formation only for RCF1a. These results indicated that the lung can process the fibres in different ways dependent on their composition.

Two studies have reported on the retention of fibres injected directly into the peritoneal cavity of rats.

Collier et al. (1994) reported that long glass fibres (> 20 μm) instilled into the lung of Fischer 344 rats underwent a reduction in diameter at a rate consistent with their exposure to a near-neutral pH environment. The same type of fibres, when injected into the peritoneal cavity, dissolved at a rate consistent with a more acidic environment. In a second study, Collier et al. (1995) compared the retention of fibres injected into the peritoneal cavity of Fischer 344 rats with the retention of the same fibres after intratracheal instillation into the lung. Long fibres (> 20 μm) were retained to a greater extent in the peritoneal cavity than in the lung. At doses in excess of 1.5 mg injected into the peritoneal cavity, most of the injected material was found in nodules which were either free in the peritoneal cavity or loosely bound to the peritoneal organs. The fibres tested had a mean diameter of 2.0 μm. [The Working Group noted that if biopersistence assays are conducted in serosal cavities, doses that do not induce a bolus effect should be chosen.]

4.2 Toxic effects in humans

4.2.1 *Adverse health effects other than respiratory cancer*

(a) *Mortality data*

Data covering all non-malignant respiratory diseases (NMRD) and the subgroup bronchitis, emphysema and asthma (BEA) associated with exposure to MMVFs are available from the death certificates of historical cohorts of MMVF production workers. Descriptions of the cohorts are given in section 2. The mortality data on NMRD are summarized in Table 66.

In the most recent update of the European cohort study, Sali et al. (1999) reported a non-significant excess of all NMRD among glass-wool workers (standardized mortality ratio (SMR), 1.18; 95% CI, 0.98–1.40). For the subgroup BEA, i.e. NMRD excluding influenza and pneumonia, the corresponding value was slightly lower (SMR, 1.12; 95% CI, 0.82–1.49). Interestingly, short-term MMVF workers (employed for < 1 year) in the study by Sali et al. (1999) of the European cohort (Boffetta et al., 1998) had significantly higher SMRs (SMR, 1.79 [95% CI, 1.8–2.44]) for BEA than long-term workers (employed for ≥ 1 year) (SMR, 0.95 [95% CI, 0.67–1.32]. [The Working Group noted that information on smoking history was not given in these studies.]

In the 1992 update of the US cohort of glass-fibre workers, the SMR for NMRD (excluding influenza and pneumonia) (national rates) was reported to vary according to the type of glass fibre produced, i.e. 0.80 for filament, 0.89 for wool and filament and 1.10 for mostly wool. These values were not significantly different from unity. The same pattern was observed using local rates (Marsh et al., 2001a). In the 1989 update of the

Table 66. Mortality data for non-malignant respiratory diseases (NMRD) in MMVF production workers and end-users

Reference	Study type	Country	No. of subjects	Industry	Relative risk (95% CI)	Observed	Comments
Glass (excluding continuous filament)							
Chiazze et al. (1992)	Nested case–control	USA	102	Production	Odds ratio, 1.5 (0.5–4.1)		Same odds ratio for fine fibres
Sali et al. (1999)	Cohort	Europe	5275	Production	SMR, 1.18 (0.98–1.40)	127	SMR (BEA), 1.12 (95% CI, 0.82–1.49)
Marsh et al. (2001a)	Cohort	USA	10 961	Production	SMR (all workers), 1.10 (0.91–1.32) SMR (long-term workers), 0.94 (0.72–1.20)	115 63	Production of mostly glass wool; SMR for NMRD excluding influenza and pneumonia
Rock (stone) and slag wool							
Sali et al. (1999)	Cohort	Europe	4616	Production	SMR, 0.97 (0.76–1.23)	71	SMR (BEA), 0.96 (95% CI, 0.66–1.35)
Marsh et al. (1996)	Cohort	USA	3035 N-cohort[a]	Production	SMR (LR), 1.07 NS	49	NMRD excluding influenza: SMR (LR), 1.27 (NS)
Marsh et al. (1996)	Cohort	USA	443 O-cohort[a]	Production	SMR (LR), 1.80 $p < 0.01$	29	NMRD excluding influenza: SMR (LR), 1.83 ($p < 0.05$)

Table 66 (contd)

Reference	Study type	Country	No. of subjects	Industry	Relative risk (95% CI)	Observed	Comments
Gustavsson et al. (1992)	Cohort	Sweden	2807	End-users (manufacture of prefabricated wooden houses)	SMR, 0.95 (0.65–1.35)	32 (33.6 expected)	Mixture of mineral wool insulation SMR (BEA), 0.98 (95% CI, 0.49–1.76)
Continuous glass filament							
Shannon et al. (1990)	Cohort	Canada	1465	Production	SMR (men), 0.43 [0.05–1.54][b] SMR (women), 1.62 [0.04–9.29][b]	2 men 1 woman	Based on a total of 96 deaths; SMR (LR) (BEA), 0.917 (95% CI, 0.479–2.842)
Chiazze et al. (1997)	Cohort	USA	2933 white men	Production	SMR (LR), 1.03 (0.68–1.48)	28 (27.28 expected)	Based on 437 deaths
Sali et al. (1999)	Cohort	Europe	1482	Production	1.05 (0.48–2.00)	9	SMR (BEA), 0.70 (95% CI, 0.14–2.03)
Marsh et al. (2001a)	Cohort	USA	5431	Production	SMR (all workers), 0.80 (0.61–1.03) SMR (long-term workers), 0.82 (0.59–1.12)	59 40	NMRD excluding influenza and pneumonia
Watkins et al. (1997)	Cohort	USA	1074 white women 494 black men	Production	SMR (LR) women, 0.82 (0.30–1.78) SMR (LR) men, 0.21 (0.005–1.17)	6 women 1 man	SMR (LR) (BEA) women, 1.18 (95% CI, 0.38–2.74) SMR (LR) (BEA) men, 0.38 (95% CI, 0.01–2.14)

BEA, bronchitis, emphysema, asthma; LR, local rate; NR, national rate; NS, not significant
[a] See section 2 for explanation of 'N-cohort' and 'O-cohort'
[b] Calculated by the Working Group

US cohort of rock (stone)/slag wool workers, the SMR (national rates) for NMRD (excluding influenza and pnemonia) in the five-plant group was 1.31 (95% CI, 0.93–1.79). Using local rates, the SMR was 1.27 (95% CI, 0.90–1.74). In another plant that had a documented history of asbestos use, the SMRs were statistically significant [2.07; 95% CI, 1.26–3.20; and 1.83; 95% CI, 1.12–2.83; using national and local rates, respectively] (Marsh et al., 1996). Although the US nested case–control study on cancer of the respiratory system suggested that all of the excesses in respiratory cancers may be explained by cigarette smoking, this has yet to be documented for NMRD excluding influenza and pneumonia. Moreover, smoking habits seem to differ slightly between the male workers in the glass fibre subgroup and in the rock (stone)/slag wool subgroup (Marsh et al., 2001b): this should be kept in mind when considering the mortality ratios for NMRD.

In a study of workers engaged in the manufacture of prefabricated wooden houses, no excess of NMRD was observed (SMR, 0.95; 95% CI, 0.65–1.35). However, this cohort had only a short follow-up period (Gustavsson et al., 1992).

(b) *Morbidity data*

(i) *Pneumoconiosis*

The data available on pneumoconiosis mainly come from radiological studies and are limited to standard thoracic X-rays that have low sensitivity and low specificity. No data are available from radiological examinations made by computerized tomography scanning. Recent studies are summarized in Table 67.

Most of the studies found no significant excess of small opacities on chest autoradiographs except the two studies of glass fibre production workers (Kilburn & Warshaw, 1991) and end-users (Kilburn et al., 1992). However, as stated by the authors, previous exposure to asbestos could have affected the observed results. Moreover, severe methodological limitations were reported (Rossiter, 1993; Weill & Hughes, 1996). [The Working Group acknowledged and agreed with the published comments concerning these limitations.]

Hughes et al. (1993) found no significant relation between small opacities observed in the chest autoradiographs of workers at seven MMVF production plants and exposure to MMVF as compared with controls. However, it should be noted that cumulative exposure to fibres was extremely low as stated in the initial publication by Weill et al. (1983) (less than 10 fibres–years/cm^3 in the group exposed to the finest fibres, < 1 μm in diameter; and less than 1 fibre–years/cm^3 in the group exposed to fibres > 3 μm in diameter). This study also demonstrated the notable variability of radiographic assessment. Table 68 summarizes the results of two consecutive readings from the two glass fibre plants where the highest level of exposure occurred (in the other three glass fibre plants, the cumulative level of exposure was estimated at less than 1 fibre–years/cm^3 (Weill et al., 1983)).

Table 67. Pneumoconiosis: radiographic data from MMVF production workers and end-users

Reference	Country	Industry	X-ray reading method	No. of exposed subjects	Percentage of abnormalities ≥ 1/0 among exposed subjects	No. of controls	Percentage of abnormalities ≥ 1/0 among controls	Comments
Glass (excluding continuous filament)								
Hughes et al. (1993)	USA	Production	5 ILO readers (median)	1252	1.8	272	0.7	NS
Kilburn & Warshaw (1991)	USA	Production	2 ILO readers (consensus)	175	12.6	–	–	Uninterpretable data (no control group; 137 subjects with previous exposure to asbestos)
Kilburn et al. (1992)	USA	End-users (insulation of refrigerators)	2 ILO readers (consensus)	284	8.1	–	–	Uninterpretable data (no control group; previous exposure to asbestos)
Rock (stone) and slag wool								
Hughes et al. (1993)	USA	Production	5 ILO readers (median)	183	0	33	0	
Yano & Karita (1998)	Japan	Production	1 ILO reader	440	0.5	544	0.2	No details for small opacities
Refractory ceramic fibre								
Lemasters et al. (1994)	USA	Production	3 ILO readers (median)	847	0	–	–	21 pleural plaques
Rossiter et al. (1994)	Europe	Production	3 ILO readers (median)	543	7	–	–	No relation with cumulative exposure to refractory ceramic fibres
Trethowan et al. (1995)	Europe	Production	[no details given]	592	13% ≥ 0/1 3.3% ≥ 1/1	– –	– –	No relation with cumulative exposure to refractory ceramic fibres

Table 67 (contd)

Reference	Country	Industry	X-ray reading method	No. of exposed subjects	Percentage of abnormalities ≥ 1/0 among exposed subjects	No. of controls	Percentage of abnormalities ≥ 1/0 among controls	Comments
Lockey et al. (1996); Lawson et al. (2001)	USA	Production	3 ILO readers (median)	652	0.5	–	–	20 pleural plaques
Lockey et al. (2002)	USA	Production	3 ILO readers (median)	794	1.3	–	–	Non-significant excess risk with exposure to refractory ceramic fibres; significant relation with age and tobacco smoking
Cowie et al. (2001)	Europe	Production	3 ILO readers (median)	774	8	–	–	Non-significant excess risk with exposure to refractory ceramic fibres; significant relation with age and tobacco smoking; increase of opacities ≥ 1/0 (2%) on new reading of 1987 films

ILO, International Labour Organization; NS, not significant

Table 68. Autoradiographic assessments of small opacities in the lungs of workers in two glass fibre plants

	Cumulative exposure	No. of subjects	Percentage ≥ 1/0 1st radiographic assessment	Percentage ≥ 1/0 2nd radiographic assessment
Plant 3 (glass fibre diameter > 1 μm)	4.3 fibres–years/cm^3	220	5.9	3.2
Plant 5 (glass fibre diameter < 1 μm)	7.4 fibres–years/cm^3	122	6.6	0

From Hughes *et al.* (1993); data on cumulative exposure from Weill *et al.* (1983)

An Australian study reported morbidity data from 687 production workers in seven plants manufacturing both glass and rock (stone) wool. Small opacities (classified as 1/0 and 1/1) were detected on the chest radiographs of seven workers (only one ILO reader). However, no data on exposure were presented in the paper, and there was no control group (Brown *et al.*, 1996). Similar findings were reported in a Japanese study of 493 male workers in nine rock (stone) wool production plants: the prevalence of radiographic abnormalities among 440 of these workers did not differ significantly from that in 544 control subjects or in a comparable population of 401 asbestos workers (Yano & Karita, 1998).

No data have been published concerning chest radiographs of production workers in the continuous glass filament industry.

Conflicting results were obtained on workers involved in the production of refractory ceramic fibres from the two most recent updates of the US and European cohorts. In the US study (Lockey *et al.*, 2002), the incidence of irregular opacities (classified ≥ 1/0) showed a non-significantly elevated odds ratio for exposure to refractory ceramic fibres (workers with > 135 fibres–months/cm^3 cumulative exposure) (odds ratio, 4.7; 95% CI, 0.95–23.7). The same was observed for workers with > 10 years in refractory ceramic fibre production (odds ratio, 4.3; 95% CI, 0.9–28.3). As expected, small irregular opacities were significantly related to exposure to tobacco smoke and age. Similar results were observed in the European study (Cowie *et al.*, 2001), which found an association between opacities (category ≥ 1/0) and exposure subdivided by calendar time period: a positive association was observed only with exposure that occurred before 1971, although this finding was based on few data (only 51 exposed workers).

(ii) *Pleural fibrosis and related findings*

Until now the non-malignant pleural effects of MMVFs in humans have been documented using only standardized thoracic radiography. As described for pneumoconiosis this technique has serious limitations: pleural plaques are correctly classified mainly when calcified; diffuse pleural thickening is well-recognized mainly when associated with parenchymal bands. All other aspects are non-specific and can be related to other anatomical features. Moreover, other etiological factors having a definite association with pleural fibrosis are frequently present: mainly co-exposure to asbestos, but also the sequelae of infectious pleural diseases. Asbestos exposure is the most important confounding factor, since there is a strong co-linearity with exposure to MMVFs. Recent studies are summarized in Table 69.

The studies by Kilburn and Warshaw (1991) and Kilburn et al. (1992) reported the occurrence of pleural abnormalities in 7.4% and 7.0% of radiographs, respectively. [The Working Group noted that these studies are difficult to interpret for two main reasons: i.e. the absence of a control group and considerable exposure to asbestos which was not taken into account.]

In a study of rock (stone) wool production, Järvholm et al. (1995) reported a small, statistically nonsignificant, excess of circumscribed pleural thickening (pleural plaques). However, a nested case–control study failed to demonstrate a relationship with exposure to fibres (odds ratio for workers with ≥ 15 years of exposure, 1.2 [95% CI, 0.3–4.2]). [The Working Group noted that the members of this cohort were young; 57% of men and 86% of women were aged less than 40 years.] The data from the Australian (Brown et al., 1996) and Japanese (Yano & Karita, 1998) cohorts did not show any excess of pleural disease. [The Working Group noted that no excess of pleural abnormalities was observed in a comparison group of Japanese asbestos workers.]

A significant excess of pleural plaques was identified radiographically among a European cohort of refractory ceramic fibre production workers. The results presented in an initial report were inconclusive (Trethowan et al., 1995), since pleural anomalies did not show a dose–response. In contrast, the two publications of results from the US cohort (Lemasters et al., 1994; Lockey et al., 1996) demonstrated a significant relation with time since beginning of exposure and with cumulated duration of exposure. Moreover, the difference was still significant after adjustment for exposure to asbestos. The most recent updates of the cohorts in Europe and the USA contributed further information. The longitudinal data from the US cohort reinforced the initial results, showing a significant association (after adjustment for exposure to asbestos and age) with duration, latency and cumulative exposure (Lockey et al., 2002). For the highest classes of exposure duration (> 20 years), latency (> 20 years) and cumulative exposure (> 135 fibres/cm^3–months), the odds ratios were: 3.7 (95% CI, 1.1–11.8), 6.1 (95% CI, 1.9–27.1) and 6.0 (95% CI, 1.4–33.1), respectively. In the European cohort, the findings on pleural abnormalities were more difficult to interpret. Of a total of 774 subjects included in the update, 158 had been exposed to asbestos while employed in the refractory ceramic fibre industry and 351 (including 98 who were also exposed within the refractory ceramic fibre industry) were potentially exposed to asbestos outside the industry. The prevalences of pleural changes were lower among the 355 subjects not exposed to asbestos (Cowie et al., 2001) than the cohort as a whole (9% and 11% for pleural changes and 3% and 5% for pleural plaques, respectively). However, the 355 subjects not exposed to asbestos showed a significant association of pleural changes

Table 69. Pleural fibrosis: radiographic data from MMVF production workers and end-users

Reference	Country	Industry	X-ray reading method	No. of exposed subjects	Percentage of pleural abnormalities among exposed subjects	No. of controls	Percentage of pleural abnormalities among controls	Comments
Glass fibre (excluding continuous filament)								
Kilburn & Warshaw (1991)	USA	Production	2 ILO readers (consensus)	175	7.4	–	–	Uninterpretable data (no control group; 137 subjects with previous exposure to asbestos)
Kilburn et al. (1992)	USA	End-users	2 ILO readers (consensus)	284	7.0	–	–	Uninterpretable data (no control group; previous exposure to asbestos)
Rock (stone) and slag wool								
Hughes et al. (1993)	USA	Production	5 ILO readers (median)	183	0	33	0	
Järvholm et al. (1995)	Sweden	Production	1 reader (radiophotography)	933	1.3% PP	865	0.3% PP	No relation with exposure to rock (stone) wool
Yano & Karita (1998)	Japan	Production	1 ILO reader	440	2.7	544	2.9	No details on pleural opacities
Refractory ceramic fibres								
Lemasters et al. (1994)	USA	Production	3 ILO readers (median)	686	3.1% PP 0.3 % DPT	–	–	Significant relation with exposure to refractory ceramic fibres (latency and duration) even after adjustment for exposure to asbestos
Rossiter et al. (1994)	Europe	Production	3 ILO readers (median)	543	2.7% PP or DPT; 0.9% bilateral	–	–	No relation with exposure to refractory ceramic fibres

Table 69 (contd)

Reference	Country	Industry	X-ray reading method	No. of exposed subjects	Percentage of pleural abnormalities among exposed subjects	No. of controls	Percentage of pleural abnormalities among controls	Comments
Refractory ceramic fibres (contd)								
Lockey et al. (1996)	USA	Production	3 ILO readers (median)	438	4.1% PP or DPT	214	0.9% PP or DPT	Significant relationship with exposure to refractory ceramic fibres (latency and duration) even after adjustment for exposure to asbestos
Lockey et al. (2002)	USA	Production	3 ILO readers (median)	794	3.4% any pleural abnormalities and 81% plaques	–	–	Significant relationship with duration, latency and cumulative exposure adjusted for exposure to asbestos and age
Cowie et al. (2001)	Europe	Production	3 ILO readers (median)	774	11% any pleural abnormalities and 5% plaques	–	–	Increased risk with time since first exposure to refractory ceramic fibres among 355 workers not exposed to asbestos

DPT, diffuse pleural thickening; ILO, International Labour Organization; PP, pleural plaque with time since first exposure to refractory ceramic fibres, after adjustment for age, smoking habits and body mass index. When no adjustment was made for age, a highly significant association existed between time since first exposure to refractory ceramic fibres and both pleural changes and pleural plaques. [The Working Group considered it desirable to extend the survey of both cohorts with a tomodensitometric examination of the thorax for at least those subjects who showed pleural changes, since it is considered as a sensitive method to detect such abnormalities.]

(iii) *Symptoms and changes in lung function*

Data from morbidity studies have been collected using questionnaires and lung function tests. The results on symptoms are summarized in Table 70 and for lung function tests in Table 71.

Four studies have been published that concern mainly glass wool: two in the production industry and two among end-users. In the French study of 524 production workers, a significant increase in symptoms in the upper respiratory tract and of dyspnoea was noted. However, a nested case–control study did not confirm the occurrence of any impairment of lung function (Moulin *et al.*, 1987). The results of the USA production study were more complicated to interpret: asbestos was a major confounding factor since 137 out of 175 workers had been previously exposed to asbestos (Kilburn & Warshaw, 1991). Two studies of end-users, one of workers installing insulation in refrigerators (Kilburn *et al.*, 1992) and the other of sheet-metal workers (Hunting & Welch, 1993) reported high rates of chronic bronchitis (10.9% and 15%, respectively). The interpretation of the results was hampered by the absence of a control group and the potential for co-exposure to air contaminants including asbestos. Moreover, the results from lung function tests did not suggest any dose–response relationship with fibre exposure.

Two studies of end-users of rock (stone) and slag wool or a mixture of mineral wools have been reported. Although an excess of symptoms in the upper respiratory tract and cough were reported from the first study in Denmark (Petersen & Sabroe, 1991) and a significant decrease of lung function (measured as 'forced expiratory volume') in the second Danish study (Clausen *et al.*, 1993), other potential sources of occupational exposure were not taken into acount. Four studies of production workers have been conducted. In the study from France (Moulin *et al.*, 1988), a high prevalence of symptoms (cough, sputum, symptoms in the upper respiratory tract) was reported to be associated with exposure to fibres: however, this result was observed in only one of the five plants investigated (a glass-wool plant where 51% of the study population worked), and where workers were also exposed to resin. A Danish study comparing 235 rock (stone) wool production workers with a random sample of 243 subjects from the general population (Hansen *et al.*, 1999) demonstrated a significant incidence of airflow obstruction related to fibre exposure (14.5% in MMVF workers, 5.3% in controls), but this effect was limited to heavy smokers: the self-administered questionnaire on disease confirmed an excess of emphysema (9 cases, as against 2 in the control group) with a

Table 70. Chronic obstructive pulmonary diseases and other related findings in MMVF workers: results from questionnaires

Reference	Country	No. of subjects	Industry	Symptoms							Relation with exposure	Comment
				Upper airways	Cough	Sputum	Chronic bronchitis	Dyspnoea	Wheezing	Asthma		
Glass fibre (excluding continuous filament)												
Moulin et al. (1987)	France	524	Production	$p < 0.05$				$p < 0.05$			?	Inconsistency between plants; co-exposure to phenolic resins
Kilburn & Warshaw (1991)	USA	175	Production	1.1%[a], 19.4%[b]							?	Data uninterpretable (no control group; 137 subjects with previous exposure to asbestos)
Kilburn et al. (1992)	USA	284	End-users (insulation for refrigerators)	5.6%[a], 19.4%[b]			10.9%			6%	?	Data uninterpretable (no control group; previous exposure to asbestos)
Hunting & Welch (1993)	USA	333	End-users (sheet-metal workers)				15%				+	Other sources of exposure?
Glass and/or rock (stone) or slag wool												
Petersen & Sabroe (1991)	Denmark	2654	End-users (construction workers): glass wools	$p < 0.05$	$p < 0.05$						+	Other sources of exposure?
Moulin et al. (1988)	France	1839[c]	Production: glass wools	1.9–9.6%	6.9%–21.9%	4.3–9.2%		4.1–12%		0.9–1.4%	+	Co-exposure with phenolic resins
Hughes et al. (1993)	USA	1030	Production: fibrous glass and rock (stone) wools		13% (ns)		6% (ns)	5% (ns)		5% (ns)	–	

Table 70 (contd)

Reference	Country	No. of subjects	Industry	Symptoms						Relation with exposure	Comment
				Upper airways	Cough	Sputum	Chronic bronchitis	Dyspnoea	Wheezing	Asthma	

Glass and/or rock (stone) or slag wool (contd)

Brown et al. (1996)	Australia	533	Production		10.7%			12%	20.4%	7.3%	?	No data on exposure
Yano & Karita (1998)	Japan	493	Production: rock (stone) wool		1.4%	4.1%		0			–	No difference compared with a control group of urban residents
Hansen et al. (1999)	Denmark	377	Production: rock (stone) wool		26% (ns)	24% (ns)	3.8% (emphysema)	20% (ns)			+	Emphysema: 0.9% in a comparable control group

Refractory ceramic fibres

| Trethowan et al. (1995) | Europe | 628 | Production | 55% | 13% | | 12% | | 18% | | + | Relation with dyspnoea significant |
| Cowie et al. (2001) | Europe | 774 | Production | | 24% (men) 20% (women) | | 3.3% (men) 2.4% (women) | 3.2% (men) 8.5% (women) | | | ? | Limited evidence of association with recent exposure to respirable fibres |

ns, not significantly different from comparable control group
[a] Nose bleed and throat irritation
[b] Throat irritation
[c] Population of three plants with 1041, 535 and 263 participating workers; the percentages give the range of values.

Table 71. Chronic obstructive pulmonary diseases and other related findings from tests of lung function in MMVF workers

Reference	Type of study	Country	No. of subjects	Industry	Lung function (percentage of expected values)				Relationship with exposure	Comments
					TPC	FVC	FEV$_1$	FEF$_{25,75}$		
Glass fibre (excluding continuous filament)										
Kilburn & Warshaw (1991)	T	USA	175	Production	114.2	94.8	91.3	80.7	?	Uninterpretable data (no control group; 137 subjects with previous exposure to asbestos)
Kilburn et al. (1992)	T	USA	214[a]	End-users (insulation for refrigerators)	107.7	92.9	90.8	85.8	?	Uninterpretable data (no control group; previous exposure to asbestos)
Glass and rock (stone) and slag wool										
Hughes et al. (1993)	T	USA	1030	Production		103 (ns)	101.1 (ns)	78 (ns)	−	
Clausen et al. (1993)	L	Denmark	340	End-users (insulators)		ns	decreased ($p < 0.001$)		+	Other co-occurring exposure
Brown et al. (1996)	T	Australia	687	Production	ns	93 (ns)	95 (ns)		−	No data on exposure
Rock (stone) wool										
Hansen et al. (1999)	T	Denmark	235			96 (ns)	92 (ns)	FEF$_{25,75}$	+	Elevated risk only in heavy smokers
					TPC	FVC	FEV$_1$	FEF$_{25,75}$		

Table 71 (contd)

Reference	Type of study	Country	No. of subjects	Industry	Lung function (percentage of expected values)				Relationship with exposure	Comments
					TPC	FVC	FEV$_1$	FEF$_{25-75}$		
Refractory ceramic fibres										
Burge et al. (1995); Trethowan et al. (1995)	T	Europe	628	Production		110.6	104.6	84.1	+	Significant relation (FEV$_1$) with exposure seen only among smokers
Lemasters et al. (1998)	L	USA	736	Production		$p < 0.01$ for male smokers; $p < 0.05$ for female non-smokers	$p < 0.01$ (male smokers)		+	Sex-specific respiratory effect
Lockey et al. (1998)	L	USA	361	Production		reduced ($p < 0.05$)	reduced ($p < 0.05$)		–	Longitudinal follow-up, with selection bias due to lower level of recent exposure
Cowie et al. (2001)	L	Europe	774	Production		reduced ($p < 0.05$)	reduced ($p < 0.05$)		+	Significant effects limited to male current smokers

T, transversal; L, longitudinal; ns, not significant; TPC, total pulmonary capacity; FVC, forced vital capacity; FEV$_1$, forced expiratory volume in 1 s; FEF$_{25-75}$, forced expiratory flow (mid-flow)
[a] Workers without pulmonary opacities

relative risk of 4.5 (95% CI, 1.0– 20.6). The absence of any alteration in single-breath carbon monoxide-diffusion capacity in this population was interpreted by the authors as absence of lung fibrosis. [The Working Group noted that chest radiographs were not available in this study. In the studies from America (Hughes *et al*., 1993), Japan (Yano & Karita, 1998) and Australia (Brown *et al*., 1996), the results (assessed by questionnaire and/or tests of lung function) showed no association with exposure.]

In contrast, the studies of the American and European cohorts of refractory ceramic fibre workers did report statistically significant numbers of abnormalities related to exposure to fibres. In the American cohort (Lemasters *et al*., 1998), the relative risk of having at least one respiratory symptom was 2.9 (95% CI, 1.4–6.2) for men and 2.4 (95% CI, 1.1–5.3) for women. In male workers, a significant loss of forced vital capacity (FVC) of the lung noted in workers exposed to refractory ceramic fibres was limited to current and ex-smokers. The decline in FEV_1 (forced expiratory volume in 1 s) was significant only in current smokers. A significant loss of FVC was observed among female non-smokers. The initial findings of the European study (Burge *et al*., 1995; Trethowan *et al*., 1995) also demonstrated obstruction of the airways related to cumulative exposure to refractory ceramic fibres that was limited to current and ex-smokers. In the most recent update of this cohort (Cowie *et al*., 2001), a significant negative association between estimated cumulative exposure to refractory ceramic fibres and lung function (FVC and FEV_1) was found. [The Working Group noted that the results of this study should be reassessed since there was no indication of pulmonary interstitial disease (no alteration in single-breath carbon monoxide-diffusion capacity and an absence of small parenchymal opacities on chest radiographs), and total lung capacity was not monitored in this study.]

(*c*) *Other respiratory findings*

Isolated case reports on lung proteinosis (1 case) and lung fibrosis (2 cases) have been published (Takahashi *et al*., 1996; Riboldi *et al*., 1999; Yamaya *et al*., 2000). [The Working Group noted the lack of an association with occupational exposure to fibres in these studies.]

In a series of 50 consecutive cases of sarcoidosis, twelve patients were described whose exposure to MMVFs had been assessed either by job history or by mineralogical lung analysis (Drent *et al*., 2000a,b). [The Working group noted that the information on exposure history and the mineralogical data from analysis of lung samples were incomplete.]

Only one study was intended to test an initial event of pulmonary fibrosis through determination of serum type III procollagen aminoterminal propeptide. In this series the concentration of this propeptide in 56 male workers exposed to rock (stone) wool during production (0.05–0.75 fibres/cm^3 for up to 20 years) did not differ from that of a group of 20 controls (Cavalleri *et al*., 1992).

4.2.2 *Other toxic effects*

(a) *Mortality data*

Data on other non-malignant, non-respiratory causes of death after exposure to MMVFs are available from the US and European cohorts. In the 1985 follow-up of the US cohort, Marsh *et al.* (1990) noted a slight increase in the incidences of nephritis and nephrosis for the whole cohort (SMR (national rate), 1.46 ($p < 0.01$)). The US rock (stone) and slag wool cohort had an elevated risk for nephritis and nephrosis (N-cohort SMR (local rate), 2.02; $p < 0.05$; O-cohort SMR (local rate), 2.58; not significant; see section 2) and no risk for cirrhosis of the liver (N-cohort SMR (local rate), 0.90; O-cohort SMR (local rate), 0.91) (Marsh *et al.*, 1996). A case–control study of workers at one plant from the US cohort reported no association between exposure to glass fibre and nephritis, nephrosis or cirrhosis of the liver (Chiazze *et al.*, 1999). The corresponding SMRs for the total US cohort of glass fibre workers (Marsh *et al.*, 2001a) were 1.04 (local rate) [95% CI, 0.81–1.32] and 0.88 (local rate) [95% CI, 0.75–1.04], respectively. In the European cohort, an excess of cirrhosis was observed only among the continuous filament production workers (SMR (national rate), 2.12 [95% CI, 1.10–3.71] for 12 observed cases) (Sali *et al.*, 1999).

(b) *Morbidity data*

Most of the literature concerning the irritating effects of MMVFs on the skin and mucosa was published before the last IARC evaluation (IARC, 1988). More recent publications have reported on observations in contaminated buildings (Lockey & Ross, 1994; Thriene *et al.*, 1996; Bergamaschi *et al.*, 1997): the role of fibres with diameter > 4 μm and direct contact with deposited fibres was emphasized.

Recent epidemiological studies are scarce. In a plant manufacturing and processing mineral wool for insulation, 25% of 259 workers presented with a skin disease, resulting from an allergy related to MMVF additives (Kiec-Swierczynska & Szymczk, 1995). Dermal irritation is a common symptom of the sick-building syndrome. Forty-six of 103 white collar workers in a building in which the atmosphere was contaminated by mineral fibres from ceiling panels had symptoms of dermal irritation (Thriene *et al.*, 1996). Similarly, it was reported that 32% of 66 subjects investigated for sick-building syndrome demonstrated positive patch tests with mineral fibres (Thestrup-Pedersen *et al.*, 1990). Among 2654 Danish construction workers, a statistically significant increase in the frequency of symptoms of irritation of the skin, eyes and respiratory tract were reported (Petersen & Sabroe, 1991).

4.3 Toxic effects in experimental systems

This section covers selected toxic effects of fibres in experimental systems that are believed to be potentially important in relation to the carcinogenic process. These endpoints include in-vivo effects such as inflammation and fibrosis, as well as selected in-vitro assessments including cytotoxicity, oxidant production and alterations to the cell cycle including proliferation and apoptosis. Genetic toxicology end-points are reviewed in section 4.5.

It is important to appreciate the degree to which biopersistence plays a role in the different studies and end-points under review, as this property of fibres is thought to be critical in determining chronic toxicity and carcinogenic outcome in humans and in experimental animal systems. In-vitro assays are invariably short-term (i.e. from hours to days), and the effect of fibre durability is unlikely to be detected in such assays. [The Working Group noted that endotoxin is a potent environmental contaminant and its presence in fibre samples could enhance their ability to cause acute inflammation. The presence of endotoxin or the steps taken to inactivate it, were not always reported.] Therefore, short-term tests could give a misleading impression of possible long-term biological effects. This will most likely become manifest as a false-positive result in an in-vitro assay for long, non-biopersistent fibres. For a non-biopersistent fibre, the effects seen *in vitro* may apply only to the time interval *in vivo* before the fibre begins to undergo dissolution or breakage. In contrast, a durable fibre may show the effects much more slowly and is more likely to give rise to pathological change.

4.3.1 *Continuous glass filament and glass wool*

Pathological change related to inflammation and fibrosis in rat lungs exposed to fibres is commonly quantified by the Wagner scale (see Table 72). The original Wagner scale was modified by McConnell *et al.* (1984). Wagner scores have been given here as the score preceded by the abbreviation Wag, e.g. Wag 4. Pathological effects might be expected to occur in rats with scores of Wag 3 and higher, while the presence of fibrotic lesions is identified by scores of Wag 4 and higher.

Table 72. Wagner pathology grading scale[a]

		Cellular change
Normal	1	No lesion
Minimal	2	Macrophage response
Mild	3	Bronchiolization, inflammation
		Fibrosis
Minimal	4	Minimal
Mild	5	Linking of fibrosis
Moderate	6	Consolidation
Severe	7	Marked fibrosis and consolidation
	8	Complete obstruction of most airways

From Hesterberg *et al.* (1993)
[a] According to guidelines given by McConnell *et al.* (1984)

(a) *Inflammation and fibrosis*

Hesterberg *et al.* (1996b) exposed Fischer 344 rats for 6 h per day, on five days per week, for 13 weeks to MMVF10 at the following airborne mass concentrations (exposure by inhalation is generally expressed as the airborne mass concentration followed by the fibre concentrations in parentheses): 3.2 mg/m^3 (36 WHO fibres/cm^3, 14 fibres > 20 μm long/cm^3), 16.5 mg/m^3 (206 WHO fibres/cm^3, 81 fibres > 20 μm long/cm^3), 30.5 mg/m^3 (316 WHO fibres/cm^3, 135 fibres > 20 μm long/cm^3), 44.5 mg/m^3 (552 WHO fibres/cm^3, 223 fibres > 20 μm long/cm^3) and 62.2 mg/m^3 (714 WHO fibres/cm^3, 343 fibres > 20 μm long/cm^3). Inflammation, measured as increased amounts of polymorphonuclear neutrophils and protein in bronchoalveolar lavage, was produced by exposure to MMVF10 at concentrations of 16 mg/m^3 and above. An influx of alveolar macrophages in the bronchioles and alveoli was seen at the end of the study in all dose groups.

Fischer 344 rats treated with RCF1 and positive controls treated with chrysotile asbestos were compared with rats treated for 6 h per day, on five days per week for 24 months, with glass fibre types MMVF10 and MMVF11 at the following airborne mass concentrations: MMVF10, 3.1 mg/m^3 (29 WHO fibres/cm^3), 17.1 mg/m^3 (145 WHO fibres/cm^3) and 27.8 mg/m^3 (232 WHO fibres/cm^3); MMVF11, 4.8 mg/m^3 (41 WHO fibres/cm^3), 15.8 mg/m^3 (153 WHO fibres/cm^3) and 28.3 mg/m^3 (246 WHO fibres/cm^3). Positive control rats were exposed to chrysotile at 10.1 mg/m^3 (10 600 WHO fibres/cm^3) and rats treated with RCF1 received 29.1 mg/m^3 (187 WHO fibres/cm^3). Animals exposed to chrysotile had scores of Wag 4 after three months of exposure; the scores remained at the same level for up to 24 months of continued exposure; there was no decrease over up to 18 months of post-treatment recovery. Rats exposed to RCF1 (30 mg/m^3) did not reach a score of Wag 4 until six months after the start of treatment; the scores remained at Wag 4 for the following 18 months of exposure and during the period of post-exposure recovery. There were clear time- and exposure-dependent responses in the pathological scores for MMVF10 and MMVF11, which peaked at Wag 3 later in rats exposed to 16 mg/m^3 and earlier after exposure to 30 mg/m^3. There was clear evidence of a decrease in the Wag score during recovery after short-term exposure, but not when the duration of exposure (to MMVF10) had been ≥ 18 months (Hesterberg *et al.*, 1993).

Hesterberg *et al.* (1998a) exposed Fischer 344 rats by inhalation for 6 h per day on five days per week for up to two years to one of the following: the relatively-soluble glass fibre X-607 at 30 mg/m^3 (174 WHO fibres/cm^3, 47 fibres > 20 μm long/cm^3); RCF1 at 29 mg/m^3 (187 WHO fibres/cm^3, 101 fibres > 20 μm long/cm^3); and chrysotile at 10 mg/m^3 (10 600 WHO fibres/cm^3; 0 fibre > 20 μm long/cm^3). The Wagner scores were as follows: controls exposed to clean air, Wag 1; treatment with X-607, average Wag 2.5; treatment with RCF1 and chrysotile, average Wag 4.0. In animals allowed to recover after cessation of exposure, the Wag score decreased in the rats exposed to X-607, but not in the animals exposed to RCF1. Pathological indices demonstrated the absence of bronchoalveolar or pleural collagen after treatment with X-607, but varying scores of up to Wag 2.7 after exposure to RCF1 and chrysotile. These scores generally decreased during periods of post-exposure recovery. The data suggest that soluble fibres are less pathogenic than more persistent fibres and that their pathogenic effects are reversible to some extent.

In the study by Everitt et al. (1994), Fischer 344 rats were exposed by intrapleural injection to RCF1 at 0.56×10^6 WHO fibres/cm^3 or to MMVF10 at 10.5×10^6 WHO fibres/cm^3. The preparation of RCF1 contained 1.75×10^6 particles (fibres with length/diameter < 3)/cm^3, the MMVF10 sample contained 1.17×10^6 particles/cm^3. At 28 days post-exposure there was little difference between the two groups in mediastinal weight or pleural thickness. [This should be viewed against the much greater exposure to MMVF10 on a per fibre basis.]

Drew et al. (1987) examined the effects in Fischer 344 rats of 10 repeated intratracheal instillations (1 h/week for 10 weeks; 0.5 mg per treatment) of specially made and sized glass fibres obtained from the Thermal Insulation Manufacturers' Association (TIMA). The average fibre dimensions were 1.5×5 μm or 1.5×60 μm. Another group of rats was treated with unsized UICC (Union internationale contre le Cancer [International Union Against Cancer]) crocidolite. The lesions seen after the 10 weekly instillations of glass fibres were described as aggregations of macrophages and foreign-body granulomas; no striking differences were observed between the results of the treatment with long and short glass fibres. Rats given five weekly instillations of crocidolite asbestos showed not only macrophage responses, but also accumulation of haemosiderin within the macrophages and thickening and fibrosis of the alveolar wall. [The Working Group noted that these pathological changes are consistent with a bolus effect due to instillation of a large quantity of foreign material.]

Cullen et al. (2000) studied the pathogenicity of E-glass fibres (code 104/E) and microfibres (code 100/475), with amosite asbestos as a control. Wistar rats were exposed by inhalation for 7 h per day on five days per week for one year to E-glass (1022 WHO fibres/cm^3, 72 fibres > 20 μm long/cm^3), microfibres (code 100/475) (1119 WHO fibres/cm^3, 38 fibres > 20 μm long/cm^3) and amosite (981 WHO fibres/cm^3, 89 fibres > 20 μm long/cm^3). One year after the end of exposure, the mean level of advanced fibrosis (percentage of lung area affected, rather than Wagner score) was determined. The results were as follows: controls, 0.08; treatment with E-glass, 8.0; treatment with microfibres, 0.2; and exposure to amosite, 7.6. The authors attributed their findings to the higher number of long fibres of E-glass and their greater biopersistence compared with that of other fibre types.

Syrian golden hamsters received MMVF10a by nose-only inhalation for 6 h per day, on five days per week for 13, 26 and 52 weeks at an airborne mass concentration of 30 mg/m^3 (323 WHO fibres/cm^3, 151 fibres > 20 μm long/cm^3), while MMVF33 (475 glass wool), a special-purpose fibre that is more durable than MMVF10a, was delivered at a concentration of 37 mg/m^3 (283 WHO fibres/cm^3, 106 fibres > 20 μm long/cm^3). Other groups of animals were exposed to amosite at three concentrations:

0.8 mg/m^3 (33 WHO fibres/cm^3, 9 fibres > 20 µm long/cm^3), 3.7 mg/m^3 (157 WHO fibres/cm^3, 37 fibres > 20 µm long/cm^3) and 7.3 mg/m^3 (255 WHO fibres/cm^3, 67 fibres > 20 µm long/cm^3). The Wagner pathology scores showed time-dependent increases for all types of exposure tested. At 52 weeks of exposure, the scores were as follows: controls, Wag 0; treatment with MMVF10a, Wag 2.3; treatment with MMVF33, Wag, 4.0; and treatment with amosite (at 0.8 mg/m^3), Wag 4.0. Exposure to the two higher concentrations of amosite produced the following Wag scores: amosite at 3.7 mg/m^3, Wag 5.3; amosite at 7.3 mg/m^3, Wag 6.0. The authors concluded that there was a clear effect of increased durability causing a greater pathological response, and that the pathological effects could be related to the numbers of long fibres (> 20 µm) persisting in the lungs (Hesterberg et al., 1997).

Hesterberg et al. (1999) exposed Syrian golden hamsters by inhalation to MMVF10.1 (a thinner fibre than MMVF10) and MMVF10a (a batch of MMVF10 that had a slightly lower fluorine content and was slightly thinner than MMVF10) for 6 h per day, on five days per week for 13 weeks with or without a period of 10 weeks of post-exposure recovery. The airborne mass concentrations of MMVF10.1 used ranged from 3.2–62.2 mg/m^3 (62.2 mg/m^3 contained 714 WHO fibres/cm^3; 343 fibres > 20 µm long/cm^3). The pathology score never exceeded Wag 2 for any exposure group and decreased to Wag 1 after recovery following the lower levels of exposure. Bronchoalveolar lavage analysis revealed that only exposure to airborne concentrations ≥ 16.5 mg/m^3 (206 WHO fibres/cm^3, 81 fibres > 20 µm long/cm^3) caused inflammation, as judged by a significant increase in the number of neutrophils.

In a chronic study by McConnell et al. (1999), Syrian golden hamsters were exposed for 6 h per day, on five days per week for 78 weeks to MMVF10a at 29.6 mg/m^3 (339 WHO fibres/cm^3, 134 fibres > 20 µm long/cm^3), MMVF33, a more durable special-purpose glass, at 37.0 mg/m^3 (310 WHO fibres/cm^3, 109 fibres > 20 µm long/cm^3) or amosite at three concentrations (highest dose, 7.1 mg/m^3, 263 WHO fibres/cm^3, 69 fibres > 20 µm/cm^3). Both visceral and parietal fibrosis were described. Visceral pleural fibrosis (expressed as the fibrosis index, i.e. the area of polarizable pleural collagen per length of pleural basement membrane, mm^2/mm) was estimated at 78 weeks in the periphery of the entire lobes and averaged 3.1 in the controls, 3.83 in hamsters exposed to MMVF10a, 3.97 in hamsters exposed to MMVF33 (these values are not significantly different from those of the controls) and 8.86 ($p < 0.05$) in hamsters exposed to the high dose of amosite. After six weeks of recovery, these figures showed a decrease in animals exposed to MMVF10a, but a further increase in the animals exposed to MMVF33 and to the high dose of amosite.

Rutten et al. (1994) injected Fischer 344 rats and Syrian golden hamsters intra-pleurally (injection volumes, 1.4 mL and 1.0 mL, respectively) with RCF1 (6–312 µg) or MMVF10 (43–2000 µg); MMVF10 contained about three times more WHO fibres than RCF1. The objective was to have the same number of long fibres in both groups. There was little difference between the effects of these two treatments in rats, but hamsters had more thickening of the visceral pleura (fibrosis) after treatment with

MMVF10; in contrast, more proliferation of mesothelial cells was observed in both the visceral and parietal pleura in hamsters treated with RCF1.

Donaldson *et al.* (1995a) exposed rats by inhalation for 7 h to 1000 WHO fibres/cm^3 long-fibre amosite asbestos or to the more soluble code 100/475 glass fibres which had comparable length distributions, and assessed cell proliferation by means of incorporation of bromodeoxyuridine 24 h later. Increased cell proliferation was seen throughout the lung after exposure to amosite, but not after exposure to code 100/475 glass fibres. This result after short-term exposure suggests that effects other than solubility, which was unlikely to be a factor, can determine the short-term proliferative response to inhaled fibres; a difference in surface reactivity between the fibres used is a possible explanation.

Donaldson *et al.* (1993) used a number of fibre types from the repository of the TIMA to assess the ability of the fibres to cause short-term inflammation in the peritoneal cavity of the C57BL6 mouse, when equal numbers of each type of fibre were used. Inflammation was not related to the mass instilled, which varied greatly depending on the type of fibre tested. A similar degree of inflammation was caused by MMVF10, MMVF11, MMVF21 and MMVF22; UICC crocidolite was most inflammogenic, probably reflecting its greater content of long fibres. These data highlight the limited usefulness of short-term assays in discriminating between fibres of variable biopersistence.

(*b*) *Cell toxicity*

The results of many studies on short-term cytotoxic effects have been published. The reports evaluated in this section have been chosen according to their ability to meet the following criteria:
- The dose was expressed as number of fibres given, in the way that exposure to fibres is regulated, and allowing potency to be compared.
- Fibre length was specified so that false-negative results from preparations of short fibres could be excluded.
- Adequate documentation of the fibre source was supplied to allow cross-reference to longer-term studies and to define composition.
- Studies involving instillation of fibres directly into the lungs were screened to exclude those in which an excessive dose might confound the results (overload effects).
- Where possible, control fibres were included or different categories of fibre length were used.

Hart *et al.* (1994) used Chinese hamster ovary cells, which are not direct target cells, to assess the cytostatic effects of glass fibre MvL 901 (Schuller International, Denver, CO) of various lengths and thicknesses at equal fibre number. Cell proliferation was inhibited to approximately 25% of control values by longer fibres (average length, 25 μm), whereas short fibres (average length, 3.5 μm) did not inhibit cell proliferation. Fibre thickness (tested with MvL 475) had a modest effect, with thinner fibres (0.3 μm) being more effective inhibitors of cell proliferation than thicker fibres (7 μm) at equal fibre number. This study showed that long fibres are toxic *per se*, in addition to their ability to accumulate because of their slower clearance from the lung. [The Working Group noted that, on a number basis, longer fibres also have a larger surface area than shorter fibres.]

Code 100 glass fibres were separated by dielectrophoresis into five mean length categories: 33, 17, 7, 4 and 3 μm. In general, the longer fibres were thicker; the average diameter was 0.75 μm for fibres in the longest category versus 0.35 μm in the shortest category. Four doses of fibres with equal fibre numbers for each length category were tested for their ability to inhibit lucigenin chemiluminescence and to cause release of lactate dehydrogenase from rat alveolar macrophages. The fibres showed length-related toxicity; the 17-μm- and 33-μm-long fibres had a similar high potency, and fibres ≤ 7 μm long had a significantly lower potency (Blake *et al.*, 1998). The authors noted that their study pointed to a direct cellular basis for the increased toxicity of long fibres, namely frustrated phagocytosis, leading to leakage of oxidants and enzymes from a macrophage that is trying to engulf a fibre.

(*c*)   *Cell activation*

Brown *et al.* (1998a) tested a series of fibres that included MMVF10 and code 100/475 for their ability to activate release of superoxide anions by rat alveolar macrophages at equal WHO fibre number. Uncoated (naked) fibres of both types inhibited superoxide release, but after coating with IgG the MMVF10 fibres, but not the code 100/475 fibres, showed stimulation. Amosite and RCF1 showed a similar pattern of response. Thus, there was no discrimination in this assay between fibres defined as either pathogenic or non-pathogenic by some studies of chronic toxicity and oncogenicity in animals (see secton 3).

Two subsets of the size-separated code 100 glass fibres used by Blake *et al.* (1998) with average lengths of 17 and 7 μm, respectively, were tested in cultured RAW 264.7 mouse macrophage cells. The production of tumour necrosis factor α and the activation of the transcription factor were more strongly induced by 17-μm-long fibres than by 7-μm-long fibres on a per fibre basis (Ye *et al.*, 1999).

(*d*)   *Other effects*

Brown *et al.* (2000b) examined the ability of equal numbers of WHO fibres (MMVF10, MMVF11 and JM code 100/475) to deplete antioxidants from solution in comparison with refractory ceramic fibres and amosite. The ability to deplete antioxidants did not correlate with the pathogenic or non-pathogenic properties of fibres (as defined by some studies of their chronic toxicity and oncogenicity in animals). The MMVF10 and code 100/475 fibres were more active in depleting glutathione and vitamin C than amosite asbestos.

Jensen and Watson (1999) used the SV-40-immortalized human mesothelial cell line MeT-5A and the rhesus monkey kidney epithelial cell line LLC-MK$_2$ to test the ability of MMVF10a, crocidolite, chrysotile and RCF1 (0.3 or 1.0 mg/mL) to impair separation of cells at the end of mitosis (cytokinesis). By use of high-resolution time-lapse microscopy, all fibres longer than about 15 µm, regardless of composition, were seen to be trapped in the cleavage furrow of the dividing cells. This caused failure of cytokinesis resulting in the formation of multinucleate cells.

Gilmour et al. (1995) tested the ability of MMVF10 and MMVF11 at equal numbers of WHO fibres to cause oxidative cleavage of bacterial plasmid supercoiled φX174RF$_1$ DNA. The MMVF10 and MMVF11 fibres showed little or no activity when compared with amosite and crocidolite asbestos, and this was not related to the ability of the fibres to release bioavailable iron.

Fisher et al. (1998) measured the release of bioavailable (soluble) iron from various types of man-made fibres (MMVF10, code 100/475, RCF1 and silicon carbide) and from long-fibre amosite asbestos. When equal numbers of WHO fibres were tested, more iron was released by MMVF10 than by code 100/475, which released about the same amount of iron as amosite; generally, more iron was released under acid conditions (pH 4.5) than under neutral conditions (pH 7.2) and there was more release into surfactant than into saline. The authors concluded that simple quantification of the release of soluble (bioavailable) iron from fibres under these in-vitro conditions cannot discriminate between pathogenic and non-pathogenic fibres.

Zoller and Zeller (2000) tested a sample of a respirable glass wool fibre (code A) and a sample of rock (stone) wool (code G) to examine whether fibres lost their biological activity when incubated in unbuffered saline. Both untreated fibres and fibres that had been incubated for four weeks in saline solution were tested for their ability to affect the production of reactive oxygen species by cultured human HL-60 cells, as assessed by luminol-enhanced chemiluminescence. Incubation in saline solution reduced the ability of these glass wool and rock (stone) wool fibres to cause chemiluminescence mediated by reactive oxygen species, suggesting that leaching or shortening *in vivo* might also occur, decreasing the long-term reactivity of the fibre. These effects were not seen with the rock (stone) wool fibres MMVF21 and HT-N.

4.3.2 *Rock (stone) and slag wool*

(a) *Inflammation and fibrosis*

Creutzenberg et al. (1997) exposed Wistar rats by inhalation for one or three weeks to MMVF21 at 38.7 mg/m^3 (695 WHO fibres/cm^3) or 43.8 mg/m^3 (879 WHO fibres/cm^3), or for three weeks to RCF1 (51.2 mg/m^3, 679 WHO fibres/cm^3); there were similar proportions of long fibres in the aerosol clouds of both fibre types. Following a three-day recovery period, inflammation, as measured by increased numbers of polymorphonuclear leukocytes in bronchoalveolar lavage, was observed after treatment with both 38.7 mg/m^3 and 43.8 mg/m^3 of MMVF21: 4.7% and 13.4%, respectively, compared with 1.1% in the lavage of rats exposed to clean air. In contrast, the rats exposed to refractory ceramic fibres had 18.9% polymorphonuclear leukocytes. During the extended post-exposure recovery period, the levels of polymorphonuclear leukocytes in the rats exposed to MMVF21 had returned to those of control animals by three months (0.5% and 0.8% in the groups treated with 38.7 mg/m^3 and 43.8 mg/m^3, respectively); however, at this time-point, the rats treated with RCF1 still had 13.5% polymorphonuclear leukocytes, although by 12 months this had decreased to 3.4%, which was not different from controls.

McConnell *et al.* (1994) exposed Fischer 344/N rats for 6 h per day, on five days per week for up to 24 months to MMVF21 at 3.1 mg/m^3 (34 WHO fibres/cm^3, 13 fibres > 20 μm long/cm^3), 16.1 mg/m^3 (150 WHO fibres/cm^3, 74 fibres > 20 μm long/cm^3) and 30.4 mg/m^3 (243 WHO fibres/cm^3, 114 fibres > 20 μm long/cm^3). Other groups of rats were exposed to MMVF22 at 3.1 mg/m^3 (30 WHO fibres/cm^3, 10 fibres > 20 μm long/cm^3), 16.1 mg/m^3 (131 WHO fibres/cm^3, 50 fibres > 20 μm long/cm^3) or 29.9 mg/m^3 (213 WHO fibres/cm^3, 99 fibres > 20 μm long/cm^3). Crocidolite (exposure period, 10 months; dose, 10 mg/m^3; 1610 WHO fibres/cm^3; 236 fibres > 20 μm long/cm^3) was used as a positive control. After various times up to 24 months, lung pathology was assessed and expressed as Wagner scores. The negative controls scored Wag 1 and animals exposed to crocidolite had a score of Wag 4 from three months of exposure onwards. Exposure to 16.1 mg/m^3 and 30.4 mg/m^3 MMVF21 resulted in scores of Wag 4 after 18 months which did not decrease after a period of recovery. Animals that received 29.9 mg/m^3 MMVF22 scored Wag 3 from six months onwards, and these scores decreased to Wag 2 or Wag 2.5 during recovery.

Kamstrup *et al.* (1998) exposed Fischer 344 rats to MMVF21 at a concentration of 16.1 mg/m^3 (150 WHO fibres/cm^3, 74 fibres > 20 μm long/cm^3) or 30.4 mg/m^3 (243 WHO fibres/cm^3, 114 fibres > 20 μm long/cm^3) or MMVF34 (soluble HT rock (stone) wool), at 30.5 mg/m^3 (288 WHO fibres/cm^3, 86 fibres > 20 μm long/cm^3) for 6 h per day, on five days per week for up to 18 months. At this time, the Wagner pathology score was Wag 4 in rats exposed to both the 16.1 mg/m^3 and 30.4 mg/m^3 doses of MMVF21 and 2.8 for rats exposed to MMVF34. The authors concluded that the lower biopersistence of the MMVF34 provided the most likely explanation for the observed differences. The same study reported the percentage of lung parenchyma affected with fibrosis or showing interstitial collagen in a morphometric assay (Kamstrup *et al.*, 2001). A maximum of 0.5% collagen deposition (expressed as percentage of lung parenchyma affected) was reported for rats exposed to MMVF34 at 24 months, whilst in rats exposed to MMVF21 the scores were 1.68% after exposure to 16.1 mg/m^3 and 3.84% after exposure to 30.4 mg/m^3 at the same time-point. The authors concluded that this study provided further evidence that biopersistence is important in the pathogenicity of fibres.

Donaldson (1994) attempted to adapt short-term inflammation assays in the mouse peritoneal cavity to detect differences between fibres of variable biopersistence by including a pre-treatment step in which the fibres were incubated for four weeks at neutral or mild acid pH. Pre-treatment of MMVF21 and RCF1 fibres at pH 5.0 decreased the degree of inflammation caused by MMVF21, but not RCF1 when compared with the effect seen after pre-treatment at pH 7.0. The author considered these findings confirmed that MMVF21 is not biopersistent at pH 5.0 and either shortens or dissolves. [The Working Group noted that the treated fibres were not sized or counted to confirm this contention.]

(b) Cell activation

Hill et al. (1996) assessed the ability of rat alveolar macrophages to release superoxide anions in response to treatment with uncoated or immunoglobulin (IgG)-coated MMVF21, code 100/475, RCF1 and silicon carbide fibres at equal numbers of WHO fibres. There was little release of superoxide in response to uncoated fibres of any type but, after incubation with IgG, the ability of both MMVF21 and RCF1 to cause superoxide release was enhanced. This effect appeared to be related to the high affinity of these fibres for the IgG, which is present in lung lining fluid and could also be adsorbed to the fibres *in vivo* to cause this enhancing effect.

Zoller and Zeller (2000) assessed the production of reactive oxygen species (ROS) by cultured human HL-60 cells using three types of rock (stone) wool fibre: code G, MMVF21 and HT-N (origin of the fibres specified in the paper). All samples were respirable and had the following proportions of long fibres: code G, 34% > 15 µm; MMVF21, 10% > 5 µm; HT-N, 50% > 15 µm. Both the untreated fibres and fibres that had been incubated at 37 °C in unbuffered saline solution for four weeks were tested. Pre-incubation reduced the ability of the code G rock (stone) wool fibres to cause chemiluminescence mediated by reactive oxygen species, but had no effect on MMVF21 or HT-N fibres.

(c) Other effects

Gilmour et al. (1995) assessed the ability of MMVF21 and MMVF22 at equal numbers of fibres to cause scission of supercoiled plasmid DNA in a cell-free system. These two fibres had little or none of this activity when compared with amosite asbestos. Tests on a series of fibres revealed no relationship between the ability of a fibre to release bioavailable iron and its pathogenicity.

4.3.3 Refractory ceramic fibres

(a) Inflammation

Refractory ceramic (aluminosilicate) fibres have been studied *in vivo* in rats and in Syrian golden hamsters and were reported to induce inflammation in the lung following inhalation.

In long-term inhalation studies (McConnell et al., 1995), Syrian golden hamsters were exposed for 18 months by nose-only inhalation for 6 h per day on five days per week to 'size-selected' kaolin-based RCF1 at an airborne mass concentration of

~30 mg/m³ (approximately 220 fibres/cm³; diameter, 1 μm; length, 25 μm). Inflammation of the lung was noted histologically as macrophage infiltration and microgranuloma formation. This study was designed to assess carcinogenicity (see section 3).

In chronic inhalation studies similar to those described above, the same kaolin-based RCF1 fibre preparation together with three other refractory ceramic fibres, *viz* an aluminium zirconium silica fibre (RCF2), a high-purity fibre (RCF3), and an 'after-service' heated kaolin-based fibre (RCF4) (all preparations contained an appreciable number of non-fibrous particles), were administered to Fischer 344 rats by nose-only inhalation (Mast *et al.*, 1995b). The animals were exposed for 6 h per day, on five days per week for 24 months to the size-selected (1 μm diameter and ~20 μm length; 220 fibres/cm³) refractory ceramic fibres at an airborne concentration of ~30 mg/m³. Inflammation of the lung was noted histologically as macrophage infiltration and microgranuloma formation and was found to be treatment-related in all groups exposed to refractory ceramic fibres beginning at three months of exposure. In a follow-up dose–response study of RCF1, Mast *et al.* (1995a) used the same exposure regimen for airborne fibre concentrations of 3, 9 or 16 mg/m³ (~36, 91 and 162 fibres/cm³, with equal or larger numbers of non-fibrous particles (≤ 3 μm) and observed mild macrophage infiltration and microgranuloma formation by 12 months in rats in all concentration groups. The carcinogenicity data from this study have been reviewed in section 3.

In a re-analysis of the above-mentioned experiments, the authors of more recent studies of well-characterized RCF1 and RCF1a fibres, containing 25% and 2% non-fibrous particles, respectively, concluded that the non-fibrous particulate fraction of the aerosol used in these chronic studies was likely to be partially responsible for the persistent inflammation that resulted in the long-term effects in the lungs of rats exposed to the 30-mg/m³ concentration (Brown *et al.*, 2000a; Mast *et al.*, 2000b).

Analysis of bronchoalveolar lavage fluid provided cytological and biochemical evidence that subchronic inhalation of RCF1 fibres causes pulmonary inflammation. Creutzenberg *et al.* (1997) exposed rats for three weeks to RCF1 at a single concentration of 51.2 mg/m³ (679 WHO fibres/cm³) and noted increased numbers of polymorphonuclear leukocytes for up to three months following cessation of exposure. Gelzleichter *et al.* (1999) conducted a 12-week inhalation study (treatment for 4 h per day, on 5 days/week) in both Fischer 344 rats and Syrian golden hamsters with a single RCF1 concentration of ~46 mg/m³ (~300 WHO fibres/cm³, with about 32% non-fibrous particles) and found cytological and biochemical indications of inflammation in bronchoalveolar lavage and pleural fluid in both species.

(*b*)   *Fibrosis*

In the long-term studies mentioned above (Mast *et al.*, 1995b; McConnell *et al.*, 1995), RCF1 fibres were demonstrated to be fibrogenic in Fischer 344 rats and Syrian golden hamsters. When both species were exposed long-term by nose-only inhalation for 6 h per day on five days per week to RCF1 at an airborne mass concentration of ~30 mg/m³, with approximately 220 fibres/cm³ (size-selected fibres: 1 μm in diameter and ~25 μm in length), both rats and Syrian golden hamsters were shown by histological analysis to have deposits of collagen at the bronchiolo–alveolar duct junctions. In addition to pulmonary parenchymal fibrosis, rats and hamsters exposed to RCF1 had pleural fibrosis (McConnell *et al.*, 1995). In hamsters chronically exposed to RCF1, pulmonary fibrosis was less widespread than in control animals treated with chrysotile while pleural fibrosis was more marked.

In the long-term inhalation study of refractory ceramic fibres (Mast *et al.*, 1995b), rats were exposed to RCF1, RCF2, RCF3 and RCF4 fibres. Fibrotic changes in the lungs developed after six months of exposure to RCF1 and RCF3. Collagen deposition was noted by histopathology after 24 months of exposure in response to all four preparations of refractory ceramic fibres. In addition, minimal to mild multifocal deposition of collagen was noted in the pleura of rats exposed to RCF1, RCF2 and RCF3 as well as in positive control animals exposed to chrysotile asbestos.

The long-term studies described above suggest that Syrian golden hamsters are more predisposed to the pleural effects of inhaled RCF1 than Fischer 344 rats. This finding is supported by other studies (Everitt *et al.*, 1997; Gelzleichter *et al.*, 1999) in which Fischer 344 rats and Syrian golden hamsters were exposed under the same experimental conditions, namely by nose-only inhalation to RCF1 for 4 h per day on five days per week at a single concentration (~46 mg/m^3, ~300 WHO fibres/cm^3) for 12 weeks. At 12 weeks post-exposure, the hamsters, but not the rats, had significant increases in deposition of pleural collagen, which correlated with an increased pleural mesothelial reaction, as assessed by cell proliferation.

(*c*)   *Cell proliferation*

Rutten *et al.* (1994) demonstrated proliferation of mesothelial cells seven and 28 days after direct intrapleural instillation of RCF1 into Fischer 344 rats and Syrian golden hamsters. Hamsters demonstrated more extensive proliferation of mesothelial cells than did rats exposed to similar numbers of MMVF10 glass or RCF1 refractory ceramic fibres: RCF1 fibres induced a higher rate of proliferation in both species than MMVF10 glass fibres.

More recent studies by Gelzleichter *et al.* (1999) involving exposure by inhalation of both Fischer 344 rats and Syrian golden hamsters to a single concentration of RCF1 (~46 mg/m^3, ~300 WHO fibres/cm^3) for 12 weeks demonstrated proliferation of visceral and parietal mesothelial cells. Stronger and more persistent effects were seen in the hamsters than in the rats. [The Working Group noted that this study employed only one exposure concentration and did not compare RCF1 with any other fibrous particulates.]

(*d*)   *Cellular toxicity*

Dopp *et al.* (1995) tested asbestos (chrysotile, amosite and crocidolite) fibres and refractory ceramic fibres [type unspecified] for their ability to induce apoptosis in cultured Syrian hamster embryo cells. Treatment with 0.5–10 μg/cm^2 refractory ceramic fibre or amosite for 16–72 h showed a small (3–3.4%), but not dose-related, development of apoptotic nuclei. This is in contrast to a significantly higher rate of apoptosis observed in Syrian hamster embryo cells treated with chrysotile (5 µg/cm^2) (up to 33%). [The Working Group noted that the dose levels, as expressed in number of fibres, were different.]

Hart *et al.* (1994) used Chinese Hamster ovary cells to assess the toxic effects of a large variety of fibres of various lengths and thicknesses, including crocidolite, chrysotile and various synthetic vitreous fibres. The primary effect of all the compositions tested was a concentration-dependent decrease in cell proliferation as a result of the physical disruption of cell division. These effects were strongly correlated with fibre length, but not with diameter. The effects did not differ between fibre types, which led the authors to conclude that this in-vitro system could not discriminate between fibres of different pathogenicity.

Refractory ceramic fibre [type unspecified] was found to be cytotoxic to cultured rat alveolar macrophages (24 h of exposure to 100, 300, or 1000 µg/mL), as assessed by the release of lactate dehydrogenase, but this effect was small in comparison with that observed after exposure to silica at a similar mass dose (Leikauf *et al.*, 1995). Fujino *et al.* (1995) treated rat alveolar macrophages with refractory ceramic fibres [type unspecified] (12.5–200 µg/mL) and determined the release of tumour necrosis factor, lactate dehydrogenase and β-glucuronidase. In comparison with a series of natural mineral and synthetic vitreous fibres, the refractory ceramic fibres exhibited relatively low levels of biological activity. This was in contrast to the findings of earlier studies on an aluminium silicate fibre (Nadeau *et al.*, 1987) which under certain experimental conditions elicited a greater release of lactate dehydrogenase and β-glucuronidase from macrophages than did chrysotile. [The Working Group noted that differences in fibre type, preparations, dimensions and dose may explain these results.]

(*e*)   *Cell activation*

Refractory ceramic fibres have been demonstrated to activate polymorphonuclear neutrophils through an increase in intracellular calcium (Ruotsalainen *et al.*, 1999). In this study, the production of reactive oxygen species induced by refractory ceramic fibres [not specified] in polymorphonuclear neutrophils was markedly higher than with other synthetic vitreous fibres, although the study did not use fibres of equivalent length distribution.

Treadwell *et al.* (1996) compared the abilities of RCF1 and asbestos to cause adherence of neutrophils to monolayers of endothelial cells at non-cytotoxic concentrations. Crocidolite and chrysotile asbestos, but not RCF1, activated endothelial cells and induced adherence of neutrophils through the expression of endothelial adhesion molecules such as intercellular adhesion molecule-1. In similar studies Barchowsky *et al.* (1997) found that crocidolite and chrysotile, but not RCF1, activated tyrosine kinases, increased cell motility and caused marked changes in endothelial cell morphology.

(f)   Oxidant generation

Brown et al. (1998b) compared MMVF10, code 100/475, RCF1, RCF4, silicon carbide and amosite asbestos for their ability to generate free radicals using two assays: the assay of DNA φX174 plasmid scission and a salicylate assay of hydroxyl radical activity. In the plasmid assay, only long-fibre amosite asbestos produced free radical activity, while in the salicylate assay, both RCF1 and amosite caused release of hydroxyl radicals.

Hill et al. (1996) assessed the ability of rat alveolar macrophages to release superoxide anions in response to treatment with uncoated and immunoglobulin (IgG)-coated synthetic vitreous fibres. The studies used equal numbers of fibres of MMVF21, code 100/475, RCF1 and silicon carbide. There was little release of superoxide in response to uncoated fibres of any type but, after incubation with IgG, the activities of both MMVF21 and RCF1 fibres were enhanced in this respect. This appeared to be related to the high affinity of these fibres for the IgG, which is present in lung lining fluid and which could be adsorbed *in vivo* to cause this enhancing effect.

(g)   Other effects

Brown et al. (2000b) examined the ability of equal numbers of fibres of different types to deplete antioxidants from lung lining fluid and from solutions of individual antioxidants. Included in the series of MMVFs tested were RCF1 and RCF4. The ability to deplete antioxidants did not correlate with differences between pathogenic and non-pathogenic fibres (as defined in long-term rodent bioassays). In comparison with MMVF10 glass fibres, RCF1 and RCF4 were not found to be especially active in depleting glutathione in these studies.

4.4   Effects on gene expression

4.4.1   *Continuous glass filament and glass wool*

Driscoll (1996) reviewed the role of inflammatory cytokines, growth factors and proto-oncogenes in asbestos-induced carcinogenesis. Churg et al. (2000) highlighted the role of reactive oxygen species in fibre-induced activation of signalling pathways controlling the expression of pro-inflammatory cytokines and growth factors.

In a sub-chronic inhalation study, the exposure of male Syrian golden hamsters to 901 glass fibre (MMVF10.1) at a concentration of ≥ 16 mg/m^3 for 6 h per day on five days per week for 13 weeks resulted in proliferation of epithelial cells (measured by immunostaining of proliferating cell nuclear antigen) and inflammation at bifurcations of the bronchoalveolar duct (Hesterberg et al., 1999). However, the staining intensity returned to normal after a 10-week recovery period and the proliferative response to glass fibres was lower than that for inhalation of amosite. This difference in the magnitude and duration of the proliferative activity of amosite asbestos in comparison with glass fibre (code 100/475) was also demonstrated by the use of bromodeoxyuridine-labelling in male Wistar rats 16 h after a single 7-h period of exposure to fibres (Donaldson et al., 1995a). The authors concluded that glass fibres on a mass basis may be less potent activators of cell proliferation *in vivo* than amosite asbestos.

The fibre-induced expression of DNA damage-inducible genes, proto-oncogenes or tumour suppressor genes has recently been investigated. Johnson and Jaramillo (1997) reported that, unlike crocidolite, code 100 glass microfibres failed to enhance the expression of DNA damage-inducible genes (*p53*, *Cip 1* and *Gadd 153*) in an alveolar epithelial type II tumour cell line (A549).

Exposure to MMVF10 at 25 µg/cm^2 did increase the concentrations of mRNA for c-*fos* and c-*jun* in rat pleural mesothelial cells, but it was less potent than crocidolite as it affected these proto-oncogene mRNA levels only at doses which caused cytotoxicity. High concentrations of MMVF10 also increased mRNA levels for ornithine decarboxylase in hamster tracheal epithelial cells, indicating induction of cell proliferation (Janssen et al., 1994).

The mammalian activator protein-1 family comprises homodimers and heterodimers of the Jun, Fos and ATF (activating transcription factor) subgroups of transcription factors (Shaulian & Karin, 2001). Activator protein-1 regulates the expression and function of regulators of the cell cycle, such as cyclin D1, p53, p21$^{Cip1/Waf1}$, p19^{ARF} and p16 and, therefore, plays an important role in cell proliferation. Mitogen-activated protein kinases (MAPK) play an important role in the signalling pathway for activator protein-1: extracellular signal-regulated kinase (ERK) can activate c-*fos* while Jun N-terminal kinase (JNK) activates c-*jun* and ATF (Karin, 1995). An increase of c-Jun protein has been shown in BALB/c-3T3 cells transformed by AAA-10 glass fibres (Gao et al., 1997). In addition, exposure of rat alveolar macrophages to code 100 glass fibres induced the phosphorylation of p38 and ERK MAPK (Ye et al., 2001). Exposure to code 100 glass fibres also stimulated the DNA binding activity of activator protein-1. Activation of activator protein-1 was partially inhibited by SB203580 (an inhibitor of p38) or PD98059 (which prevents the phosphorylation of ERK), indicating a role for MAPK in the signalling pathway to activator protein-1. Glass fibres (MMVF10) have also been shown to activate activator protein-1 in rat alveolar macrophages, but the strength of this effect was only about a third of that observed for amosite (Gilmour et al., 1997).

Tumour necrosis factor α (TNF-α) is an inflammatory mediator that has also been implicated to play a role in cell proliferation and apoptosis. Treatment of rat tracheal epithelial explants with TNF-α has been shown to enhance the binding of MMVF10 fibres to the cell surface. This enhancement was mediated through the activation of nuclear factor-κB (NF-κB) (Xie et al., 2000). Indeed, binding of TNF-α to the TNF type 1 receptor has been linked to phosphorylation of JNK which results in activation of activator protein-1 and cell proliferation (Karin & Delhase, 1998). Brass et al. (1999) have reported that the fibroproliferative response at the junctions of the bronchiolar–alveolar duct induced by inhalation of chrysotile is directly related to fibre-induced expression of mRNA and protein for TNF-α and transforming growth factor β1

(TGF-β_1); C57BL/6 mice were found to be susceptible to development of this lesion while 129 inbred mice were resistant.

Since TNF-α has been linked to fibre binding and cell proliferation, it is important to determine the effect of glass fibres on its production. To date, conflicting results have been reported. Cullen *et al.* (1997) showed that, unlike amosite or crocidolite, glass fibres (MMVF10 and MMVF11) failed to induce TNF-α production in cultures of rat alveolar macrophages. Similarly, Code 100/475 or MMVF10 glass fibres failed to significantly stimulate the release of TNF-α from rat alveolar macrophages, human blood monocytes, THP-1 human macrophage cell line or J774.2 mouse macrophage cell line in culture (Fisher *et al.*, 2000). Similarly, glass fibres [not further specified] did not significantly stimulate TNF-α production by J774 cells (Murata-Kamiya *et al.*, 1997). In contrast, significant stimulation by glass fibres [not further specified] of the release of TNF-α from rat alveolar macrophages was reported by Fujino *et al.* (1995), with a potency that was only slightly less than that of crocidolite, chrysotile or amosite asbestos.

A significant increase in TNF-α production was reported after in-vitro exposure for 6 h and 16 h to code 100 glass fibres (size-selected by dielectrophoresis) at a fibre:cell ratio of 5:1 in the mouse peritoneal monocyte cell line RAW 264.7 and in NR 8383 rat alveolar macrophages. Glass fibres 17 µm long were found to be two to four times more potent than short fibres (7 µm) in stimulating release of TNF-α at equal fibre numbers. Code 100 microfibres also activated the TNF-α gene promoter in RAW 264.7 monocyte and NR8383 alveolar macrophage cell lines; the 17-µm-long fibres were two to three times more potent than the 7-µm fibres. Several transcription factors have been identified as being involved in the expression of TNF-α induced by glass fibres. That is, the release of TNF-α in response to exposure to glass fibres was almost completely inhibited by an NF-κB inhibitor (SN50), by 70% with an inhibitor of p38 (SB203580) and by about 50% with an inhibitor of ERK (PD98059), which suggested the involvement of both NF-κB and activator protein-1. Similarly, fibre-induced TNF-α gene promoter activity was also inhibited by blockers of NF-κB and mitogen-activated protein kinase phosphorylation. In addition, transfection of NR8383 rat alveolar macrophage cells with an inactivated cAMP response element inhibited the glass-fibre-induced activation of the TNF-α gene promoter by about 80% (Ye *et al.*, 1999, 2001).

In addition to the proto-oncogenes discussed above, increased cell proliferation has been linked to the secretion of inflammatory cytokines and growth factors from lung cells exposed to asbestos (Driscoll, 1996). The transcription factor NFκB is involved in the activation pathway for several such cytokines and growth factors induced by various particles and fibres (Schins & Donaldson, 2000). Brown *et al.* (1999) reported that amosite, the refractory ceramic fibre RCF1 and silicon carbide fibres were effective stimulants of NF-κB activation in lung type II epithelial cells (A549 tumour cell line). In contrast, twice the number of fibres of MMVF10 were required to activate NF-κB while code 100/475 glass fibres were inactive even at this high dose. Similarly, Gilmour *et al.* (1997) showed that amosite asbestos activated

NF-κB in rat alveolar macrophages *in vitro*, but MMVF10 failed to cause a significant activation of this transcription factor. In contrast, in-vitro treatment for 2 h of RAW 264.7 mouse monocytes with code 100 glass fibres at a fibre:cell ratio of 5:1 caused a substantial increase in the DNA-binding activity of NF-κB; the 17-μm fibres caused a threefold increase and the 7-μm fibres a twofold increase. The inhibition of the translocation of NF-κB from the cytoplasm to the nucleus by the NF-κB inhibitor (SN50) virtually eliminated the TNF-α production induced by glass fibres. There appears to be a positive feedback loop between NF-κB and TNF-α. Not only does activation of NF-κB stimulate the TNF-α gene promoter and TNF-α production (Ye *et al.*, 1999), but TNF-α has been shown to cause a 10-fold increase in NF-κB activity in rat lung epithelial alveolar type II cells and to deplete the concentrations of IκB-α, an endogenous inhibitor of NF-κB activity (Janssen-Heininger *et al.*, 1999).

As stated above, NF-κB has been implicated in the upregulation of several pro-inflammatory cytokines and growth factors observed after exposure to various particles and fibres (Schins & Donaldson, 2000). Driscoll *et al.* (1996) reported that in-vitro exposure of rat lung type II epithelial cells to crocidolite for 6 h increased mRNA concentrations for chemokines, macrophage inflammatory protein-2 and cytokine-induced neutrophil chemoattractant, but, under similar conditions, MMVF10 glass fibres did not. In an immunofluorescence assay, neither MMVF10 nor crocidolite asbestos fibres (length, < 60 μm) affected epidermal growth factor–receptor protein staining intensity and distribution in an SV-40-immortalized human mesothelial cell line (Met5A) (Pache *et al.*, 1998), whereas crocidolite fibres > 60 μm in length increased the staining intensity and altered the distribution of this protein. It was noted by the authors that the MMVF10 glass fibres used in this study were only 20 μm in length, which may explain their ineffectiveness. [The Working Group noted that fibres in this length category usually do show effects in other cellular assays.] Glass fibres [not specified] reduced the plating efficiency of B14F28 Chinese hamster fibroblasts at concentrations > 2 μg/cm^2, but increased colony formation at concentrations < 0.33 μg/cm^2 (Fischer *et al.*, 1998). This proliferative effect appears to be due to fibre-induced production of growth factors by these cells, since conditioned medium (supernatant medium of fibre-treated colonies) was also effective in stimulating colony growth. Glass fibres were somewhat less potent than amosite or crocidolite asbestos in this assay. In contrast, Koshi *et al.* (1991) observed an inhibition of colony formation in Chinese hamster lung cells exposed to code 100, code 104, code 108A and code 108B fibres, with TD$_{50}$ values (concentrations required for 50% inhibition) of 10, 11, 18 and 27 μg/mL, respectively.

Fibres and particles have been reported to activate signalling pathways for NF-κB and activator protein-1 (Ding *et al.*, 1999; Schins & Donaldson, 2000). Fibre-induced generation of reactive oxygen species has been implied from the results of a φX174 RF plasmid DNA assay to detect oxidant-induced DNA damage (Gilmour *et al.*, 1995; Donaldson *et al.*, 1996). In this assay, amosite and crocidolite asbestos were shown to generate significant amounts of reactive species in solution, whereas glass fibres (MMVF10 and MMVF11) were virtually ineffective. In contrast, code 100 glass fibres and glass wool (Owens Corning) generated reactive species (hydroxyl radicals) in solutions containing hydrogen peroxide at rates higher than those observed with crocidolite and amosite asbestos, as detected by a salicylate assay (Maples & Johnson, 1992). Code 100/475 and MMVF10 glass fibres were also able to decrease the concentrations of glutathione and ascorbate in cell-free solutions, suggesting that glass fibres can generate reactive oxygen species under these conditions (Brown et al., 2000b). Similarly, in-vitro treatment with MMVF10 decreased the concentrations of glutathione in rat alveolar macrophages, although to a lesser extent than treatment with amosite asbestos (Gilmour et al., 1997). Wang et al. (1999a) reported that glass fibres were effective in stimulating the release of superoxide anions from guinea-pig alveolar macrophages in vitro, their potency being about 70% of that of chrysotile. However, glass fibres and chrysotile asbestos were equally potent in stimulating hydrogen peroxide production in these macrophages. This level of oxidant production was associated with a depletion of glutathione in these phagocytes, but glass fibres were not as effective as chrysotile in this respect. Superoxide release was also observed to occur in rat and hamster alveolar macrophages exposed in vitro to code 100 glass microfibres, or to MMVF21 coated with IgG (Hansen & Mossman, 1987; Mossman & Sesko, 1990; Hill et al., 1996). By means of chemoluminescence detection, the release of reactive oxygen species was measured in human polymorphonuclear neutrophils exposed in vitro to MMVF10, MMVF11, MMVF22, various types of refractory ceramic fibre or glass wool and in human monocytes exposed to MG-1 microglass fibres (Leandersen & Tagesson, 1992; Luoto et al., 1997; Ruotsalainen et al., 1999; Ohyama et al., 2000).

Oxidant production is believed to play an important role in mediating the proliferative response of lung cells to fibres. Churg (1996) reported that antioxidants decreased the uptake of asbestos fibres by explants of rat trachea. Antioxidants such as curcumin and pyrrolidine dithiocarbamate have been shown to inhibit the amosite-induced translocation of NF-κB to the nucleus in the lung epithelial tumour cell line A549 while aspirin inhibited crocidolite-induced activation of activator protein-1 in the mouse JB6 epidermal cell line (Brown et al., 1999; Ding et al., 1999). Aspirin was also shown to be effective in inhibiting crocidolite-induced activation of activator protein-1 in the bronchiolar tissue of mice, when given 30 min before intratracheal instillation of the fibres (Ding et al., 1999). In addition, exposure of rat lung epithelial alveolar type II cells to 1 mM 3-morpholinosydnonimine (which releases nitric oxide and superoxide which then react to yield peroxynitrite) caused an eightfold increase in activity of NF-κB. This oxidant-induced activation of NF-κB can be enhanced synergistically by TNF-α (Janssen-Heininger et al., 1999). Ye et al. (1999) have also shown that oxidants play a role in the activation of NF-κB in mouse monocytes (RAW 264.7 cell line) exposed in vitro to code 100 glass fibres (length, 17 μm). Treatment with N-acetyl-L-cysteine (a non-specific antioxidant) completely inhibited glass fibre-induced activation of NF-κB, activation of the TNF-α gene promoter and TNF-α production.

Increases in free intracellular calcium have been associated with cell proliferation through induction of c-*fos* and activation of activator protein-1 (Karin et al., 1997).

Glass wool (not specified) has been reported to increase the intracellular calcium concentration in human polymorphonuclear neutrophils exposed *in vitro* (Ruotsalainen *et al.*, 1999). Similarly, microglass fibres (not specified) mobilized intracellular calcium in guinea-pig alveolar macrophages *in vitro* (Wang *et al.*, 1999a). In both these studies, the potency of glass fibres to increase free intracellular calcium was found to be significantly lower than that of refractory ceramic fibres or chrysotile.

4.4.2 Rock (stone) wool, slag wool and refractory ceramic fibres

(a) Pulmonary epithelial cells

Proto-oncogene (*c-fos*, *c-jun*) mRNA levels were studied *in vitro* in hamster tracheal epithelial cells exposed to 5–25 $\mu g/cm^2$ crocidolite, MMVF10 or RCF1 (Janssen *et al.*, 1994). Changes in *c-fos* expression were not detected after exposure to any of these three fibre types. Upregulation of *c-jun* was detected in response to crocidolite, MMVF10 and RCF1, but the increase caused by MMVF10 and RCF1 was less pronounced than with crocidolite.

Reactive oxygen and nitrogen species mediate cell proliferation and the production of pro-inflammatory cytokines through signal transduction and transcription factor activation. Brown *et al.* (1999) exposed A549 human lung epithelial type II tumour cells *in vitro* to equal numbers of fibres of long-fibre amosite, silicon carbide, RCF1 and RCF4 (8.24×10^6 fibres/mL) and investigated the induction of nuclear translocation of NF-κB. After exposure to amosite, silicon carbide or RCF1 the translocation of NF-κB to the nucleus was increased; a similar effect was observed after exposure to MMVF10 at 16.5×10^6 fibres/mL. No increase was detected after exposure to RCF4. Nuclear translocation of NF-κB was inhibited by antioxidants, such as curcumin.

Marks-Konczalik *et al.* (1998) exposed SV-40-immortalized human bronchial epithelial (BEAS 2B) cells to crocidolite, rock (stone) wool (115-4) and silica at concentrations of 0–50 $\mu g/cm^2$ and measured manganese superoxide dismutase gene expression and activity. At high concentrations (above 25 $\mu g/cm^2$), a decrease in the activity of manganese superoxide dismutase was seen in response to crocidolite and silica, but not to rock (stone) wool. Upregulation was seen in cells exposed to low concentrations (2 $\mu g/cm^2$) of all three fibre types. The decreased activity of manganese superoxide dismutase coincided with increased cytotoxicity.

Ljungman *et al.* (1994) exposed rat alveolar macrophages to equal masses of a range of mineral fibres and measured mRNA expression and activity of TNF-α. Following exposure for 90 min, an increase in TNF-α mRNA expression was observed. The results, from highest to lowest, were: chrysotile B > chrysotile A > crocidolite > RCF1 > silicon carbide > MMVF21. No increase in expression was seen after exposure to MMVF22. Upregulation of TNF-α activity was seen in response to the two types of chrysotile, MMVF21 and crocidolite, but not with MMVF22, RCF1 or silicon carbide.

Fisher *et al.* (2000) exposed rat alveolar macrophages and monocytes from human peripheral blood to a variety of mineral fibres (3×10^6 fibres/mL) for 16 h and measured the production of TNF-α protein. Rat macrophages showed increased concentrations of TNF-α protein following exposure to silicon carbide, MMVF10 and RCF1. In human monocytes, increased concentrations of TNF-α protein were observed following exposure to silicon carbide or RCF1.

Cullen et al. (1997) added equal numbers of mineral fibres (8.2×10^6/well) to rat alveolar macrophages and measured the increase in production of TNF-α protein. The order of activity was: silicon carbide > crocidolite > long amosite > code 100/475 > MMVF22 > RCF1 and RCF2 > MMVF21 > RCF3 and RCF4.

Gilmour et al. (1997) exposed rat alveolar macrophages to amosite asbestos, RCF1 and MMVF10 each at a concentration of 8.24×10^6 fibres/mL and measured the activation of NF-κB and activator protein-1. Nuclear binding of activator protein-1 transcription factor was upregulated by 37%, 9% and 12%, by amosite asbestos, RCF1 and MMVF10, respectively. Only amosite increased the activity of NF-kB.

(b)   *Mesothelial cells*

Janssen et al. (1994) exposed rat pleural mesothelial cells to various concentrations (up to 25 μg/cm^2 area of culture dish) of crocidolite, MMVF10 and RCF1, and determined the steady-state levels of *c-fos* and *c-jun* proto-oncogene mRNA. Changes in *c-fos* and *c-jun* expression induced by crocidolite were seen in response to exposure to concentrations as low as 2.5 μg/cm^2. Exposure to MMVF10 or RCF1 increased expression only at concentrations of 25 μg/cm^2. The potency of MMVF10 and RCF1 was less on a per mass basis than that of crocidolite in this assay.

4.5   Genetic and related effects

4.5.1   *Continuous glass filament and glass wool*

No data were available to the Working Group on the genetic effects of continuous glass filaments.

Asbestos and glass fibres are not direct mutagens in assays with *Escherichia coli* (strain WP2*uvr*A) or *Salmonella typhimurium* (strain TA1538) (Chamberlain & Tarmy, 1977; Barrett et al., 1989). However, asbestos fibres have been shown to generate reactive oxygen species and cause DNA damage (strand breaks) in cell-free systems (Gilmour et al., 1995; Donaldson et al., 1996). In a Syrian hamster embryo cell culture system, exposure to asbestos fibres resulted in accumulation of fibres in the perinuclear region of the cell. Although no gene mutations were detected at two specific loci in these cells (Oshimura et al., 1984), it has been proposed that the physical presence of such fibres interferes with chromosome segregation during mitosis, which could result in aneuploidy or chromosomal translocations leading to the activation of proto-oncogenes and/or the inhibition of suppressor genes. The proliferation of such transformed cells could perpetuate genetic errors and lead to oncogenesis (Barrett et al., 1989).

Studies on glass fibres have demonstrated a positive response in genotoxicity tests in non-human mammalian cells, as shown by the induction of DNA damage, micronuclei and binucleated and multinucleated cells following exposure to fibres of different origin (Table 73 and references therein). Negative results were obtained with very long and thick fibres; for instance, no genotoxicity was observed in V79 hamster fibroblasts treated with general-purpose building insulation fibres (median length and diameter, 98 μm and 7.3 μm, respectively), whereas treatment with fibres of AAA-10 (median length and diameter, 2 μm and 0.18 μm, respectively) and JM 100 (median length and diameter, 3.5 μm and 0.2 μm, respectively) did produce DNA and chromosomal damage in the same assay. The dependence of micronucleus formation and chromosomal damage on fibre dimensions has been emphasized by numerous independent studies.

Nuclear abnormalities (micronucleus, polynuclei) were investigated in Chinese hamster ovary cells exposed to 17 different samples of glass fibres with average dimensions ranging between 3.5 and 27 μm in length and 0.3 and 7.0 μm in diameter. The effects were directly related to fibre length, whereas diameter had little impact on the induction of nuclear abnormalities. The induction of abnormal anaphases or telophases in rat pleural mesothelial cells was correlated with the presence of fibres — described as the most carcinogenic by Stanton et al. (1977) (length, > 8 μm; diameter, ≤ 0.25 μm) — with a non-observable effect level estimated at 2.5×10^5 fibres/cm^2 area of culture dish. Fibres of these dimensions were poorly represented in the MMVF samples used in this assay, possibly accounting for the absence of any activity of MMVF10 or MMVF11.

Formation of micronuclei was observed in Syrian hamster embryo cells exposed to 1 μg/cm^2 of code 100 fibres; cell transformation was also seen under these conditions. Reduction of fibre length by milling, without affecting fibre diameter, diminished the transforming potency of these fibres. Morphological transformation was also found in BALBc-3T3 cells exposed to code 100 fibres.

Formation of binucleated and multinucleated cells was detected after treatment of either untransformed or SV40-immortalized Met5A human mesothelial cells with thin glass wool (median length and diameter, 6.1 μm and 0.29 μm, respectively). Moreover, DNA damage and structural and numerical chromosomal aberrations were observed in human mesothelial cells exposed to glass fibres [not specified].

Many studies have focused on the production of reactive oxygen species by rodent alveolar macrophages or human polymorphonuclear leukocytes treated with glass wool (Table 74 and references). These effects are not necessarily associated with genotoxicity; however, they are cited here because reactive oxygen species may damage DNA in other target cells of the lung (Kane et al., 1996). All studies reported production of reactive oxygen species after exposure to glass fibres. In-vitro cell-free assays have been used to investigate the intrinsic ability of fibres to produce reactive oxygen species. Two types of end-point have been used: measurement of guanine hydroxylation in DNA or in deoxyguanosine; or scission of plasmid DNA, as assessed by agarose gel electrophoresis. These assays indicate the formation of hydroxyl radical (• OH) and DNA

Table 73. Genetic and related effects of glass wool

Test system	Result	Dose[a]	Fibre type[b]	Reference
DNA damage (comet assay), Chinese hamster V79 cells	+	1.7 µg/cm^2	Owens Corning AAA-10	Zhong et al. (1997)
Formation of micronuclei, Syrian hamster embryo cells	+	1 µg/cm^2 (single dose)	Code 100 (unmilled)	Hesterberg et al. (1986
Formation of bi- and multinucleated cells, rat liver epithelial cells	+	1 µg/cm^2 (no dose–response; higher doses had no effect	Thin glass wool	Pelin et al. (1995)
Formation of bi- and multinucleated cells, Met-5A and PL102 human mesothelial cells	+	1 µg/cm^2 (higher doses also active)	Thin glass wool	Pelin et al. (1995)
Formation of micronuclei and multinucleated cells, Chinese hamster V79 cells	+	10 µg/mL	AAA-10 (Owens Corning); code JM 100	Ong et al. (1997)
Formation of micronuclei and multinucleated cells, Chinese hamster V79 cells	–	160 µg/mL	General-purpose building insulation (Owens Corning)	Ong et al. (1997)
Nuclear abnormalities (micronucleus and ploidy), Chinese hamster ovary cells	+	Not applicable: data not provided individually	MMVF10; MMVF11; MvL 475/code 90; MvL 475/code 108; MvL 475/code 110; MvL 475/code 112; MvL 475/475TK; MvL 901/FG3; MvL 901/FG5; MvL 901/FG9; MvL 901/FG15; MvL 901/FG19; MvL 901/FG22; MvL 901/FG25; MvL 901/FG31; X607 thick; X607 thin	Hart et al. (1994)

Table 73 (contd)

Test system	Result	Dose[a]	Fibre type[b]	Reference
Chromosomal aberrations (structural), Chinese hamster lung cells	−	300 μg/mL	Code 100; code 104; code 108A	Koshi et al. (1991)
Chromosomal aberrations (numerical; polyploidy), Chinese hamster lung cells	+	10 μg/mL	Code 100; code 104	Koshi et al. (1991)
Chromosomal aberrations (numerical; polyploidy), Chinese hamster lung cells	+	100 μg/mL	Code 108A	Koshi et al. (1991)
Cell transformation, Syrian hamster embryo cells	+	2 μg/cm^2	Code 110 (thick)	Hesterberg & Barrett (1984)
Cell transformation, Syrian hamster embryo cells	+[c]	0.5 μg/cm^2	Code 100 (thin)	Hesterberg & Barrett (1984)
Cell transformation, Syrian hamster embryo cells	+[c]	1 μg/cm^2 (single dose)	Code 100 (unmilled)	Hesterberg et al. (1986)
Micronucleus formation, Syrian hamster embryo cells	+[c]	1 μg/cm^2 (single dose)	Code 100 (unmilled)	Hesterberg et al. (1986)
Cell transformation (foci), anchorage-independent growth of transfectants and transforming potency of transfectants on NIH-3T3, BALB/c-3T3 cells	+	1 μg/cm^2	AAA-10 (Owens Corning)	Gao et al. (1995)
Cell transformation (foci), anchorage-independent growth of transfectants and transforming potency of transfectants on NIH-3T3, BALB/c-3T3 cells	+	38 μg/cm^2	General building insulation (Owens Corning)	Gao et al. (1995)
Cell transformation (foci), anchorage-independent growth of transfectants and transforming potency of transfectants on NIH-3T3, BALB/c-3T3 cells	+	10 μg/cm^2	Code 100	Gao et al. (1995)
8-OH-dG formation, mouse reticulum-cell sarcoma cell line (J774)	−	27 μg/cm^2	Glass	Murata-Kamiya et al. (1997)
Anaphase/telophase abnormalities in rat pleural mesothelial cells	−	< 2.5 × 10^5 Stanton fibres/cm$^{2\,d}$	MMVF10; MVVF11	Yegles et al. (1995)

Table 73 (contd)

Test system	Result	Dose[a]	Fibre type[b]	Reference
DNA damage (comet assay), human Hel 299 lung fibroblasts	+	3.4 µg/cm^2	Owens Corning AAA-10 microfibres	Zhong et al. (1997)
Induction of DNA breakage, repair and interstrand cross-linking, human lung epithelial (A549) cell line	+	40 µg/cm^2 (single dose)	MG1 micro glass fibres (JFMRA); GW1 glass wool fibres (JFMRA)	Wang et al. (1999b)
Formation of bi- and multinucleated cells, human MeT-5A mesothelial cells and human primary mesothelial cells from pleural effusions	+	1 µg/cm^2	Thin glass wool	Pelin et al. (1995)
Transfection of plasmid and DNA replication, human MeT-5A mesothelial cells	–	1.33 µg/cm^2 (single dose)	Glass fibres	Gan et al. (1993)
Chromosomal aberrations in human embryo lung cells	+	1.0 µg/cm^2	MG1 micro glass fibres (JFMRA); GW1 glass wool fibres (JFMRA)	Wang et al. (1999b)
Gene amplification and mutation, glass fibre-induced transformed BALB/c-3T3 cells (H-*ras*, K-*ras*, c-*myc* and c-*fos*) (*p53* and *K-ras*)			AAA-10 (Owens Corning)	Whong et al. (1999)

8-OH-dG, 8-hydroxydeoxyguanosine
[a] Lowest effective dose or highest ineffective dose
[b] According to the authors
[c] Milling of the fibres strongly reduced the effect.
[d] Defined according to Stanton's criteria: length > 8 µm, diameter ≤ 0.25 µm (Stanton et al., 1977)

Table 74. Other effects of glass wool on cells *in vivo*

Test system	Result	Dose[a]	Fibre type[b]	Reference
Superoxide production in rat alveolar macrophages (fibres either uncoated or coated with rat IgG)	+	3 million fibres (single dose)	Code 100/475	Donaldson et al. (1995b)
Superoxide production in rat alveolar macrophages (fibres either uncoated or coated with rat IgG)	+	25 µg[c]	Code 100/475	Hill et al. (1996)
Production of reactive oxygen species, superoxide anions, hydrogen peroxide, and reduction of cellular glutathione content in guinea-pig alveolar macrophages	+[d]	200 µg/mL (single dose)	MG1 micro glass fibres (JFMRA); GW1 glass wool fibres (JFMRA)	Wang et al. (1999a)
Production of reactive oxygen species, human polymorphonuclear leukocytes	+	200 µg/mL	MMVF11	Luoto et al. (1997)
Production of reactive oxygen species, human polymorphonuclear leukocytes	+	100 µg/mL	MMVF10	Luoto et al. (1997)
Release of superoxide anions (cytochrome C reduction), hamster alveolar macrophages (+ zymosan)	+	2.5 µg/mL	Code 100	Hansen & Mossman (1987)
Release of superoxide anions (cytochrome C reduction), rat alveolar macrophages	+	5 µg/cm^2	Code 100	Hansen & Mossman (1987); Mossman & Sesko (1990)
Production of hydrogen peroxide, human polymorphonuclear leukocytes	+	200 µg/mL (single dose)	Glass wool	Leanderson & Tagesson (1992)
8-OH-dG formation (from dG added in the mixture), human polymorphonuclear leukocytes	+	500 µg/mL	Glass wool	Leanderson & Tagesson (1992)
Production of reactive oxygen species (chemiluminescence), human monocytes	+	5×10^5 fibres[c]	GW1 glass wool fibres (JFMRA)	Ohyama et al. (2000)

Table 74 (contd)

Test system	Result	Dose[a]	Fibre type[b]	Reference
Production of reactive oxygen species (chemiluminescence), human monocytes	+	35×10^5 fibres[c]	MG1 micro glass fibres (JFMRA)	Ohyama et al. (2000)
Production of reactive oxygen species (chemiluminescence), human polymorphonuclear leukocytes	+	500 µg/mL	Glass wool 2; glass wool 3	Ruotsalainen et al. (1999)

JFMRA, Japan Fibrous Material Research Association; 8-OH-dG, 8-hydroxydeoxyguanosine; dG, deoxyguanosine
[a] Lowest effective dose or highest ineffective dose
[b] According to the authors
[c] Dose estimated from paper
[d] GW1 fibres did not significantly reduce glutathione content.

breakage, respectively. The results obtained with glass wool show formation of 8-hydroxydeoxyguanosine. The data obtained with plasmids are more difficult to interpret since both positive and negative results have been reported (Table 75).

Hesterberg and Barrett (1984) studied the ability of code 100 (thin) and code 110 (thick) glass fibres to induce cell transformation in Syrian hamster embryo cells. When compared on a per weight basis, thick glass fibres (average diameter, 0.8 ± 0.06 μm) were 20 times less potent than thin fibres (average diameter, 0.13 ± 0.005 μm) in inducing cell transformation. Fibre length was also found to be critical, since milling glass fibres to lengths < 1 μm resulted in a more than 10-fold decrease in the transforming potency. Examination by microscopy indicated that fibres entered the cells and accumulated in the perinuclear regions (Barrett *et al.*, 1989). These glass fibres altered chromosomal segregation by blocking cytokinesis (Jensen & Watson, 1999) and gave rise to binucleation or polynucleation in Syrian hamster embryo cells, human mesothelial cells, monkey epithelial cells, Chinese hamster ovary cells and V79 Chinese hamster lung fibroblasts (Oshimura *et al.*, 1984; Hart *et al.*, 1994; Ong *et al.*, 1997; Jensen & Watson, 1999). Binucleation was reported to be directly dependent on the length of the glass fibres and inversely dependent on fibre diameter (Hart *et al.*, 1994). Gao *et al.* (1995) demonstrated that transformation of BALB/c-3T3 cells with code 100 or AAA-10 glass fibres resulted in gene transformation, since DNA isolated from fibre-transformed cells showed transforming potency upon transfection into naive cells. It was also demonstrated that thin glass fibres, but not thick fibres, induced cell transformation.

Gao *et al.* (1997) reported that BALB/c-3T3 cells morphologically transformed by AAA-10 glass microfibres overexpressed c-*jun*. Similarly, BALB/c-3T3 cells transformed by AAA-10 glass microfibres showed amplification of the genes H-*ras*, K-*ras*, c-*myc* and c-*fos* in 100, 56, 56 and 67% of the cells examined, respectively (Whong *et al.*, 1999). Mutations in the *p53* gene were also noted. These results suggest that activation of proto-oncogenes and inactivation of the *p53* suppressor gene may play a mechanistic role in cell transformation induced by glass fibres.

4.5.2 Rock (stone) wool and slag wool

Few data are available on the genotoxic effects of rock (stone) wool and slag wool (Table 76). Increased numbers of revertants were found in *Salmonella typhimurium* TA100 exposed to MMVF21 and the number of revertants was greater in the glutathione-deficient strains TA100/NG-54 and TA100/NG-57. DNA adducts were detected in strain TA104.

Both rock (stone) wool (MMVF21) and slag wool (MMVF22) produced chromosomal abnormalities in a Chinese hamster ovary cell line. Induction of DNA damage (breakage, cross-links) was observed in a human lung cancer cell line, A549 (one dose tested), and chromosomal aberrations were found in human embryonic lung cells.

Increased production of reactive oxygen species has been noted in most of the studies carried out with slag wool and rock (stone) wool and rodent alveolar macro-

Table 75. Other effects of glass wool in vitro

Test system	Result	Dose[a]	Fibre type[b]	Reference
Guanine hydroxylation in DNA (potentiation by hydrogen peroxide)	+	10 mg/mL (single dose)	Glass wool	Leanderson & Tagesson (1989)
Guanine hydroxylation in DNA (negative without hydrogen peroxide; potentiation by $FeSO_4$, EDTA and mannitol)	+	2.5 mg/mL	Glass wool	Adachi et al. (1992)
Guanine hydroxylation, calf thymus DNA and in dG solution (potentiation by $FeCl_2$ and hydrogen peroxide)	+	10 mg/mL (single dose)	Glass wool	Leanderson et al. (1989)
Plasmid φX174, scission of supercoiled DNA	–	46.25×10^6 fibres/mL (single dose)	Code 100/475	Brown et al. (1998b)
Plasmid φX174, scission of supercoiled DNA	–	30.8×10^6 WHO fibres/mL (single dose)	MMVF10; MMVF11	Donaldson et al. (1995c)
Plasmid φX174, scission of supercoiled DNA	–	61.7×10^6 WHO fibres/mL	MMVF10; MMVF11	Gilmour et al. (1995)
Plasmid φX174, scission of supercoiled DNA	+	46.5×10^6 fibres/mL	MMVF10	Gilmour et al. (1997)
Plasmid φX174, scission of supercoiled DNA	–	46.25×10^6 fibres/mL (single dose)	MMVF10	Brown et al. (1998b)
Hydroxylation of deoxyguanosine	+	NR	Not specified	Leanderson et al. (1988)
Hydroxylation of guanine in DNA in vitro	+	20 mg (single dose)	Not specified	Leanderson et al. (1988)
Hydroxyl radical release in salicylate solution in the presence of hydrogen peroxide	+[c]	1 mg/mL (single dose)	GW Owens Corning; Code 100	Maples & Johnson (1992)
Hydroxyl radical release in salicylate solution	–	8.24×10^7 fibres/mL	Code 100/475	Brown et al. (1998b)
Glutathione depletion from rat lung lining fluid	+	8.24×10^6 fibres/mL	MMVF10	Brown et al. (2000b)

Table 75 (contd)

Test system	Result	Dose[a]	Fibre type[b]	Reference
Glutathione depletion from rat lung lining fluid	+	4.12×10^7 fibres/mL	Code 100/475	Brown et al. (2000b)
Glutathione depletion from pure solution	+	4.12×10^7 fibres/mL	Code 100/475; MMVF10	Brown et al. (2000b)
Ascorbate depletion from rat lung lining fluid	+	4.12×10^7 fibres/mL	Code 100/475; MMVF10	Brown et al. (2000b)
Ascorbate depletion from pure solution	+	4.12×10^7 fibres/mL	Code 100/475; MMVF10	Brown et al. (2000b)

NR, not reported
[a] Lowest effective dose or highest ineffective dose
[b] According to the authors
[c] No effect without hydrogen peroxide

Table 76. Genetic and related effects of slag wool and rock (stone) wool

Test system	Result	Dose[a]	Fibre type[b]	Reference
Salmonella typhimurium TA100, reverse mutation, without exogenous metabolic system	(+)[c]	0.1–0.25 mg/plate	MMVF21	Howden & Faux (1996a)
Formation of fluorescent DNA adducts in *Salmonella typhimurium* TA104	+	40 µg/mL	MMVF21	Howden & Faux (1996b)
Nuclear abnormalities (micronucleus and ploidy), Chinese hamster ovary cells	+	NR	MMVF21; MMVF22	Hart et al. (1994)
Induction of DNA breakage, inhibition of DNA repair and DNA interstrand cross-linking, human lung epithelial (A549) cells	+	40 µg/cm^2 (single dose)	RW1 rock (stone) wool (slag wool, RW1; JFMRA)	Wang et al. (1999b)
Chromosomal aberrations, human embryo lung cells	+	1.0 µg/cm^2	RW1 rock (stone) wool (slag wool, RW1; JFMRA)	Wang et al. (1999b)

NR, not reported
[a] Lowest effective dose or highest ineffective dose
[b] According to the authors
[c] Higher number of revertants in glutathione-deficient strains

phages or human polymorphonuclear leukocytes (Table 77). Moreover, lipid peroxidation was detected in *S. typhimurium* TA104 and rat lung fibroblasts exposed to MMVF21. Glass wool, slag wool and rock (stone) wool all induced guanosine hydroxylation and a low level of DNA breakage (Table 78). Inconclusive results were obtained in DNA plasmid scission tests.

4.5.3 *Refractory ceramic fibres*

Genetic effects of refractory ceramic fibres have been reported (see Table 79 for details and references). Fluorescent malondialdehyde–DNA adducts were observed in *S. typhimurium* TA104 exposed to RCF-1. Aneuploidy was observed in *Drosophila melanogaster* fed with different samples of refractory ceramic fibres. However, no dose–response relationships were reported in these assays. In studies in non-human mammalian cells, nuclear abnormalities were observed after exposure of Chinese hamster ovary cells to RCF1, RCF2, RCF3 and RCF4. As with glass wool, the effect on nuclear abnormalities (micronuclei and polynuclei) was directly related to fibre length. In this assay, the average length of the refractory ceramic fibres ranged between 9.2 μm and 24.3 μm and the average diameter from 1.0 μm–1.4 μm.

Negative results were obtained in studies in which other non-human mammalian cells were exposed to refractory ceramic fibres. No abnormalities of anaphase or telophase were observed in rat pleural mesothelial cells treated with different samples of refractory ceramic fibres. As mentioned for glass wool, this absence of genotoxic effects may be related to the small number of Stanton fibres (length > 8 μm; diameter ≤ 0.25 μm) present in these samples. No deoxyguanosine hydroxylation was detected in a reticulum-cell sarcoma cell line (J774) exposed to a sample of refractory ceramic fibres and no mutagenicity was detected at the *Hprt* or *S1* locus in human–hamster hybrid A_L cells following exposure to RCF1. In this experiment, the total numbers of fibres at the highest ineffective dose (20 or 40 μg/cm^2) were 2×10^5 fibres and 4×10^5 fibres [according to the data provided in the paper]. As the percentage of Stanton fibres is probably small, the cells in this study may have been exposed to an insufficient number of effective fibres.

The refractory ceramic fibres RCF1, RCF2 and RCF3 caused DNA damage (breakage and cross-links) in a human lung tumour epithelial cell line A549 (one dose investigated). Micronucleus formation and structural and numerical chromosomal aberrations were detected in human amniotic fluid cells and structural chromosomal aberrations were noted in human embryonic lung cells.

All studies conducted on rodent alveolar macrophages and human polymorphonuclear leukocytes exposed to refractory ceramic fibres demonstrated the production of reactive oxygen species (Table 80). Results of the hydroxyl radical generation assay were reported in two studies in which only one dose was tested (Table 81). The results obtained with the DNA plasmid scission assay are inconclusive since both negative and weakly positive effects have been reported.

Table 77. Other effects of slag wool and rock (stone) wool in cultured cells

Test system	Result	Dose[a]	Fibre type[b]	Reference
Lipid peroxidation, formation of malondialdehyde-DNA adducts, *Salmonella typhimurium* TA104	+	40 μg/mL	MMVF21	Howden & Faux (1996b)
Production of reactive oxygen species, superoxide anion production, hydrogen peroxide production in guinea-pig alveolar macrophages	+	200 μg/mL (single dose)	RW1 rock (stone) wool (slag wool, JFMRA)	Wang et al. (1999a)
Lipid peroxidation, rat lung RFL-6 fibroblasts	+	2 μg/cm²	MMVF21	Howden & Faux (1996b)
Superoxide anion production, rat alveolar macrophages (fibres either coated or not coated with rat IgG)	+[c]	3 million fibres (single dose)	MMVF21	Donaldson et al. (1995b)
Superoxide anion production, rat alveolar macrophages (fibres either coated or not coated with rat IgG)	+[c]	12.5 μg	MMVF21	Hill et al. (1996)
Haemolysis, sheep erythrocytes	+	2.5 mg/mL	MMVF21; MMVF22	Luoto et al. (1997)
Viability of rat alveolar macrophages in suspension (LDH release)	+	1 mg/mL (single dose)	MMVF21; MMVF22	Luoto et al. (1997)
Production of reactive oxygen species, human polymorphonuclear leukocytes	+	100 μg/ml	MMVF21; MMVF22	Luoto et al. (1997)
Hydrogen peroxide production by human polymorphonuclear leukocytes	+	200 μg/mL (single dose)	Rockwool I; rockwool II	Leanderson & Tagesson (1992)
8-OH-dG formation (from dG added in the mixture) from polymorphonuclear leukocytes	+	500 μg/mL	Rockwool I; rockwool II	Leanderson & Tagesson (1992)
Production of reactive oxygen species (chemiluminescence) by human monocyte-derived macrophages	+	5×10^5 fibres	RW1 rock (stone) wool (JFMRA)	Ohyama et al. (2000)

Table 77 (contd)

Test system	Result	Dose[a]	Fibre type[b]	Reference
Production of reactive oxygen species (chemiluminescence) by human polymorphonuclear leukocytes	+	500 μg/mL	Rockwool 4; Rockwool 5; Rockwool 6	Ruotsalainen et al. (1999)
Manganese superoxide dismutase (MnSOD) mRNA induction, MnSOD enzyme activity induction, SV40-transformed human bronchial epithelial (BEAS 2B) cells	+	2 μg/cm^2 [d]	Rockwool 115-4	Marks-Konczalik et al. (1998)

LDH, lactate dehydrogenase; 8-OH-dG, 8-hydroxydeoxyguanosine; dG, deoxyguanosine

[a] Lowest effective dose or highest ineffective dose
[b] According to the authors
[c] Coated fibres were much more active than uncoated fibres.
[d] A decrease is observed at fibre concentrations ≥ 25 μg/cm^2.

Table 78. Other effects of slag wool and rock (stone) wool *in vitro*

Test system	Result	Dose[a]	Fibre type[b]	Reference
Guanine hydroxylation	+	NR	Rock (stone) wool	Leanderson et al. (1988)
Guanine hydroxylation in calf thymus DNA	+	10 mg/mL (single dose)	Rock (stone) wool	Leanderson et al. (1988)
Guanine hydroxylation (potentiation by hydrogen peroxide and $FeCl_2$), calf thymus DNA and in dG solution	+	10 mg/mL (single dose)	Rock (stone) wool	Leanderson et al. (1989)
Guanine hydroxylation in calf thymus DNA and in dG solution	+	10 mg/mL (single dose)	Slag wool	Leanderson et al. (1989)
Plasmid φX174, scission of supercoiled DNA	–	30.8×10^6 WHO fibres/mL (single dose)	MMVF21; MMVF22	Donaldson et al. (1995c)
Plasmid φX174, scission of supercoiled DNA	–	61.7×10^6 WHO fibres/mL (single dose)	MMVF21; MMVF22	Gilmour et al. (1995)
Plasmid φX174, scission of supercoiled DNA	(+)	NR	MMVF21; MMVF22	Donaldson et al. (1996)
Formation of 8-OH-dG, calf thymus DNA	+	0.5 mg/mL	MMVF21	Howden & Faux (1996a)
Oxidative potential (ethylene formation from α-keto-γ-methiol butyric acid)	+	4 mg/mL (single dose)	MMVF13	Hippeli et al. (1997)

dG, deoxyguanosine; 8-OH-dG, 8-hydroxydeoxyguanosine; NR, not reported
[a] Lowest effective dose or highest ineffective dose
[b] According to the authors

Table 79. Genetic and related effects of refractory ceramic fibres

Test system	Result	Dose[a]	Fibre type[b]	Reference
Formation of malondialdehyde-DNA adducts (lipid peroxidation), Salmonella typhimurium TA104	+	40 µg/mL (single dose)	RCF1	Howden & Faux (1996b)
Aneuploidy, Drosophila melanogaster, adult females	+	25 mg/mL in feed (single dose)	RCF1 (MTC); RCF2 (MTC); RCF3 (MTC); RCF4 (MTC)	Osgood (1994)
Aneuploidy, Drosophila melanogaster, larvae	+	250 mg/bottle (single dose)	RCF1 (MTC); RCF3 (MTC)	Osgood (1994)
Aneuploidy, Drosophila melanogaster, larvae	–	250 mg/bottle (single dose)	RCF2 (MTC); RCF4 (MTC)	Osgood (1994)
8-OH-dG formation, mouse reticulum sarcoma (J774) cell line	–	27 µg/cm^2	Refractory ceramic fibres	Murata-Kamiya et al. (1997)
Formation of malondialdehyde-DNA adducts (lipid peroxidation), rat lung (RFL-6) fibroblasts	–	5 µg/cm^2	RCF1	Howden & Faux (1996b)
Gene mutation, human–hamster hybrid A$_L$ cells, HPRT locus	–	40 µg/cm^2	RCF1 (TIMA)	Okayasu et al. (1999)
Gene mutation, human–hamster hybrid A$_L$ cells, S1 locus	–	20 µg/cm^2	RCF1 (TIMA)	Okayasu et al. (1999)
Micronucleus formation and apoptosis, Syrian hamster embryo cells	+	10 µg/cm^2	Refractory ceramic fibres	Dopp et al. (1995)
Nuclear abnormalities (micronucleus and polynucleus formation), Chinese hamster ovary (K1) cells	+	5 µg/cm^2	RCF1; RCF2; RCF3; RCF4	Hart et al. (1992, 1994)
Anaphase/telophase abnormalities in rat pleural mesothelial cells	–	> 2.5 × 10^5 Stanton fibres/cm^2 (threshold dose)[c]	RCF1; RCF3; RCF4 (TIMA)	Yegles et al. (1995)

Table 79 (contd)

Test system	Result	Dose[a]	Fibre type[b]	Reference
Micronucleus formation, human amniotic fluid cells or Syrian hamster embryo cells	+	0.5 µg/cm^2	Refractory ceramic fibres	Dopp et al. (1997); Dopp & Schiffmann (1998)
Chromosome breakage and hyperdiploidy, human amniotic fluid cells	+	5.0 µg/cm^2 (single dose)	Refractory ceramic fibres	Dopp et al. (1997)
Induction of DNA breakage, DNA repair and DNA interstrand cross-linking, human lung epithelial tumour (A549) cell line	+	40 µg/cm^2 (single dose)	RF1, RF2, RF3 (JFMRA)	Wang et al. (1999b)
Chromosomal aberrations, human embryo lung cells	+	1.0 µg/cm^2	RF1, RF2, RF3 (JFMRA)	Wang et al. (1999b)

8-OH-dG, 8-hydroxydeoxyguanosine

[a] Lowest effective dose or highest ineffective dose
[b] According to the authors; MTC, Mountain Technical Center, Littleton, CO; TIMA, Thermal Insulation Manufacturers' Association
[c] Defined according to Stanton's criteria: length > 8 µm, diameter ≤ 0.25 µm (Stanton et al., 1977)

Table 80. Other effects of refractory ceramic fibres in cultured cells

Test system	Result	Dose[a]	Fibre type[b]	Reference
Superoxide production, rat alveolar macrophages (fibres either uncoated or coated with rat IgG)	+	3×10^6 (single dose)	RCF1	Donaldson et al. (1995b)
Superoxide production, rat alveolar macrophages (fibres either uncoated or coated with rat IgG)	+	125 µg	RCF1	Hill et al. (1996)
Production of reactive oxygen species, human polymorphonuclear leukocytes	+	100 µg/mL	RCF1, RCF2, RCF3	Luoto et al. (1997)
Production of reactive oxygen species in human polymorphonuclear leukocytes	+	200 µg/mL	RCF4	Luoto et al. (1997)
Production of reactive oxygen species, superoxide anion production, guinea-pig alveolar macrophages	+	200 µg/mL (single dose)	RF1, RF2, RF3 (JFMRA)	Wang et al. (1999a)
Reduction of cellular glutathione content, guinea-pig alveolar macrophages	+	200 µg/mL (single dose)	RF2 (JFMRA)	Wang et al. (1999a)
Reduction of cellular glutathione content, guinea-pig alveolar macrophages	–	200 µg/mL (single dose)	RF1, RF3 (JFMRA)	Wang et al. (1999a)
Reduction of cellular glutathione content, rat alveolar macrophages	+	8.24×10^6 fibres/mL (single dose)	RCF1	Gilmour et al. (1997)
Regulation of transcription factors (gel mobility shift assay) for activator protein-1, rat alveolar macrophages	+	8.24×10^6 fibres/mL (single dose)	RCF1	Gilmour et al. (1997)
Regulation of transcription factors (gel mobility shift assay) for NFκB, rat alveolar macrophages	–	8.24×10^6 fibres/mL (single dose)	RCF1	Gilmour et al. (1997)
Hydrogen peroxide production, human polymorphonuclear leukocytes	+	200 µg/mL (single dose)	Refractory ceramic fibre (Kerlane®)	Leanderson & Tagesson (1992)
8-OH-dG formation (from dG added in the mixture), human polymorphonuclear leukocytes	+	500 µg/mL	Refractory ceramic fibre (Kerlane®)	Leanderson & Tagesson (1992)

Table 80 (contd)

Test system	Result	Dose[a]	Fibre type[b]	Reference
Production of reactive oxygen species (chemiluminescence), human polymorphonuclear leukocytes	+	500 µg/mL	Ceramic fibre 1; ceramic fibre 7	Ruotsalainen et al. (1999)
Production of reactive oxygen species, human monocyte-derived macrophages	+	10×10^5 fibres	RF1, RF2 (JFMRA)	Ohyama et al. (2000)
Production of reactive oxygen species, human monocyte-derived macrophages	+	30×10^5 fibres	RF3 (JFMRA)	Ohyama et al. (2000)

8-OH-dG, 8-hydroxydeoxyguanosine; dG, deoxyguanosine
[a] Lowest effective dose or highest ineffective dose
[b] According to the authors

Table 81. Other effects of refractory ceramic fibres *in vitro*

Test system	Result	Dose[a]	Fibre type[b]	Reference
Plasmid φX174, scission of supercoiled DNA	−	30.8×10^6 WHO fibres/mL (single dose)	RCF1, RCF2, RCF3, RCF4	Donaldson et al. (1995c)
Plasmid φX174, scission of supercoiled DNA	(+)	61.7×10^6 fibres/mL	RCF1, RCF3, RCF4	Gilmour et al. (1995)
Plasmid φX174, scission of supercoiled DNA	−	61.7×10^6 fibres/mL (single dose)	RCF2	Gilmour et al. (1995)
Plasmid φX174, scission of supercoiled DNA	(+)	NR	RCF1, RCF2, RCF3, RCF4	Donaldson et al. (1996)
Plasmid φX174, scission of supercoiled DNA	+	46.5×10^6 fibres/mL (single dose)	RCF1	Gilmour et al. (1997)
Plasmid φX174, scission of supercoiled DNA	−	46.25×10^6 fibres/mL (single dose)	RCF1, RCF4	Brown et al. (1998b)
Hydroxyl radical release in salicylate solution	+	8.24×10^7 fibres/mL (single dose)	RCF1	Brown et al. (1998b)
Hydroxyl radical release in salicylate solution	−	8.24×10^7 fibres/mL (single dose)	RCF4	Brown et al. (1998b)

NR, not reported
[a] Lowest effective dose and highest ineffective dose
[b] According to the authors

5. Summary of Data Reported and Evaluation

5.1 Exposure data

Significant commercial production of man-made vitreous fibres began in the early twentieth century. In 2001, it was estimated that over 9 million tonnes of man-made vitreous fibres (MMVFs) were produced annually in over 100 factories around the world. Most of the man-made vitreous fibre produced is used as thermal or acoustical insulation. Usage for this purpose is divided about equally between glass wool (~ 3 million tonnes, used predominantly in North America) and rock (stone) and slag wool (~ 3 million tonnes, used predominantly in Europe and the rest of the world). In recent years, high-alumina, low-silica wools (~ 1 million tonnes) have been increasingly replacing rock (stone) and slag wools in this application. Special-purpose glass fibres are limited-production, small-diameter fibre products that are typically used for purposes other than insulation as in filtration media and batteries. Continuous glass filaments (~ 2 million tonnes) are generally used in the reinforcement of plastics and in textiles. Refractory ceramic fibres, first produced commercially in the 1950s, are widely used (~ 150 thousand tonnes) in high-temperature applications such as furnace insulation. The more recently developed alkaline earth silicate wools (~ 10 thousand tonnes) are replacing refractory ceramic fibres in some applications.

Man-made vitreous fibre products can release airborne respirable fibres during their production, use and removal. In general, as the nominal diameter of man-made vitreous fibre products decreases, both the concentration of respirable fibres and the ratio of respirable to total fibres increase. Although exposure to man-made vitreous fibres during their production, processing and use is thought to have been higher in the past, current average exposure levels are generally less than 0.5 respirable fibre/cm^3 (500 000 respirable fibres/m^3) as an 8-h time-weighted average. Higher levels have been measured in production of special-purpose glass fibres and refractory ceramic fibres, installation of loose-fill insulation without binder, and removal of insulation products.

The concentrations of man-made vitreous fibres measured in outdoor and indoor air in non-occupational settings have been found to be much lower than in occupational settings related to their production, use or removal.

5.2 Human carcinogenicity data

Two large cohort studies and case–control studies nested within these cohorts from the USA and Europe provide most of the epidemiological evidence concerning potential risk for respiratory and other cancers associated with occupational exposure to glass wool, continuous glass filament and rock (stone)/slag wool during manufacture. The United States cohort study included 16 plants, extended the follow-up to 1992 and expanded a previous cohort to include women and non-white workers. This study included information on smoking habits and a new assessment of historical workplace exposure to respirable fibres and several sources of co-exposure including asbestos, formaldehyde and silica. The European cohort extended the follow-up to 1990 in 13 plants.

Glass wool

The findings of the United States cohort study provided no evidence of excess mortality from all causes combined or from all cancers combined, using local rates. A statistically significant 6% excess in respiratory cancer (primarily trachea, bronchus and lung) mortality was observed. When analysis was restricted to long-term workers, the excess was reduced and was no longer statistically significant. Adjustment for smoking based on a random sample of workers suggests that smoking may account for the excesses in respiratory cancer observed in the male glass fibre cohort (glass wool and continuous glass filament combined). The standardized mortality ratios for respiratory cancer were related neither to duration of employment among the total cohort or among long-term workers nor to duration of exposure, cumulative exposure or average intensity of exposure to respirable glass fibre (glass wool and continuous glass filament combined). Analysis by product group showed a statistically significant excess of respiratory cancer for all workers from plants grouped as 'mostly glass wool', but this excess risk for the 'mostly-glass-wool' product group was reduced and no longer statistically significant when the cohort was limited to long-term workers ($\geq$ 5 years of employment). There was no evidence of an excess of mesothelioma or non-respiratory cancers.

The case–control study of respiratory cancer nested within the United States cohort enabled control of plant co-exposure and a more detailed control for confounding by smoking. Duration of exposure, cumulative exposure, average intensity of exposure and the time since first exposure to respirable glass fibre were not associated with an increased risk for respiratory cancer. These results were not altered by using different characterizations of categorized respirable fibre exposure or by alternative models for continuous exposure data.

The European cohort study of glass wool workers demonstrated an increased mortality from lung cancer (trachea, bronchus and lung) but no trend with time since first hire or duration of employment. One death from mesothelioma was observed in this cohort. This study did not estimate fibre exposure, but used surrogate measures such as 'technological phase at first employment'. No information was available either on co-exposure or on smoking habits.

Continuous glass filament

Two of the plants of the United States cohort study manufactured only continuous glass filament. For all workers and for long-term workers from these two plants, no evidence of excess mortality from respiratory cancer was found when compared with local rates. Adjustment for smoking had little effect on the standardized mortality ratio for respiratory cancer. A nested case–control study that included adjustments for smoking and co-exposure also provided no consistent evidence of excess mortality from respiratory cancer. The exposure–response analyses that combined exposure to continuous glass filament and to glass wool are reported in the section on glass wool.

The European cohort study reported few data to evaluate cancer risks among workers exposed to continuous glass filament. This study provided no convincing evidence of an elevated risk for lung cancer.

Results were also available from two smaller cohort studies in the USA and Canada. The United States cohort study on one continuous glass filament plant, which included a nested case–control study, with information on smoking and co-exposure, provided no consistent evidence of an excess risk for lung cancer. The Canadian cohort study of one continuous glass filament plant did not include an assessment of smoking or co-exposure. This study also provided no consistent evidence of an excess risk for lung cancer.

Rock (stone) and slag wool

The present evaluation relies mainly on cohort and nested case–control studies, in which exposure to rock (stone) wool and exposure to slag wool were not considered separately.

The extended follow-up of the rock (stone)/slag wool cohort from the USA indicated an overall elevated risk of respiratory cancer when either national or local comparison rates were used. However, no association was found with duration of exposure or with time since first exposure. Standardized mortality ratios were no longer elevated when indirect adjustment for smoking was made. The nested case–control study showed no association between respiratory cancer and estimated cumulative exposure to respirable fibres, with or without adjustment for possible confounding by smoking and other sources of occupational exposure. Another nested case–control study partially overlapping with the study in the USA showed no increased risk for respiratory cancer in association with exposure to slag wool.

The extended follow-up of the European cohort study indicated an overall elevated risk for lung cancer when national comparison rates were used. This study showed an increasing risk with years since first exposure. The highest standardized mortality ratio was found among workers with the longest time since first employment and among those first employed in the 'early technological phase', i.e. before the introduction of oil and binders and use of the batch-processing method. However, in a case–control study that included detailed information on exposure to fibres, individual smoking habits and potential occupational confounders, no increased risk of lung cancer with increasing fibre exposure was reported.

The results from these studies provide no evidence of an increased risk for pleural mesotheliomas or any other tumours.

Refractory ceramic fibres

Preliminary results from a United States epidemiological mortality study of refractory ceramic fibre workers were available. However, the limited epidemiological data do not permit an adequate evaluation of the cancer risk associated with exposure to refractory ceramic fibres.

Man-made vitreous fibres (not otherwise specified)

A number of studies did not separate exposure to glass wool from exposure to rock (stone) and slag wool or other fibre types, or had limited ability to distinguish between these different fibre types. Since much more information was available from epidemiological studies in the fibre production industries, no separate evaluation is made for the studies of mixed exposure. The results of these studies were, however, taken into consideration for the evaluation of the distinct fibre types.

A cohort study of Swedish wooden house industry workers exposed to man-made vitreous fibres demonstrated a decreased risk for lung cancer and no positive trend in standardized mortality ratios for lung cancer with duration of employment. An increased risk for stomach cancer was found, but the risk did not increase with duration of employment.

Two population-based case–control studies in Germany were combined in a pooled analysis that suggested an association between lung cancer and occupational exposure to man-made vitreous fibres. Odds ratios were adjusted for smoking and exposure to asbestos, but exposure to man-made vitreous fibres and asbestos may not have been separated well enough to rule out residual confounding as an explanation of the results. A low response rate in one of the reference groups adds to the uncertainty of the validity of this study.

A population-based case–control study from Canada found no association between lung cancer and occupational exposure to glass wool or rock (stone) and slag wool.

A German case–control study suggested an association between mesothelioma and exposure to man-made vitreous fibres adjusted for asbestos exposure. However, several limitations constrain the interpretation of the reported results, particularly the potential for misclassification of exposure to asbestos and man-made vitreous fibres and the small number of cases and controls classified as ever having been exposed to man-made vitreous fibres without exposure to asbestos.

An increased risk for laryngeal and hypopharyngeal cancer in association with exposure to man-made vitreous fibres was reported in a case–control study from France, but this was an isolated finding not observed in other studies.

Conclusion

Results from the most recent cohort and nested case–control studies of United States workers exposed to glass wool and continuous glass filament and of European workers exposed to rock (stone) and slag wool have not provided consistent evidence of an association between exposure to fibres and risk for lung cancer or mesothelioma.

These studies, like all epidemiological investigations, have limitations that must be borne in mind when interpreting their results. Although the exposure assessment methods used in these studies are far better than in most epidemiological studies, there is still the potential for exposure misclassification. Notably these studies were not able to examine fully the risks to workers exposed to more durable fibres. Information on smoking and on the other potential confounders that were adjusted for in these studies are also subject to measurement error, which may have influenced the validity of the adjustments made for these factors. Underascertainment and misclassification of mesothelioma may also be a concern in these studies, which primarily relied upon death certificate information. Finally, although these studies are very large by epidemiological standards, their sensitivity may be limited by the fact that fibre exposure levels were low for a large proportion of the study population.

Of some concern are risks for workers in industries that use or remove these products (e.g. construction), who may have experienced higher, but perhaps more intermittent, exposure to man-made vitreous fibres. The data available to evaluate cancer risks from exposure to man-made vitreous fibres in these populations are very limited.

Results on mortality among refractory ceramic fibre workers have also been published since the previous *IARC Monographs* evaluation (1988). However, the epidemiological evidence for refractory ceramic fibres is still extremely limited. Radiographic evidence indicating pleural plaques has been reported for refractory ceramic fibre workers. Although the prognostic significance of pleural plaques is unclear, such plaques are also a common finding among asbestos-exposed workers.

5.3 Animal carcinogenicity data

Continuous glass filament

In experiments in which three types of continuous glass filament of relatively large diameter (> 3 μm) were administered intraperitoneally to rats, no significant increase in tumour response was observed.

Insulation glass wool

Insulation glass wools were tested in well-designed, long-term inhalation studies in rats and hamsters. No significant increase in lung tumours and no mesotheliomas were observed in rats and no lung tumours or mesotheliomas were observed in hamsters exposed to insulation glass wool. Two different asbestos types used as positive controls produced increases in lung tumours and mesotheliomas.

Two insulation glass wools that produced no increase in tumours when administered by inhalation did induce mesotheliomas when injected at high doses (approximately 10^9 fibres) into the peritoneal cavity of rats.

Special-purpose glass fibres

A number of chronic inhalation studies of special-purpose glass fibres have been conducted in rats, hamsters and guinea-pigs. Early inhalation studies demonstrated no significant increases in lung tumours or mesotheliomas. In some of these studies, asbestos did not induce tumours in the controls, which was probably related to use of short fibres in the aerosols. More recent studies of special-purpose glass fibres, using improved methods of fibre preparation and delivery, resulted in significant increases in lung tumours and mesotheliomas in rats (E-glass fibre) and in a single mesothelioma in hamsters ('475' fibre).

Many intraperitoneal studies of special-purpose glass fibres have been conducted, most of which have examined the tumorigenic potential of two compositions of special-purpose glass fibres ('475' and E-glass fibres) after injection or surgical implantation of fibres at high doses (approximately 10^9 fibres) into the peritoneal cavity of rats. All of these studies reported an increase in peritoneal tumours.

Special-purpose glass fibres were tested by intratracheal instillation in two experiments in rats and two in hamsters. A significant increase in lung tumours was observed in one of the rat studies and increases in lung tumours and mesotheliomas were observed in one of the hamster studies. The other two studies showed no increase in either tumour type.

Rock (stone) wool

In a well-designed, long-term inhalation study in which rats were exposed to rock (stone) wool, no significant increase in lung tumour incidence and no mesotheliomas were observed. Crocidolite asbestos was used as the positive control and led to high lung tumour incidence and one mesothelioma.

After intratracheal instillation of rock (stone) wool in two studies, no significant increase in the incidence of lung tumours or mesotheliomas was found. Tremolite asbestos was used as a positive control and induced lung tumours.

In several studies of intraperitoneal injection of high doses (approximately 10^9 fibres), rock (stone) wool induced a significant increase in mesothelioma incidence.

The more biopersistent rock (stone) wool fibres produced a higher incidence of tumours than fibres with lower biopersistence.

Slag wool

In a well-designed, long-term inhalation study of slag wool in rats, no statistically significant increase in the incidence of lung tumours and no mesotheliomas were observed. Crocidolite asbestos was used as a positive control and led to high lung tumour incidence. In two intraperitoneal studies, a high dose (approximately 10^9 fibres) of slag wool induced a statistically significant increase in the incidence of mesotheliomas.

Refractory ceramic fibres

In a well-designed, long-term inhalation study with refractory ceramic fibres in rats, a statistically significant increase in the incidence of lung tumours and a few mesotheliomas were observed. In a well-designed, long-term inhalation study of refractory ceramic fibres in hamsters, a significant increase in the incidence of mesotheliomas was observed.

After intratracheal instillation, two studies reported no excess in tumour incidence in rats. In three intrapleural studies in rats, no significant increase in tumour incidence was observed. In intraperitoneal studies in rats and hamsters, tumour incidence was related to fibre length and dose.

Newly developed wools

Two newly developed, less biopersistent fibres (an alkaline earth silicate (X-607) wool and a high-alumina, low-silica (HT) wool) have been tested in well-designed, long-term inhalation studies in rats and produced no significant increase in the incidence of lung tumours and no mesotheliomas.

In a study in rats of less biopersistent high-alumina, low-silica (HT) wool administered by intraperitoneal injection at a high dose (approximately 10^9 fibres), no abdominal tumours were observed. Four other less biopersistent fibres (A, C, F and G) have been tested by intraperitoneal injection at a high dose (approximately 10^9 fibres) in rats and produced no significant increase in the incidence of abdominal tumours.

One more biopersistent fibre type (H) was tested by intraperitoneal injection at a high dose (approximately 10^9 fibres) in rats and produced abdominal tumours.

5.4 Other relevant data

Deposition and retention

The deposition of inhaled fibres in the respiratory tract is mainly governed by their aerodynamic behaviour, including deposition by impaction, sedimentation and interception. In addition, deposition by diffusional displacement is induced by Brownian motion. Model calculations show that the respirability of fibres, i.e. their penetration to the alveolar region, differs between rodents and humans. A larger fraction of inhaled long fibres is deposited in the alveolar region of humans than in that of rats.

Chemical composition, fibre size and the deposited dose of fibres in the lung are determinants of their retention kinetics. The main mechanisms of mechanical fibre clearance include mucociliary movement in the nasopharyngeal and tracheobronchial regions and alveolar macrophage phagocytosis in the alveolar region with subsequent removal towards the mucociliary escalator. Macrophage-mediated clearance becomes negligible for long fibres, i.e. fibres with lengths approaching 20 μm and longer, which cannot be completely phagocytosed by alveolar macrophages. Alveolar macrophage-mediated clearance is significantly slower in humans than in rats, with retention half-times of several hundred days in humans and about 70 days in rats. In addition to these mechanisms, chemical dissolution and leaching, as well as breakage, can occur. These processes are important and lead to more rapid elimination of fibres deposited in the respiratory tract, thereby lowering the potential for inducing long-term adverse effects. Because the retention half-time due to mechanical clearance is much longer in humans than in rats, higher fibre solubility reduces persistence more in the human lung than in the rat lung.

Few data are available on retention of man-made vitreous fibres in human lungs. In the one available study, the lung burden of man-made vitreous fibres did not differ between workers in glass, rock (stone) and slag wool production compared between themselves or with controls. The interpretation of this difference is limited by the long delay between the end of exposure and sampling. In other studies, refractory ceramic fibres, some with morphological or chemical alterations, have been recovered from the lungs of both production workers and end-users.

Fibre biopersistence

The biopersistence of fibres deposited in the respiratory tract results from a combination of physiological clearance processes (mechanical translocation/removal) and physico-chemical processes (chemical dissolution and leaching, mechanical breaking). Long and short fibres differ in the way in which their elimination from the respiratory tract is affected by each of these mechanisms. Short fibres are taken up by macrophages and subjected to chemical dissolution/leaching within an acidic milieu while at the same time they are actively removed by these phagocytic cells. In contrast, long fibres which can be incompletely phagocytosed by several macrophages are not efficiently removed by physical translocation but may be subjected to chemical dissolution/leaching at variable pH. Since long fibres are most potent with respect to carcinogenicity, the focus of an animal biopersistence assay is on long-fibre retention kinetics in the lung. A number of studies in rats have suggested a correlation between the biopersistence of long fibres (> 20 µm) and their pathogenicity with respect to lung fibrosis and thoracic tumours.

In-vitro dissolution

The physico-chemical mechanisms whereby fibres may degrade in the lung have been examined in a variety of cell-free systems. The basic process by which dissolution of man-made vitreous fibres occurs is via attack of water molecules on the surface of fibres leading to dissolution and subsequent disruption of the fibre structure. The dissolution rate of any fibre is determined primarily by its composition. The most informative studies employ flow-through systems using balanced salt solutions at physiological pHs likely to be encountered in the intrapulmonary environment. The results from such studies have shown correlations with rates of removal of long fibres from the lung in short-term biopersistence assays. While considerable variation occurs between laboratories, the rank order of the durability of tested fibres is generally consistent.

The experimental dissolution rates of tested fibres have been reported to span over five orders of magnitude. Such a range may predict that fibres could persist in lung tissue from a few days to several years.

In-vitro studies of man-made vitreous fibres using cell culture techniques allow estimation of dissolution of fibres in the presence of lung cells. These studies provide information on the joint effects of cells and fluid on different types of man-made vitreous fibre that is helpful in comparing the dissolution rates of a given fibre and then assessing the rank order of the relative dissolution of different man-made vitreous fibres. The results of these studies are consistent with those of studies on the solubility of man-made vitreous fibres in cell-free systems.

Toxic effects in humans

With the exception of a single rock (stone)/slag wool plant in the United States cohort study which had a documented history of asbestos use, none of the mortality studies demonstrated a significant risk for non-malignant respiratory diseases. No mortality data were available on workers exposed to refractory ceramic fibres.

No convincing data for an excess of small parenchymal opacities in chest radiographs compatible with pneumoconiosis have been published. No pleural changes related to any glass fibre type or to rock (stone)/slag wool have been observed. However, an excess of pleural changes, particularly pleural plaques, has consistently been demonstrated in the cohorts of workers in the USA and Europe involved in the production of refractory ceramic fibres.

No indications of a significant excess of respiratory symptoms or of a significant decrease in lung function have been reported for glass fibre workers. The results for rock (stone) wool workers are more conflicting due to a possible interaction between fibre exposure and smoking. In contrast, a small exposure-related effect has been observed in the cohorts of workers involved in the production of refractory ceramic fibres in both the USA and Europe.

A few well-designed studies have supported previous findings of mechanical irritative effects on the skin, eyes and upper respiratory tract associated with coarse fibres.

Low levels of exposure were estimated in most production worker cohorts. With similar low cumulative exposure to asbestos, lung fibrosis would not have been detected in epidemiological studies using standard chest radiography. Limited interpretable data are available from end-users (particularly workers involved in the removal or modification of materials containing man-made vitreous fibres).

Toxic effects in experimental systems

The most important end-points that have been associated with exposure to man-made vitreous fibres include chronic persistent inflammation, fibrosis and cell proliferation in the lungs and mesothelial lining. In general, for a range of man-made vitreous fibres, the data support the contention that long, biopersistent fibres cause prolonged inflammation and fibrosis. Although mechanistically they are not conclusively linked, pulmonary and occasionally pleural fibrosis is found with conditions of exposure to man-made vitreous fibres that are carcinogenic in laboratory animals.

Because biopersistence is believed to be an important factor in the toxicity of man-made vitreous fibres, there are limitations inherent in short-term in-vitro assays of fibre toxicity.

Effect on gene expression

Mutation and/or activation of proto-oncogenes, inhibition of tumour suppressor genes and activation of transcription factors controlling the production of inflammatory cytokines and growth factors have been proposed to play a role in asbestos-induced carcinogenesis. The evidence indicates that glass fibres enter cells and cause genetic modification by physically interfering with chromosomal segregation during mitosis. Glass fibres also generate oxidants and/or mobilize intracellular calcium to activate signalling pathways controlling transcription factor activity. This interaction of glass fibres or refractory ceramic fibres with cells has been reported to induce proto-oncogenes, activate transcription factors, increase tumour necrosis factor α production, induce cell transformation and enhance cell growth. The potency of glass fibres is generally lower than that of asbestos on a per unit mass basis.

Several caveats can be raised about these in-vitro studies: (i) these assays are short-term and do not address issues related to fibre dissolution or biopersistence; and (ii) relatively high levels of man-made vitreous fibres on a mass basis have been studied, and the relevance to in-vivo exposure levels is questionable.

Genetic effects

Genotoxic effects of man-made vitreous fibres have been demonstrated in several cultured cell types, including human cells, and in cell-free assays. Many glass wool samples have been found to produce DNA damage, chromosomal aberrations, nuclear abnormalities and cell transformation. The effects were observed to depend on fibre dimensions, with long fibres being more active than shorter fibres. A few rock (stone) wool and slag wool samples have been investigated. Both DNA damage and chromosomal and nuclear aberrations have been observed, as well as mutations in bacterial test systems. The studies on refractory ceramic fibres have so far mostly been limited to the RCF1, 2, 3 and 4 samples and, to a lesser extent, to Japanese standard reference samples. Findings were similar to those with rock (stone) and slag wool.

The occurrence of mutations and some forms of DNA/chromosomal damage may be related to the production of activated oxygen species which have been detected in cell-free systems and in cells exposed to man-made vitreous fibres. Chromosomal and nuclear abnormalities may also be related to the impairment of cell division by the fibres. While reactive oxygen species are produced by either non-fibrous or fibrous particles, cell cycle-associated chromosomal and nuclear abnormalities appear to be a specific response to exposure to fibres. Despite the fact that in-vitro assessment of genetic effects does not address issues related to fibre dissolution or biopersistence, these assays can determine whether a fibre has the potential to be directly genotoxic.

A major gap in the current database is the absence of any studies that correlate genotoxic end-points with the pathogenic effects of man-made vitreous fibres in the same experimental animal system.

Mechanistic considerations

Man-made vitreous fibres deposit in the lungs where they are phagocytosed by macrophages, either completely or incompletely, depending on fibre length. Incomplete phagocytosis is a potent pro-inflammatory stimulus for the release of a cascade of mediators and reactive oxygen and nitrogen species, leading to genotoxicity and proliferation of lung cells. *In vitro*, the direct entry of fibres into cells followed by, or associated with, cell division can produce chromosomal/nuclear abnormalities and genetic changes which may lead to cell transformation and dysregulated proliferation. Animal studies have shown a range of severity of inflammation and fibrosis which has been related to more biopersistent fibres in the lungs. There is a consistent relationship between persistent inflammation, fibrosis and tumour development in animal models.

Refractory ceramic fibres, unlike other man-made vitreous fibres, have the ability to induce pleural plaques in humans. Although pleural plaques in themselves are probably not directly related to cancer development, either in the pleura or the lung, concern over potential carcinogenic effects in the pleura seems valid for refractory ceramic fibres, in view of the ability of asbestos to induce both plaques and pleural cancer.

5.5   Evaluation[1]

There is *inadequate evidence* in humans for the carcinogenicity of glass wool.

There is *inadequate evidence* in humans for the carcinogenicity of continuous glass filament.

There is *inadequate evidence* in humans for the carcinogenicity of rock (stone) wool/slag wool.

There is *inadequate evidence* in humans for the carcinogenicity of refractory ceramic fibres.

There is *sufficient evidence* in experimental animals for the carcinogenicity of special-purpose glass fibres including E-glass and '475' glass fibres.

There is *sufficient evidence* in experimental animals for the carcinogenicity of refractory ceramic fibres.

There is *limited evidence* in experimental animals for the carcinogenicity of insulation glass wool.

There is *limited evidence* in experimental animals for the carcinogenicity of rock (stone) wool.

There is *limited evidence* in experimental animals for the carcinogenicity of slag wool.

There is *limited evidence* in experimental animals for the carcinogenicity of certain newly developed, more biopersistent fibres including fibre H.

There is *inadequate evidence* in experimental animals for the carcinogenicity of continuous glass filament.

There is *inadequate evidence* in experimental animals for the carcinogenicity of certain newly developed, less biopersistent fibres including the alkaline earth silicate (X-607) wool, the high-alumina, low-silica (HT) wool and fibres A, C, F and G.

---

[1] Observers/representatives from the industry (B.C. Brown, J.G. Hadley, O. Kamstrup, L.D. Maxim and C.E. Rossiter) were not present during the evaluations.

Overall evaluation

Special-purpose glass fibres such as E-glass and '475' glass fibres are *possibly carcinogenic to humans (Group 2B)*.

Refractory ceramic fibres are *possibly carcinogenic to humans (Group 2B)*.

Insulation glass wool, continuous glass filament, rock (stone) wool and slag wool are *not classifiable as to their carcinogenicity to humans (Group 3)*.

The Working Group elected not to make an overall evaluation of the newly developed fibres designed to be less biopersistent such as the alkaline earth silicate or high-alumina, low-silica wools. This decision was made in part because no human data were available, although such fibres that have been tested appear to have low carcinogenic potential in experimental animals, and because the Working Group had difficulty in categorizing these fibres into meaningful groups based on chemical composition.

6. References

ACGIH (American Conference of Governmental Industrial Hygienists) (2001) *2001 TLVs® and BEIs® Threshold Limit Values for Chemical Substances and Physical Agents and Biological Exposure Indices*, Cincinnati, OH

Adachi, S., Takemoto, K. & Kimura, K. (1991) Tumorigenicity of fine man-made fibers after intratracheal administrations to hamsters. *Environ. Res.*, **54**, 52–73

Adachi, S., Kawamura, K., Yoshida, S. & Takemoto, K. (1992) Oxidative damage on DNA induced by asbestos and man-made fibers *in vitro*. *Int. Arch. occup. environ. Health*, **63**, 553–557

Alexander, I.C., Jubb, G.A. & Penn, J.A. (1997) Fibre sizing for classification. *Br. Ceram. Transact.*, **96**, 74–79

APFE (European Glass Fibre Producers Association) (2001) *Continuous Filament Glass Fibre and Human Health*, Brussels

Arbejdstilsynet (Danish Working Environmental Authority) (2001) *At-Vejledning C.O.1. Graensevaerdier for Stoffer og Matierialer* [Limit Values for Substances and Materials], Copenhagen (in Danish)

Asgharian, B. & Anjilvel, S. (1998) A multiple-path model of fiber deposition in the rat lung. *Toxicol. Sci.*, **44**, 80–86

Asgharian, B. & Yu, C.P. (1988) Deposition of inhaled fibrous particles in the human lung. *J. Aerosol Med.*, **1**, 37–50

Asgharian, B. & Yu, C.P. (1989) Deposition of fibers in the rat lung. *J. Aerosol Sci.*, **20**, 355–366

ASTM (American Society for Testing and Materials) (2000) *Standard Specification for Glass Fiber Strands* (ASTM D578-00), Philadephia, PA

Assuncao, J. & Corn, M. (1975) The effects of milling on diameters and lengths of fibrous glass and chrysotile asbestos fibers. *Am. ind. Hyg. Assoc. J.*, **36**, 811–819

Australian National Occupational Health and Safety Commission (1999) *List of Designated Hazardous Substances* (NOHSC: 10005)

Australian National Occupational Health & Safety Commission (2001) *National Standard for Synthetic Mineral Fibres* (http://www.nohsc.gov.au)

Axelson, O. & Steenland, K. (1988) Indirect methods of assessing the effects of tobacco use in occupational studies. *Am. J. ind. Med.*, **13**, 105–118

Baier, R., Meyer, A., Glaves-Rapp, D., Axelson, E., Forsberg, R., Kozak, M. & Nickerson, P. (2000) The body's response to inadvertent implants: Respirable particles in lung tissues. *J. Adhesion*, **74**, 103–124

Bailey, M.R., Fry, F.A. & James, A.C. (1982) The long-term clearance kinetics of insoluble particles from the human lung. *Ann. occup. Hyg.*, **26**, 273–290

Bailey, M.R., Fry, F.A. & James, A.C. (1985) Long-term retention of particles in the human respiratory tract. *J. Aerosol Sci.*, **16**, 295–305

Balzer, J.L. (1976) Environmental data: Airborne concentrations found in various operations. In: *Occupational Exposure to Fibrous Glass, Proceedings of a Symposium, College Park, Maryland, June 26–27, 1974*, Washington, DC, U.S. Department of Health, Education and Welfare, pp. 83–89

Balzer, J.L., Cooper, W.C. & Fowler, D.P. (1971) Fibrous glass-lined air transmission systems: An assessment of their environmental effects. *Am. ind. Hyg. Assoc. J.*, **32**, 512–518

Barchowsky, A., Lannon, B.M., Elmore, L.C. & Treadwell, M.D. (1997) Increased focal adhesion kinase- and urokinase-type plasminogen activator receptor-associated cell signaling in endothelial cells exposed to asbestos. *Environ. Health Perspect.*, **105** (Suppl. 5), 1131–1137

Barrett, J.C., Lamb, P.W. & Wiseman, R.W. (1989) Multiple mechanisms for the carcinogenic effects of asbestos and other mineral fibers. *Environ. Health Perspect.*, **81**, 81–89

Bauer, J.F. (1998a) Interaction of glass fiber with physiological fluids: The role of surface. In: *Proceedings of XVIII International Congress on Glass*, San Francisco, American Ceramic Society, pp. 1–10

Bauer, J.F. (1998b) Effect of fiberization process on glass fiber surfaces. *Ceram. Transact.*, **82**, 187–202

Bauer, J.F., Law, B.D. & Hesterberg, T.W. (1994) Dual pH durability studies of man-made vitreous fiber (MMVF). *Environ. Health Perspect.*, **102** (Suppl. 5), 61–65

Bellmann, B., Muhle, H., Creutzenberg, O., Ernst, H., Brown, R.C. & Sébastien, P. (2001) Effects of nonfibrous particles on ceramic fiber (RCF1) toxicity in rats. *Inhal. Toxicol.*, **13**, 101–125

Bergamaschi, A., Ripanucci, G., Sacco, A. & De Lorenzo, G. (1997) [Indoor pollution from airborne glass fibers (MMMF) in biomedical research laboratories.] *G. ital. med. Lav. Erg.*, **19**, 44–46 (in Italian)

Bernstein, D.M. & Riego-Sintes, J.M. (1999) *Methods for the Determination of the Hazardous Properties for Human Health of Man Made Mineral Fibres (MMMF) (Report No. EUR 18748)*, Brussels, European Commission Joint Research Centre, European Chemicals Bureau

Bernstein, D.M., Drew, R.T., Schidlowsky, G. & Kuschner, M. (1980) Experimental approaches for exposure to sized glass fibers. *Environ. Health Perspect.*, **34**, 47–57

Bernstein, D.M., Drew, R.T., Schidlowsky, G. & Kuschner, M. (1984) Pathogenicity of MMMF and the contrasts with natural fibres. *Biological Effects of Man-made Mineral Fibres*, Vol. 2, Copenhagen, World Health Organization, pp. 169–195

Bernstein, D.M., Mast, R., Anderson, R., Hesterberg, T.W., Musselman, R., Kamstrup, O. & Hadley, J. (1994) An experimental approach to the evaluation of the biopersistence of respirable synthetic fibers and minerals. *Environ. Health Perspect.*, **102** (Suppl. 5), 15–18

Bernstein, D.M., Morscheidt, C., Grimm, H.G., Thévenaz, P. & Teichert, U. (1996) Evaluation of soluble fibers using the inhalation biopersistence model, a nine-fiber comparison. *Inhal. Toxicol.*, **8**, 345–385

Bernstein, D.M., Morscheidt, C., de Meringo, A., Schumm, M., Grimm, H.G., Teichert, U., Thévenaz, P. & Mellon, L. (1997) The biopersistence of fibres following inhalation and intratracheal instillation exposure. *Ann. occup. Hyg.*, **41** (Suppl. 1), 224–230

Bernstein, D.M., Riego Sintes, J.M., Ersboell, B.K. & Kunert, J. (2001a) Biopersistence of synthetic mineral fibers as a predictor of chronic inhalation toxicity in rats. *Inhal. Toxicol.*, **13**, 823–849

Bernstein, D.M., Riego Sintes, J.M., Ersboell, B.K. & Kunert, J. (2001b) Biopersistence of synthetic mineral fibers as a predictor of chronic intraperitoneal injection tumor response in rats. *Inhal. Toxicol.*, **13**, 851–875

Berry, G. (1999) Models for mesothelioma incidence following exposure to fibers in terms of timing and duration of exposure and the biopersistence of the fibers. *Inhal. Toxicol.*, **11**, 111–130

Blake, T., Castranova, V., Schwegler-Berry, D., Baron, P., Deye, G.J., Li, C. & Jones, W. (1998) Effect of fiber length on glass microfiber cytotoxicity. *J. Toxicol. environ. Health*, **A54**, 243–259

Boffetta, P., Saracci, R., Andersen, A., Bertazzi, P.A., Chang-Claude, J., Ferro, G., Fletcher, A.C., Frentzel-Beyme, R., Gardner, M.J., Olsen, J.H., Simonato, L., Teppo, L., Westerholm, P., Winter, P. & Zocchetti, C. (1992) Lung cancer mortality among workers in the European production of man-made mineral fibers — A Poisson regression analysis. *Scand. J. Work Environ. Health*, **18**, 279–286

Boffetta, P., Saracci, R., Andersen, A., Bertazzi, P.A., Chang-Claude, J., Cherrie, J., Ferro, G., Frentzel-Beyme, R., Hansen, J., Olsen, J.H., Plato, N., Teppo, L., Westerholm, P., Winter, P.D. & Zocchetti, C. (1997) Cancer mortality among man-made vitreous fibre production workers. *Epidemiology*, **8**, 259–268

Boffetta, P., Sali, D., Kolstad, H., Coggon, D., Olsen, J., Andersen, A., Spence, A., Pesatori, A.C., Lynge, E., Frentzel-Beyme, R., Chang-Claude, J., Lundberg, I., Biocca, M., Gennaro, V., Teppo, L., Partanen, R., Welp, E., Saracci, R. & Kogevinas, M. (1998) Mortality of short-term workers in two international cohorts. *J. occup. environ. Med.*, **40**, 1120–1126

Boffetta, P., Andersen, A., Hansen, J., Olsen, J.H., Plato, N., Teppo, L., Westerholm, P. & Saracci, R. (1999) Cancer incidence among European man-made vitreous fibre production workers. *Scand. J. Work environ. Health*, **25**, 222–226

Böckler, M., Kempf, E. & Mattes, L. (1995) [MMMF concentrations when removing and mounting insulation material in thermal stations.] *Staub-Reinhalt. Luft*, **55**, 293–298 (in German)

Brass, D.M., Hoyle, G.W., Poovey, H.G., Liu, J.-Y. & Brody, A.R. (1999) Reduced tumor necrosis factor α and transforming growth factor-β_1 expression in the lungs of inbred mice that fail to develop fibroproliferative lesions consequent to asbestos exposure. *Am. J. Pathol.*, **154**, 853–862

Breysse, P.N., Cherrie, J.W., Lees, P.S.J. & Brown, P. (1994) Comparison of NIOSH 7400 'B' rules and WHO reference methods for the evaluation of airborne man-made mineral fibres. *Ann. occup. Hyg.*, **38** (Suppl. 1), 527–531

Breysse, P.N., Lees, P.S.J. & Rooney, B.C. (1999) Comparison of NIOSH method 7400 A and B counting rules for assessing synthetic vitreous fiber exposures. *Am. ind. Hyg. Assoc. J.*, **60**, 526–532

Breysse, P.N., Lees, P.S.J., Rooney, B.C., McArthur, B.R., Miller, M.E. & Robbins, C. (2001) End user exposures to synthetic vitreous fibers. II. Fabrication and installation of commercial products. *Appl. occup. environ. Hyg.*, **16**, 464–470

Brody, A.R. & Roe, M.W. (1983) Deposition pattern of inorganic particles at the alveolar level in the lungs of rats and mice. *Am. Rev. respir. Dis.*, **128**, 724–729

Brown, R.C. (2000) Influence of non-fibrous particles in the animal testing of refractory ceramic fibers. In: Rammlmair, D., Mederer, J., Oberthur, T., Heimann, R.B. & Pentinghaus, H., eds, *Applied Mineralogy in Research, Economy, Technology, Ecology and Culture*, Rotterdam, Balkema, pp. 739–742

Brown, R.C., Sara, E.A., Hoskins, J.A., Evans, C.E., Young, J., Laskowski, J.J., Acheson, R., Forder, S.D. & Rood, A.P. (1992) The effects of heating and devitrification on the structure and biological activity of aluminosilicate refractory ceramic fibers. *Ann. occup. Hyg.*, **36**, 115–129

Brown, N., Peat, J., Mellis, C. & Woolcock, A. (1996) Respiratory health of workers in the Australian glass wool and rock wool manufacturing industry. *J. occup Health Safety Aust. NZ*, **12**, 319–325

Brown, D.M., Roberts, N.K. & Donaldson, K. (1998a) Effect of coating with lung lining fluid on the ability of fibers to produce a respiratory burst in rat alveolar macrophages. *Toxicol. In Vitro*, **12**, 15–24

Brown, D.M., Fisher, C. & Donaldson, K. (1998b) Free radical activity of synthetic vitreous fibers: Iron chelation inhibits hydroxyl radical generation by refractory ceramic fiber. *J. Toxicol. environ. Health*, **A53**, 545–561

Brown, D.M., Beswick, P.H. & Donaldson, K. (1999) Induction of nuclear translocation of NF-κB in epithelial cells by respirable mineral fibres. *J. Pathol.*, **189**, 258–264

Brown, R.C., Sébastien, P., Bellmann, B. & Muhle, H. (2000a) Particle contamination in experimental fiber preparations. *Inhal. Toxicol.*, **12** (Suppl. 3), 99–107

Brown, D.M., Beswick, P.H., Bell, K.S. & Donaldson, K. (2000b) Depletion of glutathione and ascorbate in lung lining fluid by respirable fibres. *Ann. occup. Hyg.*, **44**, 101–108

Brüske-Hohlfeld, I., Möhner, M., Pohlabeln, H., Ahrens, W., Bolm-Audorff, U., Kreienbrock, L., Kreuzer, M., Jahn, I., Wichmann, H.-E. & Jöckel, K.-H. (2000) Occupational lung cancer risk for men in Germany: Results from a pooled case–control study. *Am. J. Epidemiol.*, **151**, 384–395

Buchanich, J.B., Marsh, G.M. & Youk, A.O. (2001) Historical cohort study of US man-made vitreous fiber production workers. V. Tobacco-smoking habits. *J. occup. environ. Med.*, **43**, 793–802

Buchta, T.M., Rice, C.H., Lockey, J.E., Lemasters, G.K. & Gartside, P.S. (1998) A comparative study of the National Institute for Occupational Safety and Health 7400 'A' and 'B' counting rules using refractory ceramic fibers. *Appl. occup. environ. Hyg.*, **13**, 58–61

Bundesgesetzblatt (2000) Ordinance of 25 May 2000 amending the Chemicals Regulations. *German Federal Law Gazette*, Part I, No. 24, 747–749

Burge, P.S., Calvert, I.A., Trethowan, W.N. & Harrington, J.M. (1995) Are the respiratory health effects found in manufacturers of ceramic fibres due to the dust rather than the exposure to fibres? *Occup. environ. Med.*, **52**, 105–109

Burley, C.G., Brown, R.C. & Maxim, L.D. (1997) Refractory ceramic fibres: The measurement and control of exposure. *Ann. occup. Hyg.*, **41**, 267–272

Carter, C.M., Axten, C.W., Byers, C.D., Chase, G.R., Koenig, A.R., Reynolds, J.W. & Rosinski, K.D. (1999) Indoor airborne fiber levels of MMVF in residential and commercial buildings. *Am. ind. Hyg. Assoc. J.*, **60**, 794–800

Carthew, P., Edwards, R.E., Dorman, B.M., Brown, R.C., Young, J., Laskowski, J.J. & Wagner, J.C. (1995) Intrapleural administration of vitreous high duty ceramic fibres and heated devitrified ceramic fibres does not give rise to pleural mesothelioma in rats. *Hum. exp. Toxicol.*, **14**, 657–661

Cavalleri, A., Gobba, F., Ferrari, D., Bacchella, L., Robotti, M., Mineo, F. & Pedroni, C. (1992) [The determination of serum amino-terminal procollagen type-III propeptide PIIINP) in occupational exposure to rockwool fiber.] *Med. Lav.*, **83**, 127–134 (in Italian)

CEN (European Committee for Standardization) (Comité européen de Normalisation) (1993) *Workplace Atmospheres — Size Fraction Definitions for Measurements of Airborne Particles* (European Standard EN 481), Brussels

Chamberlain, M. & Tarmy, E.M. (1977) Asbestos and glass fibers in bacterial mutation tests. *Mutat. Res.*, **43**, 159–164

Cherrie, J. & Schneider, T. (1999) Validation of a new method for structured subjective assessment of past concentrations. *Ann. occup. Hyg.*, **43**, 235–245

Cherrie, J., Dodgson, J., Groat, S. & Maclaren, W. (1986) Environmental surveys in the European man-made mineral fiber production industry. *Scand. J. Work Environ. Health*, **12** (Suppl. 1), 18–25

Cherrie, J., Krantz, S., Schneider, T., Öhberg, I., Kamstrup, O. & Linander, W. (1987) An experimental simulation of an early rock wool/slag wool production process. *Ann. occup. Hyg.*, **31**, 583–593

Cherrie, J.W., Crawford, N.P. & Dodgson, J. (1988) Problems in assessing airborne man-made mineral fibre concentrations in relation to epidemiology. *Ann. occup. Hyg.*, **32** (Suppl. 1), 715–723

Cherrie, J.W., Schneider, T., Spankie, S. & Quinn, M. (1996) A new method for structured, subjective assessments of past concentrations. *Occup. Hyg.*, **3**, 75–83

Chiazze, L., Watkins, D.K. & Fryar, C. (1992) A case–control study of malignant and non-malignant respiratory disease among employees of a fiberglass manufacturing facility. *Br. J. ind. Med.*, **49**, 326–331

Chiazze, L., Watkins, D.K., Fryar, C. & Kozono, J. (1993) A case–control study of malignant and non-malignant respiratory disease among employees of a fibreglass manufacturing facility. II. Exposure assessment. *Br. J. ind. Med.*, **50**, 717–725

Chiazze, L., Watkins, D.K. & Fryar, C. (1997) Historical cohort mortality study of a continuous filament fiberglass manufacturing plant. I. White men. *J. occup. environ. Med.*, **39**, 432–441

Chiazze, L., Watkins, D.K., Fryar, C., Fayerweather, W., Bender, J.R. & Chiazze, M. (1999) Mortality from nephritis and nephrosis in the fibreglass manufacturing industry. *Occup. environ. Med.*, **56**, 164–166

Cholak, J. & Schafer, L.J. (1971) Erosion of fibers from installed fibrous-glass ducts. *Arch. environ. Health*, **22**, 220–229

Christensen, V.R., Lund Jensen, S., Guldberg, M. & Kamstrup, O. (1994) Effect of chemical composition of man-made vitreous fibers on the rate of dissolution *in vitro* at different pHs. *Environ. Health Perspect.*, **102** (Suppl. 5), 83–86

Churg, A. (1996) The uptake of mineral particles by pulmonary epithelial cells. *Am. J. respir. crit. Care Med.*, **154**, 1124–1140

Churg, A. & Green, F.H.Y. (1998) Pathology of Occupational Lung Disease, Vancouver, Lippincott, Williams & Wilkins

Churg, A. & Stevens, B. (1995) Enhanced retention of asbestos fibers in the airways of human smokers. *Am. J. respir. crit. Care Med.*, **151**, 1409–1413

Churg, A., Wright, J.L., Hobson, J. & Stevens, B. (1992) Effects of cigarette smoke on the clearance of short asbestos fibres from the lung and a comparison with the clearance of long asbestos fibres. *Int. J. exp. Pathol.*, **73**, 287–297

Churg, A., Wright, J., Gilks, B. & Dai, J. (2000) Pathogenesis of fibrosis produced by asbestos and man-made mineral fibers: What makes a fiber fibrogenic? *Inhal. Toxicol.*, **12** (Suppl. 3), 15–26

Class, P., Deghilage, P. & Brown, R.C. (2001) Dustiness of different high-temperature insulation wools and refractory ceramic fibres. *Ann. occup. Hyg.*, **45**, 381–384

Clausen, J., Netterstrom, B. & Wolff, C. (1993) Lung function in insulation workers. *Br. J. ind. Med.*, **50**, 252–256

Collier, C.G., Morris, K.J., Launder, K.A., Humphreys, J.A., Morgan, A., Eastes, W. & Townsend, S. (1994) The behavior of glass fibers in the rat following intraperitoneal injection. *Reg. Toxicol. & Pharmacol.*, **20**, S89–S103

Collier, C.G., Morris, K.J., Launder, K.A., Humphreys, J.A., Morgan, A., Eastes, W. & Townsend, S. (1995) The durability and distribution of glass fibres in the rat following intraperitoneal injection. *Ann. occup. Hyg.*, **39**, 699–704

Consonni, D., Boffetta, P., Andersen, A., Chang, C.-J., Cherrie, J.W., Ferro, G., Frentzel-Beyme, R., Hansen, J., Olsen, J., Plato, N., Westerholm, P. & Saracci, R. (1998) Lung cancer mortality among European rock/slag wool workers: Exposure–response analysis. *Cancer Causes Control*, **9**, 411–416

Corn, M. & Sansone, E.B. (1974) Determination of total suspended particulate matter and airborne fiber concentrations at three fibrous glass manufacturing facilities. *Environ. Res.*, **8**, 37–52

Corn, M., Hammad, Y., Whittier, D. & Kotsko, N. (1976) Employee exposure to airborne fiber and total particulate matter in two mineral wool facilities. *Environ. Res.*, **12**, 59–74

Corn, M., Lees, P.S.J. & Breysse, P.N. (1992) *Characterization of End-User Exposures to Industrial (RCF) Insulation Products*, Final report for Thermal Insulation Manufacturers Association, Baltimore, MD, Johns Hopkins University

Cornett, M.J., Rice, C., Hertzberg, V.S. & Lockey, J.E. (1989) Assessment of fiber deposition on the conductive sampling cowl in the refractory ceramic fiber industry. *Appl. ind. Hyg.*, **4**, 201–204

Cowie, H.A., Wild, P., Beck, J., Auburtin, G., Piekarski, C., Massin, N., Cherrie, J.W., Hurley, J.F., Miller, B.G., Groat, S. & Soutar, C.A. (2001) An epidemiological study of the respiratory health of workers in the European refractory ceramic fibre (RCF) industry. *Occup. environ. Med.*, **58**, 800–810

Crapo, J.D., Young, S.L., Fram, E.K., Pinkerton, K.E., Barry, B.E. & Crapo, R.O. (1983) Morphometric characteristics of cells in the alveolar region of mammalian lungs. *Am. Rev. respir. Dis.*, **128**, S42–S46

Creutzenberg, O., Bellmann, B. & Muhle, H. (1997) Biopersistence and bronchoalveolar lavage investigations in rats after a subacute inhalation of various man-made mineral fibres. *Ann. occup. Hyg.*, **41** (Suppl. 1), 213–218

Cullen, R.C., Miller, B.G., Davis, J.M.G., McAllister Brown, D. & Donaldson, K. (1997) Short-term inhalation and *in vitro* tests as predictors of fiber pathogenicity. *Environ. Health Perspect.*, **105** (Suppl. 5), 1235–1240

Cullen, R.T., Searl, A., Buchanan, D., Davis, J.M., Miller, B.G. & Jones, A.D. (2000) Pathogenicity of a special-purpose glass microfibre (E glass) relative to another glass microfibre and amosite asbestos. *Inhal. Toxicol.*, **12**, 959–977

Cuypers, J.M.C., Bleumink, E. & Nater, J.P. (1975) [Dermatological aspect of glass fibre manufacture.] *Berufsdermatosen*, **23**, 143–154 (in German)

Dai, Y.T. & Yu, C.P. (1998) Alveolar deposition of fibers in rodents and humans. *J. Aerosol Med.*, **11**, 247–258

Davis, J.M.G. (1970) Further observations on the ultrastructure and chemistry of the formation of asbestos bodies. *Exper. mol. Pathol.*, **13**, 346–358

Davis, J.M.G. (1976) Pathological aspects of the injection of glass fiber into the pleural and peritoneal cavities of rats and mice. In: LeVee, W.N. & Schulte, P.A., eds, *Occupational Exposure to Fibrous Glass* (DHEW Publ. No. (NIOSH) 76-151; NTIS Publ. No. PB-258869), Cincinnati, OH, National Institute for Occupational Safety and Health, pp. 141–149

Davis, J.M.G., Addison, J., Bolton, R.E., Donaldson, K., Jones, A.D. & Wright, A. (1984) The pathogenic effects of fibrous ceramic aluminium silicate glass administered to rats by inhalation or peritoneal injection. In: *Biological Effects of Man-made Mineral Fibres* (Proceedings of a WHO/IARC Conference), Vol. 2, Copenhagen, World Health Organization, pp. 303–322

Davis, J.M.G., Jones, A.D. & Miller, B.G. (1991) Experimental studies in rats on the effects of asbestos inhalation coupled with the inhalation of titanium dioxide or quartz. *Int. J. exp. Pathol.*, **72**, 501–525

Davis, J.M.G., Brown, D.M., Cullen, R.T., Donaldson, K., Jones, A.D., Miller, B.G., McIntosh, C. & Searl, A. (1996a) A comparison of methods of determining and predicting the pathogenicity of mineral fibres. *Inhal. Toxicol.*, **8**, 747–770

Davis, J.M., Dungworth, D.L. & Boorman, G.A. (1996b) Concordance in diagnosis of mesotheliomas. *Toxicol. Pathol.*, **24**, 662–663

Dement, J.M. (1975) Environmental aspects of fibrous glass production and utilization. *Environ. Res.*, **9**, 295–312

Deutsche Forschungsgemeinschaft (DFG) (2001) *List of MAK and BAT Values. Commission for the Investigation of Health Hazards of Chemical Compounds on the Work area* (Report No. 37), Weinheim, Wiley-VCH Verlag

Ding, M., Dong, Z., Chen, F., Pack, D., Ma, W.-Y., Ye, J., Shi, X., Castranova, V. & Vallyathan, V. (1999) Asbestos induces activator protein-1 transactivation in transgenic mice. *Cancer Res.*, **59**, 1884–1889

Dodgson, J., Cherrie, J. & Groat, S. (1987a) Estimates of past exposure to respirable man-made mineral fibres in the European insulation wool industry. *Ann. occup. Hyg.*, **31**, 567–582

Dodgson, J., Harrison, G.E., Cherrie, J.W. & Sneddon, E. (1987b) *Assessment of Airborne Mineral Wool Fibres in Domestic Houses (Report TM/87/18)*, Edinburgh, UK, Environmental Branch, Institute of Occupational Medicine

Donaldson, K. (1994) Biological activity of respirable industrial fibres treated to mimic residence in the lung. *Toxicol. Lett.*, **72**, 299–305

Donaldson, K. (2000) Nonneoplastic lung responses induced in experimental animals by exposure to poorly soluble nonfibrous particles. *Inhal. Toxicol.*, **12**, 121–139

Donaldson, K., Miller, B.G., Brown, G.M., Slight, J., Addison, J. & Davis, J.M.G. (1993) Inflammation in the mouse peritoneal cavity in the investigation of factors determining the biological activity of respirable industrial fibers. In: Hurych, J., Lesage, M. & David, A., eds, *Proceedings of the 8th International Conference on Occupational Lung Diseases, Prague, Czechoslovakia, September 14–17, 1992*, pp. 531–539

Donaldson, K., Brown, D.M., Miller, B.G. & Brody, A.R. (1995a) Bromo-deoxyuridine (BrdU) uptake in the lungs of rats inhaling amosite asbestos or vitreous fibres at equal airborne fibre concentrations. *Exp. Toxicol. Pathol.*, **47**, 207–211

Donaldson, K., Hill, I.M. & Beswick, P.H. (1995b) Superoxide anion release by alveolar macrophages exposed to respirable industrial fibres: Modifying effect of fibre opsonisation. *Exp. Toxicol. Pathol.*, **47**, 229–231

Donaldson, K., Gilmour, P.S. & Beswick, P.H. (1995c) Supercoiled plasmid DNA as a model target for assessing the generation of free radicals at the surface of fibres. *Exp. Toxicol. Pathol.*, **47**, 235–237

Donaldson, K., Beswick, P.H. & Gilmour, P.S. (1996) Free radical activity associated with the surface of particles: A unifying factor in determining biological activity? *Toxicol. Lett.*, **88**, 293–298

Dopp, E. & Schiffmann, D. (1998) Analysis of chromosomal alterations induced by asbestos and ceramic fibers. *Toxicol. Lett.*, **96/97**, 155–162

Dopp, E., Nebe, B., Hahnel, C., Papp, T., Alonso, B., Simkó, M. & Schiffmann, D. (1995) Mineral fibers induce apoptosis in Syrian hamster embryo fibroblasts. *Pathobiology*, **63**, 213–221

Dopp, E., Schuler, M., Schiffmann, D. & Eastmond, D.A. (1997) Induction of micronuclei, hyperdiploidy and chromosomal breakage affecting the centric/pericentric regions of chromosomes 1 and 9 in human amniotic fluid cells after treatment with asbestos and ceramic fibers. *Mutat. Res.*, **377**, 77–87

Draeger, U. & Löffler, F.-W. (1991) [Workplace fibre concentrations in the manufacture of rock wool insulation.] *Zbl. Arbeitsmed., Arbeitssch. Prophylaxe Ergonom.*, **41**, 218–222 (in German)

Draeger, U., Teichert, U., Schneider, T. & Trappmann, J. (1998) Criteria for the identification of insulation wool fibres by microscopic evaluation of filter samples. *Gefahrstoffe Reinhalt. Luft*, **58**, 343–346

Dörger, M., Münzing, S., Allmeling, A.-M. & Krombach, F. (2000) Comparison of the phagocytic response of rat and hamster alveolar macrophages to man-made vitreous fibers *in vitro*. *Human & Experim. Toxicol.*, **19**, 635–640

Dörger, M., Münzing, S., Allmeling, A.-M., Messmer, K. & Krombach, F. (2001) Differential responses of rat alveolar and peritoneal macrophages to man-made vitreous fibers *in vitro*. *Environ. Res. Section A*, **85**, 207–214

Drent, M., Bomans, P.H.H., Van Suylen, R.J., Lamers, R.J.S., Bast, A. & Wouters, E.F.M. (2000a) Association of man-made mineral fibre exposure and sarcoidlike granulomas. *Respir. Med.*, **94**, 815–820

Drent, M., Kessels, B.L.J., Bomans, P.H.H., Wagenaar, S.Sc. & Henderson, R.F. (2000b) Sarcoidlike lung granulomatosis induced by glass fibre exposure. *Sarcoidosis Vasc. Diffuse Lung Dis.*, **17**, 86–87

Drew, R.T., Kuschner, M. & Bernstein, D.M. (1987) The chronic effects of exposure of rats to sized glass fibres. *Ann. occup. Hyg.*, **31**, 711–729

Driscoll, K.E. (1996) Effects of fibres on cell proliferation, cell activation and gene expression. In: Kane, A.B., Boffetta, P. & Wilbourn, J.D., eds, *Mechanisms of Fibre Carcinogenesis* (IARC Scientific Publication No. 140), Lyon, IARC*Press*, pp. 73–96

Driscoll, K.E., Howard, B.W., Carter, J.M., Asquith, T., Johnston, C., Detilleux, P., Kunkel, S.I. & Isfort, R.J. (1996) α-Quartz-induced chemokine expression by rat lung epithelial cells. Effects of in vivo and in vitro particle exposure. *Am. J. Pathol.*, **149**, 1627–1637

Dufresne, A., Loosereewanich, P., Armstrong, B., Infante-Rivard, C., Perrault, G., Dion, C., Massé, S. & Bégin, R. (1995) Pulmonary retention of ceramic fibers in silicon carbide (SiC) workers. *Am. ind. Hyg. Assoc. J.*, **56**, 490–498

Dumortier, P., Broucke, I. & De Vuyst, P. (2001) Pseudoasbestos bodies and fibers in bronchoalveolar lavage of refractory ceramic fiber users. *Am. J. respir. crit. Care Med.*, **164**, 499–503

Eastes, W. & Hadley, J.G. (1996) A mathematical model of fiber carcinogenicity and fibrosis in inhalation and intraperitoneal experiments in rats. *Inhal. Toxicol.*, **8**, 323–343

Eastes, W., Potter, R.M. & Hadley, J.G. (2000a) Estimating in-vitro glass fiber dissolution rate from composition. *Inhal. Toxicol.*, **12**, 269–280

Eastes, W., Potter, R.M. & Hadley, J.G. (2000b) Estimation of dissolution rate from in-vivo studies of synthetic vitreous fibers. *Inhal. Toxicol.*, **12**, 1037–1054

Eastes, W., Potter, R.M. & Hadley, J.G. (2000c) Estimating rock and slagwool fiber dissolution rate from composition. *Inhal. Toxicol.*, **12**, 1127–1139

ECA (Everest Consulting Associates) (1996) *Refractory Ceramic Fibers: A Substitute Study*. Refractory Ceramic Fibers Coalition (RCFC), Augusta, Georgia, Miller Printing Co.

ECFIA & RCFC (European Ceramic Fibre Industries Association and Refractory Ceramic Fibres Coalition) (USA) (2001) *Alkaline Earth Silicate (AES) Wools Fact Sheet*, October 2001, Paris, France/Washington, DC

EIPPCB (European Integrated Pollution Prevention and Control Bureau) (2000) *Reference Document on Best Available Techniques in the Glass Manufacturing Industry*, Seville, Directorate-General Joint Research Centre, Institute for Prospective Technological Studies (http://eippcb.jrc.es)

Eller, P.M. (1984) Asbestos and other fibers by PCM. Method 7400. In: National Institute for Occupational Safety and Health, Cincinnati, OH, *Manual of Analytical Methods*, 3rd Ed.

Ellouk, S.A. & Jaurand, M.-C. (1994) Review of animal/*in vitro* data on biological effects of man-made fibers. *Environ. Health Perspect.*, **102** (Suppl. 2), 47–61

Engholm, G., Englund, A., Fletcher, T. & Hallin, N. (1987) Respiratory cancer incidence in Swedish construction workers exposed to man-made mineral fibres and asbestos. *Ann. occup. Hyg.*, **31**, 663–675

Enterline, P.E. & Marsh, G.M. (1984) The health of workers in the MMVF industry. In: *Biological Effects of Man-Made Mineral Fibres (Proceedings of a WHO/IARC Conference)*, Vol. 1, Copenhagen, World Health Organization, pp. 311–339

Enterline, P.E., Marsh, G.M. & Esmen, N.A. (1983) Respiratory disease among workers exposed to man-made mineral fibers. *Am. Rev. respir. Dis.*, **128**, 1–7

Enterline, P.E., Marsh, G.M., Henderson, V. & Callahan, C. (1987) Mortality update of a cohort of U.S. man-made mineral fibre workers. *Ann. occup. Hyg.*, **31**, 625–656

ERM (Environmental Resources Management) (1995) *Description and Characterisation of the Ceramic Fibres Industry of the European Union*. Final Report for the European Commission (Reference 3059), Brussels

Esmen, N.A., Hammad, Y.Y., Corn, M., Whittier, D., Kotsko, N., Haller, M. & Kahn, R.A. (1978) Exposure of employees to man-made mineral fibers: Mineral wool production. *Environ. Res.*, **15**, 262–277

Esmen, N.A., Corn, M., Hammad, Y., Whittier, D. & Kotsko, N. (1979a) Summary of measurements of employee exposure to airborne dust and fiber in sixteen facilities producing man-made mineral fibers. *Am. ind. Hyg. Assoc. J.*, **40**, 108–117

Esmen, N.A., Corn, M., Hammad, Y.Y., Whittier, D., Kotsko, N., Haller, M. & Kahn, R.A. (1979b) Exposure of employees to man-made mineral fibers: Ceramic fiber production. *Environ. Res.*, **19**, 265–278

Esmen, N.A., Sheehan, M.J., Corn, M., Engel, M. & Kotsko, N. (1982) Exposure of employees to man-made vitreous fibers: Installation of insulation materials. *Environ. Res.*, **28**, 386–398

Etherington, D.J., Pugh, D. & Silver, I.A. (1981) Collagen degradation in an experimental inflammatory lesion: Studies on the role of the macrophage. *Acta biol. med. germ.*, **40**, 1625–1636

European Chemicals Bureau (1999) *Methods for the Determination of the Hazardous Properties for Human Health of Man Made Mineral Fibres (MMMF)* (EUR 18748 EN), Ispra, Italy, European Commission Joint Research Centre

European Commission (1997) Commission Directive 97/69/EC of 13 December 1997. *Off. J. Eur. Comm.*, **L343**, 19–24

European Insulation Manufacturers Association (EURIMA) (1990) *Thermal Insulation Means Environmental Protection*

Everitt, J.I., Bermudez, E., Mangum, J.B., Wong, B., Moss, O.R., Janszen, D. & Rutten, A.A.J.J.L. (1994) Pleural lesions in Syrian golden hamsters and Fischer-344 rats following intrapleural instillation of man-made ceramic or glass fibers. *Toxicol. Pathol.*, **22**, 229–236

Everitt, J.I., Gelzleichter, T.R., Bermudez, E., Mangum, J.B., Wong, B.A., Janszen, D.B. & Moss, O.R. (1997) Comparison of pleural responses of rats and hamsters to subchronic inhalation of refractory ceramic fibers. *Environ. Health Perspect.*, **105** (Suppl. 5), 1209–1213

Fallentin, B. & Kamstrup, O. (1993) Simulation of past exposure in slag wool production. *Ann. occup. Hyg.*, **37**, 419–433

Fenoglio, I., Martra, G., Coluccia, S. & Fubini, B. (2000) Possible role of ascorbic acid in the oxidative damage induced by inhaled crystalline silica particles. *Chem. Res. toxicol.*, **13**, 971–975

Feron, V.J., Scherrenberg, P.M., Immel, H.R. & Spit, B.J. (1985) Pulmonary response of hamsters to fibrous glass: Chronic effects of repeated intratracheal instillation with or without benzo[a]pyrene. *Carcinogenesis*, **6**, 1495–1499

Fischer, M. (1993) Benefits and risks from MMMF in indoor air. In: Kalliokoski, P., Jantunen, M. & Seppänen, O., *Indoor Air, 1993 Proceedings of the 6th International Conference on Indoor Air Quality and Climate,* Volume 4, Helsinki, Jyväskylä, Finland, Gummerus, Oy, pp. 27–31

Fischer, A.B., Kaw, J.L., Diemer, K. & Eikmann, T. (1998) Low dose effects of fibrous and non-fibrous mineral dusts on the proliferation of mammalian cells in vitro. *Toxicol. Lett.,* **96–97**, 97–103

Fisher, C.E., Brown, D.M., Shaw, J., Beswick, P.H. & Donaldson, K. (1998) Respirable fibres: Surfactant coated fibres release more Fe^{3+} than native fibres at both pH 4.5 and 7.2. *Ann. occup. Hyg.,* **42**, 337–345

Fisher, C.E., Rossi, A.G., Shaw, J., Beswick, P.H. & Donaldson, K. (2000) Release of TNFα in response to SiC fibres: Differential effects in rodent and human primary macrophages, and in macrophage-like cell lines. *Toxicol. in Vitro,* **14**, 25–31

Fowler, D.P. (1980) *Industrial Hygiene Surveys of Occupational Exposure to Mineral Wool* (DHHS (NIOSH) Publication No. 80-135), Cincinnati, OH, US Department of Health and Human Services, Public Health Service, Center for Disease Control, National Institute for Occupational Safety and Health, Division of Surveillance, Hazard Evaluation and Field Studies

Fowler, D.P., Balzer, L. & Cooper, W.C. (1971) Exposure of insulation workers to airborne fibrous glass. *Am. ind. Hyg. Assoc. J.,* **32**, 86–91

the Freedonia Group (2001) *World Insulation to 2004,* Cleveland, OH

Friar, J.J. & Phillips, A.M. (1989) Exposure to ceramic man-made mineral fibers. In: Bignon, J., Peto, J. & Saracci, R., eds, *Non-Occupational Exposure to Mineral Fibres* (IARC Scientific Publication No. 90), Lyon, IARC*Press,* pp. 299–303

Friedrichs, K.-H., Höhr, D. & Grover, Y.P. (1983) [Results from non-source-related emission measurements of fibres in the Federal Republic of Germany.] In: Reinisch, D., Schneider, H.W. & Birkner, K.F., eds, *Fibrous Dusts Measurement, Effects, Prevention* (VDI-Berichte 475), Dusseldorf, VDI-Verlag, pp.113–116 (in German)

Fubini, B., Aust, A.E., Bolton, R.E., Borm, P.J.A., Bruch, J., Ciapetti, G., Donaldson, K., Elias, Z., Gold, J., Jaurand, M.-C., Kane, A.B., Lison, D. & Muhle, H. (1998) Non-animal tests for evaluating the toxicity of solid xenobiotics. *ATLA,* **26**, 579–617

Fujino, A., Hori, H., Higashi, T., Morimoto, Y., Tanaka, I. & Kaji, H. (1995) In-vitro biological study to evaluate the toxic potentials of fibrous materials. *Int. J. occup. environ. Health,* **1**, 21–28

Gamble, J.L. (1967) *Chemical Anatomy, Physiology and Pathology of Extracellular Fluid,* Cambridge, MA, Harvard University Press

Gan, L., Savransky, E.F., Fasy, T.M. & Johnson, E.M. (1993) Transfection of human mesothelial cells mediated by different asbestos fiber types. *Environ. Res.,* **62**, 28–42

Gao, H.-G., Whong, W.-Z., Jones, W.G., Wallace, W.E. & Ong, T. (1995) Morphological transformation induced by glass fibers in BALB/3-BT3 cells. *Teratog. Carcinog. Mutag.,* **15**, 63–71

Gao, H., Brick, J., Ong, S.-H., Miller, M., Whong, W.-Z. & Ong, T.-M. (1997) Selective hyperexpression of c-*jun* oncoprotein by glass fiber- and silica-transformed BALB/3-3T3 cells. *Cancer Lett.,* **112**, 65–69

Gardner, M.J., Magnani, C., Pannett, B., Fletcher, A.C. & Winter, P.D. (1988) Lung cancer among glass fibre production workers: A case–control study. *Br. J. ind. Med.*, **45**, 613–618

Gaudichet, A., Petit, G., Billon-Galland, M.A. & Dufour, G. (1989) Levels of atmospheric pollution by man-made mineral fibres in buildings. In: Bignon, J., Peto, J. & Saracci, R., eds, *Non-occupational Exposure to Mineral Fibres* (IARC Scientific Publication No. 90), Lyon, IARCPress, pp. 291–298

Gehr, P., Geiser, M., Im Hof, V., Schürch, S., Waber, U. & Baumann, M. (1993) Surfactant and inhaled particles in the conducting airways: Structural, stereological and biophysical aspects. *Microsc. Res. Techn.*, **26**, 423–436

Gelzleichter, T.R., Bermudez, E., Mangum, J.B., Wong, B.A., Moss, O.R. & Everitt, J.I. (1996a) Pulmonary and pleural responses in Fischer 344 rats following short-term inhalation of a synthetic vitreous fiber. II. Pathobiologic responses. *Fundam. appl. Toxicol.*, **30**, 39–46

Gelzleichter, T.R., Bermudez, E., Mangum, J.B., Wong, B.A., Everitt, J.I. & Moss, O.R. (1996b) Pulmonary and pleural responses in Fischer 344 rats following short-term inhalation of a synthetic vitreous fiber. I. Quantitation of lung and pleural fiber burden. *Fundam. appl. Toxicol.*, **30**, 31–38

Gelzleichter, T.R., Mangum, J.B., Bermudez, E., Wong, B.A., Moss, O.R. & Everitt, J.I. (1996c) Pulmonary and pleural leukocytes from F344 rats produce elevated levels of fibronectin following inhalation of refractory ceramic fibers. *Exp. Toxicol. Pathol.*, **48**, 487–489

Gelzleichter, T.R., Bermudez, E., Mangum, J.B., Wong, B.A., Janszen, D.B., Moss, O.R. & Everitt, J.I. (1999) Comparison of pulmonary and pleural responses of rats and hamsters to inhaled refractory ceramic fibers. *Toxicol. Sci.*, **49**, 93–101

Gilmour, P.S., Beswick, P.H., Brown, D.M. & Donaldson, K. (1995) Detection of surface free radical activity of respirable industrial fibres using supercoiled φX174 RF1 plasmid DNA. *Carcinogenesis*, **16**, 2973–2979

Gilmour, P.S., Brown, D.M., Beswick, P.H., MacNee, W., Rahman, I. & Donaldson, K. (1997) Free radical activity of industrial fibers: Role of iron in oxidative stress and activation of transcription factors. *Environ. Health Perspect.*, **105** (Suppl. 5), 1313–1317

Goldberg, M.S., Parent, M.-E., Siematiatycki, J., Désy, M., Nadon, L., Richardson, L., Lakhani, R., Latreille, B. & Valois, M.-F. (2001) A case–control study of the relationship between the risk of colon cancer in men and exposures to occupational agents. *Am. J. ind. Med.*, **39**, 531–546

Goldmann, D. & Kruger, D. (1989) *Abschlussbericht über die Messungen und Ergebinisse des Imissions — Messprogrammes Berlin (West)* (Technischer Bericht nr D-89/185), Berlin, Technischer Überwachungs — Verein eV.

Government of Canada (1993) *Canadian Environmental Protection Act. Mineral Fibres (Man-Made Vitreous Fibres) Priority Substances List Assessment Report*, Canada Communication Group-Publishing, Ottawa

Grimm, H.-G. (1983) [Occupational exposure to man-made mineral fibres and their effects on health.] *Zbl. Arbeitsmed.*, **33**, 156–162 (in German)

Grimm, H.-G., Löffler, F.-W. & Mayer, P. (2000) *Retrospective Assessment of the Exposure Situation and of Health Effects for a Former Stone Wool Production Plant in Germany (Schriftenreihe Zentralblatt für Arbeitsmedizin, Vol. 19)*, Heidelberg, Curt Haefner Verb.

Guldberg, M., Christensen, V.R., Perander, M., Zoitos, B., Koenig, A.R. & Sebastian, K. (1998) Measurement of in-vitro fibre dissolution rate at acidic pH. *Ann. Occup. Hyg.*, **42**, 233–243

Guldberg, M., de Meringo, A., Kamstrup, O., Furtak, H. & Rossiter, C. (2000) The development of glass and stone wool compositions with increased biosolubility. *Regul. Toxicol. Pharmacol.*, **32**, 184–189

Guldberg, M., Jensen, S.L., Knudsen, T., Steenberg, T. & Kamstrup, O. (2002) High-alumina low-silica HT stone wool fibers: A chemical compositional range with high biosolubility. *Regul. Toxicol. Pharmacol.*, **35**, 217–226

Gustavsson, P., Plato, N., Axelson, O., Brage, H.N., Hogstedt, C., Ringbäck, G., Tornling, G. & Wingren, G. (1992) Lung cancer risk among workers exposed to man-made mineral fibers (MMMF) in the Swedish prefabricated house industry. *Am. J. ind. Med.*, **21**, 825–834

Hallin, N. (1981) *Mineral Wool Dust in Construction Sites* (Report 1981-09-01), Stockholm, Bygghälsan [The Construction Industry's Organization for Working Environment, Safety and Health]

Hammad, Y.Y. (1984) Deposition and elimination of MMMF. In: *Biological Effects of Man-made Mineral Fibres*, Copenhagen, World Health Organization, pp. 126–142

Hansen, K. & Mossman, B.T. (1987) Generation of superoxide (O_2^-) from alveolar macrophages exposed to asbestiform and nonfibrous particles. *Cancer Res.*, **47**, 1681–1686

Hansen, E.F., Rasmussen, F.V., Hardt, F. & Kamstrup, O. (1999) Lung function and respiratory health of long-term fiber-exposed stonewool factory workers. *Am. J. respir. crit. Care Med.*, **160**, 466–472

Hart, G.A., Newman, M.M., Bunn, W.B. & Hesterberg, T.W. (1992) Cytotoxicity of refractory ceramic fibres to Chinese hamster ovary cells in culture. *Toxicol. In Vitro*, **6**, 317–326

Hart, G.A., Kathman, L.M. & Hesterberg, T.W. (1994) *In vitro* cytotoxicity of asbestos and man-made vitreous fibers: Roles of fiber length, diameter and composition. *Carcinogenesis*, **15**, 971–977

Hart, G.A., Hesterberg, T.H. & Bauer, J.F. (1999) Physical and chemical degradation of synthetic vitreous fibres in the lung and in vitro (Abstract). *Toxicologist*, **48** (Suppl. 1), 133

Hartman, D.R., Greenwood, M.E. & Miller, D.M. (1996) *High Strength Glass Fibers (Owens Corning technical paper 1-PL-19025A)*, Granville, OH, Owens Corning Science & Technical Center

Head, I.W.H. & Wagg, R.M. (1980) A survey of occupational exposure to man-made mineral fibre dust. *Ann. occup. Hyg.*, **23**, 235–258

Health and Safety Executive (1988) *Man-made Mineral Fibre* (Methods for the Determination of Hazardous Substances 59), London, Occupational Medicine and Hygiene Laboratory

Hesterberg, T.W. & Barrett, J.C. (1984) Dependence of asbestos- and mineral dust-induced transformation of mammalian cells in culture on fiber dimension. *Cancer Res.*, **44**, 2170–2180

Hesterberg, T.W. & Hart, G.A. (2001) Synthetic vitreous fibers: A review of toxicology research and its impact on hazard classification. *Crit. Rev. Toxicol.*, **31**, 1–53

Hesterberg, T.W., Butterick, C.J., Oshimura, M., Brody, A.R. & Barrett, J.C. (1986) Role of phagocytosis in Syrian hamster cell transformation and cytogenetic effects induced by asbestos and short and long glass fibers. *Cancer Res.*, **46**, 5795–5802

Hesterberg, T.W., Miiller, W.C., McConnell, E.E., Chevalier, J., Hadley, J., Bernstein, D.M., Thevenaz, P. & Anderson, R. (1993) Chronic inhalation toxicity of size-separated glass fibers in Fischer 344 rats. *Fundam. appl. Toxicol.*, **20**, 464–476

Hesterberg, T.W., Miiller, W.C., Thevenaz, P. & Anderson, R. (1995) Chronic inhalation studies of man-made vitreous fibres: Characterization of fibres in the exposure aerosol and lungs. *Ann. occup. Hyg.*, **39**, 637–653

Hesterberg, T., Miiller, W., Hart, G., Bauer, J. & Hamilton, R. (1996a) Physical and chemical transformation of synthetic vitreous fibres in the lung and in vitro. *J. occup. Health Safety Aust. NZ*, **12**, 345–355

Hesterberg, T.W., McConnell, E.E., Miiller, W.C., Chevalier, J., Everitt, J., Thevenaz, P., Fleissner, H. & Oberdörster, G. (1996b) Use of lung toxicity and lung particle clearance to estimate the maximum tolerated dose (MTD) for a fiber glass chronic inhalation study in the rat. *Fundam. appl. Toxicol.*, **32**, 31–44

Hesterberg, T.W., Miiller, W.C., Musselman, R.P., Kamstrup, O., Hamilton, R.D. & Thevenaz, P. (1996c) Biopersistence of man-made vitreous fibers and crocidolite asbestos in the rat lung following inhalation. *Fundam. appl. Toxicol.*, **29**, 267–279

Hesterberg, T.W., Axten, C., McConnell, E.E., Oberdörster, G., Everitt, J., Miiller, W.C., Chevalier, J., Chase, G.R. & Thevenaz, P. (1997) Chronic inhalation study of fiber glass and amosite asbestos in hamsters: Twelve-month preliminary results. *Environ. Health Perspect.*, **105** (Suppl. 5), 1223–1229

Hesterberg, T.W., Hart, G.A., Chevalier, J., Miiller, W.C., Hamilton, R.D., Bauer, J. & Thevenaz, P. (1998a) The importance of fiber biopersistence and lung dose in determining the chronic inhalation effects of X607, RCF1, and chrysotile asbestos in rats. *Toxicol. appl. Pharmacol.*, **153**, 68–82

Hesterberg, T.W., Chase, G., Axten, C., Miiller, W.C., Musselman, R.P., Kamstrup, O., Hadley, J., Morscheidt, C., Bernstein, D. & Thevenaz, P. (1998b) Biopersistence of synthetic vitreous fibers and amosite asbestos in the rat lung following inhalation. *Toxicol. appl. Pharmacol.*, **151**, 262–275

Hesterberg, T.W., Axten, C., McConnell, E.E., Hart, G.A., Miiller, W., Chevalier, J., Everitt, J., Thevenaz, P. & Oberdörster, G. (1999) Studies on the inhalation toxicology of two fiber-glasses and amosite asbestos in the Syrian golden hamster. Part I. Results of a subchronic study and dose selection for a chronic study. *Inhal. Toxicol.*, **11**, 747–784

Hesterberg, T.W., Hart, G.A., Miiller, W.C., Chase, G., Rogers, R.A., Mangum, J.B. & Everitt, J.I. (2002) Use of short-term assays to evaluate the potential toxicity of two new biosoluble glasswool fibers. *Inhal. Toxicol.*, **14**, 217–246

Heyder, J. (1982) Particle transport onto human airway surfaces. *Eur. J. respir. Dis.*, **63** (Suppl. 119), 29–50

Hill, I.M., Beswick, P.H. & Donaldson, K. (1996) Enhancement of the macrophage oxidative burst by immunoglobulin coating of respirable fibers: Fiber-specific differences between asbestos and man-made mineral fibers. *Exp. Lung Res.*, **22**, 133–148

Hippeli, S., Dornisch, K., Kaiser, S., Dräger, U. & Elstner, E.F. (1997) Biological durability and oxidative potential of a stonewool mineral fibre compared to crocidolite asbestos fibres. *Arch. Toxicol.*, **71**, 532–535

Höfert, N. & Rödelsperger, K. (1998) [Parallel edges of artificial mineral fibers.] *Gefahrstoffe Reinhalt. Luft*, **58**, 405–406 (in German)

Höhr, D. (1985) [Investigations by means of transmission electron microscopy (TEM): Fibrous particles in ambient air.] *Reinhalt. Luft*, **45**, 171–174 (in German)

Hori, H., Higashi, T., Fujino, A., Yamato, H., Ishimatsu, S., Oyabu, T. & Tanaka, I. (1993) Measurement of airborne ceramic fibers in manufacturing and processing factories. *Ann. occup. Hyg.*, **37**, 623–629

Horie, E. (ed.) (1987) *Ceramic Fiber Insulation Theory and Practice*, The Energy Conservation Center, Osaka, Japan, The Eibun Press

Howden, P.J. & Faux, S.P. (1996a) Glutathione modulates the formation of 8-hydroxydeoxyguanosine in isolated DNA and mutagenicity in *Salmonella typhimurium* TA100 induced by mineral fibres. *Carcinogenesis*, **17**, 2275–2277

Howden, P.J. & Faux, S.P. (1996b) Fibre-induced lipid peroxidation leads to DNA adduct formation in *Salmonella typhimurium* TA104 and rat lung fibroblasts. *Carcinogenesis*, **17**, 413–419

Hsieh, T.H., Yu, C.P. & Oberdörster, G. (1999) A dosimetry model of nickel compounds in the rat lung. *Inhal. Toxicol.*, **11**, 229–248

Hughes, J.M., Jones, R.N., Glindmeyer, H.W., Haddad, Y.Y. & Weill, H. (1993) Follow up study of workers exposed to man made mineral fibres. *Br. J. ind. Med.*, **50**, 658–667

Hunting, K.L. & Welch, L.S. (1993) Occupational exposure to dust and lung disease among sheet metal workers. *Br. J. ind. Med.*, **50**, 432–442

IARC (1988) *IARC Monographs on the Evaluation of Carcinogenic Risk to Humans*, Vol. 43, *Man-made Mineral Fibres and Radon*, Lyon, IARC*Press*, pp. 33–171

Iburg, J., Marfels, H. & Spurny, K. (1987) [Measurements of fibrous dusts in ambient air of the Federal Republic of Germany. V. Measurements of three different locations in the region of Bayreuth.] *Staub-Reinhalt Luft*, **47**, 271–274 (in German)

ICF Inc. (1991) *Preliminary Use and Substitutes Analysis for Refractory Ceramic Fiber Used in Furnace Linings*, Washington, DC

ICRP (1994) *Annals of the ICRP: Human Respiratory Tract Model for Radiological Protection* (ICRP Publication 66), Vol. 24, Nos. 1–3, New York, Pergamon

ILO (International Labour Organization) (2000) *Code of Practice on Safety in the Use of Synthetic Vitreous Fibre Insulation Wools (Glass Wool, Rock wool, Slag wool)*, Geneva, International Labour Office

ILSI (2000) ILSI Risk Science Institute Workshop Participants: The relevance of the rat lung response to particle overload for human risk assessment: A workshop consensus report. *Inhal. Toxicol.*, **12**, 1–17

INRS (Institut National de Recherche et de Sécurité) (1999) [Threshold Limit Values for Occupational Exposure to Chemicals in France.] (ND 2098-174-99), Paris (http://www.inrs.fr) (in French)

INSERM (1999) *Effets sur la Santé des Fibres de Substitution à l'Amiante*, Paris, Institut National de la Santé et de la Recherche Médicale

ISO (International Organization for Standardization) (1995) *Air Quality-Particle Size Fractions Definitions for Health-related Sampling* (ISO 7708), Geneva

Jacob, T.R., Hadley, J.G., Bender, J.R. & Eastes, W. (1992) Airborne glass fiber concentrations during installation of residential insulation. *Am. ind. Hyg. Assoc. J.*, **53**, 519–523

Jacob, T.R., Hadley, J.G., Bender, J.R. & Eastes, W. (1993) Airborne glass fiber concentrations during manufacturing operations involving glass wool insulation. *Am. ind. Hyg. Assoc. J.*, **54**, 320–326

Jaffrey, T.S.A.M., Rood, A.P., Llewellyn, J.W. & Wilson, A.J. (1990) Levels of airborne man-made mineral fibres in UK dwellings. II. Fibre levels during and after some disturbance of loft insulation. *Atmosph. Environ.*, **24A**, 143–146

Janssen, Y.M.W., Heintz, N.H., Marsh, J.P., Borm, P.J.A. & Mossman, B.T. (1994) Induction of c-*fos* and c-*jun* proto-oncogenes in target cells of the lung and pleura by carcinogenic fibers. *Am. J. Respir. Cell mol. Biol.*, **11**, 522–530

Janssen-Heininger, Y.M.W., Macara, I. & Mossman, B.T. (1999) Cooperativity between oxidants and tumor necrosis factor in the activation of nuclear factor (NF)-κB. Requirement of Ras/mitogen-activated protein kinases in the activation of NF-κB by oxidants. *Am. J. Respir. Cell mol. Biol.*, **20**, 942–952

Järvholm, B., Hillerdal, G., Järliden, A.-K., Hansson, A., Lilja, B.-G., Tornling, G. & Westerholm, P. (1995) Occurrence of pleural plaques in workers with exposure to mineral wool. *Int. Arch. occup. environ. Health*, **67**, 343–346

Jaurand, M.-C. (1994) *In vitro* assessment of biopersistence using mammalian cell systems. *Environ. Health Perspect.*, **102** (Suppl. 5), 55–59

Jeffery, G.B. (1922) The motion of ellipsoidal particles immersed in a viscous fluid. *Proc. R. Soc. London*, **A102**, 161–179

Jensen, C.G. & Watson, M. (1999) Inhibition of cytokinesis by asbestos and synthetic fibres. *Cell Biol. int.*, **23**, 829–840

Johnson, N.F. & Jaramillo, R.J. (1997) *p53*, *Cip1*, and *Gadd153* expression following treatment of A549 cells with natural and man-made vitreous fibers. *Environ. Health Perspect.*, **105** (Suppl. 5), 1143–1145

Johnson, D.L., Healey, J.J., Ayer, H.E. & Lynch, J.R. (1969) Exposure to fibers in the manufacture of fibrous glass. *Am. ind. Hyg. Assoc. J.*, **30**, 545–550

Julier, F., Tiesler, H. & Zindler, G. (1993) [Measurement of fibrous dust concentrations when handling mineral-wool insulations on industrial building sites.] *Staub-Reinhalt. Luft*, **53**, 245–250 (in German)

Kamstrup, O., Davis, J.M.G., Ellehauge, A. & Guldberg, M. (1998) The biopersistence and pathogenicity of man-made vitreous fibers after short- and long-term inhalation. *Ann. occup. Hyg.*, **42**, 191–199

Kamstrup, O., Ellehauge, A., Chevalier, J., Davis, J.M.G., McConnell, E.E. & Thévenaz, P. (2001) Chronic inhalation studies of two types of stone wool fibers in rats. *Inhal. Toxicol.*, **13**, 603–621

Kamstrup, O., Ellehauge, A., Collier, C.G. & Davis, J.M.G. (2002) Carcinogenicity studies after intraperitoneal injection of two types of stone wool fibres in rats. *Annals occup. Hyg.*, **46**, 135–142

Kane, A.B., Boffetta, P., Saracci, R. & Wilbourn, J.D., eds (1996) *Mechanisms of Fibre Carcinogenesis* (IARC Scientific Publications No. 140), Lyon, IARC*Press*

Karin, M. (1995) The regulation of AP-1 activity by mitogen-activated protein kinases. *J. biol. Chem.*, **270**, 16483–16486

Karin, M. & Delhase, M. (1998) JNK or IKK, AP-1 or NF-κB, which are the targets for MEK kinase 1 action? *Proc. natl Acad. Sci. USA*, **95**, 9067–9069

Karin, M., Liu, Z.-G. & Zandi, E. (1997) AP-1 function and regulation. *Current Opin. Cell Biol.*, **9**, 240–246

Kauffer, E. & Vigneron, J.C. (1987) [Epidemiological survey in two man-made mineral fibre producing plants. I. Measurement of dust levels.] *Arch. Mal. prof.*, **48**, 1–6 (in French)

Kauffer, E., Schneider, T. & Vigneron, J.C. (1993) Assessment of man-made mineral fibre size distributions by scanning electron microscopy. *Ann. occup. Hyg.*, **37**, 469–479

Kauppinen, T., Toikkanen, J., Pedersen, D., Young, R., Ahrens, W., Boffetta, P., Hansen, J., Kromhout, H., Maqueda Blasco, J., Mirabelli, D., de la Orden-Rivera, V., Pannett, B., Plato, N., Savela, A., Vincent, R. & Kogevinas, M. (2000) Occupational exposure to carcinogens in the European Union. *Occup. environ. Med.*, **57**, 10–18

Kiec-Šwierczynska, M. & Szymczk, W. (1995) The effect of the working environment on occupational skin disease development in workers processing rockwool. *Int. J. occup. Med. environ. Health*, **8**, 17–22

Kilburn, K.H. & Warshaw, R.H. (1991) Difficulties of attribution of effect in workers exposed to fiberglass and asbestos. *Am. J. ind. Med.*, **20**, 745–751

Kilburn, K.H., Powers, D. & Warshaw, R.H. (1992) Pulmonary effects of exposure to fine fibreglass: Irregular opacities and small airways obstruction. *Br. J. ind. Med.*, **49**, 714–720

Kim, J.-Hong, Chang, H.-S., Kim, K.-Y., Park, W.-M., Lee, Y.-J., Choi, H.-C., Kim, K.-A. & Lim, Y. (1999) Environmental measurements of total dust and fiber concentration in manufacture and use of man-made mineral fibres. *Ind. Health*, **37**, 322–328

Kjaerheim, K., Boffetta, P., Hansen, J., Cherrie, J., Chang-Claude, J., Eilber, U., Ferro, G., Guldner, K., Olsen, J.H., Plato, N., Proud, L., Saracci, R., Westerholm, P. & Andersen, A. (2002) Lung cancer among rock and slag wool production workers. *Epidemiology*, **13**, 445–453

Kjuus, H., Skjaerven, R., Langård, S., Lien, J.T. & Aamodt, T. (1986) A case-referent study of lung cancer, occupational exposures and smoking. I. Comparison of title-based and exposure-based occupational information (Original articles). *Scand. J. Work Environ. Health*, **12**, 193–202

Klingholz, R. & Steinkopf, B. (1984) The reactions of MMMF in a physiological model fluid and water. In: *Biological Effects of Man-made Mineral Fibres*, Vol. 2, Copenhagen, World Health Organization, pp. 60–86

Knudsen, T., Guldberg, M., Christiansen, V.R. & Lund Jensen, S. (1996) New type of stonewool (HT fibres) with a high dissolution rate at pH = 4.5. *Glastech. Ber. Glass Sci. Technol.*, **69**, 331–337

Koenig, A.R. & Axten, C.W. (1995) Exposures to airborne fiber and free crystalline silica during installation of commercial and industrial mineral wool products. *Am. ind. Hyg. Assoc. J.*, **56**, 1016–1022

Koenig, A.R., Hamilton, R.D., Laskowski, T.E., Olson, J.R., Gordon, J.F., Christensen, V.R. & Byers, C.D. (1993) Fiber diameter measurement of bulk man-made vitreous fiber. *Anal. chim. Acta*, **280**, 289–298

Koshi, K., Kohyama, N., Myojo, T. & Fukuda, K. (1991) Cell toxicity, hemolytic action and clastogenic activity of asbestos and its substitutes. *Ind. Health*, **29**, 37–56

Krantz, S. (1988) Exposure to man-made mineral fibers at ten production plants in Sweden. *Scand. J. Work Environ. Health*, **14** (Suppl. 1), 49–51

Krantz, S., Cherrie, J.W., Schneider, T., Öhlberg, I. & Kamstrup, O. (1991) *Modelling of Past Exposure to MMMF in the European Rock/Slag Wool Industry (Arbete och Hälsa 1991:1)*, Solna, Arbetsmiljöinstitutet

Krombach, F., Münzing, S., Allmeling, A.-M., Gerlach, J.T., Behr, J. & Dörger, M. (1997) Cell size of alveolar macrophages: An interspecies comparison. *Environ. Health Perspect.*, **105** (Suppl. 5), 1261–1263

Lambré, C., Schorsch, F., Blanchard, O., Richard, J., Boivin, J.C., Hanton, D., Grimm, H. & Morscheidt, C. (1998) An evaluation of the carcinogenic potential of five man-made vitreous fibers using the intraperitoneal test. *Inhal. Toxicol.*, **10**, 995–1021

Lanting, R.W. & Den Boeft, J. (1983) Ambient air concentrations of mineral fibers in the Netherlands. In: Reinisch, D., Schneider, H.W. & Birkner, K.F., eds, *Fibrous Dusts — Measurement, Effects, Prevention* (VDI-Berichte 475), Dusseldorf, VDI-Verlag, pp. 123–128

Laskowski, J.J., Young, J., Gray, R., Acheson, R. & Forder, S.D. (1994) The identity, development and quantification of phases in devitrified, commercial-grade, aluminosilicate, refractory ceramic fibers: 'An X-ray powder diffractometry study'. *Anal. chim. Acta*, **286**, 9–23

Law, B.D., Bunn, W.B. & Hesterberg, T.W. (1990) Solubility of polymeric organic fibers and man-made vitreous fibers in Gamble's solution. *Inhal. Toxicol.*, **2**, 321–339

Law, B.D., Bunn, W.B. & Hesterberg, T.W. (1991) Dissolution of natural mineral and man-made vitreous fibers in Karnovsky's and formalin fixatives. *Inhal. Toxicol.*, **3**, 309–321

Lawson, C.C., LeMasters, M.K., Kawas LeMasters, G., Simpson Reutman, S., Rice, C.H. & Lockey, J.E. (2001) Reliability and validity of chest radiograph surveillance programs. *Chest*, **120**, 64–68

Le Bouffant, L., Henin, J.P., Martin, J.C., Normand, C., Tichoux, G. & Trolard, F. (1984) Distribution of inhaled MMMF in the rat lung — Long-term effects. In: *Biological Effects of Man-Made Mineral Fibres* (Proceedings of a WHO/IARC Conference), Vol. 2, Copenhagen, World Health Organization, pp. 143–167

Le Bouffant, L., Daniel, H., Henin, J.P., Martin, J.C., Normand, C., Thichoux, G. & Trolard, F. (1987) Experimental study on long-term effects of inhaled MMMF on the lungs of rats. *Ann. occup. Hyg.*, **31**, 765–790

Lea, C.S., Hertz-Picciotto, I., Andersen, A., Chang-Claude, J., Olsen, J.H., Pesatori, A.C., Teppo, L., Westerholm, P., Winter, P.D. & Boffetta, P. (1999) Gender differences in the healthy workers effect among synthetic vitreous fiber workers. *Am. J. Epidemiol.*, **150**, 1099–1105

Leanderson, P. & Tagesson, C. (1989) Cigarette smoke potentiates the DNA-damaging effect of manmade mineral fibers. *Am. J. ind. Med.*, **16**, 697–706

Leanderson, P. & Tagesson, C. (1992) Hydrogen peroxide release and hydroxyl radical formation in mixtures containing mineral fibres and human neutrophils. *Br. J. ind. Med.*, **49**, 745–749

Leanderson, P., Söderkvist, P., Tagesson, C. & Axelson, O. (1988) Formation of DNA adduct 8-hydroxy-2'-deoxyguanosine induced by man-made mineral fibres. In: Bartsch, H., Hemminki, K. & O'Neill, I.K., eds, *Methods for Detecting DNA Damaging Agents in Humans: Applications in Cancer Epidemiology and Prevention* (IARC Scientific Publications No. 89), Lyon, IARC*Press*, pp. 422–424

Leanderson, P., Söderkvist, P. & Tagesson, C. (1989) Hydroxyl radical mediated DNA base modification by manmade mineral fibres. *Br. J. ind. Med.*, **46**, 435–438

Lee, K.P., Barras, C.E., Griffith, F.D., Waritz, R.S. & Lapin, C.A. (1981) Comparative pulmonary responses to inhaled inorganic fibers with asbestos and fiberglass. *Environ. Res.*, **24**, 167–191

Lees, P.S.J., Breysse, P.N., McArthur, B.R., Miller, M.E., Rooney, B.C., Robbins, C.A & Corn, M. (1993) End user exposures to man-made vitreous fibers. I. Installation of residential insulation products. *Appl. occup. environ. Hyg.*, **8**, 1022–1030

Lehuédé, P. & de Meringo, A. (1994) SEM-EDS analysis of glass fibers corroded in physiological solutions by dynamic tests with variable flow rates. *Environ. Health Perspect.*, **102** (Suppl. 5), 73–75

Lehuédé, P., de Meringo, A. & Bernstein, D.M. (1997) Comparison of the chemical evolution of MMVF following inhalation exposure in rats and acellular in vitro dissolution. *Inhal. Toxicol.*, **9**, 495–523

Leidel, N.A., Bush, K.A. & Lynch, J.R. (1977) NIOSH occupational exposure sampling strategy manual (NIOSH Publication No. 77-173), Washington, DC, United States Department of Health, Education and Welfare

Leikauf, G.D., Fink, S.P., Miller, M.L., Lockey, J.E. & Driscoll, K.E. (1995) Refractory ceramic fibers activate alveolar macrophage eicosanoid and cytokine release. *J. appl. Physiol.*, **78**, 164–171

Leineweber, J.P. (1984) Solubility of fibres *in vitro* and *in vivo*. In: *Biological Effects of Man-made Mineral Fibres*, Vol. 2, Copenhagen, World Health Organization, pp. 87–101

Lemasters, G., Lockey, J., Rice, C., McKay, R., Hansen, K., Lu, J., Levin, L. & Gartside, P. (1994) Radiographic changes among workers manufacturing refractory ceramic fibre and products. *Ann. occup. Hyg.*, **38** (Suppl. 1), 745–751

Lemasters, G.K., Lockey, J.E., Levin, L.S., McKay, R.T., Rice, C.H., Horvath, E.P., Papes, D.M., Lu, J.W. & Feldman, D.J. (1998) An industry-wide pulmonary study of men and women manufacturing refractory ceramic fibers. *Am. J. Epidemiol.*, **148**, 910–919

Lemasters, G., Lockey, J., Levin, L., Yiin, J., Reutman, S., Papes, D. & Rice, C. (2001) A longitudinal study of chest radiographic changes and mortality of workers in the refractory ceramic fiber industry (Abstract at the 2001 Congress of Epidemiology) (Abstract No. 986). *Am. J. Epidemiol.*, **153** (Suppl. 264)

Ljungman, A.G., Lindahl, M. & Tagesson, C. (1994) Asbestos fibres and man made mineral fibres: Induction and release of tumour necrosis factor-α from rat alveolar macrophages. *Occup. environ. Med.*, **51**, 777–783

Lockey, J.E. & Ross, C.S. (1994) Radon and man-made vitreous fibers. *J. Allergy clin. Immunol.*, **94**, 310–317

Lockey, J.E., Lemasters, G., Rice, C., Hansen, K., Levin, L., Shipley, R., Spitz, H. & Wiot, J. (1996) Refractory ceramic fiber exposure and pleural plaques. *Am. J. respir. crit. Care Med.*, **154**, 1405–1410

Lockey, J.E., LeMasters, G.K., Levin, L., Rice, C., Yiin, J., Reutman, S. & Papes, D.M. (2002) A longitudinal study of chest radiographic changes of workers in the refractory ceramic fiber industry. *Chest*, **121**, 2044–2051

Loewenstein, K.L. (1993) *The Manufacturing Technology of Continuous Glass Fibres* (Glass Science and Technology 6), 3rd rev. Ed., Amsterdam, Elsevier

Lum, H., Tyler, W.S., Hyde, D.M. & Plopper, C.G. (1983) Morphometry of in situ and lavaged pulmonary alveolar macrophages from control and ozone-exposed rats. *Exp. Lung Res.*, **5**, 61–77

Lundborg, M., Johard, U., Johansson, A., Eklund, A., Falk, R., Kreyling, W. & Camner, P. (1995) Phagolysosomal morphology and dissolution of cobalt oxide particles by human and rabbit alveolar macrophages. *Exp. Lung Res.*, **21**, 51–66

Luoto, K., Holopainen, M., Karppinen, K., Perander, M. & Savolainen, K. (1994a) Dissolution of man-made vitreous fibers in rat alveolar macrophage culture and Gamble's saline solution: Influence of different media and chemical composition of the fibers. *Environ. Health Perspect.*, **102** (Suppl. 5), 103–107

Luoto, K., Holopainen, M. & Savolainen, K. (1994b) Scanning electron microscopic study on the changes in the cell surface morphology of rat alveolar macrophages after their exposure to man-made vitreous fibers. *Environ. Res.*, **66**, 198–207

Luoto, K., Holopainen, M., Kangas, J., Kalliokoski, P. & Savolainen, K. (1995) The effect of fiber length on the dissolution by macrophages of rockwool and glasswool fibers. *Environ. Res.*, **70**, 51–61

Luoto, K., Holopainen, M., Perander, M., Karppinen, K. & Savolainen, K.M. (1996) Cellular effects of particles — Impact of dissolution on toxicity of man-made mineral fibers. *Centr. Eur. J. public Health*, **4**, 29–32

Luoto, K., Holopainen, M., Sarataho, M. & Savolainen, K. (1997) Comparison of cytotoxicity of man-made vitreous fibers. *Ann. occup. Hyg.*, **41**, 31–50

Luoto, K., Holopainen, M., Kangas, J., Kalliokoski, P. & Savolainen, K. (1998) Dissolution of short and long rockwool and glasswool fibers by macrophages in flowthrough cell culture. *Environ. Res.*, **A78**, 25–37

Maltoni, C. & Minardi, F. (1989) Recent results of carcinogenicity bioassays of fibers and other particulate materials. In: Bignon, J., Peto, J. & Saracci, R., eds, *Non-Occupational Exposure to Mineral Fibres* (IARC Scientific Publications No. 90), Lyon, IARCPress, pp. 46–53

Maples, K.R. & Johnson, N.F. (1992) Fiber induced hydroxyl radical formation: Correlation with mesothelioma induction in rats and humans. *Carcinogenesis*, **13**, 2035–2039

Marchand, J.L., Luce, D., Leclerc, A., Goldberg, P., Orlowski, E., Bugel, I. & Brugère, J. (2000) Laryngeal and hypopharyngeal cancer and occupational exposure to asbestos and man-made vitreous fibers: Results of a case–control study. *Am. J. ind. Med.*, **37**, 581–589

Marchant, G.E., Amen, M.A., Bullock, C.H., Carter, C.M., Johnson, K.A., Reynolds, J.W., Connelly, F.R. & Crane, A.E. (2002) A synthetic vitreous fibre (SVF) occupational exposure database: Implementing the SVF health and safety partnership program. *Appl. occup. environ. Hyg.*, **17**, 276–285

Marks-Konczalik, J., Gillissen, A., Jaworska, M., Löseke, S., Voss, B., Fisseler-Eckhoff, A., Schmitz, I. & Schultze-Werninghaus, G. (1998) Induction of manganese superoxide dismutase gene expression in bronchoepithelial cells after rockwool exposure. *Lung*, **176**, 165–180

Marsh, G.M., Enterline, P.E., Stone, R.A. & Henderson, V.L. (1990) Mortality among a cohort of US man-made mineral fiber workers: 1985 Follow-up. *J. occup. Med.*, **32**, 594–604

Marsh, G.M., Stone, R.A., Owens, A.D., Henderson, V.L., Smith, T.J. & Quinn, M.M. (1993) *A Descriptive Analysis of the New Fiber and Co-exposure Data for the US Cohort Study of Man-made Vitreous Fiber Production (Proceedings of the 24th International Congress on Occupational Health, Nice, France)*

Marsh, G., Stone, R., Youk, A., Smith, T., Quinn, M., Henderson, V., Schall, L., Wayne, L. & Lee, K. (1996) Mortality among United States rock wool and slag wool workers: 1989 Update. *J. occup. Health Safety Austr. N.Z.*, **12**, 297–312

Marsh, G.M., Youk, A.O., Stone, R.A., Buchanich, J.M., Gula, M.J., Smith, T.J. & Quinn, M.M. (2001a) Historical cohort study of US man-made vitreous fiber production workers. I. 1992 fiberglass cohort follow-up: Initial findings. *J. occup. environ. Med.*, **43**, 741–756

Marsh, G.M., Buchanich, J.M. & Youk, A.O. (2001b) Historical cohort study of US man-made vitreous fiber production workers. VI. Respiratory system cancer standardized mortality ratios adjusted for the confounding effect of cigarette smoking. *J. occup. environ. Med.*, **43**, 803–808

Marsh, G.M., Gula, M.J., Youk, A.O., Buchanich, J.M., Churg, A. & Colby, T.V. (2001c) Historical cohort study of US man-made vitreous fiber production workers. II. Mortality from mesothelioma. *J. occup. environ. Med.*, **43**, 757–766

Martin, J.-C., Imbernon, E., Goldberg, M., Chevalier, A. & Bonenfant, S. (2000) Occupational risk factors for lung cancer in the French electricity and gas industry. A case–control survey nested in a cohort of active employees. *Am. J. Epidemiol.*, **151**, 902–912

Mast, R.W., Hesterberg, T.W., Glass, L.R., McConnell, E.E., Anderson, R. & Bernstein, D.M. (1994) Chronic inhalation and biopersistence of refractory ceramic fiber in rats and hamsters. *Environ. Health Perspect.*, **102** (Suppl. 5), 207–209

Mast, R.W., McConnell, E.E., Hesterberg, T.W., Chevalier, J., Kotin, P., Thévenaz, P., Bernstein, D.M., Glass, L.R., Miiller, W.C. & Anderson, R. (1995a) Multiple-dose chronic inhalation toxicity study of size-separated kaolin refractory ceramic fiber in male Fischer 344 rats. *Inhal. Toxicol.*, **7**, 469–502

Mast, R.W., McConnell, E.E., Anderson, R., Chevalier, J., Kotin, P., Bernstein, D.M., Thévenaz, P., Glass, L.R., Miiller, W.C. & Hesterberg, T.W. (1995b) Studies on the chronic toxicity (inhalation) of four types of refractory ceramic fiber in male Fischer 344 rats. *Inhal. Toxicol.*, **7**, 425–467

Mast, R.W., Yu, C.P., Oberdörster, G., McConnell, E.E. & Utell, M.J. (2000a) A retrospective review of the carcinogenicity of refractory ceramic fiber in two chronic Fischer 344 rat inhalation studies: An assessment of the MTD and implications for risk assessment. *Inhal. Toxicol.*, **12**, 1141–1172

Mast, R.W., Maxim, L.D., Utell, M.J. & Walker A.M. (2000b) Refractory ceramic fiber: Toxicology, epidemiology, and risk analyses — A review. *Inhal. Toxicol.*, **12**, 359–399

Mattson, S.M. (1994) Glass fiber dissolution in simulated lung fluid and measures needed to improve consistency and correspondence to *in vivo* dissolution. *Environ. Health Perspect.*, **102** (Suppl. 5), 87–90

Maxim, L.D., Kelly, W.P., Walters, T. & Waugh, R. (1994) A multiyear workplace-monitoring program for refractory ceramic fibers. *Regulat. Toxicol. Pharmacol.*, **20**, S200–S215

Maxim, L.D., Allshouse, J.N., Kelly, W.P., Walters, T. & Waugh, R. (1997) A multiyear workplace-monitoring program for refractory ceramic fibers: Findings and conclusions. *Regul. Toxicol. Pharmacol.*, **26**, 156–171

Maxim, L.D., Allshouse, J.N., Deadman, J.E., Kleck, C., Kostka, M., Webster, D., Class, P. & Sébastien, P. (1998) CARE — A European programme for monitoring and reducing refractory ceramic fibre dust at the workplace: Initial results. *Gefahrstoffe Reinhalt. Luft*, **58**, 97–103

Maxim, L.D., Venturin, D. & Allshouse, J.N. (1999a) Respirable crystalline silica exposure associated with the installation and removal of RCF and conventional silica-containing refractories in industrial furnaces. *Regul. Toxicol. Pharmacol.*, **29**, 44–63

Maxim, L.D., Mast, R.W., Utell, M.J., Yu, C.P., Boymel, P.M., Zoitos, B.K. & Cason, J.E. (1999b) Hazard assessment and risk analysis of two new synthetic vitreous fibers. *Regul. Toxicol. Pharmacol.*, **30**, 54–74

Maxim, L.D., Allshouse, J.N., Chen, S.H., Treadway, J.C. & Venturin, D.E. (2000a) Workplace monitoring of refractory ceramic fiber in the United States. *Regul. Toxicol. Pharmacol.*, **32**, 293–309

Maxim, L.D., Allshouse, J.N. & Venturin, D.E. (2000b) The random-effects model applied to refractory ceramic fiber data. *Regul. Toxicol. Pharmacol.*, **32**, 190–199

McClellan, R.O. (1997) Use of mechanistic data in assessing human risks from exposure to particles. *Environ. Health Perspect.*, **105** (Suppl. 5), 1363–1372

McConnell, E.E. (1994) Synthetic vitreous fibers — Inhalation studies. *Regul. Toxicol. Pharmacol.*, **20**, S22–S34

McConnell, E.E., Wagner, J.C., Skidmore, J.W. & Moore, J.A. (1984) A comparative study of the fibrogenic and carcinogenic effects of UICC Canadian chrysotile B asbestos and glass microfibre (JM 100). In: *Biological Effects of Man-made Mineral Fibres* (Proceedings of a WHO/IARC Conference), Vol 2, Copenhagen, World Health Organization, pp. 234–252

McConnell, E.E., Kamstrup, O., Musselman, R., Hesterberg, T.W., Chevalier, J., Miiller, W.C. & Thévenaz, P. (1994) Chronic inhalation study of size-separated rock and slag wool insulation fibers in Fischer 344/N rats. *Inhal. Toxicol.*, **6**, 571–614

McConnell, E.E., Mast, R.W., Hesterberg, T.W., Chevalier, J., Kotin, P., Bernstein, D.M., Thévenaz, P., Glass, L.R. & Anderson, R. (1995) Chronic inhalation toxicity of a kaolin-based refractory ceramic fiber (RCF) in Syrian golden hamsters. *Inhal. Toxicol.*, **7**, 503–532

McConnell, E.E., Axten, C., Hesterberg, T.W., Chevalier, J., Miiller, W.C., Everitt, J., Oberdörster, G., Chase, G.R., Thévenaz, P. & Kotin, P. (1999) Studies on the inhalation toxicology of two fiberglasses and amosite asbestos in the Syrian golden hamster. Part II. Results of chronic exposure. *Inhal. Toxicol.*, **11**, 785–835

McDonald, J.C., Case, B.W., Enterline, P.E., Henderson, V., McDonald, A.D., Plourde, M. & Sébastien, P. (1990) Lung dust analysis in the assessment of past exposure of man-made mineral fibre workers. *Ann. occup. Hyg.*, **34**, 427–441

de Meringo, A., Morscheidt, C., Thélohan, S. & Tiesler, H. (1994) *In vitro* assessment of biodurability: Acellular systems. *Environ. Health Perspect.*, **102** (Suppl. 5), 47–53

Miller, B.G., Searl, A., Davis, J.M.G., Donaldson, K., Cullen, R.T., Bolton, R.E., Buchanan, D. & Soutar, C.A. (1999) Influence of fibre length, dissolution and biopersistence on the production of mesothelioma in the rat peritoneal cavity. *Ann. occup. Hyg.*, **43**, 155–166

Mohr, J.G. & Rowe, W.P. (1978) *Fiber Glass*, New York, Van Nostrand Reinhold

Monchaux, G., Bignon, J., Jaurand, M.C., Lafuma, J., Sebastien, P., Masse, R., Hirsch, A. & Goni, J. (1981) Mesotheliomas in rats following inoculation with acid-leached chrysotile asbestos and other mineral fibres. *Carcinogenesis*, **2**, 229–236

Monopolies and Mergers Commission (1991) *The Morgan Crucible Company plc and Manville Corporation, a Report on the Merger Situation*, London, Her Majesty's Stationery Office

Moolgavkar, S.H., Turim, J., Brown, R.C. & Luebeck, E.G. (2001a) Long man-made fibers and lung cancer risk. *Regul. Toxicol. Pharmacol.*, **33**, 138–146

Moolgavkar, S.H., Brown, R.C. & Turim, J. (2001b) Biopersistence, fiber length, and cancer risk assessment for inhaled fibers. *Inhal. Toxicol.*, **13**, 755–772

Moorman, W.J., Mitchell, R.T., Mosberg, A.T. & Donofrio, D.J. (1988) Chronic inhalation toxicology of fibrous glass in rats and monkeys. *Ann. occup. Hyg.*, **32** (Suppl. 1), 757–767

Morgan, A. (1994) *In vivo* evaluation of chemical biopersistence of man-made mineral fibers. *Environ. Health Perspect.*, **102** (Suppl. 5), 127–131

Morgan, A. & Holmes, A. (1986) Solubility of asbestos and man-made mineral fibers *in vitro* and *in vivo*: Its significance in lung disease. *Environ. Res.*, **39**, 475–484

Morgan, A., Holmes, A. & Davison, W. (1982) Clearance of sized glass fibres from the rat lung and their solubility *in vivo*. *Ann. occup. Hyg.*, **25**, 317–331

Morrow, P.E. (1984) Pulmonary clearance. In: Esman & Mehlman, eds, *Advances in Modern Environmental Toxicology*, Vol. VIII, *Occupational and Industrial Hygiene: Concepts and Methods*, Princeton, NJ, Princeton Scientific Publishers, pp. 183–202

Morrow, P.E. (1988) Possible mechanisms to explain dust overloading of the lungs. *Fundam. appl. Toxicol.*, **10**, 369–384

Morrow, P.E. & Yu, C.P. (1985) Models of aerosol behavior in airways. In: Moren, F., Newhouse, M.T. & Dolovich, M.B., eds, *Aerosols in Medicine. Principles, Diagnosis and Therapy*, Amsterdam, Elsevier, pp. 149–191

Mossman, B.T. & Sesko, A.M. (1990) In vitro assays to predict the pathogenicity of mineral fibers. *Toxicology*, **60**, 53–61

Moulin, J.J., Pham, Q.T., Mur, J.M., Meyer-Bisch, C., Caillard, J.F., Massin, N., Wild, P., Teculescu, D., Delepine, P., Hunzinger, E., Perreaux, J.P. & Muller, J. (1987) [Epidemiological study in two factories producing artificial mineral fibres: II. Respiratory symptoms and lung function.] *Arch. Mal. prof.*, **48**, 7–16 (in French)

Moulin, J.J., Mur, J.M., Wild, P., Perreaux, J.P. & Pham, Q.T. (1986) Oral cavity and laryngeal cancers among man-made mineral fiber production workers. *Scand. J. Work Environ. Health*, **12**, 27–31

Moulin, J.J., Wild, P., Mur, J.M., Caillard, J.F., Massin, N., Meyer-Bisch, C., Toamain, J.P., Hanser, P., Liet, S., DuRoscoat, M.N. & Segala, A. (1988) Respiratory health assessment by questionnaire of 2024 workers involved in man-made mineral fiber production. *Int. Arch. occup. environ. Health*, **61**, 171–178

Muhle, H. & Bellmann, B. (1995) Biopersistence of man-made vitreous fibres. *Ann. occup. Hyg.*, **39**, 655–660

Muhle, H. & Bellmann, B. (1997) Significance of the biodurability of man-made vitreous fibers to risk assessment. *Environ. Health Perspect.*, **105** (Suppl. 5), 1045–1047

Muhle, H., Pott, F., Bellmann, B., Takenaka, S. & Ziem, U. (1987) Inhalation and injection experiments in rats to test the carcinogenicity of MMMF. *Ann. occup. Hyg.*, **31**, 755–764

Muhle, H., Bellmann, B. & Pott, F. (1994) Comparative investigations of the biodurability of mineral fibers in the rat lung. *Environ. Health Perspect.*, **102** (Suppl. 5), 163–168

Murata-Kamiya, N., Tsutsui, T., Fujino, A., Kasai, H. & Kaji, H. (1997) Determination of carcinogenic potential of mineral fibers by 8-hydroxydeoxyguanosine as a marker of oxidative DNA damage in mammalian cells. *Int. Arch. occup. environ. Health*, **70**, 321–326

Musselman, R.P., Miiller, W.C., Eastes, W., Hadley, J.G., Kamstrup, O., Thevenaz, P. & Hesterberg, T.W. (1994a) Biopersistences of man-made vitreous fibers and crocidolite fibers in rat lungs following short-term exposures. *Environ. Health Perspect.*, **102** (Suppl. 5), 139–143

Musselman, R.P., Miiller, W.C., Eastes, W., Hadley, J.G., Kamstrup, O., Thevenaz, P. & Hesterberg, T.W. (1994b) Biopersistence of crocidolite versus man-made vitreous fibers in rat lungs after brief exposures. In: Mohr, U., Dungworth, D.L., Mauderly, J.L. & Oberdörster, G., eds, *Toxic and Carcinogenic Effects of Solid Particles in the Respiratory Tract*, Washington, DC, ILSI Press, pp. 451–454

Nadeau, D., Fouquette-Couture, L., Paradis, D., Khorami, J., Lane, D., Dunnigan, J. (1987) Cytotoxicity of respirable dusts from industrial minerals: Comparison of two naturally occurring and two man-made silicates. *Drug Chem. Toxicol.*, **10**, 49–86

NAIMA (North American Insulation Manufacturers' Association) and EURIMA (European Insulation Manufacturers' Association) (2001) *Background Material Prepared for IARC by NAIMA, EURIMA, and FARIMA (Fiberglass and Rockwool Insulation Manufacturers' Association of Australia)*

National Research Council (2000) *Review of The U.S. Navy's Exposure Standard for Manufactured Vitreous Fibers*, Washington, DC, National Academy Press, pp. 1–73

Nielsen, O. (1987) Man made mineral fibers in the indoor climate caused by ceiling of man-made mineral wool. In: Seifert, B., Esdorn, H., Fisher, M., Rüden, H. & Wegner, J., eds, *Indoor Air, 1987, Proceedings of the 4th International Conference on Indoor Air Quality and Climate*, Vol. 1, Berlin, Institute of Water, Soil and Air Hygiene, pp. 580–583

Nikitina, O.V., Kogan, F.M., Vanchugova, N.N. & Frash, V.N. (1989) [Comparative oncogenicity of basalt fibers and chrysotile asbestos.] *Gig. Tr. prof. Zbl.*, **4**, 7–11 (in Russian)

NIOSH (National Institute for Occupational Safety and Health) (1994) Asbestos and other fibers by PCM. In: *NIOSH Manual of Analytical Methods* (NMAM) (4th Ed.) (www.cdc.gov/niosh), Method 7400, Issue 2, Cincinnati, OH

Noack, K.H. & Böckler-Klusemann, M. (1993) [Asbestos in the Environment.] In: Grohmann, D. & Sonnenschein, G., eds, Asbest (Special Edition of the Newsletter 'Working Safety'), Maschinenbau- und Metall-Berufsgenossenschaft, Bielefeld, Bielefelder Verlagsanstalt (in German)

Oberdörster, G. (1988) Lung clearance of inhaled insoluble and soluble particles. *J. Aerosol Med.*, **1**, 289–330

Oberdörster, G. (1989) Combined effects of tobacco smoke and asbestos fibers in the lung: Synergism or increased dose to target sites? In: *Biological Interaction of Inhaled Mineral Fibers and Cigarette Smoke, Proceedings of an International Symposium-Workshop*, Columbus, OH, Battelle Press, pp. 195–209

Oberdörster, G. (1996) Evaluation and use of animal models to assess mechanisms of fibre carcinogenicity. In: Kane, A.B., Boffetta, P. & Wilbourn, J.D., eds., *Mechanisms of Fibre Carcinogenesis* (IARC Scientific Publication No. 140), Lyon, IARC*Press*, pp. 107–125

Oberdörster, G. (2000) Determinants of the pathogenicity of man-made vitreous fibers (MMVF). *Int. Arch. occup. environ. Health*, **73** (Suppl.), S60–S68

Oberdörster, G. (2002) Toxicokinetics and effects of fibrous and nonfibrous particles. *Inhal. Toxicol.*, **14**, 29–56

Oberdörster, G., Morrow, P.E. & Spurny, K. (1988) Size dependent lymphatic short term clearance of amosite fibers in the lung. *Ann. occup. Hyg.*, **32**, 149–156

Oberdörster, G., Ferin, J. & Lehnert, B.E. (1994) Correlation between particle size, *in vivo* particle persistence and lung injury. *Environ. Health Perspect.*, **102** (Suppl. 5), 173–179

Ohyama, M., Otake, T. & Morinaga, K. (2000) The chemiluminescent response from human monocyte-derived macrophages exposed to various mineral fibers of different sizes. *Ind. Health*, **38**, 289–293

Okayasu, R., Wu, L. & Hei, T.K. (1999) Biological effects of naturally occurring and man-made fibres: *In vitro* cytotoxicity and mutagenesis in mammalian cells. *Br. J. Cancer*, **79**, 1319–1324

Ong, T., Liu, Y., Zhong, B.-Z., Jones, W.G. & Whong, W.-Z. (1997) Induction of micronucleated and multinucleated cells by man-made fibers in vitro in mammalian cells. *J. Toxicol. environ. Health*, **50**, 409–414

Osgood, C.J. (1994) Refractory ceramic fibers (RCFs) induce germline aneuploidy in Drosophila oocytes. *Mutat. Res.*, **324**, 23–27

Oshimura, M., Hesterberg, T.W., Tsutsui, T. & Barrett, J.C. (1984) Correlation of asbestos-induced cytogenetic effects with cell transformation of Syrian hamster embryo cells in culture. *Cancer Res.*, **44**, 5017–5022

Ottery, J., Cherrie, J.W., Dodgson, J. & Harrison, G.E. (1984) A summary report on the environmental conditions at 13 European MMMF plants. In: *Biological Effects of Man-Made Mineral Fibres (Proceedings of a WHO/IARC Conference)*, Vol. 1, Copenhagen, World Health Organization, pp. 83–117

Pache, J.-C., Janssen, Y.M.W., Walsh, E.S., Quinlan, T.R., Zanella, C.L., Low, R.B., Taatjes, D.J. & Mossman, B.T. (1998) Increased epidermal growth factor-receptor protein in a human mesothelial cell line in response to long asbestos fibers. *Am. J. Pathol.*, **152**, 333–340

Pelin, K., Kivipensas, P. & Linnainmaa, K. (1995) Effects of asbestos and man-made vitreous fibers on cell division in cultured human mesothelial cells in comparison to rodent cells. *Environ. mol. Mutag.*, **25**, 118–125

Perrault, G., Dion, C. & Cloutier, Y. (1992) Sampling and analysis of mineral fibers on construction sites. *Appl. occup. environ. Hyg.*, **7**, 323–326

Petersen, R. & Sabroe, S. (1991) Irritative symptoms and exposure to mineral wool. *Am. J. ind. Med.*, **20**, 113–122

Pigott, G.H. & Ishmael, J. (1992) The effects of intrapleural injections of alumina and aluminosilicate (ceramic) fibres. *Int. J. exp. Pathol.*, **73**, 137–146

Plato, N., Krantz, S., Anderson, L., Gustavsson, P. & Lundgren, L. (1995a) Characterization of current exposure to man-made vitreous fibres (MMVF) in the prefabricated house industry in Sweden. *Ann. occup. Hyg.*, **39**, 167–179

Plato, N., Krantz, S., Gustavsson, P., Smith, T.J. & Westerholm, P. (1995b) Fiber exposure assessment in the Swedish rock wool and slag wool production industry in 1938–1990. *Scand. J. Work Environ. Health*, **21**, 345–352

Plato, N., Westerholm, P., Gustavsson, P., Hemmingsson, T., Hogstedt, C. & Krantz, S. (1995c) Cancer incidence, mortality and exposure–response among Swedish man-made vitreous fiber production workers. *Scand. J. Work Environ. Health*, **21**, 353–361

Pohlabeln, H., Jöckel, K.-H., Brüske-Hohlfeld, I., Möhner, M., Ahrens, W., Bolm-Audorff, U., Arhelger, R., Römer, W., Kreienbrock, L., Kreuzer, M., Jahn, I. & Wichmann, H.-E. (2000) Lung cancer and exposure to man-made vitreous fibers: Results from a pooled case-control study in Germany. *Am. J. ind. Med.*, **37**, 469–477

Pott, F. (1989) Carcinogenicity of fibres in experimental animals — Data and evaluation. In: Bates, D.V., Dungworth, D.L., Lee, P.N., McClellan, R.O. & Roe, F.J.C., eds, *Assessment of Inhalation Hazards* (ILSI Monographs), Berlin, Springer-Verlag, pp. 243–253

Pott, F. (1995) Detection of mineral fibre carcinogenicity with the intraperitoneal test — Recent results and their validity. *Ann. occup. Hyg.*, **39**, 771–779

Pott, F., Huth, F. & Friedrichs, K.H. (1974) Tumorigenic effect of fibrous dusts in experimental animals. *Environ. Health Perspect.*, **9**, 313–315

Pott, F., Friedrichs, K.H. & Huth, F. (1976) [Results of animal experiments concerning the carcinogenic effect of fibrous dusts and their interpretation with regard to the carcinogenesis in humans.] *Zbl. Bakteriol. Orig. B*, **162**, 467–505 (in German)

Pott, F., Huth, F. & Spurny, K (1980) Tumour induction after intraperitoneal injection of fibrous dust. In: Wagner, J.C., ed., *Biological Effects of Mineral Fibres* (IARC Scientific Publications No. 30), Lyon, IARC*Press*, pp. 337–342

Pott, F., Schlipköter, H.W., Ziem, U., Spurny, K. & Huth, F. (1984a) New results from implantation experiments with mineral fibres. In: *Biological Effects of Man-Made Mineral Fibres* (Proceedings of a WHO/IARC Conference), Vol. 2, Copenhagen, World Health Organization, pp. 286–302

Pott, F., Ziem, U. & Mohr, U. (1984b) *Lung carcinomas and mesotheliomas following intratracheal instillation of glass fibres and asbestos.* In: *Proceedings of the VIth International Pneumoconiosis Conference, Bochum, Federal Republic of Germany, 20–23 September 1983*, Vol. 2, Geneva, International Labour Office, pp. 746–756

Pott, F., Ziem, U., Reiffer, F.-J., Huth, F., Ernst, H. & Mohr, U. (1987) Carcinogenicity studies on fibres, metal compounds, and some other dusts in rats. *Exp. Pathol.*, **82**, 129–152

Pott, F., Matscheck, A., Ziem, U., Muhle, H. & Huth, F. (1988) Animal experiments with chemically treated fibres. *Ann. occup. Hyg.*, **32** (Suppl. 1), 353–359

Pott, F., Roller, M., Ziem, U., Reiffer, F.-J., Bellmann, B, Rosenbruch, R. & Huth, F. (1989) Carcinogenicity studies on natural and man-made fibres with the intraperitoneal test in rats. In: Bignon, J., Peto, J. & Saracci, R., eds, *Non-Occupational Exposure to Mineral Fibres* (IARC Scientific Publications No. 90), Lyon, IARC*Press*, pp. 173–179

Pott, F., Roller, M., Rippe, R.M., Germann, P.-G. & Bellmann, B. (1991) Tumours by the intraperitoneal and intrapleural routes and their significance for the classification of mineral fibres. In: Brown, R.C., Hoskins, J.A. & Johnson, N.F., eds, *Mechanisms in Fibre Carcinogenesis* (NATO ASI Series 223), New York, Plenum Press, pp. 547–565

Pott, F., Roller, M., Althoff, G.H., Kamino, K., Bellmann, B. & Ulm, K. (1993) [Estimation of the carcinogenicity of inhaled fibres.] *VDI Berichte*, **No. 1075**, 17–77 (in German)

Pott, F., Dungworth, D.L., Heinrich, U., Muhle, H., Kamino, K., Germann, P.-G., Roller, M., Rippe, R.M. & Mohr, U. (1994) Lung tumours in rats after intratracheal instillation of dusts. *Ann. occup. Hyg.*, **38**, 357–363

Potter, R.M. & Mattson, S.M. (1991) Glass fiber dissolution in a physiological saline solution. *Glastech. Ber.*, **64**, 16–28

Quinn, M.M., Smith, T.J., Youk, A.O., Marsh, G.M., Stone, R.A., Buchanich, J.M. & Gula, M.J. (2001) Historical cohort study of US man-made vitreous fiber production workers. VIII. Exposure-specific job analysis. *J. occup. environ. Med.*, **43**, 824–834

Riboldi, L., Rivolta, G., Barducci, M., Errigo, G. & Picchi, O. (1999) [Respiratory disease caused by MMVF fibres and yarn.] *Med. Lav.*, **90**, 53–66 (in Italian)

Rice, C., Lockey, J., Lemasters, G., Levin, L. & Gartside, P. (1996) Identification of changes in airborne fibre concentrations in refractory ceramic fibre manufacture related to process or ventilation modifications. *Occup. Hyg.*, **3**, 85–90

Rice, C.H., Lockey, J.E., Lemasters, G.K., Levin, L.L., Staley, P. & Hansen, K.R. (1997) Estimation of historical and current employee exposure to refractory ceramic fibers during manufacturing and related operations. *Appl. occup. environ. Hyg.*, **12**, 54–61

Rindel, A.C., Bach, E., Breum, N.O., Hugod, E. & Schneider, T. (1987) Correlating health effect with indoor air quality in kindergartens. *Int. Arch. Occup. Environ. Health*, **59**, 363–373

Rockwool International (2001) *Statement for IARC, 2001-10-10*, Rockwool International A/S, Hedenhusene, Denmark

Rödelsperger, K., Teichert, U., Marfels, H., Spurny, K., Arhelger, R. & Woitowitz, H.-J. (1989) Measurement of inorganic fibrous particulates in ambient air and indoors with the scanning electron microscope. In: Bignon, J., Peto, J. & Saracci, R., eds, *Non-Occupational Exposure to Mineral Fibres* (IARC Scientific Publication No. 90), Lyon, IARC*Press*, pp. 361–366

Rödelsperger, K., Barbisan, P., Teichert, U., Arhelger, R. & Woitowitz, H.-J. (1998) Indoor emission of mineral wool products. *VDI Berichte*, **1417**, 337–354

Rödelsperger, K., Jöckel, K.-H., Pohlabeln, H., Römer, W. & Woitowitz, H.-J. (2001) Asbestos and man-made vitreous fibers as risk factors for diffuse malignant mesothelioma: Results from a German hospital-based case–control study. *Am. J. ind. Med.*, **39**, 262–275

Roller, M. & Pott, F. (1998) Carcinogenicity of man-made fibres in experimental animals and its relevance for classification of insulation wools. *Eur. J. Oncol.*, **3**, 231–239

Roller, M., Pott, F., Kamino, K., Althoff, G.-H. & Bellmann, B. (1996) Results of current intraperitoneal carcinogenicity studies with mineral and vitreous fibers. *Exp. Toxicol. Pathol.*, **48**, 3–12

Roller, M., Pott, F., Kamino, K., Althoff, G.-H. & Bellmann, B. (1997) Dose-response relationship of fibrous dusts in intraperitoneal studies. *Env. Health Persp.*, **105** (Suppl. 5), 1253–1256

Rood, A.P. & Streeter, R.R. (1985) Size distribution of airborne superfine man-made mineral fibers determined by transmission electron microscopy. *Am. ind. Hyg. Assoc. J.*, **46**, 257–261

Rossiter, C.E. (1993) Pulmonary effects of exposure to fine fibreglass: Irregular opacities and small airways obstruction. *Br. J. ind. Med.*, **50**, 382–384 (correspondence)

Rossiter, C.E., Gilson, J.C., Sheers, G., Thomas, H.F., Trenthowan, W.N., Cherrie, J.W. & Harrington, J.M. (1994) Refractory ceramic fiber production workers: Analysis of radiograph readings. *Ann. occup. Hyg.*, **38**, 731–738

Ruotsalainen, M., Hirvonen, M.-R., Luoto, K. & Savolainen, K.M. (1999) Production of reactive oxygen species by man-made vitreous fibers in human polymorphonuclear leukocytes. *Hum. exp. Toxicol.*, **18**, 354–362

Rutten, A., Bermudez, E., Mangum, J., Wong, B., Moss, O. & Everitt, J. (1994a) Pleural pathology in hamsters and rats exposed to instilled man-made vitreous fibers. In: Dungworth, D.L., Mauderly, J.L. & Oberdörster, G., eds, *Toxic and Carcinogenic Effects of Solid Particles in the Respiratory Tract* (ILSI Monographs), Washington, DC, ILSI Press, pp. 595–598

Rutten, A.A.J.J.L., Bermudez, E., Mangum, J.B., Wong, B.A., Moss, O.R. & Everitt, J.I. (1994b) Mesothelial cell proliferation induced by intrapleural instillation of man-made fibers in rats and hamsters. *Fundam. appl. Toxicol.*, **23**, 107–116

Sali, D., Boffetta, P., Andersen, A., Cherrie, J.W., Chang Claude, J., Hansen, J., Olsen, J.H., Pesatori, A.C., Plato, N., Teppo, L., Westerholm, P., Winter, P. & Saracci, R. (1999) Non-neoplastic mortality of European workers who produce man made vitreous fibers. *Occup. environ. Med.*, **56**, 612–617

Saracci, R., Simonato, L., Acheson, E.D., Andersen, A., Bertazzi, P.A., Claude, J., Charnay, N., Estève, J., Frentzel-Beyme, R.R., Gardner, M.J., Jensen, O.M., Maasing, R., Olsen, J.H., Teppo, L.H.I., Westerholm, P. & Zocchetti, C. (1984a) The IARC mortality and cancer incidence study of MMMF production workers. In: *Biological Effects of Man-Made Mineral Fibres (Proceedings of a WHO/IARC Conference)*, Vol. 1, Copenhagen, World Health Organization, pp. 279–310

Saracci, R., Simonato, L., Acheson, E.D., Andersen, A., Bertazzi, P.A., Claude, J., Charnay, N., Estève, J., Frentzel-Beyme, R.R., Gardner, M.J., Jensen, O.M., Maasing, R., Olsen, J.H., Teppo, L.H.I., Westerholm, P. & Zocchetti, C. (1984b) Mortality and incidence of cancer of workers in the man made vitreous fibres producing industry: An international investigation at 13 European plants. *Br. J. ind. Med.*, **41**, 425–436

Schins, R.P.F. & Donaldson, K. (2000) Nuclear factor kappa-B activation by particles and fibers. *Inhal. Toxicol.*, **12** (Suppl. 3), 317–326

Schlesinger, R.B., Ben-Jebria, A., Dahl, A.R., Snipes, M.D. & Ultman, J. (1997) Disposition of inhaled toxicants. In: Massaro, E.J., ed., *Handbook of Human Toxicology*, Boca Raton, FL, CRC Press, pp. 493–550

Schneider, T. (1979a) Exposures to man-made mineral fibres in user industries in Scandinavia. *Ann. occup. Hyg.*, **22**, 153–162

Schneider, T. (1979b) The influence of counting rules on the number and on the size distribution of fibers. *Ann. occup. Hyg.*, **21**, 341–350

Schneider, T. (1984) Review of surveys in industries that use MMMF. In: *Biological Effects of Man-Made Mineral Fibres (Proceedings of a WHO/IARC Conference)*, Vol. 1, Copenhagen, World Health Organization., pp. 178–190

Schneider, T. (1986) Man-made mineral fibres and other fibres in the air and settled dust. *Environ. int.*, **12**, 61–65

Schneider, T. (2000) Synthetic vitreous fibers. In: Spengler, J.D., Samet, J.M. & McCarthy, J.F., eds, *Indoor Air Quality Handbook*, New York, McGraw-Hill, pp. 39.1–39.29

Schneider, T. & Holst, E. (1983) Man-made mineral fibre size distributions utilizing unbiased and fibre length biased counting methods and the bivariate log-normal distribution. *J. Aerosol. Sci.*, **14**, 139–146

Schneider, T. & Stokholm, J. (1981) Accumulation of fibres in the eyes of workers handling man-made mineral fibre products. *Scand. J. Work Environ. Health*, **7**, 271–276

Schneider, T., Holst, E. & Skotte, J. (1983) Size distributions of airborne fibres generated from man-made mineral fibre products. *Ann. occup. Hyg.*, **27**, 157–171

Schneider, T., Skotte, J. & Nissen, P. (1985) Man-made mineral fibre size fractions and their interrelation. *Scand. J. Work Environ. Health*, **11**, 117–122

Schneider, T., Nielsen, O., Bredsdorff, P. & Linde, P. (1990) Dust in buildings with man-made mineral fiber ceiling boards. *Scand. J. Work Environ. Health*, **16**, 434–439

Schneider, T., Burdett, G., Martinon, L., Brochard, P., Guillemin, M., Teichert, U. & Draeger, U. (1996) Ubiquitous fiber exposure in selected sampling sites in Europe. *Scand. J. Work Environ. Health*, **22**, 274–284

Schnittger, J. (1993) [Measurement and identification of inorganic fibres in ambient air.] In: *Faserförmige Stäube Vorschriften, Wirkungen, Messung, Minderung* (VDI Berichte No. 1075), Düsseldorf, Verein Deutscher Ingenieure Verlag, pp. 283–295 (in German)

Scholze, H. & Conradt, R. (1987) An *in vitro* study of the chemical durability of siliceous fibres. *Ann. occup. Hyg.*, **31**, 683–692

Schupp, M. (1990) A global view of the RCF market. *Ceram. Bull.*, **69**, 63–64

Schürch, S., Gehr, P., Im Hof, V., Geiser, M. & Green, F. (1990) Surfactant displaces particles toward the epithelium in airways and alveoli. *Respir. Physiol.*, **80**, 17–32

Searl, A., Buchanan, D., Cullen, R.T., Jones, A.D., Miller, B.G. & Soutar, C.A. (1999) Biopersistence and durability of nine mineral fibre types in rat lungs over 12 months. *Ann. occup. Hyg.*, **43**, 143–153

Sébastien, P. (1994) Biopersistence of man-made vitreous silicate fibers in the human lung. *Environ. Health Perspect.*, **102** (Suppl. 5), 225–228

Sebring, R.J. & Lehnert, B.E. (1992) Morphometric comparisons of rat alveolar macrophages, pulmonary interstitial macrophages, and blood monocytes. *Exp. Lung Res.*, **18**, 479–496

Shannon, H.S., Hayes, M., Julian, J.A. & Muir, D.C.F. (1984) Mortality experience of glass fibre workers. *Br. J. ind. Med.*, **41**, 35–38

Shannon, H.S., Jamieson, E., Julian, J.A., Muir, D.C.F. & Walsh, C. (1987) Mortality experience of Ontario glass fibre workers — Extended follow-up. *Ann. occup. Hyg.*, **31**, 657–662

Shannon, H.S., Jamieson, E., Julian, J.A. & Muir, D.C.F. (1990) Mortality of glass filament (textile) workers. *Br. J. ind. Med.*, **47**, 533–536

Shaulian, E. & Karin, M. (2001) AP-1 in cell proliferation and survival. *Oncogene*, **20**, 2390–2400

Siemiatycki, J., ed. (1991) *Risk Factors for Cancer in the Workplace*, Boca Raton, FL, CRC Press

Simonato, L., Fletcher, A.C., Cherrie, J., Andersen, A., Bertazzi, P.A., Charnay, N., Claude, J., Dodgson, J., Estève, J., Frentzel-Beyme, R.R., Gardner, M.J., Jensen, O.M., Olsen, J., Saracci, R., Teppo, L.H.I., Westerholm, P., Winkelmann, R., Winter, P.D. & Zocchetti, C. (1986a) The man-made mineral fibre European historical cohort study: Extension of the follow-up. *Scand. J. Work environ. Health*, **12** (Suppl. 1), 34–47 (Corrigendum in *Scand. J. Work environ. Health*, **13**, 192)

Simonato, L., Fletcher, A.C., Cherrie, J., Andersen, A., Bertazzi, P.A., Charnay, N., Claude, J., Dodgson, J., Estève, J., Frentzel-Beyme, R.R., Gardner, M.J., Jensen, O.M., Olsen, J., Saracci, R., Teppo, L.H.I., Westerholm, P., Winkelmann, R., Winter, P.D. & Zocchetti, C. (1986b) Updating lung cancer mortality among a cohort of man-made mineral fibre production workers in seven European countries. *Cancer Lett.*, **30**, 189–200

Simonato, L., Fletcher, A.C., Cherrie, J., Andersen, A., Bertazzi, P., Charnay, N., Claude, J., Dodgson, J., Estève, J., Frentzel-Beyme, R., Gardner, M.J., Jensen, O., Olsen, J., Teppo, L., Winkelmann, R., Westerholm, P., Winter, P.D., Zocchetti, C. & Saracci, R. (1987) The International Agency for Research on Cancer historical cohort study of MMMF production workers in seven European countries: Extension of the follow-up. *Ann. occup. Hyg.*, **31**, 603–623

Skov, P. & Valbjørn, O. and the Danish Indoor Climate Study Group (1987) The 'sick' building syndrome in office environment: The Danish Town Hall study. *Environ. Int.*, **13**, 339–349

Smith, D.M., Ortiz, L.W., Archuleta, R.F. & Johnson, N.F. (1987) Long-term health effects in hamsters and rats exposed chronically to man-made vitreous fibres. *Ann. occup. Hyg.*, **31**, 731–754

Smith, T.J., Quinn, M.M., Marsh, G.M., Youk, A.O., Stone, R.A., Buchanich, J.M. & Gula, M.J. (2001) Historical cohort study of US man-made vitreous fiber workers: VII. Overview of the exposure assessment. *J. occup. environ. Med.*, **43**, 809–823

Snipes, M.B. (1989) Long-term retention and clearance of particles inhaled by mammalian species. *Crit. Reviews in Toxicol.*, **20**, 175–211

Spurny, K.R. & Stöber, W. (1981) Some aspects of analysis of single fibers in environmental and biological samples. *Int. J. environ. anal. Chem.*, **9**, 265–281

Stahlhofen, W., Scheuch, G. & Bailey, M.R. (1995) Investigations of retention of inhaled particles in the human bronchial tree. *Radiat. Protect. Dosim.*, **60**, 311–319

Stanton, M.F. & Wrench, C. (1972) Mechanisms of mesothelioma induction with asbestos and fibrous glass. *J. natl Cancer Inst.*, **48**, 797–821

Stanton, M.F., Layard, M., Tegeris, A., Miller, E., May, M. & Kent, E. (1977) Carcinogenicity of fibrous glass: Pleural response in the rat in relation to fiber dimension. *J. natl Cancer Inst.*, **58**, 587–603

Stanton, M.F., Layard, M., Tegeris, A., Miller, E., May, M., Morgan, E. & Smith, A. (1981) Relation of particle dimension to carcinogenicity in amphibole asbestoses and other fibrous minerals. *J. natl Cancer Inst.*, **67**, 965–975

Steenberg, T., Hjenner, H.K., Lund Jensen, S., Guldberg, M. & Knudsen, T. (2001) Dissolution behaviour of biosoluble HT stone wool fibres. *Glastech. Ber. Glass Sci. Technol.*, **74**, 97–105

Stone, K.C., Mercer, R.R., Gehr, P., Stockstill, B. & Crapo, J.D. (1992) Allometric relationships of cell numbers and size in the mammalian lung. *Am. J. respir. Cell mol. Biol.*, **6**, 235–243

Stone, R.A., Youk, A.O., Marsh, G.M., Buchanich, J.M., McHenry, M.B. & Smith, T.J. (2001) Historical cohort study of US man-made vitreous fiber production workers. IV. Quantitative exposure-response analysis of the nested case-control study of respiratory system cancer. *J. occup. environ. Med.*, **43**, 779–792

Stöber, W., McClellan, R.O. & Morrow, P.E. (1993) Approaches to modeling disposition of inhaled particles and fibers in the lung. In: Gardner *et al.*, eds, *Toxicology of the Lung*, 2nd Ed., New York, Raven Press, pp. 527–601

Swedish National Board of Occupational Safety and Health (1996) *Provisions on Occupational Exposure Limit Values* (AFS 1996:2). Stockholm, Arbetarskyddsstyrelsen (http://www.av.se)

Switala, E.D., Harlan, R.C., Schlaudecker, D.G. & Bender, J.R. (1994) Measurement of respirable glass and total fiber concentrations in the ambient air around a fiberglass wool manufacturing facility and a rural area. *Regul. Toxicol. Pharmacol.*, **20**, S76–S88

Takahashi, T., Munakata, M., Takekawa, H., Homma, Y. & Kawakami, Y. (1996) Pulmonary fibrosis in a carpenter with long-lasting exposure to fiberglass. *Am. J. ind. Med.*, **30**, 596–600

Taylor, D.G. (1977) Asbestos fibers in air. Method No. P & CAM 289. In: *NIOSH Manual of Analytical Methods*, 2nd Ed., Cincinnati, OH, National Institute for Occupational Safety and Health

Thestrup-Pedersen, K., Bach, B. & Pedersen, R. (1990) Allergic investigation in patients with the sick building syndrome. *Contact Derm.*, **23**, 53–55

Thriene, B., Sobottka, A., Willer, H. & Weidhase, J. (1996) Man-made mineral fibre boards in buildings — Health risks caused by quality deficiencies. *Toxicol. Lett.*, **88**, 299–303

Tiesler, H. & Draeger, U. (1993) Measurement and identification of insulation product-related fibres in contrast to ubiquitous fibres. In: Kalliokoski, P., Jantunen, M. & Seppänen, O., *Indoor Air 1993 Proceedings of the 6th International Conference on Indoor Air Quality and Climate*, Vol. 4, Helsinki, Jyväskylä, Finland, Gummerus, Oy, pp. 111–116

Tiesler, H., Teichert, U. & Draeger, U. (1990) [Studies on the fibre dust exposition at building sites caused by insulation material from mineral wool.] *Zb. Arbeitsmed.*, **40**, 307–319 (in German)

Tiesler, H., Draeger, U. & Rogge, D. (1993) Emission of fibrous particles from installed insulation mineral wool products. In: Kalliokoski, P., Jantunen, M. & Seppänen, O., *Indoor Air 1993 Proceedings of the 6th International Conference on Indoor Air Quality and Climate*, Vol. 4, Helsinki, Jyväskylä, Finland, Gummerus, Oy, pp. 117–121

TIMA (Thermal Insulation Manufacturers Association) (1993) *Man-made Vitreous Fibers: Nomenclature, Chemical and Physical Properties*, 4th Ed., Eastes, W., ed., Nomenclature Committee of Thermal Insulation Manufacturers' Association, Refractory Ceramic Fibers Coalition (RCFC), Washington, DC

Timbrell, V. (1965) The inhalation of fibrous dusts. *Ann. N.Y. Acad. Sci.*, **132**, 255–273

Tran, C.L., Buchanan, D., Cullen, R.T., Searl, A., Jones, A.D. & Donaldson, K. (2000) Inhalation of poorly soluble particles. II. Influence of particle surface area on inflammation and clearance. *Inhal. Toxicol.*, **12**, 1113–1126

Treadwell, M.D., Mossman, B.T. & Barchowsky, A. (1996) Increased neutrophil adherence to endothelial cells exposed to asbestos. *Toxicol. appl. Pharmacol.*, **139**, 62–70

Trethowan, W.N., Burge, P.S., Rossiter, C.E., Harrington, J.M. & Calvert, I.A. (1995) Study of the repiratory health of employees in seven European plants that manufacture ceramic fibres. *Occup. environ. Med.*, **52**, 97–104

TRGS (Technische Regeln für Gefahrstoffe) (1999) [Limit values in the air for non-organic dust fibres.], TRGS 901, German Federal Ministry of Labour and Social Affairs, *Bundes-ArbeitsBlatt*, **4**, 42 (in German)

TRGS (Technische Regeln für Gefahrstoffe) (2001) [Justification for the Classification of Dusts from Natural and Man-Made Mineral Fibres], TRGS 905, German Federal Ministry of Labour and Social Affairs, *Bundesarbeitsblatt*, **6**, 57

United Kingdom Factories Inspectorate (1987) *Survey of Superfine Man-Made Mineral Fibre Exposure in the UK*, London, Health and Safety Executive Advisory Committee on Toxic Substances, Occupational Medicine and Hygiene Laboratories

Van der Wal, J.F., Ebens, R. & Tempelman, J. (1987) Man-made mineral fibres in homes caused by thermal insulation. *Atmos. Environ.*, **21**, 13–19

Van De Weyer, R. & Nolard, N. (1992) [Extrinsic allergic alveolitis in a fiber glass production worker.] *Rev. fr. Mal. respir.*, **Suppl. 3**, 150 (in French)

Vasama-Neuvonen, K., Pukkala, E., Paakkulainen, H., Mutanen, P., Weiderpass, E., Boffetta, P., Shen, N., Kauppinen, T., Vainio, H. & Partanen, T. (1999) Ovarian cancer and occupational exposures in Finland. *Am. J. ind. Med.*, **36**, 83–89

Vastag, E., Matthys, H., Zsamboki, G., Köhler, D. & Daikeler, G. (1986) Mucociliary clearance in smokers. *Eur. J. respir. Dis.*, **68**, 107–113

VDI (Verein Deutscher Ingenieure) (1994) *Indoor Air Pollution Measurement. Measurement of Inorganic Fibrous Particles. Measurement Planning and Procedure. Scanning Electron Microscopy Method* (VDI 3492, part 2), Berlin, Deuth Verlag

Vetrotex (2001) La Défense, France (http://www.saint-gobainvetrotex.com)

Vorwald, A.J., Durkan, T.M. & Pratt, P.C. (1951) Experimental studies of asbestosis. *AMA Arch. ind. Hyg. occup. Med.*, **3**, 1–43

Wagner, J.C. (1963) Asbestosis in experimental animals. *Br. J. ind. Med.*, **20**, 1–12

Wagner, J.C., Berry, G. & Timbrell, V. (1973) Mesotheliomata in rats after inoculation with asbestos and other materials. *Br. J. Cancer*, **28**, 173–185

Wagner, J.C., Berry, G. & Skidmore, J.W. (1976) Studies of the carcinogenic effects of fiber glass of different diameters following intrapleural inoculation in experimental animals. In: LeVee, W.N. & Schulte, P.A., eds, *Occupational Exposure to Fibrous Glass* (DHEW Publ. No. (NIOSH) 76-151; NTIS Publ. No. PB-258869), Cincinnati, OH, National Institute for Occupational Safety and Health, pp. 193–204

Wagner, J.C., Berry, G.B., Hill, R.J., Munday, D.E. & Skidmore, J.W. (1984) Animal experiments with MMM(V)F — Effects of inhalation and intrapleural inoculation in rats. In: *Biological Effects of Man-made Mineral Fibres* (Proceedings of a WHO/IARC Conference), Vol 2, Copenhagen, World Health Organization, pp. 209–233

Walker, A.M., Maxim, L.D. & Utell, M. (2002) Risk analysis for mortality from respiratory tumors in a cohort of refractory ceramic fiber workers. *Regul. Toxicol. Pharmacol.*, **35**, 95–104

Wang, Q.-E., Han, C.-H., Wu, W.-D., Wang, H.-B., Liu, S.-J. & Kohyama, N. (1999a) Biological effects of man-made mineral fibers (I). Reactive oxygen species production and calcium homeostasis in alveolar macrophages. *Ind. Health*, **37**, 62–67

Wang, Q.-E., Han, C.-H., Yang, Y.-P., Wang, H.-B., Wu, W.-D., Liu, S.-J. & Kohyama, N. (1999b) Biological effects of man-made mineral fibers (II). Their genetic damages examined by in vitro assay. *Ind. Health*, **37**, 342–347

Warheit, D.B., Overby, L.H., George, G. & Brody, A.R. (1988) Pulmonary macrophages are attracted to inhaled particles through complement activation. *Exp. Lung Res.*, **14**, 51–66

Wastiaux, A., Blanchard, O. & Honnons, S. (1994) Possible application of urinary analysis to estimate dissolution of some man-made vitreous fibers. *Environ. Health Perspect.*, **102** (Suppl. 5), 217–219

Watkins, D.K., Chiazze, L. & Fryar, C. (1997) Historical cohort mortality study of a continuous filament fiberglass manufacturing plant. II. Women and minorities. *J. occup. environ. Med.*, **39**, 548–555

Weiderpass, E., Pukkala, E., Kauppinen, T., Mutanen, P., Paakkulainen, H., Vasama-Neuvonen, K., Boffetta, P. & Partanen, T. (1999) Breast cancer and occupational exposures in women in Finland. *Am. J. ind. Med.*, **36**, 48–53

Weill, H. & Hughes, J.M. (1996) Review of epidemiological data on morbidity following exposure to man-made vitreous fibres. *J. occup. Health Safety Aust. NZ*, **12**, 313–317

Weill, H., Hughes, J.M., Hammad, Y.Y., Glindmeyer, H.W., Sharon, G. & Jones, R.N. (1983) Respiratory health of workers exposed to man-made vitreous fibers. *Am. Rev. respir. Dis.*, **128**, 104–112

WHO (1985) *Reference Methods for Measuring Airborne Man-Made Mineral Fibres (MMMF)* (Environmental Health Series 4), World Health Organization, Copenhagen

WHO (1996) *Determination of Airborne Fibre Number Concentrations. A Recommended Method, by Phase Contrast Optical Microscopy (Membrane Filter Method)*, World Health Organization, Geneva (http://www.who.int/environmental-information)

Whong, W.-Z., Gao, H.-G., Zhou, G. & Ong, T. (1999) Genetic alterations of cancer-related genes in glass fiber-induced transformed cells. *J. Toxicol. environ. Health*, **56A**, 397–404

Wojtczak, J. (1994) [Exposure to ceramic fibers in the occupational environment. I. Production, kinds of ceramic fibers, changes in structure of fibers, preliminary studies in the working environment.] *Med. Prac.*, **45**, 479–486 (in Polish)

Wojtczak, J., Lao, I. & Krajnow, A. (1996) [Exposure to ceramic fibres in the work environment. II. Occupational exposure to dust in plants producing ceramic fibre: Fibrogenic effect of the fibres.] *Med. Prac.*, **47**, 559–567 (in Polish)

Wojtczak, J., Kiec-Šwierczynska, M. & Maciejewska, A. (1997) [Exposure to ceramic fibres in the environment. III. Occupational exposure to ceramic fibres in plants which produce and apply insulation materials made of ceramic fibres.] *Med. Prac.*, **48**, 51–60 (in Polish)

Wong, O., Foliart, D. & Trent, L.S. (1991) A case–control study of lung cancer in a cohort of workers potentially exposed to slag wool fibres. *Br. J. ind. Med.*, **48**, 818–824

Xie, C., Reusse, A., Dai, J., Zay, K., Harnett, J. & Churg, A. (2000) TNFα increases tracheal epithelial asbestos and fiberglass binding via a NF-κB-dependent mechanism. *Am. J. Physiol. Lung cell. mol. Physiol.*, **279**, L608–L614

Yamato, H., Tanaka, I., Higashi, T. & Kido, M. (1994a) Clearance of inhaled ceramic fibers from rat lungs. *Environ. Health Perspect.*, **102** (Suppl. 5), 169–171

Yamato, H., Hori, H., Tanaka, I., Higashi, T., Morimoto, Y. & Kido, M. (1994b) Retention and clearance of inhaled ceramic fibres in rat lungs and development of a dissolution model. *Occup. environ. Med.*, **51**, 275–280

Yamato, H., Morimoto, Y., Tsuda, T., Ohgami, A., Kohyama, N. & Tanaka, I. (1998) Fiber numbers per unit weight of JFM standard reference samples determined with a scanning electron microscope. *Ind. Health*, **36**, 384–387

Yamaya, M., Nakayama, K., Hosoda, M., Yanai, M. & Sasaki, H. (2000) A rockwool fibre worker with lung fibrosis. *Lancet*, **355**, 1723–1724

Yano, E. & Karita, K. (1998) Prevalence of respiratory abnormalities of workers in rock/slag wool producing industries in Japan. In: Keizo, C., Yutaka, H. & Yoshiharu, A., eds, *Advances in the Prevention of Occupational Respiratory Diseases: Proceedings of the 9th International Conference on Occupational Respiratory Diseases, Kyoto, 13–16 October 1997*, Amsterdam, Elsevier, pp. 337–341

Ye, J., Shi, X., Jones, W., Rojanasakul, Y., Cheng, N., Schwegler-Berry, D., Baron, P., Deye, G.J., Li, C. & Castranova, V. (1999) Critical role of glass fiber length in TNF-α production and transcription factor activation in macrophages. *Am. J. Physiol.*, **276**, L426–L434

Ye, J., Zeidler, P., Young, S.-H., Martinez, A., Robinson, V.A., Jones, W., Baron, P., Shi, X. & Castranova, V. (2001) Activation of mitogen-activated protein kinase p38 and extracellular signal-regulated kinase is involved in glass fiber-induced tumor necrosis factor-α production in macrophages. *J. biol. Chem.*, **276**, 5360–5367

Yegles, M., Janson, X., Dong, H.Y., Renier, A. & Jaurand, M.-C. (1995) Role of fibre characteristics on cytotoxicity and induction of anaphase/telophase aberrations in rat pleural mesothelial cells *in vitro*: Correlations with *in vivo* animal findings. *Carcinogenesis*, **16**, 2751–2758

Yeung, P. & Rogers, A. (1996) A comparison of synthetic mineral fibres exposures pre- and post the NOHSC national exposure standard and code of practice. *J. occup. Health Safety Austr. NZ*, **12**, 279–288

Youk, A.O., Marsh, G.M., Stone, R.A., Buchanich, J.M. & Smith, T.J. (2001) Historical cohort study of US man-made vitreous fiber production workers. III. Analysis of exposure-weighted measures of respirable fibers and formaldehyde in the nested case-control study of respiratory system cancer. *J. occup. environ. Med.*, **43**, 767–778

Yu, C.P., Zhang, L., Oberdörster, G., Mast, R.W., Glass, L.R. & Utell, M.J. (1994) Clearance of refractory ceramic fibers (RCF) from rat lung: Development of a model. *Environ. Res.*, **65**, 243–253

Yu, C.P., Zhang, L., Oberdörster, G., Mast, R.W., Maxim, D. & Utell, M.J. (1995a) Deposition of refractory ceramic fibers (RCF) in the human respiratory tract and comparison with rodent studies. *Aerosol Sci. Technol.*, **23**, 291–300

Yu, C.P., Ding, Y.J., Zhang, L., Oberdörster, G., Mast, R.W., Glass, L.R. & Utell, M.J. (1995b) Deposition and clearance modeling of inhaled kaolin refractory ceramic fibers (RCF) in hamsters — Comparison between species. *Inhal. Toxicol.*, **7**, 165–177

Yu, C.P., Ding, Y.J., Zhang, L., Oberdörster, G., Mast, R.W., Maxim, D. & Utell, M.J. (1996) A clearance model of refractory ceramic fibers (RCF) in the rat lung including fiber dissolution and breakage. *J. Aerosol Sci.*, **27**, 151–159

Yu, C.P., Ding, Y.J., Zhang, L., Oberdörster, G., Mast, R.W., Maxim, L.D. & Utell, M.J. (1997) Retention modeling of refractory ceramic fibers (RCF) in humans. *Regul. Toxicol. Pharmacol.*, **25**, 18–25

Yu, C.P., Dai, Y.T., Boymel, P.M., Zoitos, B.K., Oberdörster, G. & Utell, M.J. (1998) A clearance model of man-made vitreous fibers (MMVFs) in the rat lung. *Inhal. Toxicol.*, **10**, 253–274

Zhong, B.-Z., Whong, W.-Z. & Ong, T.-M. (1997) Detection of mineral-dust-induced DNA damage in two mammalian cell lines using the alkaline single cell gel/comet assay. *Mutat. Res.*, **393**, 181–187

Zoitos, B.K., de Meringo, A., Rouyer, E., Thélohan, S., Bauer, J., Law, B., Boymel, P.M., Olson, J.R., Christensen, V.R., Guldberg, M., Koenig, A.R. & Perander, M. (1997) *In vitro* measurement of fiber dissolution rate relevant to biopersistence at neutral pH: An interlaboratory round robin. *Inhal. Toxicol.*, **9**, 525–540

Zoller, T. & Zeller, W.J. (2000) Production of reactive oxygen species by phagocytic cells after exposure to glass wool and stone wool fibres — Effect of fibre preincubation in aqueous solution. *Toxicol. Lett.*, **114**, 1–9

LIST OF ABBREVIATIONS USED IN THIS VOLUME

ACGIH: American Conference of Governmental Industrial Hygienists
AES: alkaline earth silicate
AI: alveolar–interstitial (region)
APFE: European Glass Fibre Producers Association
AR: alkali-resistant
ASTM: American Society for Testing and Materials
ATF: activating transcription factor
B fibre: Bayer fibre
BEA: bronchitis, emphysema and asthma
CARE: control and reduce exposure
CEN: Comité européen de Normalisation (European Committee of Standardization)
Chrome: chromium oxides
CIIT: Chemical Industry Institute of Toxicology
CVF: colloidal and vacuum formed
D: diameter
ECA: Everest Consulting Associates
ECFIA: European Ceramic Fibre Industries Association
EDXA: energy dispersive X-ray diffraction analysis
E-glass: electrical glass (i.e. developed for electrical applications)
EIPPCB: European Integrated Pollution Prevention and Control Bureau
ERK: extracellular signal-regulated kinase
ERM: Environmental Resources Management
ET: extrathoracic
EU: European Union
EURIMA: European Insulation Manufacturers' Association
FARIMA: Fiberglass and Rockwool Insulation Manufacturers' Association of Australia
FEV$_1$: forced expiratory volume in 1 s
FVC: forced vital capacity
GM: geometric mean
GMD: geometric mean diameter

GML: geometric mean length
GSD: geometric standard deviation
HEPA: high-efficiency particulate air
HID: highest ineffective dose
HSPP: Health and Safety Partnership Program
HT: high-alumina, low-silica wool
ICRP: International Committee on Radiological Protection
IgG: immunoglobulin G
ILO: International Labour Office (of the International Labour Organization)
ILSI: International Life Sciences Institute
INRS: Institut national de Recherche et Sécurité
INSERM: Institut national de la Santé et de la Recherche médicale
IPF: idiopathic pulmonary fibrosis
ISO: International Organization for Standardization
JM Fibre: Johns Manville fibre
JNK: Jun N-terminal kinase
k_{dis}: dissolution constant
L: length
LED: lowest effective dose
LR: local rate
MAPK: mitogen-activated protein kinases
Met: mesothelial
MMAD: mass median aerodynamic diameter
MMMF: man-made mineral fibre
MMVF: man-made vitreous fibre
NAIMA: North American Insulation Manufacturers' Association
NF: nuclear factor
NIOSH: National Institute for Occupational Safety and Health
NMRD: non-malignant respiratory disease
NOHSC: National Occupational Health and Safety Commission of Australia
NR: national rate
NRC: National Research Council (USA)
8-OH-dG: 8-hydroxydeoxyguanosine
OSHA: Occupational Safety and Health Administration (USA)
P & CAM: physical and chemical analytical method
PAH: polycyclic aromatic hydrocarbons
PCOM: phase-contrast optical microscopy
PLM: polarized light microscopy
PSP: Product Stewardship Program
R: retained
RCF: refractory ceramic fibre
RCFC: Refractory Ceramic Fiber Coalition (USA)

RH: Rheinstahl (slag wool)
RRF: respiratory response function
SD: standard deviation
SEM: scanning electron microscopy
SIR: standardized incidence ratio
SMR: standardized mortality ratio
TB: tracheobronchial
TEM: transmission electron microscopy
TG: transcription factor
TGF: tumour growth factor
TIMA: Thermal Insulation Manufacturers' Association (USA)
TNF: tumour necrosis factor
TPC: total pulmonary capacity
TRGS: Technische Regeln für Gefahrstoffe
TWA: time-weighted average
UICC: Union internationale contre le Cancer (International union against cancer)
VDI: Verein Deutscher Ingenieure
WHO fibre: see Glossary
WT$_{1/2}$: weighted lung retention half-time
ZI: Zimmermann (slag wool)

GLOSSARY

Aerodynamic diameter: the diameter of a sphere with the density 1 (1 g/cm^3) which sediments at the same rate as the particle in still or laminarly flowing air. This definition also applies for fibrous particles.

Aspect ratio: the ratio of length:diameter of a fibre (see Fibre, for definition)

Binder: a substance that glues otherwise loose fibres together so that the product can be shaped. It is usually a phenol–formaldehyde or urea–formaldehyde resin.

Biopersistence: the ability of a fibre to remain in the lung. Biopersistence is a function of the solubility of the fibre in the lung, and the biological ability of the lung to clear the fibre from the lung.

Breathing zone: a person's breathing zone is described by a hemisphere of 300 mm radius extending in front of the face and measured from the midpoint of an imaginary line joining the ears.

Ceramic fibre: see Refractory ceramic fibre

Clearance rate: the rate at which deposited particles are removed by various processes from the respiratory tract. (This depends on both the physical and chemical characteristics of the fibre.)

Continuous glass filament: an extruded filament usually having a relatively large diameter, greater than 6 µm, and a very narrow range of diameter distribution. Typically formed from a glass melt

Fibre: a particle with a length to width ratio of at least 3:1

Fibre glass: see Glass fibre

Gamble's solution: a complex salt solution designed to mimic the salt balance of extracellular fluid

Geometric diameter: the geometric diameter of a spherical particle multiplied by the square root of the specific density of the material gives the aerodynamic diameter. For a non-spherical particle a shape factor also needs to be considered.

Glass fibre: may refer to reinforcing glass filament, glass wool or superfine glass fibre.

Glass filament: see Continuous glass filament

Glass wool: a fibrous product formed by either blowing or spinning a molten mass of glass. The resultant fibres are collected as a tangled mat of fibrous product.

HT (rock) stone wool: a recently developed high-alumina, low-silica wool

Inhalability: ratio of the particle (fibre) concentration in the inhaled air to that in the ambient air. (The inhalable fraction of an aerosol consists of particles than can enter nose or mouth upon inhalation.)

Intercept method: diameter is measured in proportion to fibre length (length-weighted diameter).

Kaolin: naturally occurring mineral (china clay) composed mainly of alumina and silica

k_{dis}: the dissolution constant of a glass fibre, being the rate at which it dissolves *in vitro* in a salt solution such as Gamble's solution (the unit is $ng/cm^2/h$).

Mineral wool: may refer to either slag wool or rock (stone) wool depending on the raw material from which it is produced.

NIOSH fibre: length greater than 5 μm, diameter less than 3 μm, length:diameter ratio 5:1

Nominal diameter: is the median diameter to which the fibrous product is manufactured. It may be thought of as the diameter at the midpoint of a long fibre created by joining all the fibres in a sample together in order of increasing thickness.

Personal sample: a sample taken within the breathing zone of the worker

Refractory: resistant to heat

Refractory ceramic fibre: amorphous, glassy, predominantly alumino-silicate products created from molten masses of either alumina and silica or naturally occurring kaolin clays

Respirability: ratio of airborne particles (fibres) penetrating to the alveolar region of the lung to that in the ambient air

Respirable fibre: a particle with a diameter less than 3 μm and length greater than 5 μm and with a length to width ratio of greater than 3:1. These fibres can reach the deepest part of the lung.

Retention half-time $T_{1/2}$: time by which 50% of the amount of the fibres in the lungs has been eliminated by a monoexponential function

Rock (stone) wool: a fibrous product manufactured by a process of blowing or spinning from a molten mass of rock. The resultant fibres are collected as a tangled mass of fibrous product.

Shards: particles of respirable dust (from highly chopped and pulverized continuous glass filament) with aspect ratios equal to or greater than 3:1

Shot: some wool fibre formation processes can produce numerous large, rounded particles approximately 60 μm or larger in diameter. These are known as shot.

Size: see Binder

Slag wool: a fibrous product manufactured by a process of blowing or spinning from a molten mass of metallurgical furnace slag

Stanton fibres: fibres with length > 8 μm and diameter ≤ 0.25 μm

Static sample: a sample taken at a fixed location, commonly between 1 m and 2 m above floor level

Superfine fibre: an extremely fine fibre with a diameter less than 1 µm, usually made of glass for specialist applications

Time-weighted average (TWA) concentration: the concentration of a contaminant that has been weighted for the time duration of the sample. High exposure of short sample duration does not 'weigh' as heavily in the calculation as do moderate levels for extended periods.

Wagner scale: a scale for assessing the extent of inflammation and fibrosis in the lungs of rats exposed to particles or fibres

Weighted lung retention half-time: sum of the product of each halftime of a double exponential retention curve weighted by the coefficient of each exponential expressed in days

WHO fibres: any particle that has a length greater than 5 µm, a fibre diameter less than 3 µm and a length:diameter ratio larger than 3:1

SUPPLEMENTARY CORRIGENDUM TO VOLUMES 1–80

Volume 68

p. 42, table 1: *move* two last lines of this table

'Flux-calcined   68855-54-9
diatomaceous earth'

under Crystalline silica *after* Tridymite

CUMULATIVE CROSS INDEX TO *IARC MONOGRAPHS ON THE EVALUATION OF CARCINOGENIC RISKS TO HUMANS*

The volume, page and year of publication are given. References to corrigenda are given in parentheses.

A

A-α-C	*40*, 245 (1986); *Suppl. 7*, 56 (1987)
Acetaldehyde	*36*, 101 (1985) (*corr. 42*, 263); *Suppl. 7*, 77 (1987); *71*, 319 (1999)
Acetaldehyde formylmethylhydrazone (*see* Gyromitrin)	
Acetamide	*7*, 197 (1974); *Suppl. 7*, 56, 389 (1987); *71*, 1211 (1999)
Acetaminophen (*see* Paracetamol)	
Aciclovir	*76*, 47 (2000)
Acridine orange	*16*, 145 (1978); *Suppl. 7*, 56 (1987)
Acriflavinium chloride	*13*, 31 (1977); *Suppl. 7*, 56 (1987)
Acrolein	*19*, 479 (1979); *36*, 133 (1985); *Suppl. 7*, 78 (1987); *63*, 337 (1995) (*corr. 65*, 549)
Acrylamide	*39*, 41 (1986); *Suppl. 7*, 56 (1987); *60*, 389 (1994)
Acrylic acid	*19*, 47 (1979); *Suppl. 7*, 56 (1987); *71*, 1223 (1999)
Acrylic fibres	*19*, 86 (1979); *Suppl. 7*, 56 (1987)
Acrylonitrile	*19*, 73 (1979); *Suppl. 7*, 79 (1987); *71*, 43 (1999)
Acrylonitrile-butadiene-styrene copolymers	*19*, 91 (1979); *Suppl. 7*, 56 (1987)
Actinolite (*see* Asbestos)	
Actinomycin D (*see also* Actinomycins)	*Suppl. 7*, 80 (1987)
Actinomycins	*10*, 29 (1976) (*corr. 42*, 255)
Adriamycin	*10*, 43 (1976); *Suppl. 7*, 82 (1987)
AF-2	*31*, 47 (1983); *Suppl. 7*, 56 (1987)
Aflatoxins	*1*, 145 (1972) (*corr. 42*, 251); *10*, 51 (1976); *Suppl. 7*, 83 (1987); *56*, 245 (1993)
Aflatoxin B_1 (*see* Aflatoxins)	
Aflatoxin B_2 (*see* Aflatoxins)	
Aflatoxin G_1 (*see* Aflatoxins)	
Aflatoxin G_2 (*see* Aflatoxins)	
Aflatoxin M_1 (*see* Aflatoxins)	
Agaritine	*31*, 63 (1983); *Suppl. 7*, 56 (1987)
Alcohol drinking	*44* (1988)
Aldicarb	*53*, 93 (1991)
Aldrin	*5*, 25 (1974); *Suppl. 7*, 88 (1987)

Allyl chloride	*36*, 39 (1985); *Suppl. 7*, 56 (1987); *71*, 1231 (1999)
Allyl isothiocyanate	*36*, 55 (1985); *Suppl. 7*, 56 (1987); *73*, 37 (1999)
Allyl isovalerate	*36*, 69 (1985); *Suppl. 7*, 56 (1987); *71*, 1241 (1999)
Aluminium production	*34*, 37 (1984); *Suppl. 7*, 89 (1987)
Amaranth	*8*, 41 (1975); *Suppl. 7*, 56 (1987)
5-Aminoacenaphthene	*16*, 243 (1978); *Suppl. 7*, 56 (1987)
2-Aminoanthraquinone	*27*, 191 (1982); *Suppl. 7*, 56 (1987)
para-Aminoazobenzene	*8*, 53 (1975); *Suppl. 7*, 56, 390 (1987)
ortho-Aminoazotoluene	*8*, 61 (1975) (*corr. 42*, 254); *Suppl. 7*, 56 (1987)
para-Aminobenzoic acid	*16*, 249 (1978); *Suppl. 7*, 56 (1987)
4-Aminobiphenyl	*1*, 74 (1972) (*corr. 42*, 251); *Suppl. 7*, 91 (1987)
2-Amino-3,4-dimethylimidazo[4,5-*f*]quinoline (*see* MeIQ)	
2-Amino-3,8-dimethylimidazo[4,5-*f*]quinoxaline (*see* MeIQx)	
3-Amino-1,4-dimethyl-5*H*-pyrido[4,3-*b*]indole (*see* Trp-P-1)	
2-Aminodipyrido[1,2-*a*:3′,2′-*d*]imidazole (*see* Glu-P-2)	
1-Amino-2-methylanthraquinone	*27*, 199 (1982); *Suppl. 7*, 57 (1987)
2-Amino-3-methylimidazo[4,5-*f*]quinoline (*see* IQ)	
2-Amino-6-methyldipyrido[1,2-*a*:3′,2′-*d*]imidazole (*see* Glu-P-1)	
2-Amino-1-methyl-6-phenylimidazo[4,5-*b*]pyridine (*see* PhIP)	
2-Amino-3-methyl-9*H*-pyrido[2,3-*b*]indole (*see* MeA-α-C)	
3-Amino-1-methyl-5*H*-pyrido[4,3-*b*]indole (*see* Trp-P-2)	
2-Amino-5-(5-nitro-2-furyl)-1,3,4-thiadiazole	*7*, 143 (1974); *Suppl. 7*, 57 (1987)
2-Amino-4-nitrophenol	*57*, 167 (1993)
2-Amino-5-nitrophenol	*57*, 177 (1993)
4-Amino-2-nitrophenol	*16*, 43 (1978); *Suppl. 7*, 57 (1987)
2-Amino-5-nitrothiazole	*31*, 71 (1983); *Suppl. 7*, 57 (1987)
2-Amino-9*H*-pyrido[2,3-*b*]indole (*see* A-α-C)	
11-Aminoundecanoic acid	*39*, 239 (1986); *Suppl. 7*, 57 (1987)
Amitrole	*7*, 31 (1974); *41*, 293 (1986) (*corr. 52*, 513; *Suppl. 7*, 92 (1987); *79*, 381 (2001)
Ammonium potassium selenide (*see* Selenium and selenium compounds)	
Amorphous silica (*see also* Silica)	*42*, 39 (1987); *Suppl. 7*, 341 (1987); *68*, 41 (1997) (*corr. 81*, 383)
Amosite (*see* Asbestos)	
Ampicillin	*50*, 153 (1990)
Amsacrine	*76*, 317 (2000)
Anabolic steroids (*see* Androgenic (anabolic) steroids)	
Anaesthetics, volatile	*11*, 285 (1976); *Suppl. 7*, 93 (1987)
Analgesic mixtures containing phenacetin (*see also* Phenacetin)	*Suppl. 7*, 310 (1987)
Androgenic (anabolic) steroids	*Suppl. 7*, 96 (1987)
Angelicin and some synthetic derivatives (*see also* Angelicins)	*40*, 291 (1986)
Angelicin plus ultraviolet radiation (*see also* Angelicin and some synthetic derivatives)	*Suppl. 7*, 57 (1987)
Angelicins	*Suppl. 7*, 57 (1987)
Aniline	*4*, 27 (1974) (*corr. 42*, 252); *27*, 39 (1982); *Suppl. 7*, 99 (1987)

ortho-Anisidine	27, 63 (1982); *Suppl. 7*, 57 (1987); 73, 49 (1999)
para-Anisidine	27, 65 (1982); *Suppl. 7*, 57 (1987)
Anthanthrene	32, 95 (1983); *Suppl. 7*, 57 (1987)
Anthophyllite (*see* Asbestos)	
Anthracene	32, 105 (1983); *Suppl. 7*, 57 (1987)
Anthranilic acid	16, 265 (1978); *Suppl. 7*, 57 (1987)
Antimony trioxide	47, 291 (1989)
Antimony trisulfide	47, 291 (1989)
ANTU (*see* 1-Naphthylthiourea)	
Apholate	9, 31 (1975); *Suppl. 7*, 57 (1987)
para-Aramid fibrils	68, 409 (1997)
Aramite®	5, 39 (1974); *Suppl. 7*, 57 (1987)
Areca nut (*see* Betel quid)	
Arsanilic acid (*see* Arsenic and arsenic compounds)	
Arsenic and arsenic compounds	1, 41 (1972); 2, 48 (1973); 23, 39 (1980); *Suppl. 7*, 100 (1987)
Arsenic pentoxide (*see* Arsenic and arsenic compounds)	
Arsenic sulfide (*see* Arsenic and arsenic compounds)	
Arsenic trioxide (*see* Arsenic and arsenic compounds)	
Arsine (*see* Arsenic and arsenic compounds)	
Asbestos	2, 17 (1973) (*corr.* 42, 252); 14 (1977) (*corr.* 42, 256); *Suppl. 7*, 106 (1987) (*corr.* 45, 283)
Atrazine	53, 441 (1991); 73, 59 (1999)
Attapulgite (*see* Palygorskite)	
Auramine (technical-grade)	1, 69 (1972) (*corr.* 42, 251); *Suppl. 7*, 118 (1987)
Auramine, manufacture of (*see also* Auramine, technical-grade)	*Suppl. 7*, 118 (1987)
Aurothioglucose	13, 39 (1977); *Suppl. 7*, 57 (1987)
Azacitidine	26, 37 (1981); *Suppl. 7*, 57 (1987); 50, 47 (1990)
5-Azacytidine (*see* Azacitidine)	
Azaserine	10, 73 (1976) (*corr.* 42, 255); *Suppl. 7*, 57 (1987)
Azathioprine	26, 47 (1981); *Suppl. 7*, 119 (1987)
Aziridine	9, 37 (1975); *Suppl. 7*, 58 (1987); 71, 337 (1999)
2-(1-Aziridinyl)ethanol	9, 47 (1975); *Suppl. 7*, 58 (1987)
Aziridyl benzoquinone	9, 51 (1975); *Suppl. 7*, 58 (1987)
Azobenzene	8, 75 (1975); *Suppl. 7*, 58 (1987)
AZT (*see* Zidovudine)	

B

Barium chromate (*see* Chromium and chromium compounds)	
Basic chromic sulfate (*see* Chromium and chromium compounds)	
BCNU (*see* Bischloroethyl nitrosourea)	
Benz[*a*]acridine	32, 123 (1983); *Suppl. 7*, 58 (1987)
Benz[*c*]acridine	3, 241 (1973); 32, 129 (1983); *Suppl. 7*, 58 (1987)
Benzal chloride (*see also* α-Chlorinated toluenes and benzoyl chloride)	29, 65 (1982); *Suppl. 7*, 148 (1987); 71, 453 (1999)

Benz[*a*]anthracene	*3*, 45 (1973); *32*, 135 (1983); *Suppl. 7*, 58 (1987)
Benzene	*7*, 203 (1974) (*corr. 42*, 254); *29*, 93, 391 (1982); *Suppl. 7*, 120 (1987)
Benzidine	*1*, 80 (1972); *29*, 149, 391 (1982); *Suppl. 7*, 123 (1987)
Benzidine-based dyes	*Suppl. 7*, 125 (1987)
Benzo[*b*]fluoranthene	*3*, 69 (1973); *32*, 147 (1983); *Suppl. 7*, 58 (1987)
Benzo[*j*]fluoranthene	*3*, 82 (1973); *32*, 155 (1983); *Suppl. 7*, 58 (1987)
Benzo[*k*]fluoranthene	*32*, 163 (1983); *Suppl. 7*, 58 (1987)
Benzo[*ghi*]fluoranthene	*32*, 171 (1983); *Suppl. 7*, 58 (1987)
Benzo[*a*]fluorene	*32*, 177 (1983); *Suppl. 7*, 58 (1987)
Benzo[*b*]fluorene	*32*, 183 (1983); *Suppl. 7*, 58 (1987)
Benzo[*c*]fluorene	*32*, 189 (1983); *Suppl. 7*, 58 (1987)
Benzofuran	*63*, 431 (1995)
Benzo[*ghi*]perylene	*32*, 195 (1983); *Suppl. 7*, 58 (1987)
Benzo[*c*]phenanthrene	*32*, 205 (1983); *Suppl. 7*, 58 (1987)
Benzo[*a*]pyrene	*3*, 91 (1973); *32*, 211 (1983) (*corr. 68*, 477); *Suppl. 7*, 58 (1987)
Benzo[*e*]pyrene	*3*, 137 (1973); *32*, 225 (1983); *Suppl. 7*, 58 (1987)
1,4-Benzoquinone (see *para*-Quinone)	
1,4-Benzoquinone dioxime	*29*, 185 (1982); *Suppl. 7*, 58 (1987); *71*, 1251 (1999)
Benzotrichloride (*see also* α-Chlorinated toluenes and benzoyl chloride)	*29*, 73 (1982); *Suppl. 7*, 148 (1987); *71*, 453 (1999)
Benzoyl chloride (*see also* α-Chlorinated toluenes and benzoyl chloride)	*29*, 83 (1982) (*corr. 42*, 261); *Suppl. 7*, 126 (1987); *71*, 453 (1999)
Benzoyl peroxide	*36*, 267 (1985); *Suppl. 7*, 58 (1987); *71*, 345 (1999)
Benzyl acetate	*40*, 109 (1986); *Suppl. 7*, 58 (1987); *71*, 1255 (1999)
Benzyl chloride (*see also* α-Chlorinated toluenes and benzoyl chloride)	*11*, 217 (1976) (*corr. 42*, 256); *29*, 49 (1982); *Suppl. 7*, 148 (1987); *71*, 453 (1999)
Benzyl violet 4B	*16*, 153 (1978); *Suppl. 7*, 58 (1987)
Bertrandite (*see* Beryllium and beryllium compounds)	
Beryllium and beryllium compounds	*1*, 17 (1972); *23*, 143 (1980) (*corr. 42*, 260); *Suppl. 7*, 127 (1987); *58*, 41 (1993)

Beryllium acetate (*see* Beryllium and beryllium compounds)
Beryllium acetate, basic (*see* Beryllium and beryllium compounds)
Beryllium-aluminium alloy (*see* Beryllium and beryllium compounds)
Beryllium carbonate (*see* Beryllium and beryllium compounds)
Beryllium chloride (*see* Beryllium and beryllium compounds)
Beryllium-copper alloy (*see* Beryllium and beryllium compounds)
Beryllium-copper-cobalt alloy (*see* Beryllium and beryllium compounds)
Beryllium fluoride (*see* Beryllium and beryllium compounds)
Beryllium hydroxide (*see* Beryllium and beryllium compounds)
Beryllium-nickel alloy (*see* Beryllium and beryllium compounds)
Beryllium oxide (*see* Beryllium and beryllium compounds)

Beryllium phosphate (*see* Beryllium and beryllium compounds)	
Beryllium silicate (*see* Beryllium and beryllium compounds)	
Beryllium sulfate (*see* Beryllium and beryllium compounds)	
Beryl ore (*see* Beryllium and beryllium compounds)	
Betel quid	*37*, 141 (1985); *Suppl. 7*, 128 (1987)
Betel-quid chewing (*see* Betel quid)	
BHA (*see* Butylated hydroxyanisole)	
BHT (*see* Butylated hydroxytoluene)	
Bis(1-aziridinyl)morpholinophosphine sulfide	*9*, 55 (1975); *Suppl. 7*, 58 (1987)
2,2-Bis(bromomethyl)propane-1,3-diol	*77*, 455 (2000)
Bis(2-chloroethyl)ether	*9*, 117 (1975); *Suppl. 7*, 58 (1987); *71*, 1265 (1999)
N,N-Bis(2-chloroethyl)-2-naphthylamine	*4*, 119 (1974) (*corr. 42*, 253); *Suppl. 7*, 130 (1987)
Bischloroethyl nitrosourea (*see also* Chloroethyl nitrosoureas)	*26*, 79 (1981); *Suppl. 7*, 150 (1987)
1,2-Bis(chloromethoxy)ethane	*15*, 31 (1977); *Suppl. 7*, 58 (1987); *71*, 1271 (1999)
1,4-Bis(chloromethoxymethyl)benzene	*15*, 37 (1977); *Suppl. 7*, 58 (1987); *71*, 1273 (1999)
Bis(chloromethyl)ether	*4*, 231 (1974) (*corr. 42*, 253); *Suppl. 7*, 131 (1987)
Bis(2-chloro-1-methylethyl)ether	*41*, 149 (1986); *Suppl. 7*, 59 (1987); *71*, 1275 (1999)
Bis(2,3-epoxycyclopentyl)ether	*47*, 231 (1989); *71*, 1281 (1999)
Bisphenol A diglycidyl ether (*see also* Glycidyl ethers)	*71*, 1285 (1999)
Bisulfites (see Sulfur dioxide and some sulfites, bisulfites and metabisulfites)	
Bitumens	*35*, 39 (1985); *Suppl. 7*, 133 (1987)
Bleomycins (*see also* Etoposide)	*26*, 97 (1981); *Suppl. 7*, 134 (1987)
Blue VRS	*16*, 163 (1978); *Suppl. 7*, 59 (1987)
Boot and shoe manufacture and repair	*25*, 249 (1981); *Suppl. 7*, 232 (1987)
Bracken fern	*40*, 47 (1986); *Suppl. 7*, 135 (1987)
Brilliant Blue FCF, disodium salt	*16*, 171 (1978) (*corr. 42*, 257); *Suppl. 7*, 59 (1987)
Bromochloroacetonitrile (*see also* Halogenated acetonitriles)	*71*, 1291 (1999)
Bromodichloromethane	*52*, 179 (1991); *71*, 1295 (1999)
Bromoethane	*52*, 299 (1991); *71*, 1305 (1999)
Bromoform	*52*, 213 (1991); *71*, 1309 (1999)
1,3-Butadiene	*39*, 155 (1986) (*corr. 42*, 264 *Suppl. 7*, 136 (1987); *54*, 237 (1992); *71*, 109 (1999)
1,4-Butanediol dimethanesulfonate	*4*, 247 (1974); *Suppl. 7*, 137 (1987)
n-Butyl acrylate	*39*, 67 (1986); *Suppl. 7*, 59 (1987); *71*, 359 (1999)
Butylated hydroxyanisole	*40*, 123 (1986); *Suppl. 7*, 59 (1987)
Butylated hydroxytoluene	*40*, 161 (1986); *Suppl. 7*, 59 (1987)
Butyl benzyl phthalate	*29*, 193 (1982) (*corr. 42*, 261); *Suppl. 7*, 59 (1987); *73*, 115 (1999)
β-Butyrolactone	*11*, 225 (1976); *Suppl. 7*, 59 (1987); *71*, 1317 (1999)
γ-Butyrolactone	*11*, 231 (1976); *Suppl. 7*, 59 (1987); *71*, 367 (1999)

C

Cabinet-making (*see* Furniture and cabinet-making)
Cadmium acetate (*see* Cadmium and cadmium compounds)
Cadmium and cadmium compounds   2, 74 (1973); *11*, 39 (1976) (*corr.* 42, 255); *Suppl. 7*, 139 (1987); *58*, 119 (1993)

Cadmium chloride (*see* Cadmium and cadmium compounds)
Cadmium oxide (*see* Cadmium and cadmium compounds)
Cadmium sulfate (*see* Cadmium and cadmium compounds)
Cadmium sulfide (*see* Cadmium and cadmium compounds)
Caffeic acid   *56*, 115 (1993)
Caffeine   *51*, 291 (1991)
Calcium arsenate (*see* Arsenic and arsenic compounds)
Calcium chromate (see Chromium and chromium compounds)
Calcium cyclamate (*see* Cyclamates)
Calcium saccharin (*see* Saccharin)
Cantharidin   *10*, 79 (1976); *Suppl. 7*, 59 (1987)
Caprolactam   *19*, 115 (1979) (*corr.* 42, 258); *39*, 247 (1986) (*corr.* 42, 264); *Suppl. 7*, 59, 390 (1987); *71*, 383 (1999)
Captafol   *53*, 353 (1991)
Captan   *30*, 295 (1983); *Suppl. 7*, 59 (1987)
Carbaryl   *12*, 37 (1976); *Suppl. 7*, 59 (1987)
Carbazole   *32*, 239 (1983); *Suppl. 7*, 59 (1987); *71*, 1319 (1999)
3-Carbethoxypsoralen   *40*, 317 (1986); *Suppl. 7*, 59 (1987)
Carbon black   *3*, 22 (1973); *33*, 35 (1984); *Suppl. 7*, 142 (1987); *65*, 149 (1996)
Carbon tetrachloride   *1*, 53 (1972); *20*, 371 (1979); *Suppl. 7*, 143 (1987); *71*, 401 (1999)
Carmoisine   *8*, 83 (1975); *Suppl. 7*, 59 (1987)
Carpentry and joinery   *25*, 139 (1981); *Suppl. 7*, 378 (1987)
Carrageenan   *10*, 181 (1976) (*corr.* 42, 255); *31*, 79 (1983); *Suppl. 7*, 59 (1987)
Catechol   *15*, 155 (1977); *Suppl. 7*, 59 (1987); *71*, 433 (1999)
CCNU (*see* 1-(2-Chloroethyl)-3-cyclohexyl-1-nitrosourea)
Ceramic fibres (*see* Man-made vitreous fibres)
Chemotherapy, combined, including alkylating agents (*see* MOPP and other combined chemotherapy including alkylating agents)
Chloral   *63*, 245 (1995)
Chloral hydrate   *63*, 245 (1995)
Chlorambucil   *9*, 125 (1975); *26*, 115 (1981); *Suppl. 7*, 144 (1987)
Chloramphenicol   *10*, 85 (1976); *Suppl. 7*, 145 (1987); *50*, 169 (1990)
Chlordane (*see also* Chlordane/Heptachlor)   *20*, 45 (1979) (*corr.* 42, 258)
Chlordane and Heptachlor   *Suppl. 7*, 146 (1987); *53*, 115 (1991); *79*, 411 (2001)

Chlordecone	*20*, 67 (1979); *Suppl. 7*, 59 (1987)
Chlordimeform	*30*, 61 (1983); *Suppl. 7*, 59 (1987)
Chlorendic acid	*48*, 45 (1990)
Chlorinated dibenzodioxins (other than TCDD) (*see also* Polychlorinated dibenzo-*para*-dioxins)	*15*, 41 (1977); *Suppl. 7*, 59 (1987)
Chlorinated drinking-water	*52*, 45 (1991)
Chlorinated paraffins	*48*, 55 (1990)
α-Chlorinated toluenes and benzoyl chloride	*Suppl. 7*, 148 (1987); *71*, 453 (1999)
Chlormadinone acetate	*6*, 149 (1974); *21*, 365 (1979); *Suppl. 7*, 291, 301 (1987); *72*, 49 (1999)
Chlornaphazine (*see N,N*-Bis(2-chloroethyl)-2-naphthylamine)	
Chloroacetonitrile (*see also* Halogenated acetonitriles)	*71*, 1325 (1999)
para-Chloroaniline	*57*, 305 (1993)
Chlorobenzilate	*5*, 75 (1974); *30*, 73 (1983); *Suppl. 7*, 60 (1987)
Chlorodibromomethane	*52*, 243 (1991); *71*, 1331 (1999)
Chlorodifluoromethane	*41*, 237 (1986) (*corr. 51*, 483); *Suppl. 7*, 149 (1987); *71*, 1339 (1999)
Chloroethane	*52*, 315 (1991); *71*, 1345 (1999)
1-(2-Chloroethyl)-3-cyclohexyl-1-nitrosourea (*see also* Chloroethyl nitrosoureas)	*26*, 137 (1981) (*corr. 42*, 260); *Suppl. 7*, 150 (1987)
1-(2-Chloroethyl)-3-(4-methylcyclohexyl)-1-nitrosourea (*see also* Chloroethyl nitrosoureas)	*Suppl. 7*, 150 (1987)
Chloroethyl nitrosoureas	*Suppl. 7*, 150 (1987)
Chlorofluoromethane	*41*, 229 (1986); *Suppl. 7*, 60 (1987); *71*, 1351 (1999)
Chloroform	*1*, 61 (1972); *20*, 401 (1979); *Suppl. 7*, 152 (1987); *73*, 131 (1999)
Chloromethyl methyl ether (technical-grade) (*see also* Bis(chloromethyl)ether)	*4*, 239 (1974); *Suppl. 7*, 131 (1987)
(4-Chloro-2-methylphenoxy)acetic acid (*see* MCPA)	
1-Chloro-2-methylpropene	*63*, 315 (1995)
3-Chloro-2-methylpropene	*63*, 325 (1995)
2-Chloronitrobenzene	*65*, 263 (1996)
3-Chloronitrobenzene	*65*, 263 (1996)
4-Chloronitrobenzene	*65*, 263 (1996)
Chlorophenols (*see also* Polychlorophenols and their sodium salts)	*Suppl. 7*, 154 (1987)
Chlorophenols (occupational exposures to)	*41*, 319 (1986)
Chlorophenoxy herbicides	*Suppl. 7*, 156 (1987)
Chlorophenoxy herbicides (occupational exposures to)	*41*, 357 (1986)
4-Chloro-*ortho*-phenylenediamine	*27*, 81 (1982); *Suppl. 7*, 60 (1987)
4-Chloro-*meta*-phenylenediamine	*27*, 82 (1982); *Suppl. 7*, 60 (1987)
Chloroprene	*19*, 131 (1979); *Suppl. 7*, 160 (1987); *71*, 227 (1999)
Chloropropham	*12*, 55 (1976); *Suppl. 7*, 60 (1987)
Chloroquine	*13*, 47 (1977); *Suppl. 7*, 60 (1987)
Chlorothalonil	*30*, 319 (1983); *Suppl. 7*, 60 (1987); *73*, 183 (1999)

para-Chloro-*ortho*-toluidine and its strong acid salts (*see also* Chlordimeform)	*16*, 277 (1978); *30*, 65 (1983); *Suppl. 7*, 60 (1987); *48*, 123 (1990); *77*, 323 (2000)
4-Chloro-*ortho*-toluidine (see *para*-chloro-*ortho*-toluidine)	
5-Chloro-*ortho*-toluidine	*77*, 341 (2000)
Chlorotrianisene (*see also* Nonsteroidal oestrogens)	*21*, 139 (1979); *Suppl. 7*, 280 (1987)
2-Chloro-1,1,1-trifluoroethane	*41*, 253 (1986); *Suppl. 7*, 60 (1987); *71*, 1355 (1999)
Chlorozotocin	*50*, 65 (1990)
Cholesterol	*10*, 99 (1976); *31*, 95 (1983); *Suppl. 7*, 161 (1987)
Chromic acetate (*see* Chromium and chromium compounds)	
Chromic chloride (*see* Chromium and chromium compounds)	
Chromic oxide (*see* Chromium and chromium compounds)	
Chromic phosphate (*see* Chromium and chromium compounds)	
Chromite ore (*see* Chromium and chromium compounds)	
Chromium and chromium compounds (*see also* Implants, surgical)	*2*, 100 (1973); *23*, 205 (1980); *Suppl. 7*, 165 (1987); *49*, 49 (1990) (*corr. 51*, 483)
Chromium carbonyl (*see* Chromium and chromium compounds)	
Chromium potassium sulfate (*see* Chromium and chromium compounds)	
Chromium sulfate (*see* Chromium and chromium compounds)	
Chromium trioxide (*see* Chromium and chromium compounds)	
Chrysazin (*see* Dantron)	
Chrysene	*3*, 159 (1973); *32*, 247 (1983); *Suppl. 7*, 60 (1987)
Chrysoidine	*8*, 91 (1975); *Suppl. 7*, 169 (1987)
Chrysotile (*see* Asbestos)	
CI Acid Orange 3	*57*, 121 (1993)
CI Acid Red 114	*57*, 247 (1993)
CI Basic Red 9 (*see also* Magenta)	*57*, 215 (1993)
Ciclosporin	*50*, 77 (1990)
CI Direct Blue 15	*57*, 235 (1993)
CI Disperse Yellow 3 (see Disperse Yellow 3)	
Cimetidine	*50*, 235 (1990)
Cinnamyl anthranilate	*16*, 287 (1978); *31*, 133 (1983); *Suppl. 7*, 60 (1987); *77*, 177 (2000)
CI Pigment Red 3	*57*, 259 (1993)
CI Pigment Red 53:1 (*see* D&C Red No. 9)	
Cisplatin (*see also* Etoposide)	*26*, 151 (1981); *Suppl. 7*, 170 (1987)
Citrinin	*40*, 67 (1986); *Suppl. 7*, 60 (1987)
Citrus Red No. 2	*8*, 101 (1975) (*corr. 42*, 254); *Suppl. 7*, 60 (1987)
Clinoptilolite (*see* Zeolites)	
Clofibrate	*24*, 39 (1980); *Suppl. 7*, 171 (1987); *66*, 391 (1996)
Clomiphene citrate	*21*, 551 (1979); *Suppl. 7*, 172 (1987)
Clonorchis sinensis (infection with)	*61*, 121 (1994)
Coal dust	*68*, 337 (1997)
Coal gasification	*34*, 65 (1984); *Suppl. 7*, 173 (1987)
Coal-tar pitches (*see also* Coal-tars)	*35*, 83 (1985); *Suppl. 7*, 174 (1987)

Coal-tars	35, 83 (1985); *Suppl. 7*, 175 (1987)
Cobalt[III] acetate (*see* Cobalt and cobalt compounds)	
Cobalt-aluminium-chromium spinel (*see* Cobalt and cobalt compounds)	
Cobalt and cobalt compounds (*see also* Implants, surgical)	52, 363 (1991)
Cobalt[II] chloride (*see* Cobalt and cobalt compounds)	
Cobalt-chromium alloy (*see* Chromium and chromium compounds)	
Cobalt-chromium-molybdenum alloys (*see* Cobalt and cobalt compounds)	
Cobalt metal powder (*see* Cobalt and cobalt compounds)	
Cobalt naphthenate (*see* Cobalt and cobalt compounds)	
Cobalt[II] oxide (*see* Cobalt and cobalt compounds)	
Cobalt[II,III] oxide (*see* Cobalt and cobalt compounds)	
Cobalt[II] sulfide (*see* Cobalt and cobalt compounds)	
Coffee	51, 41 (1991) (*corr. 52*, 513)
Coke production	34, 101 (1984); *Suppl. 7*, 176 (1987)
Combined oral contraceptives (*see* Oral contraceptives, combined)	
Conjugated equine oestrogens	72, 399 (1999)
Conjugated oestrogens (*see also* Steroidal oestrogens)	21, 147 (1979); *Suppl. 7*, 283 (1987)
Continuous glass filament (*see* Man-made vitreous fibres)	
Contraceptives, oral (*see* Oral contraceptives, combined; Sequential oral contraceptives)	
Copper 8-hydroxyquinoline	15, 103 (1977); *Suppl. 7*, 61 (1987)
Coronene	32, 263 (1983); *Suppl. 7*, 61 (1987)
Coumarin	10, 113 (1976); *Suppl. 7*, 61 (1987); 77, 193 (2000)
Creosotes (*see also* Coal-tars)	35, 83 (1985); *Suppl. 7*, 177 (1987)
meta-Cresidine	27, 91 (1982); *Suppl. 7*, 61 (1987)
para-Cresidine	27, 92 (1982); *Suppl. 7*, 61 (1987)
Cristobalite (*see* Crystalline silica)	
Crocidolite (*see* Asbestos)	
Crotonaldehyde	63, 373 (1995) (*corr. 65*, 549)
Crude oil	45, 119 (1989)
Crystalline silica (*see* also Silica)	42, 39 (1987); *Suppl. 7*, 341 (1987); 68, 41 (1997) (*corr. 81*, 383)
Cycasin (*see also* Methylazoxymethanol)	1, 157 (1972) (*corr. 42*, 251); 10, 121 (1976); *Suppl. 7*, 61 (1987)
Cyclamates	22, 55 (1980); *Suppl. 7*, 178 (1987); 73, 195 (1999)
Cyclamic acid (*see* Cyclamates)	
Cyclochlorotine	10, 139 (1976); *Suppl. 7*, 61 (1987)
Cyclohexanone	47, 157 (1989); 71, 1359 (1999)
Cyclohexylamine (*see* Cyclamates)	
Cyclopenta[*cd*]pyrene	32, 269 (1983); *Suppl. 7*, 61 (1987)
Cyclopropane (*see* Anaesthetics, volatile)	
Cyclophosphamide	9, 135 (1975); 26, 165 (1981); *Suppl. 7*, 182 (1987)
Cyproterone acetate	72, 49 (1999)

D

2,4-D (see also Chlorophenoxy herbicides; Chlorophenoxy herbicides, occupational exposures to)	15, 111 (1977)
Dacarbazine	26, 203 (1981); Suppl. 7, 184 (1987)
Dantron	50, 265 (1990) (corr. 59, 257)
D&C Red No. 9	8, 107 (1975); Suppl. 7, 61 (1987); 57, 203 (1993)
Dapsone	24, 59 (1980); Suppl. 7, 185 (1987)
Daunomycin	10, 145 (1976); Suppl. 7, 61 (1987)
DDD (see DDT)	
DDE (see DDT)	
DDT	5, 83 (1974) (corr. 42, 253); Suppl. 7, 186 (1987); 53, 179 (1991)
Decabromodiphenyl oxide	48, 73 (1990); 71, 1365 (1999)
Deltamethrin	53, 251 (1991)
Deoxynivalenol (see Toxins derived from *Fusarium graminearum*, *F. culmorum* and *F. crookwellense*)	
Diacetylaminoazotoluene	8, 113 (1975); Suppl. 7, 61 (1987)
N,N'-Diacetylbenzidine	16, 293 (1978); Suppl. 7, 61 (1987)
Diallate	12, 69 (1976); 30, 235 (1983); Suppl. 7, 61 (1987)
2,4-Diaminoanisole and its salts	16, 51 (1978); 27, 103 (1982); Suppl. 7, 61 (1987); 79, 619 (2001)
4,4'-Diaminodiphenyl ether	16, 301 (1978); 29, 203 (1982); Suppl. 7, 61 (1987)
1,2-Diamino-4-nitrobenzene	16, 63 (1978); Suppl. 7, 61 (1987)
1,4-Diamino-2-nitrobenzene	16, 73 (1978); Suppl. 7, 61 (1987); 57, 185 (1993)
2,6-Diamino-3-(phenylazo)pyridine (see Phenazopyridine hydrochloride)	
2,4-Diaminotoluene (see also Toluene diisocyanates)	16, 83 (1978); Suppl. 7, 61 (1987)
2,5-Diaminotoluene (see also Toluene diisocyanates)	16, 97 (1978); Suppl. 7, 61 (1987)
ortho-Dianisidine (see 3,3'-Dimethoxybenzidine)	
Diatomaceous earth, uncalcined (see Amorphous silica)	
Diazepam	13, 57 (1977); Suppl. 7, 189 (1987); 66, 37 (1996)
Diazomethane	7, 223 (1974); Suppl. 7, 61 (1987)
Dibenz[a,h]acridine	3, 247 (1973); 32, 277 (1983); Suppl. 7, 61 (1987)
Dibenz[a,j]acridine	3, 254 (1973); 32, 283 (1983); Suppl. 7, 61 (1987)
Dibenz[a,c]anthracene	32, 289 (1983) (corr. 42, 262); Suppl. 7, 61 (1987)
Dibenz[a,h]anthracene	3, 178 (1973) (corr. 43, 261); 32, 299 (1983); Suppl. 7, 61 (1987)
Dibenz[a,j]anthracene	32, 309 (1983); Suppl. 7, 61 (1987)
7H-Dibenzo[c,g]carbazole	3, 260 (1973); 32, 315 (1983); Suppl. 7, 61 (1987)
Dibenzodioxins, chlorinated (other than TCDD) (see Chlorinated dibenzodioxins (other than TCDD))	
Dibenzo[a,e]fluoranthene	32, 321 (1983); Suppl. 7, 61 (1987)
Dibenzo[h,rst]pentaphene	3, 197 (1973); Suppl. 7, 62 (1987)

Dibenzo[*a,e*]pyrene	*3*, 201 (1973); *32*, 327 (1983); *Suppl. 7*, 62 (1987)
Dibenzo[*a,h*]pyrene	*3*, 207 (1973); *32*, 331 (1983); *Suppl. 7*, 62 (1987)
Dibenzo[*a,i*]pyrene	*3*, 215 (1973); *32*, 337 (1983); *Suppl. 7*, 62 (1987)
Dibenzo[*a,l*]pyrene	*3*, 224 (1973); *32*, 343 (1983); *Suppl. 7*, 62 (1987)
Dibenzo-*para*-dioxin	*69*, 33 (1997)
Dibromoacetonitrile (*see also* Halogenated acetonitriles)	*71*, 1369 (1999)
1,2-Dibromo-3-chloropropane	*15*, 139 (1977); *20*, 83 (1979); *Suppl. 7*, 191 (1987); *71*, 479 (1999)
1,2-Dibromoethane (*see* Ethylene dibromide)	
2,3-Dibromopropan-1-ol	*77*, 439 (2000)
Dichloroacetic acid	*63*, 271 (1995)
Dichloroacetonitrile (*see also* Halogenated acetonitriles)	*71*, 1375 (1999)
Dichloroacetylene	*39*, 369 (1986); *Suppl. 7*, 62 (1987); *71*, 1381 (1999)
ortho-Dichlorobenzene	*7*, 231 (1974); *29*, 213 (1982); *Suppl. 7*, 192 (1987); *73*, 223 (1999)
meta-Dichlorobenzene	*73*, 223 (1999)
para-Dichlorobenzene	*7*, 231 (1974); *29*, 215 (1982); *Suppl. 7*, 192 (1987); *73*, 223 (1999)
3,3'-Dichlorobenzidine	*4*, 49 (1974); *29*, 239 (1982); *Suppl. 7*, 193 (1987)
trans-1,4-Dichlorobutene	*15*, 149 (1977); *Suppl. 7*, 62 (1987); *71*, 1389 (1999)
3,3'-Dichloro-4,4'-diaminodiphenyl ether	*16*, 309 (1978); *Suppl. 7*, 62 (1987)
1,2-Dichloroethane	*20*, 429 (1979); *Suppl. 7*, 62 (1987); *71*, 501 (1999)
Dichloromethane	*20*, 449 (1979); *41*, 43 (1986); *Suppl. 7*, 194 (1987); *71*, 251 (1999)
2,4-Dichlorophenol (*see* Chlorophenols; Chlorophenols, occupational exposures to; Polychlorophenols and their sodium salts)	
(2,4-Dichlorophenoxy)acetic acid (*see* 2,4-D)	
2,6-Dichloro-*para*-phenylenediamine	*39*, 325 (1986); *Suppl. 7*, 62 (1987)
1,2-Dichloropropane	*41*, 131 (1986); *Suppl. 7*, 62 (1987); *71*, 1393 (1999)
1,3-Dichloropropene (technical-grade)	*41*, 113 (1986); *Suppl. 7*, 195 (1987); *71*, 933 (1999)
Dichlorvos	*20*, 97 (1979); *Suppl. 7*, 62 (1987); *53*, 267 (1991)
Dicofol	*30*, 87 (1983); *Suppl. 7*, 62 (1987)
Dicyclohexylamine (*see* Cyclamates)	
Didanosine	*76*, 153 (2000)
Dieldrin	*5*, 125 (1974); *Suppl. 7*, 196 (1987)
Dienoestrol (*see also* Nonsteroidal oestrogens)	*21*, 161 (1979); *Suppl. 7*, 278 (1987)
Diepoxybutane (*see also* 1,3-Butadiene)	*11*, 115 (1976) (*corr. 42*, 255); *Suppl. 7*, 62 (1987); *71*, 109 (1999)
Diesel and gasoline engine exhausts	*46*, 41 (1989)
Diesel fuels	*45*, 219 (1989) (*corr. 47*, 505)

Diethanolamine	77, 349 (2000)
Diethyl ether (see Anaesthetics, volatile)	
Di(2-ethylhexyl) adipate	29, 257 (1982); Suppl. 7, 62 (1987); 77, 149 (2000)
Di(2-ethylhexyl) phthalate	29, 269 (1982) (corr. 42, 261); Suppl. 7, 62 (1987); 77, 41 (2000)
1,2-Diethylhydrazine	4, 153 (1974); Suppl. 7, 62 (1987); 71, 1401 (1999)
Diethylstilboestrol	6, 55 (1974); 21, 173 (1979) (corr. 42, 259); Suppl. 7, 273 (1987)
Diethylstilboestrol dipropionate (see Diethylstilboestrol)	
Diethyl sulfate	4, 277 (1974); Suppl. 7, 198 (1987); 54, 213 (1992); 71, 1405 (1999)
N,N'-Diethylthiourea	79, 649 (2001)
Diglycidyl resorcinol ether	11, 125 (1976); 36, 181 (1985); Suppl. 7, 62 (1987); 71, 1417 (1999)
Dihydrosafrole	1, 170 (1972); 10, 233 (1976) Suppl. 7, 62 (1987)
1,8-Dihydroxyanthraquinone (see Dantron)	
Dihydroxybenzenes (see Catechol; Hydroquinone; Resorcinol)	
Dihydroxymethylfuratrizine	24, 77 (1980); Suppl. 7, 62 (1987)
Diisopropyl sulfate	54, 229 (1992); 71, 1421 (1999)
Dimethisterone (see also Progestins; Sequential oral contraceptives)	6, 167 (1974); 21, 377 (1979))
Dimethoxane	15, 177 (1977); Suppl. 7, 62 (1987)
3,3'-Dimethoxybenzidine	4, 41 (1974); Suppl. 7, 198 (1987)
3,3'-Dimethoxybenzidine-4,4'-diisocyanate	39, 279 (1986); Suppl. 7, 62 (1987)
para-Dimethylaminoazobenzene	8, 125 (1975); Suppl. 7, 62 (1987)
para-Dimethylaminoazobenzenediazo sodium sulfonate	8, 147 (1975); Suppl. 7, 62 (1987)
trans-2-[(Dimethylamino)methylimino]-5-[2-(5-nitro-2-furyl)-vinyl]-1,3,4-oxadiazole	7, 147 (1974) (corr. 42, 253); Suppl. 7, 62 (1987)
4,4'-Dimethylangelicin plus ultraviolet radiation (see also Angelicin and some synthetic derivatives)	Suppl. 7, 57 (1987)
4,5'-Dimethylangelicin plus ultraviolet radiation (see also Angelicin and some synthetic derivatives)	Suppl. 7, 57 (1987)
2,6-Dimethylaniline	57, 323 (1993)
N,N-Dimethylaniline	57, 337 (1993)
Dimethylarsinic acid (see Arsenic and arsenic compounds)	
3,3'-Dimethylbenzidine	1, 87 (1972); Suppl. 7, 62 (1987)
Dimethylcarbamoyl chloride	12, 77 (1976); Suppl. 7, 199 (1987); 71, 531 (1999)
Dimethylformamide	47, 171 (1989); 71, 545 (1999)
1,1-Dimethylhydrazine	4, 137 (1974); Suppl. 7, 62 (1987); 71, 1425 (1999)
1,2-Dimethylhydrazine	4, 145 (1974) (corr. 42, 253); Suppl. 7, 62 (1987); 71, 947 (1999)
Dimethyl hydrogen phosphite	48, 85 (1990); 71, 1437 (1999)
1,4-Dimethylphenanthrene	32, 349 (1983); Suppl. 7, 62 (1987)
Dimethyl sulfate	4, 271 (1974); Suppl. 7, 200 (1987); 71, 575 (1999)
3,7-Dinitrofluoranthene	46, 189 (1989); 65, 297 (1996)
3,9-Dinitrofluoranthene	46, 195 (1989); 65, 297 (1996)

1,3-Dinitropyrene	*46*, 201 (1989)
1,6-Dinitropyrene	*46*, 215 (1989)
1,8-Dinitropyrene	*33*, 171 (1984); *Suppl. 7*, 63 (1987); *46*, 231 (1989)
Dinitrosopentamethylenetetramine	*11*, 241 (1976); *Suppl. 7*, 63 (1987)
2,4-Dinitrotoluene	*65*, 309 (1996) (*corr. 66*, 485)
2,6-Dinitrotoluene	*65*, 309 (1996) (*corr. 66*, 485)
3,5-Dinitrotoluene	*65*, 309 (1996)
1,4-Dioxane	*11*, 247 (1976); *Suppl. 7*, 201 (1987); *71*, 589 (1999)
2,4′-Diphenyldiamine	*16*, 313 (1978); *Suppl. 7*, 63 (1987)
Direct Black 38 (*see also* Benzidine-based dyes)	*29*, 295 (1982) (*corr. 42*, 261)
Direct Blue 6 (*see also* Benzidine-based dyes)	*29*, 311 (1982)
Direct Brown 95 (*see also* Benzidine-based dyes)	*29*, 321 (1982)
Disperse Blue 1	*48*, 139 (1990)
Disperse Yellow 3	*8*, 97 (1975); *Suppl. 7*, 60 (1987); *48*, 149 (1990)
Disulfiram	*12*, 85 (1976); *Suppl. 7*, 63 (1987)
Dithranol	*13*, 75 (1977); *Suppl. 7*, 63 (1987)
Divinyl ether (*see* Anaesthetics, volatile)	
Doxefazepam	*66*, 97 (1996)
Doxylamine succinate	*79*, 145 (2001)
Droloxifene	*66*, 241 (1996)
Dry cleaning	*63*, 33 (1995)
Dulcin	*12*, 97 (1976); *Suppl. 7*, 63 (1987)

E

Endrin	*5*, 157 (1974); *Suppl. 7*, 63 (1987)
Enflurane (*see* Anaesthetics, volatile)	
Eosin	*15*, 183 (1977); *Suppl. 7*, 63 (1987)
Epichlorohydrin	*11*, 131 (1976) (*corr. 42*, 256); *Suppl. 7*, 202 (1987); *71*, 603 (1999)
1,2-Epoxybutane	*47*, 217 (1989); *71*, 629 (1999)
1-Epoxyethyl-3,4-epoxycyclohexane (*see* 4-Vinylcyclohexene diepoxide)	
3,4-Epoxy-6-methylcyclohexylmethyl 3,4-epoxy-6-methyl-cyclohexane carboxylate	*11*, 147 (1976); *Suppl. 7*, 63 (1987); *71*, 1441 (1999)
cis-9,10-Epoxystearic acid	*11*, 153 (1976); *Suppl. 7*, 63 (1987); *71*, 1443 (1999)
Epstein-Barr virus	*70*, 47 (1997)
d-Equilenin	*72*, 399 (1999)
Equilin	*72*, 399 (1999)
Erionite	*42*, 225 (1987); *Suppl. 7*, 203 (1987)
Estazolam	*66*, 105 (1996)
Ethinyloestradiol	*6*, 77 (1974); *21*, 233 (1979); *Suppl. 7*, 286 (1987); *72*, 49 (1999)
Ethionamide	*13*, 83 (1977); *Suppl. 7*, 63 (1987)
Ethyl acrylate	*19*, 57 (1979); *39*, 81 (1986); *Suppl. 7*, 63 (1987); *71*, 1447 (1999)
Ethylbenzene	*77*, 227 (2000)

Ethylene	*19*, 157 (1979); *Suppl. 7*, 63 (1987); *60*, 45 (1994); *71*, 1447 (1999)
Ethylene dibromide	*15*, 195 (1977); *Suppl. 7*, 204 (1987); *71*, 641 (1999)
Ethylene oxide	*11*, 157 (1976); *36*, 189 (1985) (*corr. 42*, 263); *Suppl. 7*, 205 (1987); *60*, 73 (1994)
Ethylene sulfide	*11*, 257 (1976); *Suppl. 7*, 63 (1987)
Ethylenethiourea	*7*, 45 (1974); *Suppl. 7*, 207 (1987); *79*, 659 (2001)
2-Ethylhexyl acrylate	*60*, 475 (1994)
Ethyl methanesulfonate	*7*, 245 (1974); *Suppl. 7*, 63 (1987)
N-Ethyl-N-nitrosourea	*1*, 135 (1972); *17*, 191 (1978); *Suppl. 7*, 63 (1987)
Ethyl selenac (*see also* Selenium and selenium compounds)	*12*, 107 (1976); *Suppl. 7*, 63 (1987)
Ethyl tellurac	*12*, 115 (1976); *Suppl. 7*, 63 (1987)
Ethynodiol diacetate	*6*, 173 (1974); *21*, 387 (1979); *Suppl. 7*, 292 (1987); *72*, 49 (1999)
Etoposide	*76*, 177 (2000)
Eugenol	*36*, 75 (1985); *Suppl. 7*, 63 (1987)
Evans blue	*8*, 151 (1975); *Suppl. 7*, 63 (1987)
Extremely low-frequency electric fields	*80* (2002)
Extremely low-frequency magnetic fields	*80* (2002)

F

Fast Green FCF	*16*, 187 (1978); *Suppl. 7*, 63 (1987)
Fenvalerate	*53*, 309 (1991)
Ferbam	*12*, 121 (1976) (*corr. 42*, 256); *Suppl. 7*, 63 (1987)
Ferric oxide	*1*, 29 (1972); *Suppl. 7*, 216 (1987)
Ferrochromium (*see* Chromium and chromium compounds)	
Fluometuron	*30*, 245 (1983); *Suppl. 7*, 63 (1987)
Fluoranthene	*32*, 355 (1983); *Suppl. 7*, 63 (1987)
Fluorene	*32*, 365 (1983); *Suppl. 7*, 63 (1987)
Fluorescent lighting (exposure to) (*see* Ultraviolet radiation)	
Fluorides (inorganic, used in drinking-water)	*27*, 237 (1982); *Suppl. 7*, 208 (1987)
5-Fluorouracil	*26*, 217 (1981); *Suppl. 7*, 210 (1987)
Fluorspar (*see* Fluorides)	
Fluosilicic acid (*see* Fluorides)	
Fluroxene (*see* Anaesthetics, volatile)	
Foreign bodies	*74* (1999)
Formaldehyde	*29*, 345 (1982); *Suppl. 7*, 211 (1987); *62*, 217 (1995) (*corr. 65*, 549; *corr. 66*, 485)
2-(2-Formylhydrazino)-4-(5-nitro-2-furyl)thiazole	*7*, 151 (1974) (*corr. 42*, 253); *Suppl. 7*, 63 (1987)
Frusemide (*see* Furosemide)	
Fuel oils (heating oils)	*45*, 239 (1989) (*corr. 47*, 505)

Fumonisin B$_1$ (*see* Toxins derived from *Fusarium moniliforme*)	
Fumonisin B$_2$ (*see* Toxins derived from *Fusarium moniliforme*)	
Furan	*63*, 393 (1995)
Furazolidone	*31*, 141 (1983); *Suppl. 7*, 63 (1987)
Furfural	*63*, 409 (1995)
Furniture and cabinet-making	*25*, 99 (1981); *Suppl. 7*, 380 (1987)
Furosemide	*50*, 277 (1990)
2-(2-Furyl)-3-(5-nitro-2-furyl)acrylamide (*see* AF-2)	
Fusarenon-X (*see* Toxins derived from *Fusarium graminearum*, *F. culmorum* and *F. crookwellense*)	
Fusarenone-X (*see* Toxins derived from *Fusarium graminearum*, *F. culmorum* and *F. crookwellense*)	
Fusarin C (*see* Toxins derived from *Fusarium moniliforme*)	

G

Gamma (γ)-radiation	*75*, 121 (2000)
Gasoline	*45*, 159 (1989) (*corr. 47*, 505)
Gasoline engine exhaust (*see* Diesel and gasoline engine exhausts)	
Gemfibrozil	*66*, 427 (1996)
Glass fibres (*see* Man-made mineral fibres)	
Glass manufacturing industry, occupational exposures in	*58*, 347 (1993)
Glass wool (*see* Man-made vitreous fibres)	
Glass filaments (*see* Man-made mineral fibres)	
Glu-P-1	*40*, 223 (1986); *Suppl. 7*, 64 (1987)
Glu-P-2	*40*, 235 (1986); *Suppl. 7*, 64 (1987)
L-Glutamic acid, 5-[2-(4-hydroxymethyl)phenylhydrazide] (*see* Agaritine)	
Glycidaldehyde	*11*, 175 (1976); *Suppl. 7*, 64 (1987); *71*, 1459 (1999)
Glycidol	*77*, 469 (2000)
Glycidyl ethers	*47*, 237 (1989); *71*, 1285, 1417, 1525, 1539 (1999)
Glycidyl oleate	*11*, 183 (1976); *Suppl. 7*, 64 (1987)
Glycidyl stearate	*11*, 187 (1976); *Suppl. 7*, 64 (1987)
Griseofulvin	*10*, 153 (1976); *Suppl. 7*, 64, 391 (1987); *79*, 289 (2001)
Guinea Green B	*16*, 199 (1978); *Suppl. 7*, 64 (1987)
Gyromitrin	*31*, 163 (1983); *Suppl. 7*, 64, 391 (1987)

H

Haematite	*1*, 29 (1972); *Suppl. 7*, 216 (1987)
Haematite and ferric oxide	*Suppl. 7*, 216 (1987)
Haematite mining, underground, with exposure to radon	*1*, 29 (1972); *Suppl. 7*, 216 (1987)
Hairdressers and barbers (occupational exposure as)	*57*, 43 (1993)
Hair dyes, epidemiology of	*16*, 29 (1978); *27*, 307 (1982);
Halogenated acetonitriles	*52*, 269 (1991); *71*, 1325, 1369, 1375, 1533 (1999)
Halothane (*see* Anaesthetics, volatile)	
HC Blue No. 1	*57*, 129 (1993)

HC Blue No. 2	*57*, 143 (1993)
α-HCH (*see* Hexachlorocyclohexanes)	
β-HCH (*see* Hexachlorocyclohexanes)	
γ-HCH (*see* Hexachlorocyclohexanes)	
HC Red No. 3	*57*, 153 (1993)
HC Yellow No. 4	*57*, 159 (1993)
Heating oils (*see* Fuel oils)	
Helicobacter pylori (infection with)	*61*, 177 (1994)
Hepatitis B virus	*59*, 45 (1994)
Hepatitis C virus	*59*, 165 (1994)
Hepatitis D virus	*59*, 223 (1994)
Heptachlor (*see also* Chlordane/Heptachlor)	*5*, 173 (1974); *20*, 129 (1979)
Hexachlorobenzene	*20*, 155 (1979); *Suppl. 7*, 219 (1987); *79*, 493 (2001)
Hexachlorobutadiene	*20*, 179 (1979); *Suppl. 7*, 64 (1987); *73*, 277 (1999)
Hexachlorocyclohexanes	*5*, 47 (1974); *20*, 195 (1979) (*corr. 42*, 258); *Suppl. 7*, 220 (1987)
Hexachlorocyclohexane, technical-grade (*see* Hexachlorocyclohexanes)	
Hexachloroethane	*20*, 467 (1979); *Suppl. 7*, 64 (1987); *73*, 295 (1999)
Hexachlorophene	*20*, 241 (1979); *Suppl. 7*, 64 (1987)
Hexamethylphosphoramide	*15*, 211 (1977); *Suppl. 7*, 64 (1987); *71*, 1465 (1999)
Hexoestrol (*see also* Nonsteroidal oestrogens)	*Suppl. 7*, 279 (1987)
Hormonal contraceptives, progestogens only	*72*, 339 (1999)
Human herpesvirus 8	*70*, 375 (1997)
Human immunodeficiency viruses	*67*, 31 (1996)
Human papillomaviruses	*64* (1995) (*corr. 66*, 485)
Human T-cell lymphotropic viruses	*67*, 261 (1996)
Hycanthone mesylate	*13*, 91 (1977); *Suppl. 7*, 64 (1987)
Hydralazine	*24*, 85 (1980); *Suppl. 7*, 222 (1987)
Hydrazine	*4*, 127 (1974); *Suppl. 7*, 223 (1987); *71*, 991 (1999)
Hydrochloric acid	*54*, 189 (1992)
Hydrochlorothiazide	*50*, 293 (1990)
Hydrogen peroxide	*36*, 285 (1985); *Suppl. 7*, 64 (1987); *71*, 671 (1999)
Hydroquinone	*15*, 155 (1977); *Suppl. 7*, 64 (1987); *71*, 691 (1999)
4-Hydroxyazobenzene	*8*, 157 (1975); *Suppl. 7*, 64 (1987)
17α-Hydroxyprogesterone caproate (*see also* Progestins)	*21*, 399 (1979) (*corr. 42*, 259)
8-Hydroxyquinoline	*13*, 101 (1977); *Suppl. 7*, 64 (1987)
8-Hydroxysenkirkine	*10*, 265 (1976); *Suppl. 7*, 64 (1987)
Hydroxyurea	*76*, 347 (2000)
Hypochlorite salts	*52*, 159 (1991)

I

Implants, surgical	*74*, 1999
Indeno[1,2,3-*cd*]pyrene	*3*, 229 (1973); *32*, 373 (1983); *Suppl. 7*, 64 (1987)

Inorganic acids (*see* Sulfuric acid and other strong inorganic acids, occupational exposures to mists and vapours from)
Insecticides, occupational exposures in spraying and application of   *53*, 45 (1991)
Insulation glass wool (*see* Man-made vitreous fibres)
Ionizing radiation (*see* Neutrons, γ- and X-radiation)
IQ   *40*, 261 (1986); *Suppl. 7*, 64 (1987); *56*, 165 (1993)

Iron and steel founding   *34*, 133 (1984); *Suppl. 7*, 224 (1987)

Iron-dextran complex   *2*, 161 (1973); *Suppl. 7*, 226 (1987)
Iron-dextrin complex   *2*, 161 (1973) (*corr. 42*, 252); *Suppl. 7*, 64 (1987)

Iron oxide (*see* Ferric oxide)
Iron oxide, saccharated (*see* Saccharated iron oxide)
Iron sorbitol-citric acid complex   *2*, 161 (1973); *Suppl. 7*, 64 (1987)
Isatidine   *10*, 269 (1976); *Suppl. 7*, 65 (1987)

Isoflurane (*see* Anaesthetics, volatile)
Isoniazid (*see* Isonicotinic acid hydrazide)
Isonicotinic acid hydrazide   *4*, 159 (1974); *Suppl. 7*, 227 (1987)
Isophosphamide   *26*, 237 (1981); *Suppl. 7*, 65 (1987)
Isoprene   *60*, 215 (1994); *71*, 1015 (1999)
Isopropanol   *15*, 223 (1977); *Suppl. 7*, 229 (1987); *71*, 1027 (1999)

Isopropanol manufacture (strong-acid process)   *Suppl. 7*, 229 (1987)
    (*see also* Isopropanol; Sulfuric acid and other strong inorganic acids, occupational exposures to mists and vapours from)
Isopropyl oils   *15*, 223 (1977); *Suppl. 7*, 229 (1987); *71*, 1483 (1999)

Isosafrole   *1*, 169 (1972); *10*, 232 (1976); *Suppl. 7*, 65 (1987)

J

Jacobine   *10*, 275 (1976); *Suppl. 7*, 65 (1987)
Jet fuel   *45*, 203 (1989)
Joinery (*see* Carpentry and joinery)

K

Kaempferol   *31*, 171 (1983); *Suppl. 7*, 65 (1987)
Kaposi's sarcoma herpesvirus   *70*, 375 (1997)
Kepone (*see* Chlordecone)
Kojic acid   *79*, 605 (2001)

L

Lasiocarpine   *10*, 281 (1976); *Suppl. 7*, 65 (1987)
Lauroyl peroxide   *36*, 315 (1985); *Suppl. 7*, 65 (1987); *71*, 1485 (1999)

Lead acetate (*see* Lead and lead compounds)

Lead and lead compounds (*see also* Foreign bodies)	*1*, 40 (1972) (*corr. 42*, 251); *2*, 52, 150 (1973); *12*, 131 (1976); *23*, 40, 208, 209, 325 (1980); *Suppl. 7*, 230 (1987)
Lead arsenate (*see* Arsenic and arsenic compounds)	
Lead carbonate (*see* Lead and lead compounds)	
Lead chloride (*see* Lead and lead compounds)	
Lead chromate (*see* Chromium and chromium compounds)	
Lead chromate oxide (*see* Chromium and chromium compounds)	
Lead naphthenate (*see* Lead and lead compounds)	
Lead nitrate (*see* Lead and lead compounds)	
Lead oxide (*see* Lead and lead compounds)	
Lead phosphate (*see* Lead and lead compounds)	
Lead subacetate (*see* Lead and lead compounds)	
Lead tetroxide (*see* Lead and lead compounds)	
Leather goods manufacture	*25*, 279 (1981); *Suppl. 7*, 235 (1987)
Leather industries	*25*, 199 (1981); *Suppl. 7*, 232 (1987)
Leather tanning and processing	*25*, 201 (1981); *Suppl. 7*, 236 (1987)
Ledate (*see also* Lead and lead compounds)	*12*, 131 (1976)
Levonorgestrel	*72*, 49 (1999)
Light Green SF	*16*, 209 (1978); *Suppl. 7*, 65 (1987)
d-Limonene	*56*, 135 (1993); *73*, 307 (1999)
Lindane (*see* Hexachlorocyclohexanes)	
Liver flukes (*see Clonorchis sinensis, Opisthorchis felineus* and *Opisthorchis viverrini*)	
Lumber and sawmill industries (including logging)	*25*, 49 (1981); *Suppl. 7*, 383 (1987)
Luteoskyrin	*10*, 163 (1976); *Suppl. 7*, 65 (1987)
Lynoestrenol	*21*, 407 (1979); *Suppl. 7*, 293 (1987); *72*, 49 (1999)

M

Magenta	*4*, 57 (1974) (*corr. 42*, 252); *Suppl. 7*, 238 (1987); *57*, 215 (1993)
Magenta, manufacture of (*see also* Magenta)	*Suppl. 7*, 238 (1987); *57*, 215 (1993)
Malathion	*30*, 103 (1983); *Suppl. 7*, 65 (1987)
Maleic hydrazide	*4*, 173 (1974) (*corr. 42*, 253); *Suppl. 7*, 65 (1987)
Malonaldehyde	*36*, 163 (1985); *Suppl. 7*, 65 (1987); *71*, 1037 (1999)
Malondialdehyde (*see* Malonaldehyde)	
Maneb	*12*, 137 (1976); *Suppl. 7*, 65 (1987)
Man-made mineral fibres (*see* Man-made vitreous fibres)	
Man-made vitreous fibres	*43*, 39 (1988); *81* (2002)
Mannomustine	*9*, 157 (1975); *Suppl. 7*, 65 (1987)
Mate	*51*, 273 (1991)
MCPA (*see also* Chlorophenoxy herbicides; Chlorophenoxy herbicides, occupational exposures to)	*30*, 255 (1983)

MeA-α-C	*40*, 253 (1986); *Suppl. 7*, 65 (1987)
Medphalan	*9*, 168 (1975); *Suppl. 7*, 65 (1987)
Medroxyprogesterone acetate	*6*, 157 (1974); *21*, 417 (1979) (*corr. 42*, 259); *Suppl. 7*, 289 (1987); *72*, 339 (1999)
Megestrol acetate	*Suppl. 7*, 293 (1987); *72*, 49 (1999)
MeIQ	*40*, 275 (1986); *Suppl. 7*, 65 (1987); *56*, 197 (1993)
MeIQx	*40*, 283 (1986); *Suppl. 7*, 65 (1987) *56*, 211 (1993)
Melamine	*39*, 333 (1986); *Suppl. 7*, 65 (1987); *73*, 329 (1999)
Melphalan	*9*, 167 (1975); *Suppl. 7*, 239 (1987)
6-Mercaptopurine	*26*, 249 (1981); *Suppl. 7*, 240 (1987)
Mercuric chloride (*see* Mercury and mercury compounds)	
Mercury and mercury compounds	*58*, 239 (1993)
Merphalan	*9*, 169 (1975); *Suppl. 7*, 65 (1987)
Mestranol	*6*, 87 (1974); *21*, 257 (1979) (*corr. 42*, 259); *Suppl. 7*, 288 (1987); *72*, 49 (1999)
Metabisulfites (*see* Sulfur dioxide and some sulfites, bisulfites and metabisulfites)	
Metallic mercury (*see* Mercury and mercury compounds)	
Methanearsonic acid, disodium salt (*see* Arsenic and arsenic compounds)	
Methanearsonic acid, monosodium salt (*see* Arsenic and arsenic compounds	
Methimazole	*79*, 53 (2001)
Methotrexate	*26*, 267 (1981); *Suppl. 7*, 241 (1987)
Methoxsalen (*see* 8-Methoxypsoralen)	
Methoxychlor	*5*, 193 (1974); *20*, 259 (1979); *Suppl. 7*, 66 (1987)
Methoxyflurane (*see* Anaesthetics, volatile)	
5-Methoxypsoralen	*40*, 327 (1986); *Suppl. 7*, 242 (1987)
8-Methoxypsoralen (*see also* 8-Methoxypsoralen plus ultraviolet radiation)	*24*, 101 (1980)
8-Methoxypsoralen plus ultraviolet radiation	*Suppl. 7*, 243 (1987)
Methyl acrylate	*19*, 52 (1979); *39*, 99 (1986); *Suppl. 7*, 66 (1987); *71*, 1489 (1999)
5-Methylangelicin plus ultraviolet radiation (*see also* Angelicin and some synthetic derivatives)	*Suppl. 7*, 57 (1987)
2-Methylaziridine	*9*, 61 (1975); *Suppl. 7*, 66 (1987); *71*, 1497 (1999)
Methylazoxymethanol acetate (*see also* Cycasin)	*1*, 164 (1972); *10*, 131 (1976); *Suppl. 7*, 66 (1987)
Methyl bromide	*41*, 187 (1986) (*corr. 45*, 283); *Suppl. 7*, 245 (1987); *71*, 721 (1999)
Methyl *tert*-butyl ether	*73*, 339 (1999)
Methyl carbamate	*12*, 151 (1976); *Suppl. 7*, 66 (1987)

Methyl-CCNU (see 1-(2-Chloroethyl)-3-(4-methylcyclohexyl)-1-nitrosourea)
Methyl chloride *41*, 161 (1986); *Suppl. 7*, 246 (1987); *71*, 737 (1999)

1-, 2-, 3-, 4-, 5- and 6-Methylchrysenes *32*, 379 (1983); *Suppl. 7*, 66 (1987)
N-Methyl-N,4-dinitrosoaniline *1*, 141 (1972); *Suppl. 7*, 66 (1987)
4,4'-Methylene bis(2-chloroaniline) *4*, 65 (1974) (*corr. 42*, 252); *Suppl. 7*, 246 (1987); *57*, 271 (1993)

4,4'-Methylene bis(N,N-dimethyl)benzenamine *27*, 119 (1982); *Suppl. 7*, 66 (1987)
4,4'-Methylene bis(2-methylaniline) *4*, 73 (1974); *Suppl. 7*, 248 (1987)
4,4'-Methylenedianiline *4*, 79 (1974) (*corr. 42*, 252); *39*, 347 (1986); *Suppl. 7*, 66 (1987)
4,4'-Methylenediphenyl diisocyanate *19*, 314 (1979); *Suppl. 7*, 66 (1987); *71*, 1049 (1999)

2-Methylfluoranthene *32*, 399 (1983); *Suppl. 7*, 66 (1987)
3-Methylfluoranthene *32*, 399 (1983); *Suppl. 7*, 66 (1987)
Methylglyoxal *51*, 443 (1991)
Methyl iodide *15*, 245 (1977); *41*, 213 (1986); *Suppl. 7*, 66 (1987); *71*, 1503 (1999)

Methylmercury chloride (see Mercury and mercury compounds)
Methylmercury compounds (see Mercury and mercury compounds)
Methyl methacrylate *19*, 187 (1979); *Suppl. 7*, 66 (1987); *60*, 445 (1994)

Methyl methanesulfonate *7*, 253 (1974); *Suppl. 7*, 66 (1987); *71*, 1059 (1999)

2-Methyl-1-nitroanthraquinone *27*, 205 (1982); *Suppl. 7*, 66 (1987)
N-Methyl-N'-nitro-N-nitrosoguanidine *4*, 183 (1974); *Suppl. 7*, 248 (1987)
3-Methylnitrosaminopropionaldehyde [see 3-(N-Nitrosomethylamino)-propionaldehyde]
3-Methylnitrosaminopropionitrile [see 3-(N-Nitrosomethylamino)-propionitrile]
4-(Methylnitrosamino)-4-(3-pyridyl)-1-butanal [see 4-(N-Nitrosomethyl-amino)-4-(3-pyridyl)-1-butanal]
4-(Methylnitrosamino)-1-(3-pyridyl)-1-butanone [see 4-(-Nitrosomethyl-amino)-1-(3-pyridyl)-1-butanone]
N-Methyl-N-nitrosourea *1*, 125 (1972); *17*, 227 (1978); *Suppl. 7*, 66 (1987)

N-Methyl-N-nitrosourethane *4*, 211 (1974); *Suppl. 7*, 66 (1987)
N-Methylolacrylamide *60*, 435 (1994)
Methyl parathion *30*, 131 (1983); *Suppl. 7*, 66, 392 (1987)

1-Methylphenanthrene *32*, 405 (1983); *Suppl. 7*, 66 (1987)
7-Methylpyrido[3,4-c]psoralen *40*, 349 (1986); *Suppl. 7*, 71 (1987)
Methyl red *8*, 161 (1975); *Suppl. 7*, 66 (1987)
Methyl selenac (see also Selenium and selenium compounds) *12*, 161 (1976); *Suppl. 7*, 66 (1987)
Methylthiouracil *7*, 53 (1974); *Suppl. 7*, 66 (1987); *79*, 75 (2001)

Metronidazole *13*, 113 (1977); *Suppl. 7*, 250 (1987)

Mineral oils *3*, 30 (1973); *33*, 87 (1984) (*corr. 42*, 262); *Suppl. 7*, 252 (1987)

Mirex	5, 203 (1974); *20*, 283 (1979) (*corr. 42*, 258); *Suppl. 7*, 66 (1987)
Mists and vapours from sulfuric acid and other strong inorganic acids	*54*, 41 (1992)
Mitomycin C	*10*, 171 (1976); *Suppl. 7*, 67 (1987)
Mitoxantrone	*76*, 289 (2000)
MNNG (*see* N-Methyl-N'-nitro-N-nitrosoguanidine)	
MOCA (*see* 4,4'-Methylene bis(2-chloroaniline))	
Modacrylic fibres	*19*, 86 (1979); *Suppl. 7*, 67 (1987)
Monocrotaline	*10*, 291 (1976); *Suppl. 7*, 67 (1987)
Monuron	*12*, 167 (1976); *Suppl. 7*, 67 (1987); *53*, 467 (1991)
MOPP and other combined chemotherapy including alkylating agents	*Suppl. 7*, 254 (1987)
Mordanite (*see* Zeolites)	
Morpholine	*47*, 199 (1989); *71*, 1511 (1999)
5-(Morpholinomethyl)-3-[(5-nitrofurfurylidene)amino]-2-oxazolidinone	*7*, 161 (1974); *Suppl. 7*, 67 (1987)
Musk ambrette	*65*, 477 (1996)
Musk xylene	*65*, 477 (1996)
Mustard gas	*9*, 181 (1975) (*corr. 42*, 254); *Suppl. 7*, 259 (1987)
Myleran (*see* 1,4-Butanediol dimethanesulfonate)	

N

Nafenopin	*24*, 125 (1980); *Suppl. 7*, 67 (1987)
1,5-Naphthalenediamine	*27*, 127 (1982); *Suppl. 7*, 67 (1987)
1,5-Naphthalene diisocyanate	*19*, 311 (1979); *Suppl. 7*, 67 (1987); *71*, 1515 (1999)
1-Naphthylamine	*4*, 87 (1974) (*corr. 42*, 253); *Suppl. 7*, 260 (1987)
2-Naphthylamine	*4*, 97 (1974); *Suppl. 7*, 261 (1987)
1-Naphthylthiourea	*30*, 347 (1983); *Suppl. 7*, 263 (1987)
Neutrons	*75*, 361 (2000)
Nickel acetate (*see* Nickel and nickel compounds)	
Nickel ammonium sulfate (*see* Nickel and nickel compounds)	
Nickel and nickel compounds (*see also* Implants, surgical)	*2*, 126 (1973) (*corr. 42*, 252); *11*, 75 (1976); *Suppl. 7*, 264 (1987) (*corr. 45*, 283); *49*, 257 (1990) (*corr. 67*, 395)
Nickel carbonate (*see* Nickel and nickel compounds)	
Nickel carbonyl (*see* Nickel and nickel compounds)	
Nickel chloride (*see* Nickel and nickel compounds)	
Nickel-gallium alloy (*see* Nickel and nickel compounds)	
Nickel hydroxide (*see* Nickel and nickel compounds)	
Nickelocene (*see* Nickel and nickel compounds)	
Nickel oxide (*see* Nickel and nickel compounds)	
Nickel subsulfide (*see* Nickel and nickel compounds)	
Nickel sulfate (*see* Nickel and nickel compounds)	
Niridazole	*13*, 123 (1977); *Suppl. 7*, 67 (1987)
Nithiazide	*31*, 179 (1983); *Suppl. 7*, 67 (1987)
Nitrilotriacetic acid and its salts	*48*, 181 (1990); *73*, 385 (1999)

5-Nitroacenaphthene	*16*, 319 (1978); *Suppl. 7*, 67 (1987)
5-Nitro-*ortho*-anisidine	*27*, 133 (1982); *Suppl. 7*, 67 (1987)
2-Nitroanisole	*65*, 369 (1996)
9-Nitroanthracene	*33*, 179 (1984); *Suppl. 7*, 67 (1987)
7-Nitrobenz[*a*]anthracene	*46*, 247 (1989)
Nitrobenzene	*65*, 381 (1996)
6-Nitrobenzo[*a*]pyrene	*33*, 187 (1984); *Suppl. 7*, 67 (1987); *46*, 255 (1989)
4-Nitrobiphenyl	*4*, 113 (1974); *Suppl. 7*, 67 (1987)
6-Nitrochrysene	*33*, 195 (1984); *Suppl. 7*, 67 (1987); *46*, 267 (1989)
Nitrofen (technical-grade)	*30*, 271 (1983); *Suppl. 7*, 67 (1987)
3-Nitrofluoranthene	*33*, 201 (1984); *Suppl. 7*, 67 (1987)
2-Nitrofluorene	*46*, 277 (1989)
Nitrofural	*7*, 171 (1974); *Suppl. 7*, 67 (1987); *50*, 195 (1990)
5-Nitro-2-furaldehyde semicarbazone (*see* Nitrofural)	
Nitrofurantoin	*50*, 211 (1990)
Nitrofurazone (*see* Nitrofural)	
1-[(5-Nitrofurfurylidene)amino]-2-imidazolidinone	*7*, 181 (1974); *Suppl. 7*, 67 (1987)
N-[4-(5-Nitro-2-furyl)-2-thiazolyl]acetamide	*1*, 181 (1972); *7*, 185 (1974); *Suppl. 7*, 67 (1987)
Nitrogen mustard	*9*, 193 (1975); *Suppl. 7*, 269 (1987)
Nitrogen mustard *N*-oxide	*9*, 209 (1975); *Suppl. 7*, 67 (1987)
Nitromethane	*77*, 487 (2000)
1-Nitronaphthalene	*46*, 291 (1989)
2-Nitronaphthalene	*46*, 303 (1989)
3-Nitroperylene	*46*, 313 (1989)
2-Nitro-*para*-phenylenediamine (*see* 1,4-Diamino-2-nitrobenzene)	
2-Nitropropane	*29*, 331 (1982); *Suppl. 7*, 67 (1987); *71*, 1079 (1999)
1-Nitropyrene	*33*, 209 (1984); *Suppl. 7*, 67 (1987); *46*, 321 (1989)
2-Nitropyrene	*46*, 359 (1989)
4-Nitropyrene	*46*, 367 (1989)
N-Nitrosatable drugs	*24*, 297 (1980) (*corr. 42*, 260)
N-Nitrosatable pesticides	*30*, 359 (1983)
N'-Nitrosoanabasine	*37*, 225 (1985); *Suppl. 7*, 67 (1987)
N'-Nitrosoanatabine	*37*, 233 (1985); *Suppl. 7*, 67 (1987)
N-Nitrosodi-*n*-butylamine	*4*, 197 (1974); *17*, 51 (1978); *Suppl. 7*, 67 (1987)
N-Nitrosodiethanolamine	*17*, 77 (1978); *Suppl. 7*, 67 (1987); *77*, 403 (2000)
N-Nitrosodiethylamine	*1*, 107 (1972) (*corr. 42*, 251); *17*, 83 (1978) (*corr. 42*, 257); *Suppl. 7*, 67 (1987)
N-Nitrosodimethylamine	*1*, 95 (1972); *17*, 125 (1978) (*corr. 42*, 257); *Suppl. 7*, 67 (1987)
N-Nitrosodiphenylamine	*27*, 213 (1982); *Suppl. 7*, 67 (1987)
para-Nitrosodiphenylamine	*27*, 227 (1982) (*corr. 42*, 261); *Suppl. 7*, 68 (1987)
N-Nitrosodi-*n*-propylamine	*17*, 177 (1978); *Suppl. 7*, 68 (1987)
N-Nitroso-*N*-ethylurea (*see* *N*-Ethyl-*N*-nitrosourea)	
N-Nitrosofolic acid	*17*, 217 (1978); *Suppl. 7*, 68 (1987)

N-Nitrosoguvacine	*37*, 263 (1985); *Suppl. 7*, 68 (1987)
N-Nitrosoguvacoline	*37*, 263 (1985); *Suppl. 7*, 68 (1987)
N-Nitrosohydroxyproline	*17*, 304 (1978); *Suppl. 7*, 68 (1987)
3-(*N*-Nitrosomethylamino)propionaldehyde	*37*, 263 (1985); *Suppl. 7*, 68 (1987)
3-(*N*-Nitrosomethylamino)propionitrile	*37*, 263 (1985); *Suppl. 7*, 68 (1987)
4-(*N*-Nitrosomethylamino)-4-(3-pyridyl)-1-butanal	*37*, 205 (1985); *Suppl. 7*, 68 (1987)
4-(*N*-Nitrosomethylamino)-1-(3-pyridyl)-1-butanone	*37*, 209 (1985); *Suppl. 7*, 68 (1987)
N-Nitrosomethylethylamine	*17*, 221 (1978); *Suppl. 7*, 68 (1987)
N-Nitroso-*N*-methylurea (*see N*-Methyl-*N*-nitrosourea)	
N-Nitroso-*N*-methylurethane (*see N*-Methyl-*N*-nitrosourethane)	
N-Nitrosomethylvinylamine	*17*, 257 (1978); *Suppl. 7*, 68 (1987)
N-Nitrosomorpholine	*17*, 263 (1978); *Suppl. 7*, 68 (1987)
N'-Nitrosonornicotine	*17*, 281 (1978); *37*, 241 (1985); *Suppl. 7*, 68 (1987)
N-Nitrosopiperidine	*17*, 287 (1978); *Suppl. 7*, 68 (1987)
N-Nitrosoproline	*17*, 303 (1978); *Suppl. 7*, 68 (1987)
N-Nitrosopyrrolidine	*17*, 313 (1978); *Suppl. 7*, 68 (1987)
N-Nitrososarcosine	*17*, 327 (1978); *Suppl. 7*, 68 (1987)
Nitrosoureas, chloroethyl (*see* Chloroethyl nitrosoureas)	
5-Nitro-*ortho*-toluidine	*48*, 169 (1990)
2-Nitrotoluene	*65*, 409 (1996)
3-Nitrotoluene	*65*, 409 (1996)
4-Nitrotoluene	*65*, 409 (1996)
Nitrous oxide (*see* Anaesthetics, volatile)	
Nitrovin	*31*, 185 (1983); *Suppl. 7*, 68 (1987)
Nivalenol (*see* Toxins derived from *Fusarium graminearum, F. culmorum* and *F. crookwellense*)	
NNA (*see* 4-(*N*-Nitrosomethylamino)-4-(3-pyridyl)-1-butanal)	
NNK (*see* 4-(*N*-Nitrosomethylamino)-1-(3-pyridyl)-1-butanone)	
Nonsteroidal oestrogens	*Suppl. 7*, 273 (1987)
Norethisterone	*6*, 179 (1974); *21*, 461 (1979); *Suppl. 7*, 294 (1987); *72*, 49 (1999)
Norethisterone acetate	*72*, 49 (1999)
Norethynodrel	*6*, 191 (1974); *21*, 461 (1979) (*corr. 42*, 259); *Suppl. 7*, 295 (1987); *72*, 49 (1999)
Norgestrel	*6*, 201 (1974); *21*, 479 (1979); *Suppl. 7*, 295 (1987); *72*, 49 (1999)
Nylon 6	*19*, 120 (1979); *Suppl. 7*, 68 (1987)

O

Ochratoxin A	*10*, 191 (1976); *31*, 191 (1983) (*corr. 42*, 262); *Suppl. 7*, 271 (1987); *56*, 489 (1993)
Oestradiol	*6*, 99 (1974); *21*, 279 (1979); *Suppl. 7*, 284 (1987); *72*, 399 (1999)
Oestradiol-17β (*see* Oestradiol)	
Oestradiol 3-benzoate (*see* Oestradiol)	
Oestradiol dipropionate (*see* Oestradiol)	
Oestradiol mustard	*9*, 217 (1975); *Suppl. 7*, 68 (1987)

Oestradiol valerate (*see* Oestradiol)
Oestriol   6, 117 (1974); *21*, 327 (1979);
                                                                            Suppl. 7, 285 (1987); *72*, 399
                                                                            (1999)

Oestrogen-progestin combinations (*see* Oestrogens,
    progestins (progestogens) and combinations)
Oestrogen-progestin replacement therapy (*see* Post-menopausal
    oestrogen-progestogen therapy)
Oestrogen replacement therapy (*see* Post-menopausal oestrogen
    therapy)
Oestrogens (*see* Oestrogens, progestins and combinations)
Oestrogens, conjugated (*see* Conjugated oestrogens)
Oestrogens, nonsteroidal (*see* Nonsteroidal oestrogens)
Oestrogens, progestins (progestogens) and combinations   6 (1974); *21* (1979); *Suppl. 7*, 272
                                                                            (1987); *72*, 49, 339, 399, 531
                                                                            (1999)

Oestrogens, steroidal (*see* Steroidal oestrogens)
Oestrone   6, 123 (1974); *21*, 343 (1979)
                                                                            (*corr. 42*, 259); *Suppl. 7*, 286
                                                                            (1987); *72*, 399 (1999)

Oestrone benzoate (*see* Oestrone)
Oil Orange SS   8, 165 (1975); *Suppl. 7*, 69 (1987)
Opisthorchis felineus (infection with)   *61*, 121 (1994)
Opisthorchis viverrini (infection with)   *61*, 121 (1994)
Oral contraceptives, combined   *Suppl. 7*, 297 (1987); *72*, 49 (1999)
Oral contraceptives, sequential (*see* Sequential oral contraceptives)
Orange I   8, 173 (1975); *Suppl. 7*, 69 (1987)
Orange G   8, 181 (1975); *Suppl. 7*, 69 (1987)
Organolead compounds (*see also* Lead and lead compounds)   *Suppl. 7*, 230 (1987)
Oxazepam   *13*, 58 (1977); *Suppl. 7*, 69 (1987);
                                                                            66, 115 (1996)
Oxymetholone (*see also* Androgenic (anabolic) steroids)   *13*, 131 (1977)
Oxyphenbutazone   *13*, 185 (1977); *Suppl. 7*, 69 (1987)

P

Paint manufacture and painting (occupational exposures in)   *47*, 329 (1989)
Palygorskite   *42*, 159 (1987); *Suppl. 7*, 117
                                                                            (1987); *68*, 245 (1997)
Panfuran S (*see also* Dihydroxymethylfuratrizine)   *24*, 77 (1980); *Suppl. 7*, 69 (1987)
Paper manufacture (*see* Pulp and paper manufacture)
Paracetamol   *50*, 307 (1990); *73*, 401 (1999)
Parasorbic acid   *10*, 199 (1976) (*corr. 42*, 255);
                                                                            Suppl. 7, 69 (1987)
Parathion   *30*, 153 (1983); *Suppl. 7*, 69 (1987)
Patulin   *10*, 205 (1976); *40*, 83 (1986);
                                                                            Suppl. 7, 69 (1987)
Penicillic acid   *10*, 211 (1976); *Suppl. 7*, 69 (1987)
Pentachloroethane   *41*, 99 (1986); *Suppl. 7*, 69 (1987);
                                                                            71, 1519 (1999)
Pentachloronitrobenzene (see Quintozene)
Pentachlorophenol (*see also* Chlorophenols; Chlorophenols,   *20*, 303 (1979); *53*, 371 (1991)
    occupational exposures to; Polychlorophenols and their sodium salts)

Permethrin	*53*, 329 (1991)
Perylene	*32*, 411 (1983); *Suppl. 7*, 69 (1987)
Petasitenine	*31*, 207 (1983); *Suppl. 7*, 69 (1987)
Petasites japonicus (*see also* Pyrrolizidine alkaloids)	*10*, 333 (1976)
Petroleum refining (occupational exposures in)	*45*, 39 (1989)
Petroleum solvents	*47*, 43 (1989)
Phenacetin	*13*, 141 (1977); *24*, 135 (1980); *Suppl. 7*, 310 (1987)
Phenanthrene	*32*, 419 (1983); *Suppl. 7*, 69 (1987)
Phenazopyridine hydrochloride	*8*, 117 (1975); *24*, 163 (1980) (*corr. 42*, 260); *Suppl. 7*, 312 (1987)
Phenelzine sulfate	*24*, 175 (1980); *Suppl. 7*, 312 (1987)
Phenicarbazide	*12*, 177 (1976); *Suppl. 7*, 70 (1987)
Phenobarbital and its sodium salt	*13*, 157 (1977); *Suppl. 7*, 313 (1987); *79*, 161 (2001)
Phenol	*47*, 263 (1989) (*corr. 50*, 385); *71*, 749 (1999)
Phenolphthalein	*76*, 387 (2000)
Phenoxyacetic acid herbicides (*see* Chlorophenoxy herbicides)	
Phenoxybenzamine hydrochloride	*9*, 223 (1975); *24*, 185 (1980); *Suppl. 7*, 70 (1987)
Phenylbutazone	*13*, 183 (1977); *Suppl. 7*, 316 (1987)
meta-Phenylenediamine	*16*, 111 (1978); *Suppl. 7*, 70 (1987)
para-Phenylenediamine	*16*, 125 (1978); *Suppl. 7*, 70 (1987)
Phenyl glycidyl ether (*see also* Glycidyl ethers)	*71*, 1525 (1999)
N-Phenyl-2-naphthylamine	*16*, 325 (1978) (*corr. 42*, 257); *Suppl. 7*, 318 (1987)
ortho-Phenylphenol	*30*, 329 (1983); *Suppl. 7*, 70 (1987); *73*, 451 (1999)
Phenytoin	*13*, 201 (1977); *Suppl. 7*, 319 (1987); *66*, 175 (1996)
Phillipsite (*see* Zeolites)	
PhIP	*56*, 229 (1993)
Pickled vegetables	*56*, 83 (1993)
Picloram	*53*, 481 (1991)
Piperazine oestrone sulfate (*see* Conjugated oestrogens)	
Piperonyl butoxide	*30*, 183 (1983); *Suppl. 7*, 70 (1987)
Pitches, coal-tar (*see* Coal-tar pitches)	
Polyacrylic acid	*19*, 62 (1979); *Suppl. 7*, 70 (1987)
Polybrominated biphenyls	*18*, 107 (1978); *41*, 261 (1986); *Suppl. 7*, 321 (1987)
Polychlorinated biphenyls	*7*, 261 (1974); *18*, 43 (1978) (*corr. 42*, 258); *Suppl. 7*, 322 (1987)
Polychlorinated camphenes (*see* Toxaphene)	
Polychlorinated dibenzo-*para*-dioxins (other than 2,3,7,8-tetrachlorodibenzodioxin)	*69*, 33 (1997)
Polychlorinated dibenzofurans	*69*, 345 (1997)
Polychlorophenols and their sodium salts	*71*, 769 (1999)
Polychloroprene	*19*, 141 (1979); *Suppl. 7*, 70 (1987)
Polyethylene (*see also* Implants, surgical)	*19*, 164 (1979); *Suppl. 7*, 70 (1987)

Poly(glycolic acid) (*see* Implants, surgical)	
Polymethylene polyphenyl isocyanate (*see also* 4,4′-Methylenediphenyl diisocyanate)	*19*, 314 (1979); *Suppl. 7*, 70 (1987)
Polymethyl methacrylate (*see also* Implants, surgical)	*19*, 195 (1979); *Suppl. 7*, 70 (1987)
Polyoestradiol phosphate (*see* Oestradiol-17β)	
Polypropylene (*see also* Implants, surgical)	*19*, 218 (1979); *Suppl. 7*, 70 (1987)
Polystyrene (*see also* Implants, surgical)	*19*, 245 (1979); *Suppl. 7*, 70 (1987)
Polytetrafluoroethylene (*see also* Implants, surgical)	*19*, 288 (1979); *Suppl. 7*, 70 (1987)
Polyurethane foams (*see also* Implants, surgical)	*19*, 320 (1979); *Suppl. 7*, 70 (1987)
Polyvinyl acetate (*see also* Implants, surgical)	*19*, 346 (1979); *Suppl. 7*, 70 (1987)
Polyvinyl alcohol (*see also* Implants, surgical)	*19*, 351 (1979); *Suppl. 7*, 70 (1987)
Polyvinyl chloride (*see also* Implants, surgical)	*7*, 306 (1974); *19*, 402 (1979); *Suppl. 7*, 70 (1987)
Polyvinyl pyrrolidone	*19*, 463 (1979); *Suppl. 7*, 70 (1987); *71*, 1181 (1999)
Ponceau MX	*8*, 189 (1975); *Suppl. 7*, 70 (1987)
Ponceau 3R	*8*, 199 (1975); *Suppl. 7*, 70 (1987)
Ponceau SX	*8*, 207 (1975); *Suppl. 7*, 70 (1987)
Post-menopausal oestrogen therapy	*Suppl. 7*, 280 (1987); *72*, 399 (1999)
Post-menopausal oestrogen-progestogen therapy	*Suppl. 7*, 308 (1987); *72*, 531 (1999)
Potassium arsenate (*see* Arsenic and arsenic compounds)	
Potassium arsenite (*see* Arsenic and arsenic compounds)	
Potassium bis(2-hydroxyethyl)dithiocarbamate	*12*, 183 (1976); *Suppl. 7*, 70 (1987)
Potassium bromate	*40*, 207 (1986); *Suppl. 7*, 70 (1987); *73*, 481 (1999)
Potassium chromate (*see* Chromium and chromium compounds)	
Potassium dichromate (*see* Chromium and chromium compounds)	
Prazepam	*66*, 143 (1996)
Prednimustine	*50*, 115 (1990)
Prednisone	*26*, 293 (1981); *Suppl. 7*, 326 (1987)
Printing processes and printing inks	*65*, 33 (1996)
Procarbazine hydrochloride	*26*, 311 (1981); *Suppl. 7*, 327 (1987)
Proflavine salts	*24*, 195 (1980); *Suppl. 7*, 70 (1987)
Progesterone (*see also* Progestins; Combined oral contraceptives)	*6*, 135 (1974); *21*, 491 (1979) (*corr. 42*, 259)
Progestins (*see* Progestogens)	
Progestogens	*Suppl. 7*, 289 (1987); *72*, 49, 339, 531 (1999)
Pronetalol hydrochloride	*13*, 227 (1977) (*corr. 42*, 256); *Suppl. 7*, 70 (1987)
1,3-Propane sultone	*4*, 253 (1974) (*corr. 42*, 253); *Suppl. 7*, 70 (1987); *71*, 1095 (1999)
Propham	*12*, 189 (1976); *Suppl. 7*, 70 (1987)
β-Propiolactone	*4*, 259 (1974) (*corr. 42*, 253); *Suppl. 7*, 70 (1987); *71*, 1103 (1999)
n-Propyl carbamate	*12*, 201 (1976); *Suppl. 7*, 70 (1987)
Propylene	*19*, 213 (1979); *Suppl. 7*, 71 (1987); *60*, 161 (1994)

Propyleneimine (see 2-Methylaziridine)
Propylene oxide   *11*, 191 (1976); *36*, 227 (1985)
                                                    (*corr. 42*, 263); *Suppl. 7*, 328
                                                    (1987); *60*, 181 (1994)
Propylthiouracil   *7*, 67 (1974); *Suppl. 7*, 329 (1987);
                                                    79, 91 (2001)
Ptaquiloside (see also Bracken fern)   *40*, 55 (1986); *Suppl. 7*, 71 (1987)
Pulp and paper manufacture   *25*, 157 (1981); *Suppl. 7*, 385
                                                    (1987)
Pyrene   *32*, 431 (1983); *Suppl. 7*, 71 (1987)
Pyridine   *77*, 503 (2000)
Pyrido[3,4-*c*]psoralen   *40*, 349 (1986); *Suppl. 7*, 71 (1987)
Pyrimethamine   *13*, 233 (1977); *Suppl. 7*, 71 (1987)
Pyrrolizidine alkaloids (see Hydroxysenkirkine; Isatidine; Jacobine;
    Lasiocarpine; Monocrotaline; Retrorsine; Riddelliine; Seneciphylline;
    Senkirkine)

Q

Quartz (see Crystalline silica)
Quercetin (see also Bracken fern)   *31*, 213 (1983); *Suppl. 7*, 71
                                                    (1987); *73*, 497 (1999)
para-Quinone   *15*, 255 (1977); *Suppl. 7*, 71
                                                    (1987); *71*, 1245 (1999)
Quintozene   *5*, 211 (1974); *Suppl. 7*, 71 (1987)

R

Radiation (see gamma-radiation, neutrons, ultraviolet radiation,
    X-radiation)
Radionuclides, internally deposited   *78* (2001)
Radon   *43*, 173 (1988) (*corr. 45*, 283)
Refractory ceramic fibres (see Man-made vitreous fibres)
Reserpine   *10*, 217 (1976); *24*, 211 (1980)
                                                    (*corr. 42*, 260); *Suppl. 7*, 330
                                                    (1987)
Resorcinol   *15*, 155 (1977); *Suppl. 7*, 71
                                                    (1987); *71*, 1119 (1990)
Retrorsine   *10*, 303 (1976); *Suppl. 7*, 71 (1987)
Rhodamine B   *16*, 221 (1978); *Suppl. 7*, 71 (1987)
Rhodamine 6G   *16*, 233 (1978); *Suppl. 7*, 71 (1987)
Riddelliine   *10*, 313 (1976); *Suppl. 7*, 71 (1987)
Rifampicin   *24*, 243 (1980); *Suppl. 7*, 71 (1987)
Ripazepam   *66*, 157 (1996)
Rock (stone) wool (see Man-made vitreous fibres)
Rubber industry   *28* (1982) (*corr. 42*, 261); *Suppl. 7*,
                                                    332 (1987)
Rugulosin   *40*, 99 (1986); *Suppl. 7*, 71 (1987)

S

Saccharated iron oxide	2, 161 (1973); *Suppl. 7*, 71 (1987)
Saccharin and its salts	22, 111 (1980) (*corr. 42*, 259); *Suppl. 7*, 334 (1987); *73*, 517 (1999)
Safrole	*1*, 169 (1972); *10*, 231 (1976); *Suppl. 7*, 71 (1987)
Salted fish	*56*, 41 (1993)
Sawmill industry (including logging) (*see* Lumber and sawmill industry (including logging))	
Scarlet Red	*8*, 217 (1975); *Suppl. 7*, 71 (1987)
Schistosoma haematobium (infection with)	*61*, 45 (1994)
Schistosoma japonicum (infection with)	*61*, 45 (1994)
Schistosoma mansoni (infection with)	*61*, 45 (1994)
Selenium and selenium compounds	9, 245 (1975) (*corr. 42*, 255); *Suppl. 7*, 71 (1987)
Selenium dioxide (*see* Selenium and selenium compounds)	
Selenium oxide (*see* Selenium and selenium compounds)	
Semicarbazide hydrochloride	*12*, 209 (1976) (*corr. 42*, 256); *Suppl. 7*, 71 (1987)
Senecio jacobaea L. (*see also* Pyrrolizidine alkaloids)	*10*, 333 (1976)
Senecio longilobus (*see also* Pyrrolizidine alkaloids)	*10*, 334 (1976)
Seneciphylline	*10*, 319, 335 (1976); *Suppl. 7*, 71 (1987)
Senkirkine	*10*, 327 (1976); *31*, 231 (1983); *Suppl. 7*, 71 (1987)
Sepiolite	*42*, 175 (1987); *Suppl. 7*, 71 (1987); *68*, 267 (1997)
Sequential oral contraceptives (*see also* Oestrogens, progestins and combinations)	*Suppl. 7*, 296 (1987)
Shale-oils	*35*, 161 (1985); *Suppl. 7*, 339 (1987)
Shikimic acid (*see also* Bracken fern)	*40*, 55 (1986); *Suppl. 7*, 71 (1987)
Shoe manufacture and repair (*see* Boot and shoe manufacture and repair)	
Silica (*see also* Amorphous silica; Crystalline silica)	*42*, 39 (1987)
Silicone (*see* Implants, surgical)	
Simazine	*53*, 495 (1991); *73*, 625 (1999)
Slag wool (*see* Man-made vitreous fibres)	
Sodium arsenate (*see* Arsenic and arsenic compounds)	
Sodium arsenite (*see* Arsenic and arsenic compounds)	
Sodium cacodylate (*see* Arsenic and arsenic compounds)	
Sodium chlorite	*52*, 145 (1991)
Sodium chromate (*see* Chromium and chromium compounds)	
Sodium cyclamate (*see* Cyclamates)	
Sodium dichromate (*see* Chromium and chromium compounds)	
Sodium diethyldithiocarbamate	*12*, 217 (1976); *Suppl. 7*, 71 (1987)
Sodium equilin sulfate (*see* Conjugated oestrogens)	
Sodium fluoride (*see* Fluorides)	
Sodium monofluorophosphate (*see* Fluorides)	
Sodium oestrone sulfate (*see* Conjugated oestrogens)	
Sodium *ortho*-phenylphenate (*see also ortho*-Phenylphenol)	*30*, 329 (1983); *Suppl. 7*, 71, 392 (1987); *73*, 451 (1999)
Sodium saccharin (*see* Saccharin)	

Sodium selenate (*see* Selenium and selenium compounds)	
Sodium selenite (*see* Selenium and selenium compounds)	
Sodium silicofluoride (*see* Fluorides)	
Solar radiation	*55* (1992)
Soots	*3*, 22 (1973); *35*, 219 (1985); *Suppl. 7*, 343 (1987)
Special-purpose glass fibres such as E-glass and '475' glass fibres (*see* Man-made vitreous fibres)	
Spironolactone	*24*, 259 (1980); *Suppl. 7*, 344 (1987); *79*, 317 (2001)
Stannous fluoride (*see* Fluorides)	
Static electric fields	*80* (2002)
Static magnetic fields	*80* (2002)
Steel founding (*see* Iron and steel founding)	
Steel, stainless (*see* Implants, surgical)	
Sterigmatocystin	*1*, 175 (1972); *10*, 245 (1976); *Suppl. 7*, 72 (1987)
Steroidal oestrogens	*Suppl. 7*, 280 (1987)
Streptozotocin	*4*, 221 (1974); *17*, 337 (1978); *Suppl. 7*, 72 (1987)
Strobane® (*see* Terpene polychlorinates)	
Strong-inorganic-acid mists containing sulfuric acid (*see* Mists and vapours from sulfuric acid and other strong inorganic acids)	
Strontium chromate (*see* Chromium and chromium compounds)	
Styrene	*19*, 231 (1979) (*corr. 42*, 258); *Suppl. 7*, 345 (1987); *60*, 233 (1994) (*corr. 65*, 549)
Styrene-acrylonitrile-copolymers	*19*, 97 (1979); *Suppl. 7*, 72 (1987)
Styrene-butadiene copolymers	*19*, 252 (1979); *Suppl. 7*, 72 (1987)
Styrene-7,8-oxide	*11*, 201 (1976); *19*, 275 (1979); *36*, 245 (1985); *Suppl. 7*, 72 (1987); *60*, 321 (1994)
Succinic anhydride	*15*, 265 (1977); *Suppl. 7*, 72 (1987)
Sudan I	*8*, 225 (1975); *Suppl. 7*, 72 (1987)
Sudan II	*8*, 233 (1975); *Suppl. 7*, 72 (1987)
Sudan III	*8*, 241 (1975); *Suppl. 7*, 72 (1987)
Sudan Brown RR	*8*, 249 (1975); *Suppl. 7*, 72 (1987)
Sudan Red 7B	*8*, 253 (1975); *Suppl. 7*, 72 (1987)
Sulfadimidine (*see* Sulfamethazine)	
Sulfafurazole	*24*, 275 (1980); *Suppl. 7*, 347 (1987)
Sulfallate	*30*, 283 (1983); *Suppl. 7*, 72 (1987)
Sulfamethazine and its sodium salt	*79*, 341 (2001)
Sulfamethoxazole	*24*, 285 (1980); *Suppl. 7*, 348 (1987); *79*, 361 (2001)
Sulfites (*see* Sulfur dioxide and some sulfites, bisulfites and metabisulfites)	
Sulfur dioxide and some sulfites, bisulfites and metabisulfites	*54*, 131 (1992)
Sulfur mustard (*see* Mustard gas)	
Sulfuric acid and other strong inorganic acids, occupational exposures to mists and vapours from	*54*, 41 (1992)
Sulfur trioxide	*54*, 121 (1992)
Sulphisoxazole (*see* Sulfafurazole)	
Sunset Yellow FCF	*8*, 257 (1975); *Suppl. 7*, 72 (1987)
Symphytine	*31*, 239 (1983); *Suppl. 7*, 72 (1987)

T

2,4,5-T (*see also* Chlorophenoxy herbicides; Chlorophenoxy herbicides, occupational exposures to)	*15*, 273 (1977)
Talc	*42*, 185 (1987); *Suppl. 7*, 349 (1987)
Tamoxifen	*66*, 253 (1996)
Tannic acid	*10*, 253 (1976) (*corr. 42*, 255); *Suppl. 7*, 72 (1987)
Tannins (*see also* Tannic acid)	*10*, 254 (1976); *Suppl. 7*, 72 (1987)
TCDD (*see* 2,3,7,8-Tetrachlorodibenzo-*para*-dioxin)	
TDE (*see* DDT)	
Tea	*51*, 207 (1991)
Temazepam	*66*, 161 (1996)
Teniposide	*76*, 259 (2000)
Terpene polychlorinates	*5*, 219 (1974); *Suppl. 7*, 72 (1987)
Testosterone (*see also* Androgenic (anabolic) steroids)	*6*, 209 (1974); *21*, 519 (1979)
Testosterone oenanthate (*see* Testosterone)	
Testosterone propionate (*see* Testosterone)	
2,2′,5,5′-Tetrachlorobenzidine	*27*, 141 (1982); *Suppl. 7*, 72 (1987)
2,3,7,8-Tetrachlorodibenzo-*para*-dioxin	*15*, 41 (1977); *Suppl. 7*, 350 (1987); *69*, 33 (1997)
1,1,1,2-Tetrachloroethane	*41*, 87 (1986); *Suppl. 7*, 72 (1987); *71*, 1133 (1999)
1,1,2,2-Tetrachloroethane	*20*, 477 (1979); *Suppl. 7*, 354 (1987); *71*, 817 (1999)
Tetrachloroethylene	*20*, 491 (1979); *Suppl. 7*, 355 (1987); *63*, 159 (1995) (*corr. 65*, 549)
2,3,4,6-Tetrachlorophenol (*see* Chlorophenols; Chlorophenols, occupational exposures to; Polychlorophenols and their sodium salts)	
Tetrachlorvinphos	*30*, 197 (1983); *Suppl. 7*, 72 (1987)
Tetraethyllead (*see* Lead and lead compounds)	
Tetrafluoroethylene	*19*, 285 (1979); *Suppl. 7*, 72 (1987); *71*, 1143 (1999)
Tetrakis(hydroxymethyl)phosphonium salts	*48*, 95 (1990); *71*, 1529 (1999)
Tetramethyllead (*see* Lead and lead compounds)	
Tetranitromethane	*65*, 437 (1996)
Textile manufacturing industry, exposures in	*48*, 215 (1990) (*corr. 51*, 483)
Theobromine	*51*, 421 (1991)
Theophylline	*51*, 391 (1991)
Thioacetamide	*7*, 77 (1974); *Suppl. 7*, 72 (1987)
4,4′-Thiodianiline	*16*, 343 (1978); *27*, 147 (1982); *Suppl. 7*, 72 (1987)
Thiotepa	*9*, 85 (1975); *Suppl. 7*, 368 (1987); *50*, 123 (1990)
Thiouracil	*7*, 85 (1974); *Suppl. 7*, 72 (1987); *79*, 127 (2001)
Thiourea	*7*, 95 (1974); *Suppl. 7*, 72 (1987); *79*, 703 (2001)
Thiram	*12*, 225 (1976); *Suppl. 7*, 72 (1987); *53*, 403 (1991)
Titanium (*see* Implants, surgical)	
Titanium dioxide	*47*, 307 (1989)

CUMULATIVE INDEX

Tobacco habits other than smoking (*see* Tobacco products, smokeless)	
Tobacco products, smokeless	37 (1985) (*corr.* 42, 263; 52, 513); *Suppl.* 7, 357 (1987)
Tobacco smoke	38 (1986) (*corr.* 42, 263); *Suppl.* 7, 359 (1987)
Tobacco smoking (*see* Tobacco smoke)	
ortho-Tolidine (*see* 3,3'-Dimethylbenzidine)	
2,4-Toluene diisocyanate (*see also* Toluene diisocyanates)	19, 303 (1979); 39, 287 (1986)
2,6-Toluene diisocyanate (*see also* Toluene diisocyanates)	19, 303 (1979); 39, 289 (1986)
Toluene	47, 79 (1989); 71, 829 (1999)
Toluene diisocyanates	39, 287 (1986) (*corr.* 42, 264); *Suppl.* 7, 72 (1987); 71, 865 (1999)
Toluenes, α-chlorinated (*see* α-Chlorinated toluenes and benzoyl chloride)	
ortho-Toluenesulfonamide (*see* Saccharin)	
ortho-Toluidine	16, 349 (1978); 27, 155 (1982) (*corr.* 68, 477); *Suppl.* 7, 362 (1987); 77, 267 (2000)
Toremifene	66, 367 (1996)
Toxaphene	20, 327 (1979); *Suppl.* 7, 72 (1987); 79, 569 (2001)
T-2 Toxin (*see* Toxins derived from *Fusarium sporotrichioides*)	
Toxins derived from *Fusarium graminearum, F. culmorum* and *F. crookwellense*	11, 169 (1976); 31, 153, 279 (1983); *Suppl.* 7, 64, 74 (1987); 56, 397 (1993)
Toxins derived from *Fusarium moniliforme*	56, 445 (1993)
Toxins derived from *Fusarium sporotrichioides*	31, 265 (1983); *Suppl.* 7, 73 (1987); 56, 467 (1993)
Tremolite (*see* Asbestos)	
Treosulfan	26, 341 (1981); *Suppl.* 7, 363 (1987)
Triaziquone (*see* Tris(aziridinyl)-*para*-benzoquinone)	
Trichlorfon	30, 207 (1983); *Suppl.* 7, 73 (1987)
Trichlormethine	9, 229 (1975); *Suppl.* 7, 73 (1987); 50, 143 (1990)
Trichloroacetic acid	63, 291 (1995) (*corr.* 65, 549)
Trichloroacetonitrile (*see also* Halogenated acetonitriles)	71, 1533 (1999)
1,1,1-Trichloroethane	20, 515 (1979); *Suppl.* 7, 73 (1987); 71, 881 (1999)
1,1,2-Trichloroethane	20, 533 (1979); *Suppl.* 7, 73 (1987); 52, 337 (1991); 71, 1153 (1999)
Trichloroethylene	11, 263 (1976); 20, 545 (1979); *Suppl.* 7, 364 (1987); 63, 75 (1995) (*corr.* 65, 549)
2,4,5-Trichlorophenol (*see also* Chlorophenols; Chlorophenols, occupational exposures to; Polychlorophenols and their sodium salts)	20, 349 (1979)
2,4,6-Trichlorophenol (*see also* Chlorophenols; Chlorophenols, occupational exposures to; Polychlorophenols and their sodium salts)	20, 349 (1979)
(2,4,5-Trichlorophenoxy)acetic acid (*see* 2,4,5-T)	
1,2,3-Trichloropropane	63, 223 (1995)
Trichlorotriethylamine-hydrochloride (*see* Trichlormethine)	
T_2-Trichothecene (*see* Toxins derived from *Fusarium sporotrichioides*)	
Tridymite (*see* Crystalline silica)	
Triethanolamine	77, 381 (2000)

Triethylene glycol diglycidyl ether	*11*, 209 (1976); *Suppl. 7*, 73 (1987); *71*, 1539 (1999)
Trifluralin	*53*, 515 (1991)
4,4',6-Trimethylangelicin plus ultraviolet radiation (*see also* Angelicin and some synthetic derivatives)	*Suppl. 7*, 57 (1987)
2,4,5-Trimethylaniline	*27*, 177 (1982); *Suppl. 7*, 73 (1987)
2,4,6-Trimethylaniline	*27*, 178 (1982); *Suppl. 7*, 73 (1987)
4,5',8-Trimethylpsoralen	*40*, 357 (1986); *Suppl. 7*, 366 (1987)
Trimustine hydrochloride (*see* Trichlormethine)	
2,4,6-Trinitrotoluene	*65*, 449 (1996)
Triphenylene	*32*, 447 (1983); *Suppl. 7*, 73 (1987)
Tris(aziridinyl)-*para*-benzoquinone	*9*, 67 (1975); *Suppl. 7*, 367 (1987)
Tris(1-aziridinyl)phosphine-oxide	*9*, 75 (1975); *Suppl. 7*, 73 (1987)
Tris(1-aziridinyl)phosphine-sulphide (*see* Thiotepa)	
2,4,6-Tris(1-aziridinyl)-*s*-triazine	*9*, 95 (1975); *Suppl. 7*, 73 (1987)
Tris(2-chloroethyl) phosphate	*48*, 109 (1990); *71*, 1543 (1999)
1,2,3-Tris(chloromethoxy)propane	*15*, 301 (1977); *Suppl. 7*, 73 (1987); *71*, 1549 (1999)
Tris(2,3-dibromopropyl) phosphate	*20*, 575 (1979); *Suppl. 7*, 369 (1987); *71*, 905 (1999)
Tris(2-methyl-1-aziridinyl)phosphine-oxide	*9*, 107 (1975); *Suppl. 7*, 73 (1987)
Trp-P-1	*31*, 247 (1983); *Suppl. 7*, 73 (1987)
Trp-P-2	*31*, 255 (1983); *Suppl. 7*, 73 (1987)
Trypan blue	*8*, 267 (1975); *Suppl. 7*, 73 (1987)
Tussilago farfara L. (*see also* Pyrrolizidine alkaloids)	*10*, 334 (1976)

U

Ultraviolet radiation	*40*, 379 (1986); *55* (1992)
Underground haematite mining with exposure to radon	*1*, 29 (1972); *Suppl. 7*, 216 (1987)
Uracil mustard	*9*, 235 (1975); *Suppl. 7*, 370 (1987)
Uranium, depleted (*see* Implants, surgical)	
Urethane	*7*, 111 (1974); *Suppl. 7*, 73 (1987)

V

Vat Yellow 4	*48*, 161 (1990)
Vinblastine sulfate	*26*, 349 (1981) (*corr. 42*, 261); *Suppl. 7*, 371 (1987)
Vincristine sulfate	*26*, 365 (1981); *Suppl. 7*, 372 (1987)
Vinyl acetate	*19*, 341 (1979); *39*, 113 (1986); *Suppl. 7*, 73 (1987); *63*, 443 (1995)
Vinyl bromide	*19*, 367 (1979); *39*, 133 (1986); *Suppl. 7*, 73 (1987); *71*, 923 (1999)
Vinyl chloride	*7*, 291 (1974); *19*, 377 (1979) (*corr. 42*, 258*); Suppl. 7*, 373 (1987)
Vinyl chloride-vinyl acetate copolymers	*7*, 311 (1976); *19*, 412 (1979) (*corr. 42*, 258); *Suppl. 7*, 73 (1987)

4-Vinylcyclohexene	*11*, 277 (1976); *39*, 181 (1986) *Suppl. 7*, 73 (1987); *60*, 347 (1994)
4-Vinylcyclohexene diepoxide	*11*, 141 (1976); *Suppl. 7*, 63 (1987); *60*, 361 (1994)
Vinyl fluoride	*39*, 147 (1986); *Suppl. 7*, 73 (1987); *63*, 467 (1995)
Vinylidene chloride	*19*, 439 (1979); *39*, 195 (1986); *Suppl. 7*, 376 (1987); *71*, 1163 (1999)
Vinylidene chloride-vinyl chloride copolymers	*19*, 448 (1979) (*corr. 42*, 258); *Suppl. 7*, 73 (1987)
Vinylidene fluoride	*39*, 227 (1986); *Suppl. 7*, 73 (1987); *71*, 1551 (1999)
N-Vinyl-2-pyrrolidone	*19*, 461 (1979); *Suppl. 7*, 73 (1987); *71*, 1181 (1999)
Vinyl toluene	*60*, 373 (1994)
Vitamin K substances	*76*, 417 (2000)

W

Welding	*49*, 447 (1990) (*corr. 52*, 513)
Wollastonite	*42*, 145 (1987); *Suppl. 7*, 377 (1987); *68*, 283 (1997)
Wood dust	*62*, 35 (1995)
Wood industries	*25* (1981); *Suppl. 7*, 378 (1987)

X

X-radiation	*75*, 121 (2000)
Xylenes	*47*, 125 (1989); *71*, 1189 (1999)
2,4-Xylidine	*16*, 367 (1978); *Suppl. 7*, 74 (1987)
2,5-Xylidine	*16*, 377 (1978); *Suppl. 7*, 74 (1987)
2,6-Xylidine (*see* 2,6-Dimethylaniline)	

Y

Yellow AB	*8*, 279 (1975); *Suppl. 7*, 74 (1987)
Yellow OB	*8*, 287 (1975); *Suppl. 7*, 74 (1987)

Z

Zalcitabine	*76*, 129 (2000)
Zearalenone (*see* Toxins derived from *Fusarium graminearum*, *F. culmorum* and *F. crookwellense*)	
Zectran	*12*, 237 (1976); *Suppl. 7*, 74 (1987)
Zeolites other than erionite	*68*, 307 (1997)
Zidovudine	*76*, 73 (2000)
Zinc beryllium silicate (*see* Beryllium and beryllium compounds)	
Zinc chromate (*see* Chromium and chromium compounds)	
Zinc chromate hydroxide (*see* Chromium and chromium compounds)	

Zinc potassium chromate (*see* Chromium and chromium compounds)
Zinc yellow (*see* Chromium and chromium compounds)
Zineb   *12*, 245 (1976); *Suppl. 7*, 74 (1987)
Ziram   *12*, 259 (1976); *Suppl. 7*, 74 (1987); *53, 423* (1991)

List of IARC Monographs on the Evaluation of Carcinogenic Risks to Humans*

Volume 1
Some Inorganic Substances, Chlorinated Hydrocarbons, Aromatic Amines, N-Nitroso Compounds, and Natural Products
1972; 184 pages (out-of-print)

Volume 2
Some Inorganic and Organometallic Compounds
1973; 181 pages (out-of-print)

Volume 3
Certain Polycyclic Aromatic Hydrocarbons and Heterocyclic Compounds
1973; 271 pages (out-of-print)

Volume 4
Some Aromatic Amines, Hydrazine and Related Substances, N-Nitroso Compounds and Miscellaneous Alkylating Agents
1974; 286 pages (out-of-print)

Volume 5
Some Organochlorine Pesticides
1974; 241 pages (out-of-print)

Volume 6
Sex Hormones
1974; 243 pages (out-of-print)

Volume 7
Some Anti-Thyroid and Related Substances, Nitrofurans and Industrial Chemicals
1974; 326 pages (out-of-print)

Volume 8
Some Aromatic Azo Compounds
1975; 357 pages

Volume 9
Some Aziridines, N-, S- and O-Mustards and Selenium
1975; 268 pages

Volume 10
Some Naturally Occurring Substances
1976; 353 pages (out-of-print)

Volume 11
Cadmium, Nickel, Some Epoxides, Miscellaneous Industrial Chemicals and General Considerations on Volatile Anaesthetics
1976; 306 pages (out-of-print)

Volume 12
Some Carbamates, Thiocarbamates and Carbazides
1976; 282 pages (out-of-print)

Volume 13
Some Miscellaneous Pharmaceutical Substances
1977; 255 pages

Volume 14
Asbestos
1977; 106 pages (out-of-print)

Volume 15
Some Fumigants, the Herbicides 2,4-D and 2,4,5-T, Chlorinated Dibenzodioxins and Miscellaneous Industrial Chemicals
1977; 354 pages (out-of-print)

Volume 16
Some Aromatic Amines and Related Nitro Compounds—Hair Dyes, Colouring Agents and Miscellaneous Industrial Chemicals
1978; 400 pages

Volume 17
Some N-Nitroso Compounds
1978; 365 pages

Volume 18
Polychlorinated Biphenyls and Polybrominated Biphenyls
1978; 140 pages (out-of-print)

Volume 19
Some Monomers, Plastics and Synthetic Elastomers, and Acrolein
1979; 513 pages (out-of-print)

Volume 20
Some Halogenated Hydrocarbons
1979; 609 pages (out-of-print)

Volume 21
Sex Hormones (II)
1979; 583 pages

Volume 22
Some Non-Nutritive Sweetening Agents
1980; 208 pages

Volume 23
Some Metals and Metallic Compounds
1980; 438 pages (out-of-print)

Volume 24
Some Pharmaceutical Drugs
1980; 337 pages

Volume 25
Wood, Leather and Some Associated Industries
1981; 412 pages

Volume 26
Some Antineoplastic and Immunosuppressive Agents
1981; 411 pages

Volume 27
Some Aromatic Amines, Anthraquinones and Nitroso Compounds, and Inorganic Fluorides Used in Drinking-water and Dental Preparations
1982; 341 pages

Volume 28
The Rubber Industry
1982; 486 pages

Volume 29
Some Industrial Chemicals and Dyestuffs
1982; 416 pages

Volume 30
Miscellaneous Pesticides
1983; 424 pages

---

*Certain older volumes, marked out-of-print, are still available directly from IARCPress. Further, high-quality photocopies of all out-of-print volumes may be purchased from University Microfilms International, 300 North Zeeb Road, Ann Arbor, MI 48106-1346, USA (Tel.: 313-761-4700, 800-521-0600).

Volume 31
Some Food Additives, Feed Additives and Naturally Occurring Substances
1983; 314 pages (out-of-print)

Volume 32
Polynuclear Aromatic Compounds, Part 1: Chemical, Environmental and Experimental Data
1983; 477 pages (out-of-print)

Volume 33
Polynuclear Aromatic Compounds, Part 2: Carbon Blacks, Mineral Oils and Some Nitroarenes
1984; 245 pages (out-of-print)

Volume 34
Polynuclear Aromatic Compounds, Part 3: Industrial Exposures in Aluminium Production, Coal Gasification, Coke Production, and Iron and Steel Founding
1984; 219 pages

Volume 35
Polynuclear Aromatic Compounds, Part 4: Bitumens, Coal-tars and Derived Products, Shale-oils and Soots
1985; 271 pages

Volume 36
Allyl Compounds, Aldehydes, Epoxides and Peroxides
1985; 369 pages

Volume 37
Tobacco Habits Other than Smoking; Betel-Quid and Areca-Nut Chewing; and Some Related Nitrosamines
1985; 291 pages

Volume 38
Tobacco Smoking
1986; 421 pages

Volume 39
Some Chemicals Used in Plastics and Elastomers
1986; 403 pages

Volume 40
Some Naturally Occurring and Synthetic Food Components, Furocoumarins and Ultraviolet Radiation
1986; 444 pages

Volume 41
Some Halogenated Hydrocarbons and Pesticide Exposures
1986; 434 pages

Volume 42
Silica and Some Silicates
1987; 289 pages

Volume 43
Man-Made Mineral Fibres and Radon
1988; 300 pages

Volume 44
Alcohol Drinking
1988; 416 pages

Volume 45
Occupational Exposures in Petroleum Refining; Crude Oil and Major Petroleum Fuels
1989; 322 pages

Volume 46
Diesel and Gasoline Engine Exhausts and Some Nitroarenes
1989; 458 pages

Volume 47
Some Organic Solvents, Resin Monomers and Related Compounds, Pigments and Occupational Exposures in Paint Manufacture and Painting
1989; 535 pages

Volume 48
Some Flame Retardants and Textile Chemicals, and Exposures in the Textile Manufacturing Industry
1990; 345 pages

Volume 49
Chromium, Nickel and Welding
1990; 677 pages

Volume 50
Pharmaceutical Drugs
1990; 415 pages

Volume 51
Coffee, Tea, Mate, Methyl-xanthines and Methylglyoxal
1991; 513 pages

Volume 52
Chlorinated Drinking-water; Chlorination By-products; Some Other Halogenated Compounds; Cobalt and Cobalt Compounds
1991; 544 pages

Volume 53
Occupational Exposures in Insecticide Application, and Some Pesticides
1991; 612 pages

Volume 54
Occupational Exposures to Mists and Vapours from Strong Inorganic Acids; and Other Industrial Chemicals
1992; 336 pages

Volume 55
Solar and Ultraviolet Radiation
1992; 316 pages

Volume 56
Some Naturally Occurring Substances: Food Items and Constituents, Heterocyclic Aromatic Amines and Mycotoxins
1993; 599 pages

Volume 57
Occupational Exposures of Hairdressers and Barbers and Personal Use of Hair Colourants; Some Hair Dyes, Cosmetic Colourants, Industrial Dyestuffs and Aromatic Amines
1993; 428 pages

Volume 58
Beryllium, Cadmium, Mercury, and Exposures in the Glass Manufacturing Industry
1993; 444 pages

Volume 59
Hepatitis Viruses
1994; 286 pages

Volume 60
Some Industrial Chemicals
1994; 560 pages

Volume 61
Schistosomes, Liver Flukes and *Helicobacter pylori*
1994; 270 pages

Volume 62
Wood Dust and Formaldehyde
1995; 405 pages

Volume 63
Dry Cleaning, Some Chlorinated Solvents and Other Industrial Chemicals
1995; 551 pages

Volume 64
Human Papillomaviruses
1995; 409 pages

Volume 65
Printing Processes and Printing Inks, Carbon Black and Some Nitro Compounds
1996; 578 pages

Volume 66
Some Pharmaceutical Drugs
1996; 514 pages

Volume 67
Human Immunodeficiency Viruses and Human T-Cell Lymphotropic Viruses
1996; 424 pages

Volume 68
Silica, Some Silicates, Coal Dust and *para*-Aramid Fibrils
1997; 506 pages

Volume 69
Polychlorinated Dibenzo-*para*-Dioxins and Polychlorinated Dibenzofurans
1997; 666 pages

Volume 70
Epstein-Barr Virus and Kaposi's Sarcoma Herpesvirus/Human Herpesvirus 8
1997; 524 pages

Volume 71
Re-evaluation of Some Organic Chemicals, Hydrazine and Hydrogen Peroxide
1999; 1586 pages

Volume 72
Hormonal Contraception and Post-menopausal Hormonal Therapy
1999; 660 pages

Volume 73
Some Chemicals that Cause Tumours of the Kidney or Urinary Bladder in Rodents and Some Other Substances
1999; 674 pages

Volume 74
Surgical Implants and Other Foreign Bodies
1999; 409 pages

Volume 75
Ionizing Radiation, Part 1, X-Radiation and γ-Radiation, and Neutrons
2000; 492 pages

Volume 76
Some Antiviral and Antineoplastic Drugs, and Other Pharmaceutical Agents
2000; 522 pages

Volume 77
Some Industrial Chemicals
2000; 563 pages

Volume 78
Ionizing Radiation, Part 2, Some Internally Deposited Radionuclides
2001; 595 pages

Volume 79
Some Thyrotropic Agents
2001; 763 pages

Volume 80
Non-Ionizing Radiation, Part 1: Static and Extremely Low-Frequency (ELF) Electric and Magnetic Fields
2002; 429 pages

Volume 81
Man-made Vitreous Fibres
2002; 418 pages

Supplement No. 1
Chemicals and Industrial Processes Associated with Cancer in Humans (*IARC Monographs*, Volumes 1 to 20)
1979; 71 pages (out-of-print)

Supplement No. 2
Long-term and Short-term Screening Assays for Carcinogens: A Critical Appraisal
1980; 426 pages (out-of-print)

Supplement No. 3
Cross Index of Synonyms and Trade Names in Volumes 1 to 26 of the *IARC Monographs*
1982; 199 pages (out-of-print)

Supplement No. 4
Chemicals, Industrial Processes and Industries Associated with Cancer in Humans (*IARC Monographs*, Volumes 1 to 29)
1982; 292 pages (out-of-print)

Supplement No. 5
Cross Index of Synonyms and Trade Names in Volumes 1 to 36 of the *IARC Monographs*
1985; 259 pages (out-of-print)

Supplement No. 6
Genetic and Related Effects: An Updating of Selected *IARC Monographs* from Volumes 1 to 42
1987; 729 pages

Supplement No. 7
Overall Evaluations of Carcinogenicity: An Updating of *IARC Monographs* Volumes 1–42
1987; 440 pages

Supplement No. 8
Cross Index of Synonyms and Trade Names in Volumes 1 to 46 of the *IARC Monographs*
1990; 346 pages (out-of-print)

All IARC publications are available directly from
IARCPress, 150 Cours Albert Thomas, F-69372 Lyon cedex 08, France
(Fax: +33 4 72 73 83 02; E-mail: press@iarc.fr).

IARC Monographs and Technical Reports are also available from the
World Health Organization Distribution and Sales, CH-1211 Geneva 27
(Fax: +41 22 791 4857; E-mail: publications@who.int)
and from WHO Sales Agents worldwide.

IARC Scientific Publications, IARC Handbooks and IARC CancerBases are also available from
Oxford University Press, Walton Street, Oxford, UK OX2 6DP (Fax: +44 1865 267782).

IARC Monographs are also available in an electronic edition,
both on-line by internet and on CD-ROM, from GMA Industries, Inc.,
20 Ridgely Avenue, Suite 301, Annapolis, Maryland, USA
(Fax: +01 410 267 6602; internet: https//www.gmai.com/Order_Form.htm)

www.ingramcontent.com/pod-product-compliance
Lightning Source LLC
Chambersburg PA
CBHW081152020426
42333CB00020B/2485